# Advances in Design Optimization

# Advances in Design Optimizaton

*Edited by*
## PROFESSOR HOJJAT ADELI
The Ohio State University, USA

**CHAPMAN & HALL**

London · Glasgow · Weinheim · New York · Tokyo · Melbourne · Madras

**Published by Chapman and Hall, 2–6 Boundary Row, London SE1 8HN, UK**

Chapman & Hall, 2–6 Boundary Row, London SE1 8HN, UK

Blackie Academic & Professional, Wester Cleddens Road, Bishopbriggs, Glasgow G64 2NZ, UK

Chapman & Hall Inc., One Penn Plaza, 41st Floor, New York NY10119, USA

Chapman & Hall Japan, Thomson Publishing Japan, Hirakawacho Nemoto Building, 6F, 1-7-11 Hirakawa-cho, Chiyoda-ku, Tokyo 102, Japan

Chapman & Hall Australia, Thomas Nelson Australia, 102 Dodds Street, South Melbourne, Victoria 3205, Australia

Chapman & Hall India, R. Seshadri, 32 Second Main Road, CIT East, Madras 600 035, India

First edition 1994

© 1994 Chapman & Hall

Typeset in $9\frac{1}{2}/11$ pt Times by Thomson Press (India) Ltd., New Delhi

Printed in Great Britain by TJ Press (Padstow) Ltd, Padstow, Cornwall

ISBN 0 412 53730 3

A catalogue record for this book is available from the British Library

**Library of Congress Cataloging-in-Publication data**

Advances in design optimization / edited by Hojjat Adeli.— 1st ed.
    p.   cm.
    Includes bibliographical references and index.
    ISBN 0-412-53730-3 (alk. paper)
    1. Engineering design—Mathematical models. I. Adeli, Hojjat, 1950–
TA174.A415   1994
620′.0042′011—dc20                                                93-32183
                                                                      CIP

∞ Printed on acid-free text paper, manufactured in accordance with ANSI/NISO Z39.48-1992 (Permanence of Paper).

# Contents

# Preface

Design being an open-ended problem, the selection of the 'best' or 'optimum' design has always been a major concern of designers. In recent years, we have seen a growing interest in the application of mathematical optimization techniques in engineering design. Optimization algorithms are no longer considered esoteric tools used for solution of theoretical problems only. They can be in fact an effective tool for design of large and complicated engineering and industrial systems. Commercial computer-aided design packages with optimization capabilities have recently become widely available.

A number of leading optimization researchers and 'practitioners' have contributed to this volume. It attempts to summarize advances in a number of fundamental areas of optimization with application in engineering design. Introductory materials are also included in each chapter in an attempt to make the book as self-contained as possible. Thus, the book can be used as a textbook in a graduate level course on design optimization.

We hope this book will generate further interest in the subject and help engineers in their quest to create more efficient designs.

Hojjat Adeli
Columbus, Ohio

# About the editor

Hojjat Adeli is Professor of Civil Engineering and a member of Center of Cognitive Science at The Ohio State University. A contributor to 38 research journals he has authored nearly 240 research and scientific publications and edited 10 books in various fields of computer science and engineering since 1976 when he received his PhD from Stanford University. He is the author of *Interactive Microcomputer-aided Structural Steel Design* and co-author of *Expert Systems for Structural Design* both published by Prentice-Hall in 1988, and *Parallel Processing in Structural Engineering* published by Chapman & Hall in 1993. He is the Editor-in-Chief of the international journal *Microcomputers in Civil Engineering* which he founded in 1986. He has been an organizer or member of advisory boards of over 30 national and international conferences and a contributor to over 80 conferences held in 23 different countries. He was a Keynote and Plenary Lecturer at international knowledge engineering and computing conferences held in Italy (1989), Mexico (1989), Japan (1991), China (1992), Canada (1992), Portugal (1992), USA (1993) and Germany (1993). He has received numerous academic, research, and leadership awards, honors, and recognitions. His recent awards include The Ohio State University College of Engineering 1990 Research Award in Recognition of Outstanding Research Accomplishments and Lichtenstein Memorial Award for Faculty Excellence. In 1990 he was selected as Man of the Year by the American Biographical Institute.

# Biographies of contributors

**JASBIR ARORA** received his PhD in the area of Structural Mechanics from the University of Iowa in 1971. He is currently Professor of Civil Engineering and Mechanical Engineering at the University of Iowa. He is also Director of the Optimal Design Laboratory in the College of Engineering. Dr Arora has published over 120 research papers. He is the author of the textbook *Introduction to Optimum Design*. McGraw-Hill, 1989, and co-author of a graduate level textbook, *Applied Optimal Design*, Wiley–Interscience, 1979. His current research interests include software development for design optimization, numerical optimization algorithms for large-scale problems, practical applications of optimization, design sensitivity analysis and optimization of nonlinear systems, optimal design under dynamic loads, and optimal control.

**SCOTT A. BURNS** received his PhD from the University of Illinois at Urbana-Champaign in 1985. He is currently an Associate Professor of General Engineering at the University of Illinois, Urbana. Dr Burns is a 1989 recipient of the National Science Foundation Presidential Young Investigator award.

**DAN M. FRANGOPOL** received his doctorate from the University of Liege, Belgium, in 1976. From 1969 to 1973 and from 1977 to 1979, he taught at the Institute of Civil Engineering, Bucharest, Romania. He was a project engineer in the Consulting Group A. Lipski (Brussels, Belgium) from 1979 to 1983. Since March 1983 he has been a professor of Civil Engineering at the University of Colorado, Boulder. He has performed research and consulting sponsored by the National Science Foundation, Federal Highway Administration, Pennsylvania Department of Transportation, Transportation Research Board, US Department of the Interior and US Army Corps of Engineers. His research interests include structural optimization, structural reliability, bridge engineering, and earthquake engineering.

**D.E. GRIERSON** received his PhD from the University of Waterloo, Canada, in 1968. He is currently Professor of Civil Engineering at the University of Waterloo. He is the co-author of structural optimization software for the design of steel structures that is used throughout North America in professional design offices and academic institutions.

**SANTIAGO HERNÁNDEZ** received his doctorate from University of Cantabria, Spain, in 1982. He is currently Professor of Structural Analysis at the University of La Coruna, Spain. He has written a book, co-authored two other books, and edited four books on structural optimization. He has organized international conferences on optimization and European training programs on intelligent design in engineering and architecture.

**KAZUTO HORIMATSU** received his Master of Engineering degree from Keio University, Tokyo, Japan, in 1988, and worked at Sony Corporation as a software engineer for design and production system division. He was a research scientist at the Computational Mechanics Laboratory at the University of Michigan during 1990–1992. He is currently with the Sony Corporation of America.

**MANOHAR KAMAT** received his PhD from Georgia Institute of Technology, Atlanta, Georgia, in 1972. From 1972 to 1985 he was on the faculty of the Department of Engineering Science and Mechanics at the Virginia Polytechnic and State University, Blacksburg, Virgina. He is currently Professor of Aerospace Engineering at Georgia Institute of Technology. He is the co-author of the text *Elements of Structural Optimization* and Editor of ASCE Journal of Aerospace Engineering. He is also the Editor of AIAA's Progress on Aeronautics and Astronautics Series volume *Structural Optimization – Status & Promise*.

**NARBEY KHACHATURIAN** received his PhD from the University of Illinois at Urbana-Champaign in 1952. He is now an Emeritus Professor of Civil Engineering at the University of Illinois, Urbana-Champaign. He was Associate Head of Department of Civil Engineering there during 1983–1989. He has performed research and published extensively on structural optimization during the last three decades. In 1992 he was elected an Honorary Member of the American Society of Civil Engineers.

**NARENDA S. KHOT** received his PhD from the University of Cincinnati in 1964. He is currently Senior Aerospace Engineer at the Flight Dynamics Directorate of Wright Patterson Air Force Base, Ohio. He has performed research and published extensively in the area of structural optimization, in particular the optimality criteria approach and structural control. He is the past associate editor of AIAA Journal. At present, he is on the editorial board of the journals of *Computational Mechanics* and *Structural Optimization*.

**NOBORU KIKUCHI** received his PhD from the Department of Aerospace Engineering and Engineering Mechanics at the University of Texas at Austin in 1977. He joined the Department of Mechanical Engineering and Applied Mechanics at the University of Michigan in 1979 where he is currently a professor. He has authored two books in the areas of contact mechanics and finite element method and over 80 journal papers on computational mechanics and structural optimization.

**JUHANI KOSKI** received his doctorate in Mechanical Engineering from Tampere University of Technology, Finland, in 1984. He is currently an Associate Professor of Structural Mechanics at Tampere University of Technology.

**LUIS MESQUITA** received his PhD from the Virginia Polytechnic Institute and State University, Blacksburg, Virginia, in 1985. He taught at the University of Nebraska at Lincoln from 1985 to 1990. Since 1991 he has been working as Project Engineer at Stevens & Associates in Woburn, Massachusetts.

**FRED MOSES** received his PhD in Civil Engineering from Cornell University in 1963 and served on the faculty of Case Western Reserve University until 1992. Since then, he has been Professor and Chairman of the Department of Civil Engineering at the University of Pittsburgh. He has served as a visiting faculty member at the Norwegian Institute of Technology, Israel Institute of Technology, Imperial College, and the Federal University of Lausanne, Switzerland. His current research interests include structural system reliability, optimization, load models for bridges, safety codes for offshore structures, design and evaluation of highway bridges, fatigue life prediction, and testing.

**PANOS Y. PAPALAMBROS** received his PhD from the Design Division of Mechanical Engineering Department at Stanford University in 1979. He joined the University of Michigan faculty in 1979 and has been teaching design there since then. He is currently Chairman of Department of Mechanical Engineering and Applied Mechanics at the University of Michigan. He has published over 50 book and journal articles primarily in the field of design automation, and has co-authored the text *Principles of Optimal Design* with D.J. Wilde. He is a past associate editor for Design Automation for the ASME Mechanical Design Transactions Journal.

**GEORGE I.N. ROZVANY** received his PhD from Monash University, Melbourne, Australia, in 1967. Since 1985 he has been Professor of Structural Design at the University of Essen, Germany. In the 1970s he was the closest research associate of the late Professor William Prager of Brown University. His over 190 publications on structural optimization include two books, three conference proceedings, and numerous principal lectures at other conferences. He has organized three international meetings on structural optimization and held visiting positions at Oxford (England), Waterloo (Canada), Stuttgart (Germany), Bangalore (India), and Warsaw (Poland). He is the Editor of journal of Structural Optimization which he founded in 1988, and Chairman of the Executive Committee of the International Society of Structural and Multidisciplinary Optimization.

**ERIC SANDGREN** received his PhD from Purdue University in 1977. He is currently Director of Design Engineering at TRW Steering and Suspension Systems in Sterling Heights, Michigan. Previously, he was an Associate Professor in the School of Mechanical Engineering at Purdue University in West Lafayette, Indiana. He has industrial experience with IBM and General Motors Corporation as well as additional academic experience at the University of Missouri in Columbia, Missouri. He was the recipient of the US National Science Foundation Presidential Young Investigator Award in 1984.

**OLE SIGMUND** received his MSc degree from the Technical University of Denmark in 1991. During 1991–1992 he performed research at the University of Essen under the sponsorship of German Research Foundation (DFG). He has published in the areas of structural optimization, dynamics, and active control. He is currently employed at the Department of Solid Mechanics at the Technical University of Denmark.

**MING ZHOU** received his Dr-Ing degree from Essen University, Germany, in 1992. He has been a research associate at the Essen University since 1988. He has published over thirty papers on different topics of structural optimization.

# Metric/imperial equivalents for common units

**Basic conversion factors**

The following equivalents of SI units are given in imperial and, where applicable, metric technical units.

| | | | | | | | |
|---|---|---|---|---|---|---|---|
| 1 mm | = 0.039 37 in | 1 in | = 25.4 mm | $1\,m^2$ | = 1.196 $yd^2$ | $1\,yd^2$ | = 0.8361 $m^2$ |
| 1 m | = 3.281 ft | 1 ft | = 0.3048 m | 1 hectare | = 2.471 arces | 1 acre | = 0.4047 hectares |
| | = 1.094 yd | 1 yd | = 0.9144 m | $1\,mm^3$ | = 0.000 061 02 $in^3$ | $1\,in^3$ | = 16 390 $mm^3$ |
| 1 km | = 0.6214 mile | 1 mile | = 1.609 km | $1\,m^3$ | = 35.31 $ft^3$ | $1\,ft^3$ | = 0.028 32 $m^3$ |
| $1\,mm^2$ | = 0.001 55 $in^2$ | $1\,in^2$ | = 645.2 $mm^2$ | | = 1.308 $yd^3$ | $1\,yd^3$ | = 0.7646 $m^3$ |
| $1\,m^2$ | = 10.76 $ft^2$ | $1\,ft^2$ | = 0.0929 $m^2$ | $1\,mm^4$ (M of I) = 0.000 002 403 $in^4$ | | $1\,in^4$ | = 416 200 $mm^4$ |

**Force**

| | | |
|---|---|---|
| 1 N | = 0.2248 lbf = 0.1020 kgf | |
| 4.448 N = 1 lbf | = 0.4536 kgf | |
| 9.807 N = 2.205 lbf | = 1 kgf | |

| | | |
|---|---|---|
| 1 kN | = 0.1004 tonf = 102.0 kgf = 0.1020 tonne f | |
| 9.964 kN = 1 tonf | = 1016 kgf = 1.016 tonne f | |
| 9.807 kN = 0.9842 tonf = 1000 kgf = 1 tonne f | | |

**Force per unit length**

| | |
|---|---|
| 1 N/m | = 0.068 52 lbf/ft = 0.1020 kgf/m |
| 14.59 N/m = 1 lbf/ft | = 1.488 kgf/m |
| 9.807 N/m = 0.672 lbf/ft | = 1 kgf/m |

| | |
|---|---|
| 1 kN/m | = 0.0306 tonf/ft = 0.1020 tonne f/m |
| 32.69 kN/m = 1 tonf/ft | = 3.333 tonne f/m |
| 9.807 kN/m = 0.3000 tonf/ft = 1 tonne f/m | |

**Force per unit area**

| | |
|---|---|
| $1\,N/mm^2$ | = 145.0 $lbf/in^2$ = 10.20 $kgf/cm^2$ |
| 0.006 895 $N/mm^2$ = 1 $lbf/in^2$ | = 0.0703 $kgf/cm^2$ |
| 0.098 07 $N/mm^2$ = 14.22 $lbf/in^2$ | = 1 $kgf/cm^2$ |

| | |
|---|---|
| $1\,N/mm^2$ | = 0.064 75 $tonf/in^2$ = 10.20 $kgf/cm^2$ |
| 15.44 $N/mm^2$ = 1 $tonf/in^2$ | = 157.5 $kgf/cm^2$ |
| 0.098 07 $N/mm^2$ = 0.006 350 $tonf/in^2$ = 1 $kgf/cm^2$ | |

| | |
|---|---|
| $1\,N/m^2$ | = 0.020 89 $lbf/ft^2$ = 0.102 $kgf/m^2$ |
| 47.88 $N/m^2$ | = 1 $lbf/ft^2$ = 4.882 $kgf/m^2$ |
| 9.807 $N/m^2$ | = 0.2048 $lbf/ft^2$ = 1 $kgf/m^2$ |

| | |
|---|---|
| $1\,N/mm^2$ | = 9.324 $tonf/ft^2$ = 10.20 $kgf/cm^2$ |
| 0.1073 $N/mm^2$ = 1 $tonf/ft^2$ | = 1.094 $kgf/cm^2$ |
| 0.098 07 $N/mm^2$ = 0.9144 $tonf/ft^2$ | = 1 $kgf/cm^2$ |

**Force per unit volume**

| | |
|---|---|
| $1\,N/m^3$ | = 0.006 366 $lbf/ft^3$ = 0.102 $kgf/m^3$ |
| 157.1 $N/m^3$ = 1 $lbf/ft^3$ | = 16.02 $kgf/m^3$ |
| 9.807 $N/m^3$ = 0.0624 $lbf/ft^3$ | = 1 $kgf/m^3$ |

| | |
|---|---|
| $1\,kN/m^3$ | = 0.002 842 $tonf/ft^3$ = 0.1020 tonne $f/m^3$ |
| 351.9 $kN/m^3$ = 1 $tonf/ft^3$ | = 35.88 tonne $f/m^3$ |
| 9.807 $kN/m^3$ = 0.027 87 $tonf/ft^3$ | = 1 tonne $f/m^3$ |

| | |
|---|---|
| $1\,kN/m^3$ | = 0.003 684 $lbf/in^3$ = 0.1020 tonne $f/m^3$ |
| 271.4 $kN\,m^3$ = 1 $lbf/in^3$ | = 27.68 tonne $f/m^3$ |
| 9.807 $kN/m^3$ = 0.036 13 $lbf/in^3$ | = 1 tonne $f/m^3$ |

**Moment**

| | |
|---|---|
| 1 Nm | = 8.851 lbf in = 0.7376 lbf ft = 0.1020 kgf m |
| 0.1130 Nm = 1 lbf in | = 0.083 33 lbf ft = 0.011 52 kgf m |
| 1.356 Nm = 12 lbf in | = 1 lbf ft = 0.1383 kgf m |
| 9.807 Nm = 86.80 lbf in = 7.233 lbf ft | = 1 kgf m |

# 1

# Mathematical theory of optimum engineering design

S. HERNÁNDEZ

## 1.1 INTRODUCTION

Mathematical formulation of optimum design in engineering is usually written as

$$\min F(X) \tag{1.1}$$

subject to

$$g_j(X) \leqslant 0 \quad i = 1, \ldots, m \tag{1.2}$$

In this expression

$X$     is the vector of problem variables
$F(X)$   is the objective function
$g_j(X)$   are the set of problem constraints

In structural optimization (Schmit, 1960) the most usual design variables are mechanical parameters of structure elements, size of cross section internal components or coordinates of nodes which define the structure. The objective function $F(X)$ is generally the structural weight, but more complicated functions are chosen in some problems. The constraints $g_j(X)$ represent state variables of structural response for each loading case: stresses, internal forces, displacements, natural frequencies or buckling loads.

All points which satisfy the constraints $g_j(X)$ are called feasible design and form the feasible region. Figure 1.1 shows several shapes of feasible regions for a two-variables design problem and the problem solution in each case.

- In this case several local minima exist inside the feasible region. All restraints are passive and the problem can be considered an unconstrained optimization (Fig. 1.1(a)).
- There are local minima due to the shape of the objective function, and some constraints $g_j(X)$ are activated (Fig. 1.1(b)).
- Local minima appear because of the geometry of constraints. The optimum is on the boundary with some $g_j(X) = 0$ (Fig.1.1(c)).
- This is a nonlinear problem having a unique constrained minimum as solution (Fig. 1.1(d)).
- This case corresponds to a linear problem, which always has only one minimum in situations related to engineering design (Fig. 1.1(e)).

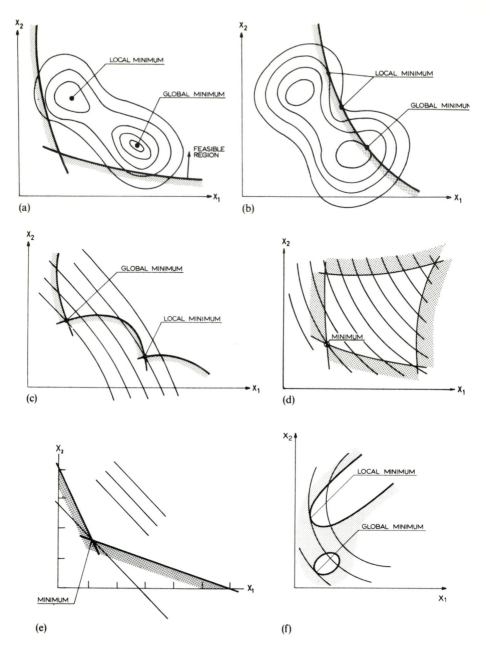

**Fig. 1.1** Shape of feasible region in minimization problems: (a) nonlinear problem with two minima in the domain (nonconvex objective function); (b) nonlinear problem with two minima on the boundary (nonconvex objective function); (c) nonlinear problem with two minima on the boundary (nonconvex feasible region); (d) nonlinear problem with one minimum; (e) linear problem with one minimum; (f) nonlinear problem with two minima (disjoint feasible regions). (Parts (a)–(c) from U. Kirsch, *Optimum Structural Design*, © 1981 McGraw-Hill, Inc. Used with permission of the publisher.)

• This is a problem with disjoint feasible regions. For each of them any of the aforementioned cases can occur (Fig. 1.1(f)).

Among situations represented in Fig. 1.1 only the case in (a) does not occur for properly posed real problems, but cases (b)–(f) are associated with many structural optimization problems. There is an important difference among them: while in cases (d) and (e) only one minimum exists, several local minima are found in cases (b) and (c) and the minimum providing the lowest value of the objective function $F(X)$ turns out to be the solution of the problem. In case (f) the feasible region is formed by various subregions, and the same considerations should be taken into account.

## 1.2 PATTERN OF SHAPE OF THE FEASIBLE REGION

In this section it is intended to classify the most important structural optimization problems according to the shape of the feasible region originated.

### 1.2.1 Feasible region bounded by straight lines

For this kind of problem the geometry of the objective function is very important. If $F(X)$ is a straight line the solution is unique and placed at a vertex of the feasible region (Fig. 1.2(a)); the optimum can be found out by using any method of linear programming such as the very well-known simplex method (Dantzig, 1963) or a more recent technique given by Karmarker (1984). By contrast, in cases when the objective function is nonlinear several local minima may exist as indicated in Fig. 1.2(b)).

A number of problems exist in structural optimization with a linear objective function and also a linear set of constraints, thus leading to a linear programming problem. A few examples follow.

### (a) Optimum design of trusses under plastic behaviour

Design variables are the cross-sectional areas of elements and the objective function chosen is the structural weight (Dorn, Gomory and Greenberg, 1964; Goble and LaPay, 1971; Russell and Reinschmidt, 1964).

An example is the symmetric four-bar truss of Fig. 1.3, with two design variables $x_1, x_2$ and loaded with two isolated forces $P$.

$$\min F(X) = \frac{4l}{3^{1/2}} x_1 + \frac{4l}{2^{1/2}} x_2 \tag{1.3}$$

subject to

$$2^{1/2} \sigma x_2 \geqslant P \tag{1.4}$$

$$3^{1/2} \sigma x_1 \geqslant P \tag{1.5}$$

$$\sigma x_1 \geqslant P \tag{1.6}$$

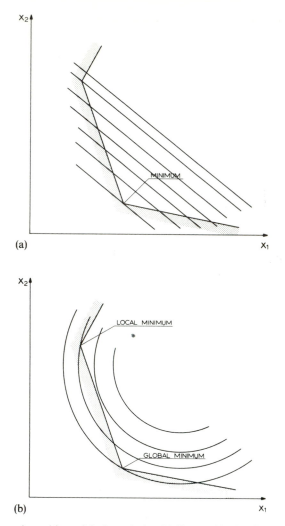

**Fig. 1.2** Feasible region with straight boundaries: (a) linear objective function; (b) nonlinear objective function.

This example is represented in Fig. 1.4 and the solution is reached at values

$$F = \left( \frac{4}{3^{1/2}} + 2 \right) \frac{Pl}{\sigma} \quad x_1 = \frac{P}{\sigma} \quad x_2 = \frac{P}{2^{1/2}\sigma} \tag{1.7}$$

**(b) Optimum design of planar frames formed by prismatic elements, considering plastic collapse related only to bending moment**

Design variables are linked to the inertia of bars and the weight is the objective function (Cohn, Ghosh and Parini, 1972; Horne and Morris, 1973). An example

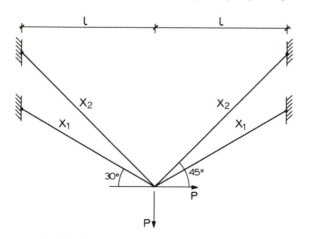

**Fig. 1.3** Four-bar truss with two design variables.

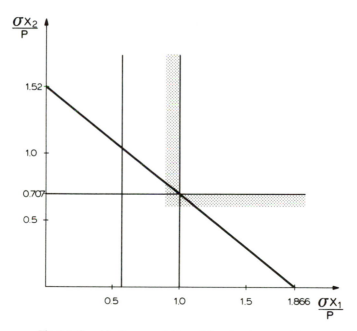

**Fig. 1.4** Graphical representation of four-bar truss problem.

is the frame shown in Fig. 1.5. A set of constraints is established by using the kinematic analysis that compares the work produced by external loads and the internal energy of plastic hinges in each collapse mode. Also, size constraints are set up in equation (1.11) to avoid the disappearance of bars.

$$F(X) = 2lx_1 + 3lx_2 \tag{1.8}$$

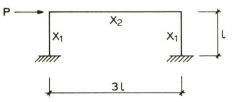

**Fig. 1.5** A simple portal frame with two design variables.

subject to

$$2x_1 + 2x_2 \geqslant 1 \tag{1.9}$$

$$4x_1 \geqslant 1 \tag{1.10}$$

$$x_1, x_2 \geqslant 0.1 \tag{1.11}$$

The problem is drawn in Fig. 1.6 and the optimal solution is

$$F = 1.1l \quad x_1 = 0.4 \quad x_2 = 0.1 \tag{1.12}$$

If optimum design of frames is carried out considering not plastic behavior but linear static analysis, the problem turns out to be nonlinear. In addition to that, many other sets of constraints need to be considered in real structures, including local buckling, lateral torsion buckling, and code-allowable stresses.

### (c) Optimization of prestressing forces and tendon configuration in continuous beams, grillages or frames

Design variables are the prestressing forces and tendon positions throughout the structure. Constraints can be related to stresses, displacements or tendon curvature (Goble and LaPay, 1971; Kirsch, 1973; Maquoi and Rondal, 1977).

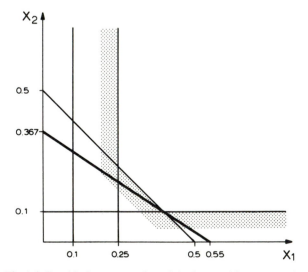

**Fig. 1.6** Graphical representation of single portal frame example.

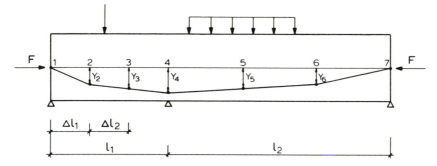

**Fig. 1.7** Prestressed beam example.

Consider a continuous beam of several spans with $i = 1, \ldots, C$ applied loading conditions (Fig. 1.7). Tendon configuration can be approximated to a polygonal line defined by the eccentricities $y_j$ at each section studied.

The prestressing force $F$ shown in Fig. 1.7 produces a bending moment $M_k$ and a displacement $u_k$ at a coordinate $x_k$ that can be written as

$$M_k = F \sum_{j=1}^{K} m_{jk} y_j \quad u_k = F \sum_{j=1}^{K} u_{jk} y_j \tag{1.13}$$

where $m_{jk}$ and $u_{jk}$ are influence coefficients of eccentricity $y_k$ at coordinate $x_k$.

The aim in this problem is to minimize the amount of prestressing force $F$. The usual formulation includes constraints for stresses $\sigma_k$ and $\tau_k$, displacement $u_k$, tendon configuration $y_k$ and curvature $\rho_k$.

Assuming Navier theory the stresses $\sigma_k$ and $\tau_k$ at a coordinate $x_k$ may be evaluated as

$$\sigma_k = F \left( \sum_{j=1}^{K} \frac{m_{jk}}{W_k} y_j - \frac{1}{A_k} \right) + \sigma_{\mathrm{P}k} \tag{1.14}$$

$$\tau_k = F \sum_{j=1}^{K} \tau_{jk} y_j + \tau_{\mathrm{P}k} \tag{1.15}$$

Similarly the displacement $u_k$ is

$$u_k = F \sum_{j=1}^{K} u_{jk} y_j + u_{\mathrm{P}k} \tag{1.16}$$

where $W_k$ and $A_k$ are the section modulus and cross-sectional area: $\tau_{jk}$ and $u_{jk}$ are influence coefficients of shear stresses and displacements; $\sigma_{\mathrm{P}k}, \tau_{\mathrm{P}k}$ and $u_{\mathrm{P}k}$ are normal stress, shear stress and displacement produced by the applied external loads.

Considering constraints for tendon eccentricity $y_j$ and curvature $\rho_j$, and setting up lower and upper limits for all constraints, the problem formulation turns out to be

$$\min F \tag{1.17}$$

subject to

$$\sigma^L \leqslant F\left(\sum_{j=1}^{K} \frac{m_{jk}}{W_k} y_j - \frac{1}{A_k}\right) + \sigma_{Pk} \leqslant \sigma^U \tag{1.18}$$

$$\tau^L \leqslant F\sum_{j=1}^{K} \tau_{jk} y_j + \tau_{Pk} \leqslant \tau^U \tag{1.19}$$

$$u^L \leqslant F\sum_{j=1}^{K} u_{jk} y_j + u_{Pk} \leqslant u^U \tag{1.20}$$

$$y^L \leqslant y_j \leqslant y^U \quad j = 1, \ldots, K \tag{1.21}$$

$$\frac{1}{\rho^U} \leqslant \frac{1}{\rho_j} \leqslant \frac{1}{\rho^L} \tag{1.22}$$

Prestressing losses through the tendon can also be included in the formulation given by equations (1.17)–(1.22). An expression which approximates quite adequately to the actual variation of prestressing force $F$ is (Hernández, 1990)

$$F = F_0 e^{-\mu x_k} \qquad 0 \leqslant x_k \leqslant L/2 \tag{1.23}$$

$$F = F_0 e^{-\mu(L - x_k)} \quad L/2 \leqslant x_k \leqslant L \tag{1.24}$$

where $F_0$ is the prestressing force at both ends of the beam, $L$ is the beam length and $\mu$ is a parameter of value $0.1 \leqslant \mu \leqslant 0.4$, depending on the prestressing procedure.

By carrying out a change of variables

$$F_0 = Y_0 \tag{1.25}$$

$$F y_j = Y_j \tag{1.26}$$

the problem formulated by equations (1.17)–(1.22) turns out to be linear, so it can be solved easily with techniques of linear programming.

### 1.2.2 Problems with a nonlinear feasible region

Most problems on structural optimization belong to this kind of situation. Problems such as optimum design of structures under elastic behavior or optimization of cross sections to obtain least weight lead to a feasible region bounded by constraints defined by curve lines. The following examples describe such situations.

The bar shown in Fig. 1.8 has the volume as objective function, and dimensions $b, e$ of its cross-section as design variables (Hernández, 1990a). Collapse modes considered are

1. yield stress $\sigma_e$,
2. buckling stress $\sigma_e/\omega$,
3. maximum value of slenderness ratio $\lambda = l/i$,
4. geometrical ratio $b/e \leqslant 30$, and
5. minimum value of $e \geqslant 0.3$.

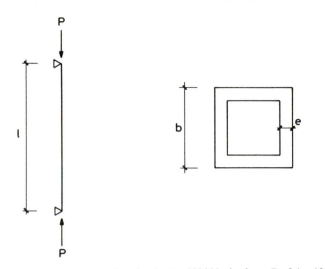

**Fig. 1.8** Single bar under compression load. $P = 800\,\text{kN}$; $l = 3\,\text{m}$; $E = 2.1 \times 10^4\,\text{kN cm}^{-2}$; $\sigma_4 = 2.6\,\text{kN cm}^{-2}$.

Buckling is incorporated via the parameter $\omega$ (Spanish code NBE-MV 103 (1978)), producing equation (1.29) where $\sigma_E$ is Euler's buckling stress

$$\sigma_E = \frac{\pi^2 EI}{l^2}$$

The slenderness ratio in equation (1.30) is limited to an upper value of 200 by the Spanish code mentioned earlier. Constraints 4 and 5 refer to design considerations in real structures.

Formulation of this problem can be written as

$$\min F = 4ebl \tag{1.27}$$

subject to

$$\frac{\rho}{4be} \leqslant \sigma_e \tag{1.28}$$

$$\frac{2P}{4be} \frac{\sigma_e}{1.3\sigma_e + \sigma_E - [(1.3\sigma_e + \sigma_E)^2 - 4\sigma_e\sigma_E]^{1/2}} \leqslant \sigma_e \tag{1.29}$$

$$2.45\frac{l}{b} \leqslant 200 \tag{1.30}$$

$$\frac{b}{e} \leqslant 30 \tag{1.31}$$

$$-e \leqslant -0.3 \tag{1.32}$$

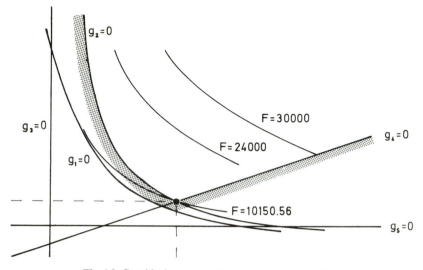

**Fig. 1.9** Graphical representation of single-bar problem.

The objective function and the set of constraints which define a feasible region with curved geometry are shown in Fig. 1.9. The solution is at the following values:

$$e = 0.53\,\text{cm} \quad b = 15.96\,\text{cm} \quad F = 10\,150.56\,\text{cm}^3$$

Another example of optimum design of structures is the grillage of Fig. 1.10. It is loaded by two distributed loads $q_1, q_2$, and an isolated load $P$ (Moses and Onoda, 1969). The objective function to be minimized is the volume of beams, considering constraints of normal stress at points of maximum bending moments. Design variables are $x_1, x_2$ and other mechanical parameters, such as the strength modulus $W$ and inertia modulus $I$, are usually linked as in equation (1.32) in

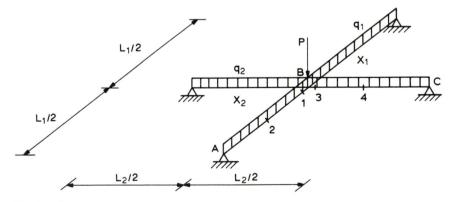

**Fig. 1.10** Four-bar grillage. (From Moses and Onoda, *International Journal of Numerical Methods in Engineering*, 1969, Vol. 1, pp. 311–331. © 1969 John Wiley & Sons Ltd. Reprinted by permission of the publisher.)

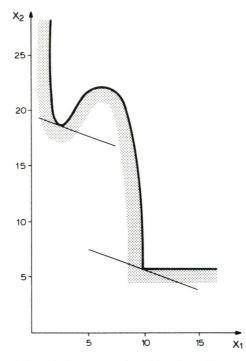

**Fig. 1.11** Graphical representation of four-bar grillage problem.

order to reduce the number of variables of the problem.

$$W_i = \left(\frac{x_i}{1.48}\right)^{1.82} \qquad I_i = 1.007\left(\frac{x_i}{1.48}\right)^{2.65}. \tag{1.33}$$

When the values are

$$-\sigma^L = \sigma^U = 20\,\text{kN}\,\text{cm}^{-2} \quad L_1 = 100\,\text{cm} \quad L_2 = 300\,\text{cm} \tag{1.34}$$

$$q_1 = q_2 = 0.1\,\text{kN}\,\text{m}^{-1} \qquad P = 1\,\text{kN} \tag{1.35}$$

the feasible region has the shape indicated in Fig. 1.11. Two local minima exist and correspond to the following values:

$$x_1 = 2.2\,\text{cm} \quad x_2 = 18.8\,\text{cm} \quad V = 5860\,\text{cm}^3 \tag{1.36}$$

$$x_1 = 9.7\,\text{cm} \quad x_2 = 6.1\,\text{cm} \quad V = 2800\,\text{cm}^3 \tag{1.37}$$

The major difference between the last two examples is the number of local minima which appear in each one. This difference is due to the shape of the feasible region.

A domain D is called convex if, given two points $X_1$ and $X_2$ contained in it, the line segment joining them is also contained completely in the domain. In other words, the domain D is convex if for $0 < \alpha < 1$ any point $X = \alpha X_1 + (1 - \alpha)X_2$ is also inside the domain.

Examples of convex and nonconvex domains are shown in Figs 1.12(a) and 1.12(b) respectively. The intersection of convex domains is also convex, as in Fig. 1.13(a); domains bounded by straight lines, as for the example in Fig. 1.13(b), are always convex. Convex or nonconvex domains may be bounded or unbounded.

A minimization problem is called convex if the objective function $F(X)$ is convex and the set of constraints $g_j(X)$ bounds a convex domain; otherwise the problem is nonconvex. In accordance with that, linear problems are always convex. If a minimization problem is convex there is only one minimum which is the solution.

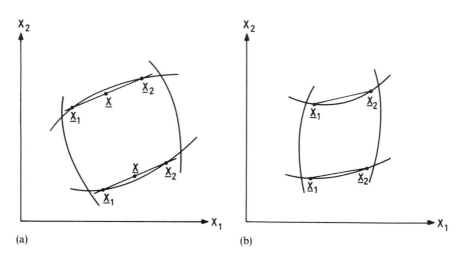

**Fig. 1.12** Examples of optimization domains: (a) convex domain; (b) nonconvex domain.

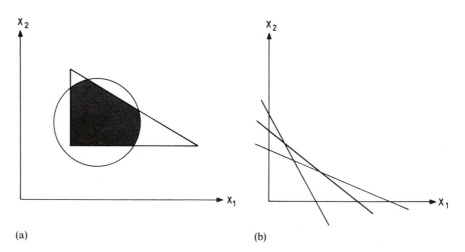

**Fig. 1.13** Examples of convex domains; (a) intersection of two convex domains; (b) unbounded convex domain.

On the contrary, in nonconvex problems several local minima may exist and the least value among them corresponds to the global minimum and the desired solution.

Some constraints considered in structural optimization problems lead to convex problems. Svanberg (1984, 1991) has identified the following cases.

- *Symmetric displacement constraint.* A constraint is said to be a symmetric displacement when the displacement vector $u_k$ of a node has the same direction as the external load $P_k$ acting at this node. Usually an upper bound is included for $u_k$.

$$u_k = \mu P_k \leqslant u_k^U \qquad (1.38)$$

- *Global displacement constraint.* This consists of imposing a maximum value limit $u^U$ to any component of the displacement vector $u$ of the structure.

$$\max u_j \leqslant u^U \quad j = 1, \ldots, N \qquad (1.39)$$

- *Lower limit of the smallest eigenvalue.* The constraint is intended to set up a lower bound for the smallest eigenvalue of the stiffness matrix $K$.

$$\lambda_m \leqslant \lambda_j \quad j = 1, \ldots, N \qquad (1.40)$$

Many other constraints, such as stress- or displacement-related constraints, produce nonconvex feasible regions. Usually, in any structural optimization problem many kinds of constraints are considered and the feasible region can be expected not to be convex, and some local minima may appear.

### 1.2.3 Problems with disjoint feasible regions

This kind of situation arises in structures undergoing dynamic loading in the absence of damping. It was first described by Cassis (1974) in planar frames subjected to horizontal vibrations at the foundation. In the case shown in Fig. 1.14 the frame is subjected to a distributed load and a harmonic vibration of ground acting during half a period. Design variables are the inertia moduli of columns $I_1$ and $I_2$ and the weight is the objective function. The feasible design domain is represented in Fig. 1.14(a) and it appears to be split into two disjoint regions, a greater, unbounded one and another that is smaller, placed as a hole in the infeasible solutions domain.

Other authors (Johnson, 1976; Johnson *et al.*, 1976; Mills-Curran and Schmit, 1985) have pointed out that in a dynamic system with $N$ degrees of freedom undergoing $C$ harmonic loads, the maximum number of disjoint feasible regions is $n(C + 1)$.

Another well-known example is a thin-walled cantilever tubular beam shown in Fig. 1.15 subjected to a torsional moment of amplitude $M_0$ and frequency $\omega_e$, acting along its length. The objective is to minimize the beam weight; the sets of constraints are the yield stress of material and the lower limits of the design values. Again, the feasible region is constituted by two regions as shown in Fig. 1.16 for two different loading values. In both cases each region has a local minimum. The solution at point B, that corresponds to a larger value of $e_1$, gives

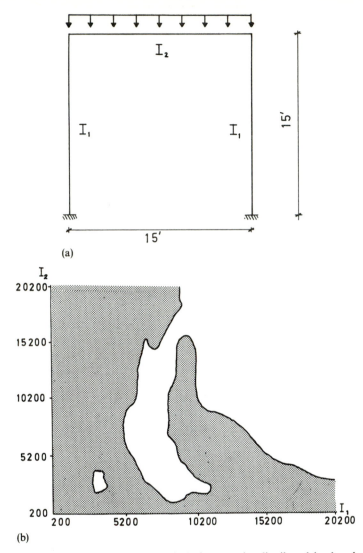

**Fig. 1.14** Single portal frame example: (a) single frame under distributed load and vibrating foundation; (b) two disjoint feasible regions arising in single portal frame example.

a structure is which the first eigenvalue $\omega_1$ satisfies $\omega_e < \omega_1$; the minimum at point A with smaller values of $e_1$ has the opposite characteristic.

## 1.3 NUMERICAL METHODS TO OBTAIN LOCAL MINIMA

There are several techniques to solve the minimization problem expressed by equations (1.1) and (1.2) but they can only guarantee that local minima are reached. A suitable division of these methods is

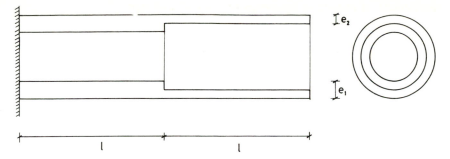

**Fig. 1.15** Cantilever tubular example.

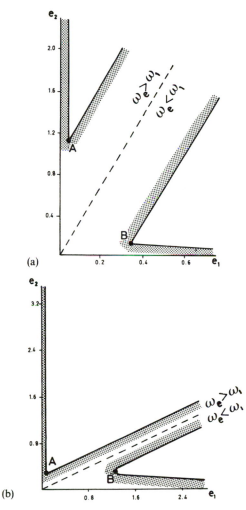

**Fig. 1.16** Feasible domains in cantilever tubular example: (a) global minimum with $\omega_e < \omega_l$; (b) global minimum with $\omega_e > \omega_l$. (From E.H. Johnson, *AIAA Journal*, 1976, Vol. 14, pp. 259–261. © 1976 American Institute of Aeronautics and Astronautics. Reprinted by permission of the publisher.)

- optimality criteria methods,
- methods of mathematical programming, and
- approximation techniques.

### 1.3.1 Optimality criteria methods

(a) Kuhn–Tucker-based methods

Let us consider a problem defined by

$$\min F(X) \tag{1.41}$$

subject to

$$g_j(X) \leqslant 0 \quad i = 1, \ldots, m \tag{1.42}$$

Kuhn–Tucker necessary conditions establish that a point $X$ is a local minimum if a set of scalars $\lambda_j$ exists such that

$$\lambda_j \geqslant 0 \quad j = 1, \ldots, J \tag{1.43}$$

and

$$\frac{\partial F}{\partial x_i} + \sum_{j=1}^{J} \lambda_j \frac{\partial g_j}{\partial x_i} = 0 \quad i = 1, \ldots, n \tag{1.44}$$

Equation (1.44) explains that at a minimum the vector $-\nabla F$ is a linear combination of the gradients of the active constraints subset with positive components for all gradients $\nabla g_j$ active at this point. In the opposite case, when some components $\lambda_j$ are nonpositive the point is not a minimum. In Fig. 1.17 two different cases of this situation are represented.

To solve the minimization problem it is necessary to find values of the Lagrange

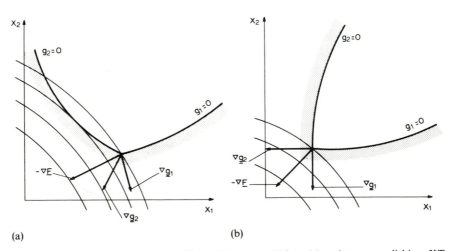

(a)                                      (b)

**Fig. 1.17** Graphical explanation of Kuhn–Tucker condition: (a) point accomplishing KT condition; (b) point not accomplishing KT condition. (From U. Kirsch, *Optimum Structural Design*, © 1981 McGraw-Hill, Inc. Used with permission of the publisher.)

multipliers $\lambda$ and design variables $X$. An iterative process is usually carried out which requires

- a recurrence relation between design variables $X$, and
- identifying the subset of active constraints at the optimum.

The recurrence relation is set up taking into account that at the solution

$$\frac{\partial F}{\partial x_i} + \sum_{j=1}^{m} \lambda_j \frac{\partial g_j}{\partial x_i} = 0 \quad i = 1, \ldots, n \tag{1.45}$$

assuming that

$$\frac{\partial F}{\partial x_i} \neq 0$$

$$- \sum_{j=1}^{m} \lambda_j \frac{\partial g_j}{\partial x_i} \Big/ \frac{\partial F}{\partial x_i} = 1 \tag{1.46}$$

Throughout the process the left-hand side of equation (1.46) has the value

$$- \sum_{j=1}^{m} \lambda_j \frac{\partial g_j}{\partial x_i} \Big/ \frac{\partial F}{\partial x_i} = T_i \quad i = 1, \ldots, n \tag{1.47}$$

and at the minimum will take the value $T_i = 1$ $(i = 1, \ldots, n)$. At the $k$th iteration equation (1.46) will be

$$- \sum_{j=1}^{m} \lambda_{jk} \frac{\partial g_{jk}}{\partial x_i} \Big/ \frac{\partial F_k}{\partial x_i} = T_{ik} \tag{1.48}$$

From equation (1.46) a recurrence relation may be established between values of any element $x_i$ of design variables $X_k$ and $X_{k+1}$ at two consecutive iterations. A commonly used relation is (Khot, Berke and Venkayya, 1979)

$$x_{ik+1} = [\alpha + (1 - \alpha) T_{ik}] x_{ik} \quad i = 1, \ldots, n \tag{1.49}$$

where $\alpha$ is a relaxation factor controlling the step size. It usually takes values $\alpha = 0.5$ or $\alpha = 0.75$.

To obtain the Lagrange multipliers $\lambda$ the following procedure is carried out. The set of active constraints $g_j(X)$ is linearized (Kinsalaas, 1972; Rizzi, 1976):

$$g_{jk+1} \approx g_{jk} + \sum_{i=1}^{n} \frac{\partial g_{jk}}{\partial x_i} (x_{ik+1} - x_{ik}) = 0 \quad j = 1, \ldots, J \tag{1.50}$$

introducing the expression for $x_{ik+1}$ given by equation (1.49)

$$g_{jk} + (1 - \alpha) \sum_{i=1}^{n} \frac{\partial g_{jk}}{\partial x_i} (T_{ik} - 1) x_{ik} = 0 \quad j = 1, \ldots, J \tag{1.51}$$

If $T_{ik}$ is substituted by its value in equation (1.47) it turns out that for $j = 1, \ldots, J$

$$\sum_{i=1}^{n} \frac{\partial g_{jk}}{\partial x_i} x_{ik} \sum_{j=1}^{m} \lambda_j \frac{\partial g_{jk}}{\partial x_i} \Big/ \frac{\partial F_k}{\partial x_i} = \frac{g_{jk}}{1 - \alpha} - \sum_{i=1}^{n} \frac{\partial g_{jk}}{\partial x_i} x_{ik} \tag{1.52}$$

Equation (1.52) is a system of linear equations with $\lambda_j (j = 1, \ldots, J)$ as unknowns. Each of them that has a nonnegative value corresponds to active constraints, according to the Kuhn–Tucker conditions.

Briefly, the whole process is carried out by repeating the following steps.

1. Define an initial design $X$.
2. Solve the system of equations to obtain the vector $\lambda$.
3. Calculate values of $T_i$ considering only the subset of constraints associated with nonnegative $\lambda_j$.
4. Go to recurrence relation (1.21) and go back to step 2.

Steps 2 through 4 should be repeated until absolute or relative convergence is achieved. If the subset of active constraints $g_j(X)$ $(j = 1, \ldots, J)$ changes frequently from one iteration to the next, convergence may become difficult. The same happens if the initial design is very dissimilar to the solution.


### (b) Method based on information theory

In this method (Templeman and Li, 1985, 1987; Templeman, 1989) the optimization problem defined by equation (1.1) is substituted by an equivalent problem:

$$\min F(X) \tag{1.53}$$

subject to

$$\sum_{j=1}^{m} \lambda_j g_j(X) = 0 \tag{1.54}$$

$$\sum_{j=1}^{m} \lambda_j = 1 \tag{1.55}$$

$$\lambda_j \geqslant 0 \tag{1.56}$$

Each multiplier $\lambda_j$ may be considered as the probability for a constraint $g_j(X)$ to be active at the optimum. Hence, for each constraint the following condition stands:

$$g_j \lambda_j = 0 \tag{1.57}$$

Consequently, the condition (1.53) is satisfied at the solution. It is well known (Everett, 1963; Brooks and Geoffrion, 1966; Gould, 1969; Greenberg and Pierskalla, 1970) that the problem defined by equations (1.53)–(1.56) has the same solution as the problem defined by equations (1.1) and (1.2).

As in the methods mentioned earlier the vectors $X$ and $\lambda$ constitute the set of unknowns. In this technique the probabilities $\lambda_j$ are obtained by using techniques not biasing the iterative process. The innovative approach taken by Templeman and coworkers is that they do this by using techniques based on information theory such as the maximum entropy formalism. This formalism, enunciated by Jaynes (1957), indicates that the least biased values for probabilities $\lambda_j$ at each iteration are those which maximize the entropy information of the process, which is measured according to the function defined by Shannon (1948):

$$\max \frac{S(\lambda)}{K} = -\sum_{j=1}^{m} \lambda_j \ln \lambda_j \tag{1.58}$$

$$\sum_{j=1}^{m} \lambda_j = 1 \tag{1.59}$$

The optimization procedure is carried out in the following way.

1. An initial design is defined and the probabilities vector $\lambda$ is obtained through the maximization problem of equations (1.58) and (1.59). The solution of the problem turns out to be

$$\lambda = \left[\frac{1}{m}, \ldots, \frac{1}{m}\right] \qquad S = K \ln m \qquad (1.60)$$

2. By solving the minimization problems of equations (1.53)–(1.56) a new design $X_2$ is obtained.
3. Beginning at iteration $k = 2$ and thereafter the elements of the vector $\lambda_k$ at the $k$th iteration are updated to $\lambda_{k+1}$ by taking in account that

$$\sum_{i=1}^{m} \lambda_{jk} g_j(X_k) = 0 \quad \text{and} \quad \sum_{i=1}^{m} \lambda_{jk+1} g_j(X_k) = \varepsilon_{k+1} \qquad (1.61)$$

where $\varepsilon_{k+1} \neq 0$ because the vectors $\lambda_{k+1}$ and $X_k$ belong to different iterations in the procedure. As the procedure approaches convergence the value of $\varepsilon_{k+1}$ should decrease until it cancels out, in theory, at the optimum. Because of that, in order not to bias the undating of $\lambda_k$, and to maintain its normality, the probability vector $\lambda_{k+1}$ is found by solving the following maximization problem:

$$\max \frac{S(\lambda)}{K} = - \sum_{j=1}^{m} \lambda_{jk+1} \ln \lambda_{jk+1} \qquad (1.62)$$

subject to

$$\sum_{j=1}^{m} \lambda_{jk+1} = 1 \qquad (1.63)$$

$$\sum_{j=1}^{m} \lambda_{jk+1} g_j(X_k) = \varepsilon_{k+1} \qquad (1.64)$$

where $\varepsilon_{k+1}$ is a user-selected parameter the expression of which must tend to zero.
4. Then, the process goes back to steps 2 and 3 until convergence is achieved. The solution usually arrives from outside the feasible region, so if the process is stopped at an intermediate design some constraints are violated, and thus it is necessary to continue the method until a good ratio of convergence is obtained.

### 1.3.2 Methods based on mathematical programming

#### (a) Methods of linear programming

The most popular method for linear problems is the simplex method (Dantzig, 1963) and it is explained in every optimization book. A different approach started by Karmarkar (1984) has been claimed to be more efficient than the simplex method for large size problems. The main reason is that the simplex method requires an effort to obtain the solution that increases exponentially for the worst case, and in Karmarkar's methods the effort increases polynomially. There is a large number of real problems of many classes solved very efficiently by the simplex method while experience with Karmarkar's method is still limited. However, it is indeed a very promising procedure and it is explained next. Let us remember that a very usual formulation for linear problems in design optimization is

$$\min F = \sum_{j=1}^{n} c_j x_j \qquad (1.65)$$

subject to

$$\sum_{j=1}^{n} a_{ij} x_j \geqslant b_i \quad i = 1, \ldots, m \qquad (1.66)$$

$$x_j \geqslant 0 \quad j = 1, \ldots, n \qquad (1.67)$$

This formulation is called a primal problem. There exists a dual problem, in the space of dual variables $\lambda_i (i = 1, \ldots, m)$, that can be formulated as

$$\max G = \sum_{i=1}^{m} b_i \lambda_i \qquad (1.68)$$

subject to

$$\sum_{i=1}^{m} a_{ji} \lambda_i \leqslant c_j \qquad (1.69)$$

$$\lambda_i \geqslant 0 \qquad (1.70)$$

It is very well known that

$$\min F = \max G \qquad (1.71)$$

The method of Karmarkar is intended to solve problems formulated as

$$\min K = \sum_{j=1}^{N+1} e'_j y'_j \qquad (1.72)$$

subject to

$$\sum_{j=1}^{N+1} k'_{ij} y'_j = 0 \quad i = 1, \ldots, M \qquad (1.73)$$

$$\sum_{j=1}^{N+1} y'_j = 1 \qquad (1.74)$$

where the objective function $K$ is zero at the optimum.

As the problem defined by equations (1.65)–(1.67) is different from the problem solved by Karmarkar, the first required step is to convert it to a formulation suitable for the method of Karmarkar. The way to do this is to combine the primal and dual formulations of equations (1.64)–(1.69) to set up a new linear problem. Using matrix notation it can be written

$$\min(F - G) = \min K = C^T X - B^T \lambda \qquad (1.75)$$

subject to

$$AX \geqslant B \qquad (1.76)$$

$$A^T \lambda \leqslant C \qquad (1.77)$$

$$X, \lambda \geqslant 0 \qquad (1.78)$$

if slack variables are introduced

$$\min(C^T X - B^T \lambda) \qquad (1.79)$$

subject to

$$AX - X' = B \tag{1.80}$$

$$A^T \lambda + \lambda' = C \tag{1.81}$$

$$X, X', \lambda\lambda' \geqslant 0 \tag{1.82}$$

and also

$$\min E^T Y \tag{1.83}$$

subject to

$$KY = 0 \tag{1.84}$$

$$Y \geqslant 0 \tag{1.85}$$

where

$$\underset{(1 \times N)}{Y^T} = \begin{bmatrix} \underset{(1 \times n)}{Y^T} & \underset{(1 \times m)}{X'^T} & \underset{(1 \times m)}{\lambda^T} & \underset{(1 \times n)}{\lambda'^T} \end{bmatrix} \tag{1.86}$$

$$\underset{(1 \times N)}{E^T} = \begin{bmatrix} \underset{(1 \times n)}{C^T} & \underset{(1 \times m)}{0} & \underset{(1 \times m)}{-B^T} & \underset{(1 \times n)}{0} \end{bmatrix} \tag{1.87}$$

$$\underset{(1 \times M)}{D^T} = \begin{bmatrix} \underset{(1 \times m)}{B^T} & \underset{(1 \times n)}{C^T} \end{bmatrix} \tag{1.88}$$

$$\underset{(M \times N)}{K} = \begin{bmatrix} \underset{(m \times n)}{A} & \underset{(m \times m)}{-I} & \underset{(m \times m)}{0} & \underset{(m \times n)}{0} \\ \underset{(n \times n)}{0} & \underset{(n \times m)}{0} & \underset{(n \times m)}{A^T} & \underset{(n \times n)}{I} \end{bmatrix} \tag{1.89}$$

Then, starting from an initial design $Y_1$, a projective transformation is carried out

$$y'_j = \frac{y_j}{y_{1j}} \bigg/ \left( 1 + \sum_{i=1}^{N} \frac{y_i}{y_{1i}} \right) \quad j = 1, \dots, N \tag{1.90}$$

and a variable $y'_{N+1}$ is defined

$$y'_{N+1} = 1 - \sum_{j=1}^{N} y'_j \tag{1.91}$$

Equation (1.90) may be written as

$$y'_j = \frac{y'_{N+1} y_j}{y_{1j}} \tag{1.92}$$

or inversely

$$y_j = \frac{y_{1j} y'_j}{y'_{N+1}} \quad j = 1, \dots, N \tag{1.93}$$

The transformation defined by equation (1.90) transforms points of an $N$-coordinate open space into points of an $(N+1)$-coordinate closed space. It can be easily proved that the point $Y_1$ is transformed into the point

$$Y'_1 = \left( \frac{1}{N+1}, \dots, \frac{1}{N+1} \right)$$

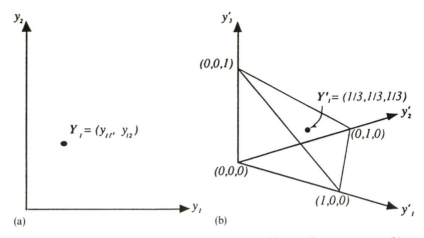

**Fig. 1.18** Projective transformation for $N = 2$: (a) space with coordinate system $y_j$; (b) space with coordinate system $y'_j$.

the center of the new coordinate system. This transformation is shown in Fig. 1.18 for an example with $N = 2$.

If we put equation (1.93) into equations (1.83)–(1.85) we obtain

$$\min \frac{e_j y_{1j} y'_j}{y'_{N+1}} \tag{1.94}$$

subject to

$$\frac{k_{ij} y_{1j} y'_j}{y'_{N+1}} = d_i \quad i = 1, \ldots, M \tag{1.95}$$

$$\sum_{j=1}^{N+1} y'_j = 1 \tag{1.96}$$

$$y'_j \geqslant 0 \quad j = 1, \ldots, N+1 \tag{1.97}$$

defining

$$e'_j = e_j y_{1j} \tag{1.98}$$

$$k'_{ij} = k_{ij} y_{1j} \quad j = 1, \ldots, N \tag{1.99}$$

$$k'_{iN+1} = -d_i \tag{1.100}$$

Examining equations (1.90) and (1.91) it can be concluded that the variable $y'_{N+1}$ always satisfies $y'_{N+1} > 0$. As the objective function $K$ is zero at the optimum, the variable $y'_{N+1}$ can be eliminated from equation (1.94). After all these changes the formulation of the problem is

$$\min K = E'^{\mathrm{T}} Y' \tag{1.101}$$

subject to

$$K' Y' = 0 \tag{1.102}$$

$$\sum_{j=1}^{N+1} y'_j = 1 \tag{1.103}$$

$$y'_j \geqslant 0 \quad j = 1, \ldots, N+1 \tag{1.104}$$

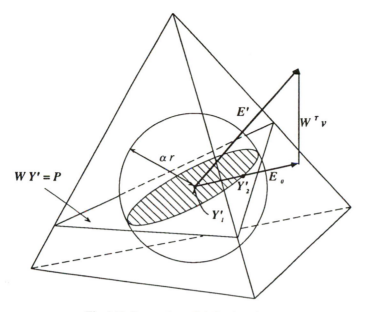

**Fig. 1.19** Karmarkar minimization phase.

that coincides with the class of problem solved by the method of Karmarkar. Constraints of equations (1.102) and (1.103) may be grouped in the form

$$WY' = \begin{bmatrix} K' \\ 1 \end{bmatrix} Y' = \begin{bmatrix} 0 \\ 1 \end{bmatrix} = P \qquad (1.105)$$

The approach of Karmarkar for obtaining the minimum is to carry out an iterative procedure that will be explained with the example that appears in Fig. 1.19. In this example $N + 1 = 4$, and thus the coordinate system is represented by a tetrahedron and the set of constraints by a plane. The projective transformation takes the initial design $Y_1$ into $Y_1'$, the center of the tetrahedron. From geometrical considerations it can be concluded that the tetrahedron inscribes a ball of radius

$$r = \frac{1}{[N(N+1)]^{1/2}}$$

Instead of moving directly to the minimum, Karmarkar tries to improve the current design $Y_1'$ by obtaining another point $Y_2'$ such that

- it is placed on the boundary of a ball centered at $Y_1'$, of radius $\alpha r$, with $0 < \alpha < 1$, and
- it decreases the objective function $K$.

As the point $Y_2'$ minimizes the objective function of equation (1.101) it will lie along $E_p'$ which is the projection of the gradient vector $-E'$ onto the null space of the constraint surface $WY' = P$, so

$$Y'_2 = Y'_1 - \alpha r \frac{E'_P}{\|E'_P\|} \qquad (1.106)$$

To identify $-E'_P$ it is useful to observe that the vector $E' - E_P$ will be in the space spanned by the gradients of $WY' = P$. Hence, there exists a vector $\boldsymbol{v}$ such that

$$W\boldsymbol{v} = E' - E'_P \qquad (1.107)$$

Premultiplying by $W^T$ gives

$$WW^T\boldsymbol{v} = WE' - WE'_P \qquad (1.108)$$

since

$$WE'_P = 0 \qquad (1.109)$$

$$\boldsymbol{v} = (WW^T)^{-1} WE' \qquad (1.110)$$

and

$$E'_P = E' - W(WW^T)^{-1}WE' = \{I - W(WW^T)^{-1}W\}^{-1}E \qquad (1.111)$$

After the point $Y'_2$ is obtained, the vector of variables $Y_2$ is calculated by reversing the projective transformation of equation (1.90). Then, the whole procedure is carried out again for as many iterations as necessary until convergence is achieved.

The four-bar structure in Fig. 1.3 may be used to explain numerically Karmarkar's procedure. Assuming numerical values

$$P = 2 \quad \sigma = 2 \quad l = 1 \qquad (1.112)$$

Equations (1.3)–(1.6) become

$$\min F(X) = 2.31x_1 + 2.82x_2 \qquad (1.113)$$

subject to

$$1.41x_2 \geqslant 1 \qquad (1.114)$$

$$1.73x_1 \geqslant 1 \qquad (1.115)$$

$$x_1 \geqslant 1 \qquad (1.116)$$

Combining the primal and dual formulation gives

$$\min 2.31x_1 + 2.82x_2 - \lambda_1 - \lambda_2 - \lambda_3 \qquad (1.117)$$

subject to

$$1.41x_2 - x_3 \qquad\qquad\qquad\qquad\qquad = 1 \qquad (1.118)$$

$$1.73x_1 \qquad\quad - x_4 \qquad\qquad\qquad\qquad = 1 \qquad (1.119)$$

$$x_1 \qquad\qquad - x_5 \qquad\qquad\qquad = 1 \qquad (1.120)$$

$$1.73\lambda_2 + \lambda_3 + \lambda_4 \quad = 2.31 \qquad (1.121)$$

$$1.41\lambda \qquad\qquad\qquad + \lambda_5 = 2.82 \qquad (1.122)$$

Choosing $Y_1$ as initial design

$$Y_1 = (2, 1.41, 2, 4.92, 2, 0.5, 0.233, 0.646, 0.5, 1) \qquad (1.123)$$

**Table 1.1** Numerical values for iterations in the Karmarkar procedure

| $N_{ite}$ | $x_1$ | $x_2$ | $K$ | $F$ |
|---|---|---|---|---|
| 1 | 1.69 | 1.15 | 0.44 | 7.15 |
| 7 | 1.01 | 0.75 | 0.098 | 4.58 |
| 15 | 1.00 | 0.71 | 0.008 | 4.33 |

and carrying out the operations indicated by equations (1.98)–(1.100) the following formulation of Karmarkar is obtained:

$$\min 4.62y_1' + 3.976y_2' - 0.5y_6' - 0.233y_7' - 0.646y_8' \qquad (1.124)$$

subject to

$$2y_2' - 2y_3' = 0 \qquad (1.125)$$
$$3.46y_1' - 4.92y_4' = 0 \qquad (1.126)$$
$$2y_1' - 2y_5' = 0 \qquad (1.127)$$
$$0.40y_7' + 0.646y_8' + 0.5y_9' = 0 \qquad (1.128)$$
$$0.70y_6' + y_{10}' = 0 \qquad (1.129)$$

Numerical values for several iterations are shown in Table 1.1.

### (b) Methods based on mathematical programming

All these methods try to obtain a minimum of the optimization problem defined by equations (1.1) and (1.2). Starting from an initial design $X_1$ the process is carried out in an iterative scheme producing intermediate designs $X_k$ which improve the objective function and bring the problem to the solution.

Design modification is done at each iteration along a search direction $S_k$ as indicated in

$$X_{k+1} = X_k + \alpha S_k \qquad (1.130)$$

Almost every method has a different approach to obtaining the direction $S_k$, and the most important techniques are explained next.

In the method of **feasible directions** (Zoutendijk, 1960; Vanderplaats, 1984a,b) the choice of $S_k$ depends on the position of point $X_k$.

1. If $X_k$ is inside the feasible region the gradient of the objective function is chosen as indicated below

$$S_k = -\nabla F_k \qquad (1.131)$$

2. If $X_k$ is on the boundary, at one or more constraints, the aim is to improve the current design $X_k$ keeping the new point $X_{k+1}$ inside the feasible region as much as possible. To do that the direction $S_k$ is calculated by solving a linear programming problem stated by equations (1.132)–(1.135):

$$\max \beta \qquad (1.132)$$

subject to

$$S_k^T \nabla g_j + \theta_j \beta \leqslant 0 \quad j = 1, \ldots, J \tag{1.133}$$

$$S_k^T \nabla F + \beta \leqslant 0 \tag{1.134}$$

$$-1 \leqslant S_k \geqslant 1 \tag{1.135}$$

There are two possibilities for placing the point $X_{k+1}$: it can be in the domain, or at a new constraint. Both possibilities are shown in Fig. 1.20. If $X_{k+1}$ is located in the domain the method proceeds according to step 1. If $X_{k+1}$ is on the boundary then step 2 is carried out.

The constraints $g_j(X)$ in equation (1.135) are those which contain the current design and thus $g_j(X) = 0$. The scalars $\theta_j$ are usually given by $\theta_j = 1$.

Other methods change the problem of equations (1.1) and (1.2) into an unconstrained optimization problem. This is how the **sequential unconstrained minimization technique** (SUMT) acts. This approach, also called the **penalty function method** (Fiacco and McCormick, 1963, 1968; Zangwill, 1967; Kavlie, 1971; Haftka and Starnes, 1976) creates a penalty function in the following way:

$$\phi(X, r) = F(X) + r \sum_{j=1}^{m} p(g_j) \tag{1.136}$$

The penalty function $P(g_j)$ is formed by using the set of constraints and the parameter $r$ is a control parameter of convergence. There are two variations of this technique.

● *Exterior penalty function.* In this formulation, equation (1.136) is written as

$$\phi(X, r) = F(X) + r \sum_{j=1}^{m} \langle g_j \rangle^\gamma \tag{1.137}$$

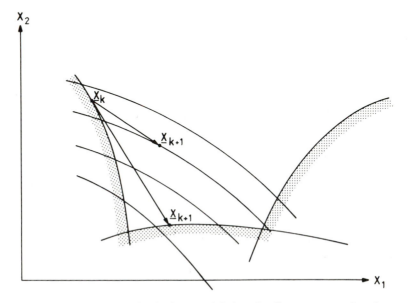

**Fig. 1.20** Possible situations for the improved design after linear programming phase.

In equation (1.137) the expression $\langle g_j \rangle$ symbolizes

$$\langle g_j \rangle = g_j \quad \text{if} \quad g_j(X) > 0 \tag{1.138}$$

$$\langle g_j \rangle = 0 \quad \text{if} \quad g_j(X) \leqslant 0 \tag{1.139}$$

and usually $\gamma = 2$.

The starting point must be an infeasible design. The function $\phi(X,r)$ is minimized repeatedly for increasing values of the control parameter $r$. This produces intermediate designs $X_k$ that reduce the values of the violated constraints at each iteration and thus leads to the solution. The starting point must be an infeasible design. Minimization of the function $\phi(X,r)$ forces each violated constraint to become smaller. The minimum $\phi(X,r)$ is also the minimum of $f(X)$ and it is reached from outside the feasible region.

- *Interior penalty function.* In this case equation (1.136) becomes

$$\phi(X,r) = F(X) - r \sum_{j=1}^{m} \frac{1}{g_j(X)} \tag{1.140}$$

In this technique the starting point must be a feasible one, so any constraint is negative. The function defined by equation (1.140) is minimized iteratively for decreasing values of $r$. This causes intermediate designs to be obtained that produce lower bounds of every constraint in order to reduce $\phi(X,r)$. As in the previous technique the minimum of $\phi(X,r)$ is the same as the minimum of the problem of equations (1.1) and (1.2). The solution is obtained from inside the feasible region, so each intermediate design is a valid one.

In both categories of penalty function methods it is necessary to solve several times an unconstrained optimization problem, which is accomplished according to equation (1.130). There are three groups of techniques to obtain $S_k$.

- *Order zero techniques.* One of the most useful algorithms is the **conjugate directions** (Powell, 1964) approach. This method requires a set of conjugate directions to be identified. A set of directions are called conjugate when they satisfy the condition

$$S_i^T A S_j = 0 \quad \text{for} \quad i \neq j \tag{1.141}$$

$A$ is a positive definite quadratic form. This technique only needs to evaluate the objective function and, when the objective function of the unconstrained problem is quadratic, the minimum is obtained by optimizing once along each direction of the conjugate directions set. In any other case it could be necessary to repeat this technique several times before reaching the optimum.

- *Order one techniques.* These are usually called gradient methods (Fletcher and Reeves, 1964) and they require the gradient of the objective function to be calculated. The direction $S_k$ is defined as

$$S_k = -\nabla F_k + \beta_k S_{k-1} \tag{1.142}$$

The value of $\beta_k$ depends on each specific formulation. A pair of more well-known cases are **steepest descent**, that is

$$\beta_k = 0 \tag{1.143}$$

and **conjugate gradient**, that is

$$\beta_k = \frac{\nabla F^T \nabla F_k}{\nabla F_{k-1}^T \nabla F_{k-1}} \tag{1.144}$$

- *Order two methods.* These are denominated Newton methods, and, in addition to the objective function $F(X)$ and its gradient, they require the hessian matrix $H(X)$ to be evaluated. Each direction $S_k$ is defined as

$$S_k = -H_k^{-1} \nabla F_k \tag{1.145}$$

Some variants of this method (Davidon, 1959; Fletcher and Powell, 1963) obtain the matrix $H(X)$ by using approximation schemes.

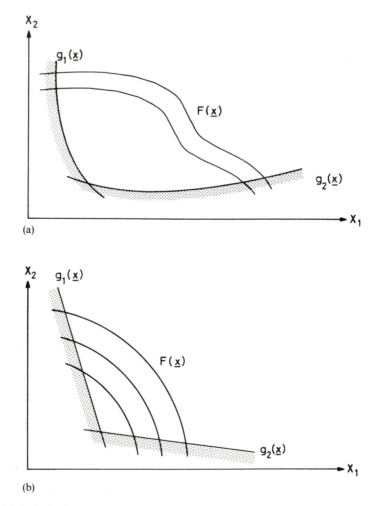

(a)

(b)

**Fig. 1.21** Optimization problem and quadratic approximation: (a) actual optimization problem; (b) quadratic approximation.

### 1.3.3 Approximation techniques

Finally, another type of method solves the optimization problem by writing equations (1.1) and (1.2) using a Taylor series expansion, which is truncated after two or three terms, thus creating an approximate problem. The most efficient techniques are as follows.

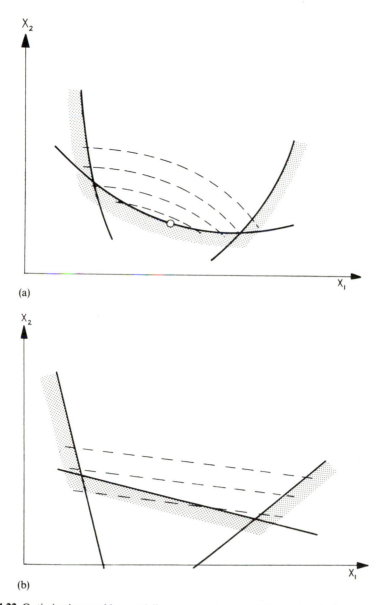

(a)

(b)

**Fig. 1.22** Optimization problem and linear approximation: (a) actual optimization problem; (b) linear approximation. (From E. Atrek, R.H. Gallagher, K.M. Ragsdell and O.C. Zienkiewicz (eds), 1984, *New Directions in Optimum Structural Design*, John Wiley & Sons. Reprinted by permission of John Wiley & Sons, Ltd.)

- *Sequence of quadratic problems.* Once an initial design $X_1$ is chosen, the problem of equations (1.1) and (1.2) is approximated to

$$F(X) \approx F(X_1) + \nabla F(X_1)^T(X - X_1) + \tfrac{1}{2}(X - X_1)^T H(X_1)(X - X_1) \quad (1.146)$$

$$g_j(X) \approx g_j(X_1) + \nabla g_j^T(X_1)(X - X_1) \leqslant 0 \quad j = 1, \ldots, m \quad (1.147)$$

By solving equations (1.146)–(1.147) a new design $X_2$ is given. Then, these expressions are again calculated at the new point $X_2$ and the optimization is carried out repeatedly until convergence is obtained (Fig. 1.21).

- *Sequence of linear problems.* In this approach (Fig. 1.22) the formulation is

$$\min F(X) \approx F(X_1) + \nabla F_1^T(X - X_1) \quad (1.148)$$

$$g_j(X) \approx g_j(X_1) + \nabla g_j^T(X_1)(X - X_1) \leqslant 0 \quad (1.149)$$

Several variants of this technique exist (Cheney and Goldstein, 1959; Kelley,

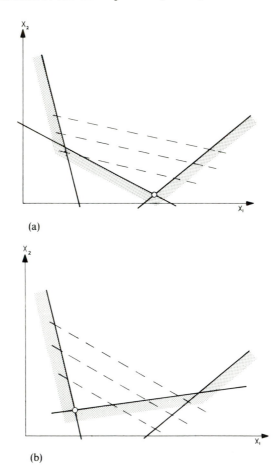

(a)

(b)

**Fig. 1.23** Solution of a linear approximation problem oscillating between two points. (From E. Atrek, R.H. Gallagher, K.M. Ragsdell and O.C. Zienkiewicz (eds), 1984, *New Directions in Optimum Structural Design*, John Wiley & Sons. Reprinted by permission of John Wiley & Sons, Ltd.)

1960), and sometimes a few inconveniences arise in this method. If the solution of the real problem formulated by equations (1.1) and (1.2) is placed on a unique constraint, the linear approximation may oscillate indefinitely between the two vertices nearer the real optimum, as shown in Fig. 1.23. There are improvements on this method which can overcome this drawback (Baldur, 1972; Hernández and Ceron, 1987) by stating some **move limits** to set bounds to the variation of the design variables at each iteration.

These move limits are intended to avoid the oscillating behavior of linear approximations by preventing any design variable from changing too much between two consecutive iterations. Let us suppose that $x_i^L, x_i^U$ are lower and upper limits of variable $x_i$; if this variable has the value $x_{ik}$ at the $k$th iteration and a move limit $\Delta x_{ik}$ it set up, the value of the design variable $x_{ik+1}$ would be limited by

$$\max(x_i^L, x_{ik} - \Delta x_{ik}) \leqslant x_{ik+1} \leqslant \min(x_i^U, x_{ik} + \Delta x_{ik}) \qquad (1.150)$$

Move limits can be set up independently for each variable with different values for lower and upper limits. Additionally, they can also vary between iterations.

## 1.4 THE TUNNELING METHOD IN GLOBAL OPTIMIZATION

Given a function $F(X)$ each optimization technique mentioned previously starts by defining an initial design $X_1$ and tries to obtain another point $X_1^*$ which satisfies

$$F(X_1^*) \leqslant F(X_1) \qquad (1.151)$$

$$F(X_1^*) \leqslant F(X_1^* + \Delta X) \qquad (1.152)$$

where $\Delta X$ is an increment of $X_1^*$.

If the problem considered is a convex one the value of $F(X_1^*)$ satisfying equations (1.151) and (1.152) is the optimal solution and the global minimum of $F(X)$, but in the opposite case several points $X_i^*$ may be found under conditions (1.151) and (1.152), so there is a set of local minima containing the global one. If points corresponding to local minima are ordered according to their objective function value the following series can be written:

$$F(X_1^*) \geqslant F(x_2^*) \geqslant \cdots \geqslant F(X_n^*) \qquad (1.153)$$

In this section a method for obtaining the global minimum in nonconvex problems is presented, and numerical results corresponding to practical examples are included to show the performance of the method.

### 1.4.1 The tunneling method of unconstrained optimization

This method has been developed by Gómez and Levy (1982) and Levy and Gómez (1984) and is intended to identify, in a systematic way, each of the local minima of a nonconvex function. Along the procedure each new minimum obtained produces an equal or lower objective function value $F(X)$ than the previous one.

Given a function $F(X)$, the aim of the method is obtain a point $X^*$ that for any vector $Y$ satisfies

$$F(X^*) \leqslant F(X) \tag{1.154}$$

$$\nabla F(X^*) = 0 \quad Y^T H(X^*) Y > 0 \tag{1.155}$$

The method has two different phases.

1. *Minimization phase.* The purpose of this phase is to obtain a local minimum satisfying equation (1.154). Starting from an initial point $X_1$ the first local minimum $X_1^*$ is identified as shown in Fig. 1.24. In order to do it, any numerical optimization technique already mentioned may be used.
2. *Tunneling phase.* Departing from $X^*$, this phase aims to identify a point $X_2$, indicated in Fig. 1.24, having $F(X_1^*) = F(X_2)$ and not being a local minimum.

After the tunneling phase the process returns to phase 1. By carrying out these two phases repeatedly a set of local minima producing decreasing objective function values is obtained, and eventually the global minimum $X_G$ is reached.

The method is conceptually similar to a miner excavating a horizontal tunnel until finding a sloping cavern to descend into, and having to proceed several

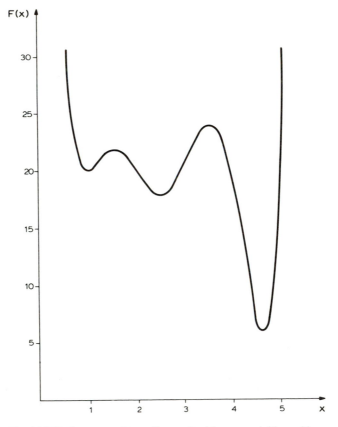

**Fig. 1.24** Performance of tunneling method in one variable problem.

times repeating the same task until arriving at the lowest point. Because of that this technique is called the **tunneling method**.

Phase 1 is not worth specific comment as it is solved by techniques discussed before. However, we need to explain how the method works in the tunneling phase. Given a local minimum $X_i^*$ the following tunneling function is defined:

$$T(X, X_i^*, s_i) = \frac{F(X) - F(X_i^*)}{[(X - X_i^*)^T (X - X_i^*)]^{s_i}} \tag{1.156}$$

In equation (1.156) the parameter $s_i$ is the order of the root of the function $F(X) - F(X_i^*) = 0$ at $X = X_i^*$. In consequence, the tunneling function is $T(X, X_i^*, s_i) \neq 0$ at this point and thus the function $T(X, X_i^*, s_i)$ can be used to obtain the new root at $X = X_2$. The parameter $s_i$ is not known and must be identified. In order to do this the function $T$ is evaluated several times at $X = X_i^*$ for increasing values of $s_i$; it is useful to begin with $s_i = 0$ and to add increments $\Delta_s$ such as $0.01 \leqslant \Delta_s \leqslant 0.05$. When for a value $s_i$ it is found that

$$T(X_i^*, X_i^*, s_i) \neq 0 \text{ and } T(X_i^* + \Delta X, X_i^*, s_i) \leqslant T(X_i^*, X_i^*, s_i) \tag{1.157}$$

the current value of the parameter $s_i$ is the root order. The procedure proceeds by trying to calculate another root of the function $T$ which will provide the

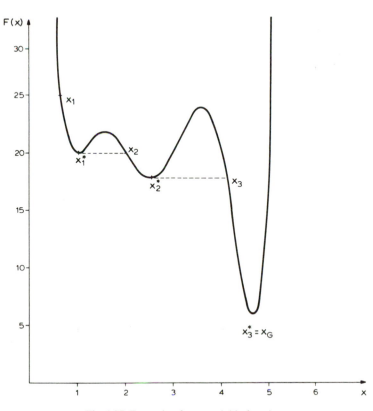

**Fig. 1.25** Example of one-variable function.

design $X_{i+1}$ desired, because according equation (1.156) if

$$T(X_{i+1}, X_i^*, s_i) = 0 \qquad (1.158)$$

$$F(X_{i+1}) = F(X_i^*) \qquad (1.159)$$

An example of a function $F(x)$ with one variable may be used to explain the method (Fig. 1.25):

$$F(x) = x^6 - 16x^5 + 100x^4 - 310x^3 + 499x^2 - 394x + 140$$

The function $F(x)$ has a local minimum at $x = 1$ where $F(1) = 20$. The function $T(x, 1, s_1)$ is represented in Figs. 1.26 and 1.27 for the cases $s_1 = 0.5$ and $s_1 = 1.0$.

$$T(x, 1, 0.5) = \frac{F(x) - F(1)}{[(x-1)(x-1)]^{0.5}} = x^5 - 15x^4 + 85x^3 - 225x^2 + 274x - 120$$

$$T(x, 1, 1) = \frac{F(x) - F(1)}{(x-1)(x-1)} = x^4 - 14x^3 + 71x^2 - 154x + 120$$

It turns out that the parameter $s_1$ has the value 1.0. The root of the function $T(x, 1, 1)$ is obtained at $x_2 = 2.0$ as shown in Fig. 1.27. By carrying out from this point another minimization phase, a local minimum at $x_2^* = 2.517$ will be reached (Fig. 1.28).

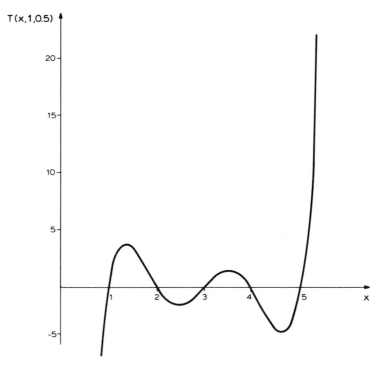

**Fig. 1.26** Tunneling function $T(x, 1, 0.5)$.

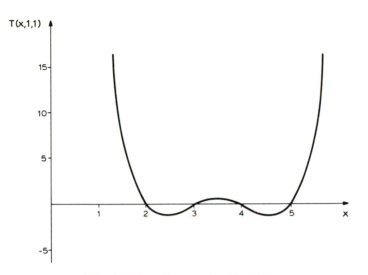

**Fig. 1.27** Tunneling function $T(x, 1, 1)$.

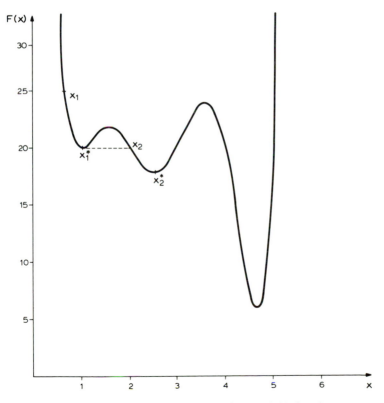

**Fig. 1.28** Tunneling phase in example of one-variable function.

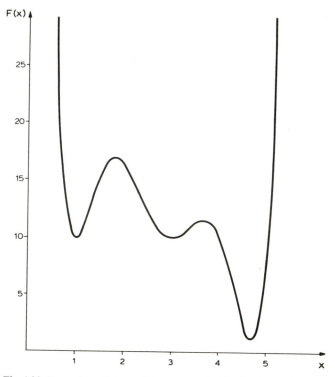

**Fig. 1.29** Example with two minima at equal objective function values.

To avoid in practice the division by zero associated with equation (1.156) the tunneling function is evaluated in the neighborhood of the current local minimum, namely at point $X_i = X_i^* + \varepsilon$ where $\varepsilon$ is a small increment.

It may happen during the process that departing from a local minimum $X_i^*$ the next phase gives a point $X_{i+1} = X_{i+1}^*$ whose function value is $F(X_i^*) = F(X_{i+1}^*)$. This situation means that there are two local minima with equal objective function values, as occurs in the function shown in Fig. 1.29.

In this case it is necessary to define an enhanced tunneling function

$$T(X, X_i^*, s_i, X_{i+1}^*, s_{i+1}) = \frac{F(X) - F(X_i^*)}{[(X - X_i^*)^{\mathrm{T}}(X - X_i^*)]^{s_i}[(X - X_{i+1}^*)^{\mathrm{T}}(X - X_{i+1}^*)]^{s_{i+1}}}$$

(1.160)

In this function only the parameter $s_{i+1}$ needs to be identified by the procedure already mentioned. If more local minima having equal objective function values are found the tunneling function is again modified to include the term linked to each minimum and the new parameter is obtained by the same procedure.

As the number of local minima of the function $F(X)$ is usually unknown, a user decision is necessary in order to complete the method. Levy and Gómez propose to do that by taking into account the computer time consumed to find a new local minimum from the previous one. Generally, as the process goes on it becomes more difficult to obtain new minima, because in the tunneling phase it is more difficult to identify points with the objective function value of the current minimum. Hence, it is suggested that the procedure be ended when no local minimum is found in computer time $10t$, $t$ being the time needed to obtain the current one.

### 1.4.2 The tunneling method in structural optimization

The tunneling method may also be a useful tool to solve constrained optimization problems by linking it with penalty function techniques. This approach may be implemented as follows.

1. From the constrained problem defined in equations (1.1) and (1.2),

$$\min F(X) \tag{1.161}$$

subject to

$$g_j(X) \leqslant 0 \quad i = 1, \ldots, m \tag{1.162}$$

an unconstrained problem via a penalty function is created:

$$\phi(X, r) = F(X) + r \sum_{j=1}^{m} P(g_j) \tag{1.163}$$

In the problem defined in equation (1.163) a local minimum $X_1^*$ may be obtained by using any mentioned method.
2. The tunneling function $T$ is defined at point $X_1^*$ and the tunneling phase is worked out to find a point $X_2$ having

$$F(X_1^*) = F(X_2) \tag{1.164}$$

3. Steps 1 and 2 are repeated until it is not possible to find a new local minimum as indicated in the previous section.

The optimum design of the grillage shown in Fig. 1.10 has been solved as indicated, by using the quadratic extended interior penalty function. This technique extends the penalty function defintion to the infeasible region (Haftka and Starnes, 1976) and improves performance of the interior penalty function method near the optimum.

$$\phi(X, r) = F(X) - r \sum_{j=1}^{m} P(g_j) \tag{1.165}$$

$$P(g_j) = \begin{cases} 1/g_j & \text{if } g_j \leqslant \varepsilon \tag{1.166} \\ [(g_j/\varepsilon)^2 - 3g_j/\varepsilon + 3]/\varepsilon & \text{if } g_j > \varepsilon \end{cases} \tag{1.167}$$

The results obtained are the following:

| | | |
|---|---|---|
| initial design | $X_1 = (5, 25)$ | $F_1 = 8000$ |
| local minimum | $X_1^* = (2.25, 19.52)$ | $F_1^* = 6082$ |
| tunneling phase | $X_2 = (9.16, 13.98)$ | $F_2 = 6076$ |
| global minimum | $X_2^* = (9.72, 6.07)$ | $F_2^* = 2795$ |

The point $X_2^*$ is the global minimum of the problem. Other examples of structural optimization may be found in Hernández (1990b, 1991).

## REFERENCES

Baldur, R. (1972) Structural optimization by inscribed hyperspheres. *Proceedings ASCE*, **98**, (EM3), 503–18.

Brooks, R. and Geoffrion, A. (1966) Finding Everett's Lagrange multipliers by lineal programming. *Operations Research*, **14**(61), 1149–53.

Cassis, J.H. (1974) Optimum Design of Structures Subjected to Dynamic Loads. *UCLA-ENG-7451*, UCLA School of Engineering and Applied Science, CA.

Cheney, E.W. and Goldstein, A.A. (1959) Newton's method for convex programming on Tchebycheff approximation. *Numerische Mathematik*, **1**, 253–68.

Cohn, N.Z., Ghosh, S.K. and Parini, S.R. (1972) Unified approach to the theory of plastic structure. *Journal of Engineering Mechanical Division, ASCE*, **98** (EM5), 1133–58.

Dantzig, G. (1963) *Linear Programming and Extensions*, Princeton University Press, Princeton, N.J.

Davidon, W.C. (1959) Variable Metric Method for Minimization. *ANL-5990 Rev.*, Argonne National Laboratory.

Dorn, W.S., Gomory, R.E. and Greenberg, H.J. (1964) Automatic design of optimal structure. *Journal de Mecanique*, **3**(1), 25–52.

Everett, H. (1963) Generalized Lagrange multiplier method for solving problems of optimum allocation of resources. *Operations Research*, **11**(2), 397–417.

Fiacco, A.V. and McCormick, G.P. (1963) Programming under Nonlinear Constraints by Unconstrained Minimization: A Primal-Dual Method. *Tech. Paper RAC-TP-96*, Research Analysis Corporation, Bethesda, MD.

Fiacco, A.V. and McCormick, G.P. (1968) *Nonlinear Programming Sequential Unconstrained Minimization Techniques*. Wiley, New York.

Fletcher, R. and Powell, M.J.D. (1963) A rapidly convergent method for minimization. *Computer Journal*, **6**(2), 163–8.

Fletcher, R. and Reeves, C.M. (1964) Function minimization by conjugate gradients. *Computer Journal*, **7**(2), 149–54.

Goble, G. and LaPay, W.S. (1971) Optimum design of prestressed beams. *Journal ACI*, **68**, 712–18.

Gómez, S. and Levy, A.V. (1982) The tunneling method for solving the constrained global optimization problems with several nonconnected feasible regions, in *Lecture Notes in Mathematics no. 909* (ed. J.P. Hennart), Springer, Berlin, Heidelberg.

Gould, F.J. (1969) Extensions of Lagrange multipliers in nonlinear programming. *SIAM Journal of Applied Mathematics*, **13**(6), 1280–97.

Greenberg, H.J. and Pierskalla, W.P. (1970) Surrogate mathematical programming. *Operations Research*, **18**(5), 924–39.

Haftka, R.T. and Starnes, J.H. (1976) Application of a quadratic extended interior penalty function for structural optimization. *AIAA Journal*, **14**, 718–24.

Hernández, S. (1990a) *Metodos de diseño óptimo de estructuras*, Colegio de Ingenieros de Caminos, Canales y Puertos (in Spanish).

Hernández, S. (1990b) El método de tunelización. Una técnica de optimización global en problemas no convexos en la ingeniería. IX Congreso de Ingeniería Mecánica, Zaragoza.

Hernández, S. (1991) A new mathematical procedure for global optimization of nonconvex problems. NATO/DFG ASI Seminar on Optimization of Large Structural Systems, Berchtesgaden, Germany.

Hernández, S. and Ceron, C. (1987) A sequence of linear problems with move limits for shape optimization of truss structures. SIAM Conference on Optimization, Houston, TX.

Horne, M.R. and Morris, L.I. (1973) Optimum design of multistorey rigid frames, in *Optimum Structural Design* (eds. R.H. Gallagher and O.C. Zienkiewicz), Wiley, New York.

Jaynes, E.T. (1957) Information theory and statistical mechanics. *The Physical Review*, **106**, 620–30, **108**, 171–90.

Johnson, E.H. (1976) Disjoint design spaces in the optimization of harmonically excited structures. *AIAA Journal*, **14**(2), 259–61.

Johnson, E.H., Rizzi, P., Ashley, H., *et al.* (1976) Optimization of continuous one-dimensional structures under steady harmonic excitation. *AIAA Journal*, **14**(2), 1690–8.

Karmarkar, N. (1984) A new polynomial-time algorithm for linear programming. *Combinatórica*, **4**, 373–95.

Kavlie, D. (1971) Optimum design of statically indeterminate structures. Ph.D. Thesis, University of California, Berkeley, CA.

Kelley, H.J. (1960) The cutting plane method for solving complex programs. *SIAM Journal*, **8**, 703–12.

Khot, N.S., Berke, L. and Venkayya, V.B. (1979) Comparison of optimality criteria. Algorithms for minimum wieght design of structures. *AIAA Journal*, **17**, 182–90.

Kinsalaas, J. (1972) Minimum Weight Design of Structures via Optimality Criteria. *NASA TND-7115*.

Kirsch, U. (1973) Optimized prestressing by linear programming. *International Journal of Numerical Methods in Engineering*, **7**, 125–6.

Kuhn, H.W. and Tucker, A.W. (1951) Nonlinear programming, in *Proceedings of 2nd Berkeley Symposium on Mathematical Statistics and Probability*, University of California Press, pp. 481–90.

LaPay, W.S. and Goble, G.G. (1971) Optimum design of trusses for ultimate loads. *Journal of Structural Division ASCE*, **97**(ST1), 157–74.

Levy, A.V. and Gómez, S. (1984) The tunneling method applied to global optimization, in *Numerical Optimization 1984*, (eds. P.T. Boggs, R.H. Byrd and R.B. Schnabel), SIAM Conference.

Maquoi, R. and Rondal, J. (1977) Approache realiste du dimensionnement optimal des ponts précontraints hyperstatiques. *Annales des Travaux Publics de Belgique*, (3), 197–215.

Mills-Curran, W.C. and Schmit, L.A. (1985) Structural optimization with dynamic behaviour constraints. *AIAA Journal*, **23**(1), 136–8.

Moses, F. and Onoda, S. (1969) Minimum weight design of structures with application to elastic grillages. *International Journal of Numerical Methods in Engineering*, **1**, 311–31.

NBE-MV103 (1978) *Cálculo de las estructuras de acero laminado en la edificación* (in Spanish).

Powell, M.J.D. (1964) An efficient method of finding the minimum of a function of several variables without calculating derivatives. *Computer Journal*, **7**(4), 303–7.

Rizzi, P. (1976) Optimization of multiconstrained structures based on optimality criteria. *17th AIAA/ASME/SAE Conference*, King of Prussia, PA.

Russell, A.D. and Reinschmidt, K.F. (1964) Discussion of optimum design of trusses for ultimate loads. *Journal of Structural Division ASCE*, **3**(1), 25–52.

Schmit, L.A. (1960) Structural design by systematic synthesis. Second Conference on Electronic Computation, ASCE, Pittsburgh, PA, pp. 105–32.

Shannon, C.E. (1948) A mathematical theory of communication. *Bell System Techanical Journal*, **27**(3), 379–428.

Svanberg, K. (1984) On local and global minima in structural optimization, in *New Directions in Optimum Design* (eds. E. Atrek *et al.*), Wiley, New York.

Svanberg, K. (1991) On local and global optima. NATO/DFG ASI Seminar on Optimization of Large Structural Systems, Berchtesgaden, Germany.

Templeman, A.B. (1989) Entropy-based minimax applications in shape-optimal design, in *Lecture Notes on Engineering, No. 42: Discretization Methods and Structural Optimization – Procedures and Applications* (eds. H.A. Eschenauer and G. Thierauf), Springer, Berlin, Heidelberg, pp. 335–42.

Templeman, A.B. and Li, X. (1985) Entropy duals. *Engineering Optimization*, **9**, 107–20.

Templeman, A.B. and Li, X. (1987) A maximum entropy approach to constrained nonlinear programming. *Engineering Optimization*, **12**, 191–205.

Vanderplaats, G.N. (1984a) An efficient feasible directions algorithm for design synthesis, *AIAA Journal*, **22**(11), 1633–40.

Vanderplaats, G.N. (1984b) *Numerical Optimization Techniques for Engineering Design: with Applications*, McGraw-Hill, New York.

Zangwill, W.I. (1967) Nonlinear programming via penalty functions. *Management Sciences*, **13**(5), 344–58.

Zoutendijk, G. (1960) *Methods of Feasible Directions*, Elsevier, Amsterdam.

# 2

# Optimality criteria methods for large discretized systems

**G.I.N. ROZVANY and M. ZHOU**

## 2.1 INTRODUCTION: WHY OPTIMALITY CRITERIA METHODS?

Research into new optimality crieria methods was motivated by the currently existing discrepancy between analysis capability and optimization capability in structural design, which was pointed out repeatedly by Berke and Khot (e.g. Berke and Khot, 1987). Reasons for this discrepancy can be understood better if we consider the quantities that determine the optimization capability of various methods used in structural design, which are summarized in Table 2.1 in the context of static problems.

Whereas in primal mathematical programming (MP) methods the critical quantity is the number of variables ($N$), in dual and traditional discretized optimality criteria (DOC) methods it is the number of active behavioral constraints which, for static problems, includes the number ($m_d$) of active deflection constraints and number ($m_s$) of active stress constraints. Finally, in optimality criteria methods combined with fully stressed design for stress constraints (DOC–FSD) and in the new optimality criteria methods (DCOC) introduced recently (Zhou and Rozvany, 1992; Zhou, 1992), the critical quantity is the number ($m_d$) of active deflection (or other global) constraints. It is important to note, however, that the DOC–FSD method is known to lead in general to a nonoptimal solution, although the error is often relatively small. For this reason, DOC–FSD should be disqualified in a comparison of correct methods of structural optimization. In the case of very large systems, many thousand stress constraints may be active, whereas the number of active global constraints (such as deflection, natural frequency or system stability constraints, involving the entire system) can be rather small. For this reason, the proposed DCOC method achieves an improvement in our optimization capability by several orders of magnitude, if typical large structural systems are being optimized.

## 2.2 HISTORICAL BACKGROUND AND GENERAL FEATURES OF THE DCOC ALGORITHM

Historically, the DCOC algorithm can be regarded as a unification of discretized optimality criteria (DOC) techniques (e.g. Berke, 1970; Venkayya, Khot and

**Table 2.1** Quantities determining the optimization capability of various methods

| | Primal MP | DOC, Dual | DOC–FSD | COC DCOC |
|---|---|---|---|---|
| Critical quantity | $N$ | $m_d + m_s$ | $m_d$ | $m_d$ |
| Solution | Optimal | Optimal | Non-optimal | Optimal |

Berke, 1973; for reviews see for example Berke and Khot, 1974, 1987, 1988) and of continuum-based optimality criteria (COC) methods derived by the analytical school of structural optimization (e.g. Prager and Shield, 1967; Prager and Taylor, 1968; for a comprehensive review see a book by Rozvany, 1989). The latter were applied indirectly to large finite element systems under the term 'iterative discretized COC methods' (Rozvany *et al.*, 1989; Rozvany, Zhou and Gollub, 1990; Rozvany and Zhou, 1991), in which the optimality criteria were derived for continua in terms of stress resultants for each cross section and then discretization was carried out for computations by FE methods. Moreover, iterative COC methods used expressions which were exact only for a continuous variation of the design variable along the members or elements involved. For these reasons, it represented a considerable improvement of the continuum-type approach when Zhou (1992) reformulated the COC method directly for discretized systems in the matrix notation of finite element analysis. This new discretized continuum-type optimality criteria technique has been termed the DCOC method for historical reasons and it is hoped that it will help in bridging over the communication gap between the numerical and analytical schools of structural optimization.

Before discussing DCOC in detail, we shall summarize briefly the main differences between the formulation of this technique and that of earlier discretized optimality criteria (DOC) methods (Table 2.2). In DCOC the variables in the formulation include not only design variables but also the real forces and the virtual forces used in work equations for the deflection constraints. Whereas in DOC the stress constraints are converted into global (displacement) constraints involving the entire system, in DCOC they are expressed in terms of real forces

**Table 2.2** Main differences between the DOC and COC–DCOC methods

| | DOC | COC–DCOC |
|---|---|---|
| Variables | Design variables | Design variables, real and virtual forces |
| Formulation of stress constraints | Through equivalent displacements | Directly in terms of real forces |
| Equilibrium | Implicit | Explicit equality constraints |
| Compatibility | Implicit | Not included, implied by optimality |
| Calculation of Lagrangians for stress constraints | Iteratively, at system level | Explicitly, at element level |

involving only the element concerned. Whilst in DOC equilibrium and compatibility are included in the formulation implicitly through the structural analysis, in DCOC the equilibrium constraint is included explicitly, but compatibility (kinetic admissibility) is relaxed initially and turns out to be an optimality condition, if any one displacement condition is active. This feature of DCOC simplifies the formulation considerably. The most important advantage of DCOC lies in the calculation of the Lagrange multipliers for stress constraints. In DOC the latter have been converted into relative displacement constraints and hence the corresponding Lagrange multipliers (i) require an analysis for a virtual load for each stress constraint and (ii) are coupled with each other at the system level. The evaluation of the Lagrangians in DOC therefore involves a complicated and expensive iterative procedure. In DCOC, these Lagrange multipliers can be calculated explicitly at the element level for certain simpler types of structures (e.g. trusses, frames and membranes). Another important advantage of DCOC is the fact that some of the optimality criteria are interpreted in terms of the so-called adjoint structure, which has the same analysis equations as the real structure. Analysis of the adjoint structure replaces the usual sensitivity analysis of other methods and involves only a simple substitution of the decomposed stiffness matrix, already computed in the analysis of the real structure. This means that the adjoint analysis in DCOC requires a relatively insignificant amount of computer time.

The difference between DCOC and DOC–FSD is that the adjoint system (virtual force, i.e. dummy load system) in DCOC is subject to prestrains (initial relative displacements) due to active stress constraints. This ensures correctness of the optimal solution, whereas DOC–FSD usually leads to an incorrect solution for statically indeterminate structures. The above statements will be demonstrated on test examples later (section 2.5).

Because DCOC is more useful than COC for practical applications, the formulation and derivation of this method is given in detail in section 2.3, and its computational implementation discussed in section 2.4. Illustrative examples are presented in section 2.5. The COC method, which is important for understanding the theoretical background and necessary for the derivation of exact analytical optimal layouts, will be discussed in section 2.6 and illustrated with examples in section 2.7.

Although the computational implementation of DCOC uses the stiffness method, derivation of the optimality criteria is based on the flexibility method, because this simplifies considerably the mathematical procedure involved.

## 2.3 DERIVATION OF OPTIMALITY CRITERIA FOR THE DCOC ALGORITHM

For didactic reasons, the optimality criteria for the DCOC method will be first derived in section 2.3.1 for a very simple class of problems, namely for the case of stress constraints and one displacement constraint. This type of derivation was originally proposed by Zhou (1992). Various extensions of DCOC are derived in sections 2.3.2 to 2.3.5.

### 2.3.1 Stress constraints and one displacement constraint

(a) Problem formulation

The considered problem can be stated as follows:

$$\min \Phi = \sum_{e=1}^{E} w^e(\{\mathbf{x}^e\}) \tag{2.1}$$

subject to the following constraints. The displacement constraint is

$$\{\hat{\mathbb{F}}\}^T[\mathbf{D}]\{\mathbb{F}\} + \{\hat{\mathbb{F}}\}^T\{\mathbf{\Delta}^*\} - t \leqslant 0 \tag{2.2}$$

The stress constraints are

$$\theta^e_{hr}(\{\mathbf{n}^e_h\}) - \sigma^e_{hr} = \theta^e_{hr}([\mathbf{\omega}^e_h]\{\mathbf{ff}^e\} + \{\mathbf{n}^e_{oh}\}) - \sigma^e_{hr} \leqslant 0$$

$$(h = 1,\ldots,H^e; r = 1,\ldots,R^e_h; e = 1,\ldots,E) \tag{2.3}$$

The side constraints are

$$-x^e_i + x^e_i\!\downarrow \leqslant 0 \quad (i = 1,\ldots,I^e; e = 1,\ldots,E) \tag{2.4}$$

$$x^e_i - x^e_i\!\uparrow \leqslant 0 \quad (i = 1,\ldots,I^e; e = 1,\ldots,E) \tag{2.5}$$

The equilibrium condition for the real forces is

$$\{\mathbf{P}\} - [\mathbf{B}]\{\mathbb{F}\} = \mathbf{0} \tag{2.6}$$

The equilibrium condition for the virtual forces is

$$\{\hat{\mathbf{P}}\} - [\mathbf{B}]\{\hat{\mathbb{F}}\} = \mathbf{0} \tag{2.7}$$

In relations (2.1)–(2.7), $\Phi$ is the objective function, which in this case can be the total weight of the structure; $e = 1,\ldots,E$ identify the elements; $w^e$ is the objective function (weight) for the element $e$; $\{\mathbf{X}\} = \lfloor \mathbf{x}^1,\ldots,\mathbf{x}^e,\ldots,\mathbf{x}^E \rfloor^T$ with $\mathbf{x}^e = \lfloor x^e_1,\ldots, x^e_i,\ldots,x^e_{I^e} \rfloor^T$ the design variables (cross-sectional dimensions); $\hat{\mathbb{F}}$ denotes the virtual nodal forces in flexibility formulation, equilibrating the virtual loads $\{\hat{\mathbf{P}}\}$ (unit dummy load for a deflection constraint at one point or loads corresponding to the weighting factors if a weighted combination of deflections is prescribed), $[\mathbf{D}]$ is the flexibility matrix, $\{\mathbb{F}\}$ denotes the real nodal forces in flexibility formulation, $\{\mathbf{\Delta}^*\} = \lfloor \mathbf{\delta}^{1*},\ldots,\mathbf{\delta}^{e*},\ldots,\mathbf{\delta}^{E*} \rfloor^T$ are the real initial relative displacements caused by (i) loads within an element, (ii) temperature strains or (iii) prestrains; $t$ is the permissible value of the displacement; $\theta^e_{hr}$ is a stress condition for the type of stress or stress location $r$ at the cross section $h$ of the element $e$; $\{\mathbf{n}^e_h\} = \lfloor n^e_{h1},\ldots,n^e_{hv},\ldots,n_{hV^e_h} \rfloor^T$ are the stress resultants at the cross section $h$ of the element $e$; $[\mathbf{\omega}^e_h]$ is a matrix converting the nodal forces $\{\mathbf{ff}^e\}$ to stress resultants at the cross section $h$; $\{\mathbf{n}^e_{oh}\}$ are the stress resultants caused by the loads acting on the interior of the element $e$; $\sigma_{hr}$ is the permissible value of the $r$ type stress at the cross section $h$; $x^e_i\!\downarrow$ and $x^e_i\!\uparrow$ are the lower and upper limits on the value of the design variable $x^e_i$; $\{\mathbf{P}\}$ are the real nodal loads at the free degrees of freedom; $[\mathbf{B}]$ is the statics matrix. In general, upper-case letters denote quantities at the system level and lower-case letters quantities at the element level.

The relaxed problem formulation in equations (2.1)–(2.7) requires statical admissibility of the real and virtual forces by relations (2.6) and (2.7) but includes no provision for compatibility conditions. It will be seen later that compatibility for the real forces is ensured by the Kuhn–Tucker conditions. In representing a displacement via a work equation in relation (2.2), the virtual forces need to be statically admissible only. However, the Kuhn–Tucker conditions will also introduce compatibility conditions for the strain caused by the virtual forces, making them thereby also kinematically admissible. The virtual force system, together with compatibility conditions and initial relative displacements due to active stress constraints, will represent a fictitious structure which will be termed the 'adjoint structure'.

The stress constraint in relation (2.3) is often linear with respect to the stress resultants, in which case it can be represented as

$$\{s_{hr}^e\}^T\{n_h^e\} - \sigma_{hr}^e = \{s_{hr}^e\}([\omega_h^e]\{ff^e\} + \{n_{oh}^e\}) - \sigma_{hr}^e \leqslant 0 \qquad (2.8)$$

where the vector $\{s_{hr}^e\}$ involves design variables and converts the stress resultants at the cross section $h$ of element $e$ into a stress of type and/or location $r$ at the same cross section.

(b) Example illustrating the problem formulation: plane frame element

Since the above formulation may be unfamiliar to some of the readers, we shall use a simple example to illustrate the concepts involved.

Figure 2.1(a) shows a plane frame element and its rectangular cross section of constant width $b$ and variable depth $x_1^e$ is indicated in Fig. 2.1(b). The element objective function shall be in this case the element weight

$$w^e(x_1^e) = bL\rho x_1^e \qquad (2.9)$$

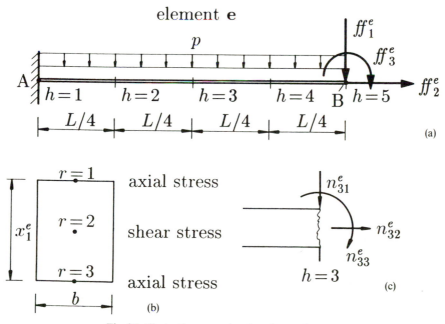

Fig. 2.1 Illustrative example; plane frame element.

where $L$ is the length of the element and $\rho$ is the specific weight of the frame material. In this example, there is only one design variable per element.

Considering the displacement constraint in equation (2.2), the following should be clarified. In the flexibility formulation, each element must be supported in a stable, statically determinate manner (Fig. 2.1(a)). For a plane frame element, it is most convenient to clamp the element at the left end (A) and leave the right end (B) free. Then the three nodal forces $\{\mathbf{ff}^e\} = \lfloor ff^e_1, ff^e_2, ff^e_3 \rfloor^T$ represent the vertical and horizontal forces and the moment at the right end (B) of the beam element (Fig. 2.1(a)). The virtual forces $\{\hat{\mathbf{ff}}^e\} = \lfloor \hat{ff}^e_1, \hat{ff}^e_2, \hat{ff}^e_3 \rfloor^T$ are similar but they equilibrate the virtual loads instead of the real loads.

The element flexibility matrix $[\mathbf{d}^e]$ maps the nodal forces $\{\mathbf{ff}^e\}$ of the element $e$ into relative displacements $\{\boldsymbol{\delta}^e\}$ which have the same location and direction as the former:

$$\{\boldsymbol{\delta}^e\} = [\mathbf{d}^e]\{\mathbf{ff}^e\} + \{\boldsymbol{\delta}^{*e}\} \tag{2.10}$$

where $\{\boldsymbol{\delta}^{*e}\}$ are the initial relative nodal displacements.

It is easy to check by elementary statical considerations that the frame element flexibility matrix is

$$[\mathbf{d}^e] = \frac{1}{E}
\begin{bmatrix}
\dfrac{L^3}{3I} & 0 & \dfrac{L^2}{2I} \\[2ex]
0 & \dfrac{L}{A} & 0 \\[2ex]
\dfrac{L^2}{2I} & 0 & \dfrac{L}{I}
\end{bmatrix}
= \frac{1}{Eb}
\begin{bmatrix}
\dfrac{4L^3}{(x^e_1)^3} & 0 & \dfrac{6L^2}{(x^e_1)^3} \\[2ex]
0 \cdot & \dfrac{L}{x^e_1} & 0 \\[2ex]
\dfrac{6L^2}{(x^e_1)^3} & 0 & \dfrac{12L}{(x^e_1)^3}
\end{bmatrix} \tag{2.11}$$

where $I$ is the moment of inertia and $A$ is the area of the cross section.

The initial relative displacement $\{\boldsymbol{\delta}^{*e}\}$ for the element in Fig. 2.1(a) are the three displacement components at the right end B of the element caused by the distributed load $p$ and have the values

$$\{\boldsymbol{\delta}^{*e}\} = \lfloor pL^4/8EI, 0, pL^3/6EI \rfloor^T = \lfloor 3pL^4/2Eb(x^e_1)^3, 0, 2pL^3/Eb(x^e_1)^3 \rfloor^T \tag{2.12}$$

If the entire element were subject to a temperature change of $\tau$, then the second entry of $\{\boldsymbol{\delta}^{*e}\}$ in relation (2.12) would change from zero to $\tau\alpha L$, where $\alpha$ is the coefficient of thermal expansion.

Considering the stress constraints in equation (2.3), the cross sections for which stress values are constrained are $h = 1, 2, 3, 4, 5$ in Fig. 2.1(a) and the type of stress and location within a cross section are $r = 1, 2, 3$ in Fig. 2.1(b). For the considered element, the stress constraints are linear in terms of the stress resultants and hence equation (2.8) can be used instead of the general formulation (2.3). The stress resultants $\{\mathbf{n}^e_h\} = \lfloor n^e_{h1}, n^e_{h2}, n^e_{h3} \rfloor^T$ represent the shear force, axial force and bending moment, respectively (Fig. 2.1(c)). For calculating the stresses at the three points $r = 1, 2, 3$ in Fig. 2.1(b) from the stress resultant, we have the following transformation vectors:

$$\{\mathbf{s}^e_{h1}\} = \left\lfloor 0, \frac{1}{A}, \frac{1}{Q} \right\rfloor^T, \quad \{\mathbf{s}^e_{h2}\} = \left\lfloor \frac{1.5}{bx^e_1}, 0, 0 \right\rfloor^T, \quad \{\mathbf{s}^e_{h3}\} = \left\lfloor 0, \frac{1}{A}, -\frac{1}{Q} \right\rfloor^T \tag{2.13}$$

where $A$ is the cross-sectional area $(A = bx_1^e)$ and $Q$ is the section modulus for calculating flexural stresses in the extreme fiber $(Q = b(x_1^e)^2/6)$. The factor $1.5/bx_1^e$ converts a shear force into a shear stress at the neutral axis $(r = 2)$.

The transformation matrix $[\boldsymbol{\omega}_h^e]$ for calculating the stress resultants $\{\mathbf{n}_h^e\}$ from the nodal forces $\{\mathbf{ff}^e\}$ takes the following form for $h = 3$, for example:

$$[\boldsymbol{\omega}_3^e] = \begin{bmatrix} 1 & 0 & 0 \\ 0 & 1 & 0 \\ \dfrac{L}{2} & 0 & 1 \end{bmatrix} \tag{2.14}$$

since the moment $n_3^e$ at the cross section $h = 3$ is the sum of the vertical nodal force $ff_1^e$ multiplied by the lever arm $L/2$ and the nodal moment $ff_3^e$.

Finally, the stress resultants $\{\mathbf{n}_{oh}^e\} = \lfloor n_{oh1}^e, n_{oh2}^e, n_{oh3}^e \rfloor^{\mathrm{T}}$ caused by the distributed transverse load $p$ are for $h = 3$

$$\{\mathbf{n}_{o3}^e\} = \lfloor pL/2, 0, pL^2/8 \rfloor^{\mathrm{T}} \tag{2.15}$$

In the above example, the stress constraint was linear with respect to the stress resultants. An example of a nonlinear stress constraint is a membrane element (plate in plane stress) for which the Mises stress is constrained:

$$\sigma_1^2 + \sigma_2^2 - \sigma_1\sigma_2 - \sigma_{hr}^2 = \sigma_x^2 + \sigma_y^2 - \sigma_x\sigma_y + 3\tau_{xy}^2 - \sigma_{hr}^2 \leqslant 0 \tag{2.16}$$

where $\sigma_{hr}$ is the Mises stress for the considered point. Denoting the axial and shear forces (usually $N_x, N_y, N_{xy}$) at a cross section (point) $h$ of the membrane element by $\{\mathbf{n}_h^e\} = \lfloor n_{h1}^e, n_{h2}^e, n_{h3}^e \rfloor^{\mathrm{T}}$, we have the stress constraint

$$[(n_{h1}^e)^2 + (n_{h2}^e)^2 - n_{h1}^e n_{h2}^e + 3(n_{h3}^e)^2]/x_1^e - \sigma_{hr}^2 \leqslant 0 \tag{2.17}$$

where $x_1^e$ is the thickness of the membrane element.

## (c) Necessary conditions for optimality

The Lagrange multipliers for the constraints in relations (2.2)–(2.7) will be denoted by

$$v, \lambda_{hr}^e, \beta_i^e\downarrow, \beta_i^e\uparrow, \{\bar{\mathbf{L}}\}, \{\mathbf{L}\} \tag{2.18}$$

For the problem in relations (2.1)–(2.8), we have the following Kuhn–Tucker conditions with respect to the variables stated below.

*Design variables*

$$\frac{\partial w^e}{\partial x_i^e} + v\{\widehat{\mathbf{ff}}^e\}^{\mathrm{T}}\left[\frac{\partial \mathbf{d}^e}{\partial x_i^e}\right]\{\mathbf{ff}^e\} + v\{\widehat{\mathbf{ff}}^e\}^{\mathrm{T}}\left\{\frac{\partial \boldsymbol{\delta}^{*e}}{\partial x_i^e}\right\} + \sum_{h=1}^{H^e}\sum_{r=1}^{R_h^e}\lambda_{hr}^e\frac{\partial \theta_{hr}^e(\{\mathbf{n}_h^e\})}{\partial x_i^e} - \beta_i^e\downarrow + \beta_i^e\uparrow = 0$$

$$(i = 1, \dots, I^e, e = 1, \dots, E) \tag{2.19}$$

*Real nodal forces* $\mathbb{F}$

$$v[\mathbf{D}]\{\widehat{\mathbb{F}}\} + \{\bar{\boldsymbol{\Delta}}^*\} - [\mathbf{B}]^{\mathrm{T}}\{\bar{\mathbf{L}}\} = \mathbf{0} \tag{2.20}$$

with

$$\{\bar{\boldsymbol{\Delta}}^*\} = \lfloor \bar{\boldsymbol{\delta}}^{*1}, \dots, \bar{\boldsymbol{\delta}}^{*e}, \dots, \bar{\boldsymbol{\delta}}^{*E} \rfloor^{\mathrm{T}} \tag{2.21}$$

and

$$\{\bar{\boldsymbol{\delta}}^{*e}\} = \sum_{h=1}^{H^e} \sum_{r=1}^{R_h^g} \lambda_{hr}^e \{\omega_h^e\}^{\mathrm{T}} \left\{ \frac{\partial \theta_{hr}^e}{\partial \mathbf{n}_h^e} \right\} \tag{2.22}$$

where the partial derivatives in curly brackets can also be represented as $\mathrm{grad}_{\mathbf{n}_h^e} \theta_{hr}^e$.

For linear stress constraints of the type given in equation (2.8), we have

$$\{\bar{\boldsymbol{\delta}}^{*e}\} = \sum_{h=1}^{H^e} \sum_{r=1}^{R_h^e} \lambda_{hr}^e [\omega_h^e]^{\mathrm{T}} \{\mathbf{s}_{hr}^e\} \tag{2.23}$$

*Virtual (adjoint) nodal forces* $\hat{\mathbb{F}}$

$$v([\mathbf{D}]\{\mathbb{F}\} + \{\boldsymbol{\Delta}^*\}) - [\mathbf{B}]^{\mathrm{T}}\{\mathbf{L}\} = 0 \tag{2.24}$$

Moreover, the Lagrangians, $\lambda_{hr}^e, v, \beta_i^e \downarrow$ and $\beta_i^e \uparrow$ must be nonnegative, and they are nonzero only if the corresponding constraint is satisfied as an equality. Since the expression in parentheses after $v$ in equation (2.24) represents the real relative displacements $\{\boldsymbol{\Delta}\}$ and $[\mathbf{B}]^{\mathrm{T}}$ is the kinematics matrix mapping the nodal displacements $\{\mathbf{U}\}$ into relative displacements $\{\boldsymbol{\Delta}\}$, we can see that $\{\mathbf{L}\}/v$ represents the real nodal displacements:

$$\frac{1}{v}\{\mathbf{L}\} = \{\mathbf{U}\} \tag{2.25}$$

Then it follows from equation (2.24) that kinematic admissibility, which was not included in the relaxed formulation (2.1)–(2.7), is automatically enforced as an optimality condition, provided that the displacement condition (2.2) is active. The fact that optimality with respect to an active displacement constraint ensures kinematic admissibility for any statically admissible real and virtual stress fields was also shown much earlier by Shield and Prager (1970) and Huang (1971).

Moreover, by equation (2.20) we can interpret the Lagrange multipliers $\{\bar{\mathbf{L}}\}$ as the nodal displacements of a fictitious structure termed the adjoint structure

$$\{\bar{\mathbf{L}}\} = \{\bar{\mathbf{U}}\} \tag{2.26}$$

and the corresponding relative displacements in equation (2.20), i.e.

$$\{\bar{\boldsymbol{\Delta}}\} = v([\mathbf{D}]\{\hat{\mathbb{F}}\} + \{\bar{\boldsymbol{\Delta}}^*\}) \tag{2.27}$$

as the relative displacements of the adjoint structure. We can see from the foregoing that for stress constraints and a simple displacement constraint the adjoint structure has the following properties:

- the internal forces $\{\bar{\mathbb{F}}\}$ equilibrate the virtual loads multiplied by $v$, that is $\{\bar{\mathbf{P}}\} = v\{\hat{\mathbf{P}}\}$;
- the elements are subject to initial relative displacements given by relation (2.22) or (2.23) for elements with active stress constraints.

Necessary conditions for optimality of a discretized elastic system with stress constraints, one displacement constraint and one loading are summarized in Fig. 2.2. This representation is called a "conceptual scheme" because for the computational implementation the necessary conditions are modified somewhat with a view to using a stiffness method.

## (d) Solution algorithm

Since the above equations cannot be solved simultaneously for most problems of a practical nature, it is convenient to adopt an iterative procedure consisting of the following two major steps in each iteration:

- analysis of the real and adjoint structures;
- updating the Lagrange multipliers and the design variables $\{x\}$ on the basis of the relations at the top of Fig. 2.2.

Further details of the computational implementation of DCOC will be explained in section 2.4.

### 2.3.2 Stress constraints, several displacement constraints and one load condition

If we consider several displacement constraints, identified by the subscripts $k = 1, \ldots, K$, then equation (2.2) is replaced by

$$\{\hat{\mathbb{F}}_k\}^{\mathrm{T}}[\mathbf{D}]\{\mathbb{F}\} + \{\hat{\mathbb{F}}_k\}^{\mathrm{T}}\{\mathbf{\Delta}^*\} - t_k \leqslant 0 \quad (k = 1, \ldots, K) \tag{2.28}$$

where $\{\hat{\mathbb{F}}_k\}$ equilibrates the virtual load associated with the $k$th displacement constraints and $t_k$ is the limiting value of the $k$th displacement. In addition, the equilibrium condition for the virtual forces becomes

$$\{\hat{\mathbf{P}}_k\} - [\mathbf{B}]\{\hat{\mathbb{F}}_k\} = \mathbf{0} \tag{2.29}$$

Moreover, instead of the Lagrange multiplier $v$ and $\{\mathbf{L}\}$ in equation (2.18) we have

$$v_k, \{\mathbf{L}_k\} \quad (k = 1, \ldots, K) \tag{2.30}$$

and then the second and third terms in equation (2.19) become

$$\cdots + \sum_{k=1}^{K} v_k \{\hat{\mathbf{ff}}^e\}^{\mathrm{T}} \left[\frac{\partial \mathbf{d}^e}{\partial x_i^e}\right] \{\mathbf{ff}^e\} + \sum_{k=1}^{K} v_k \{\hat{\mathbf{ff}}_k^e\}^{\mathrm{T}} \left\{\frac{\partial \boldsymbol{\delta}^{*e}}{\partial x_i^e}\right\} + \cdots \tag{2.31}$$

Finally, equations (2.20) and (2.24) are replaced by

$$\sum_{k=1}^{K} v_k [\mathbf{D}]\{\hat{\mathbb{F}}_k\} + \{\bar{\mathbf{\Delta}}^*\} - [\mathbf{B}]^{\mathrm{T}}\{\bar{\mathbf{L}}\} = \mathbf{0} \tag{2.32}$$

$$v_k([\mathbf{D}]\{\mathbb{F}\} + \{\mathbf{\Delta}^*\}) - [\mathbf{B}]^{\mathrm{T}}\{\mathbf{L}_k\} = \mathbf{0} \quad (k = 1, \ldots, K) \tag{2.33}$$

Introducing

$$\frac{1}{v_k}\{\mathbf{L}_k\} = \{\mathbf{U}\} \tag{2.34}$$

equation (2.33) with any one $k$ value implies kinematical admissibility of the real nodal displacements and relative displacements. By equation (2.32) the adjoint

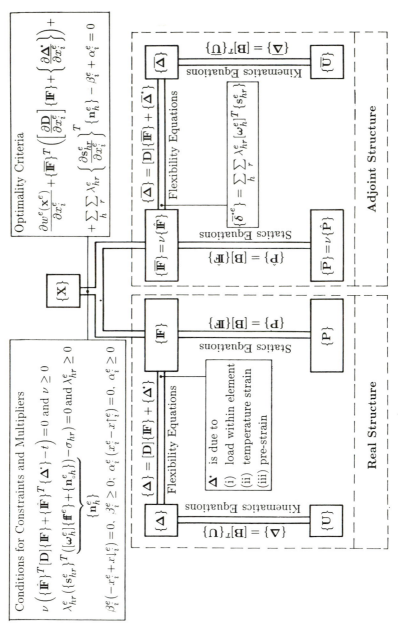

**Fig. 2.2** DCOC algorithm for linearly elastic structures with stress constraints, one displacement constraint and one load condition: conceptual scheme.

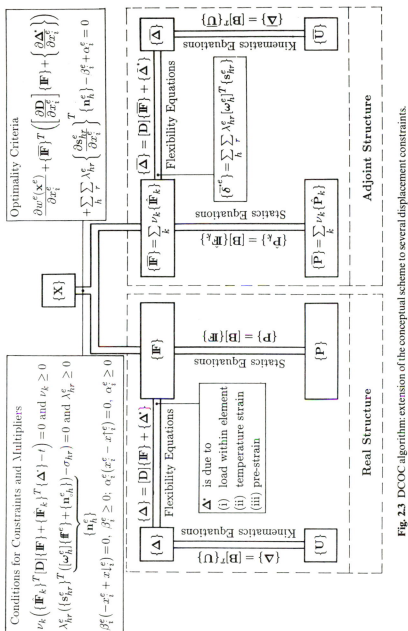

**Fig. 2.3** DCOC algorithm: extension of the conceptual scheme to several displacement constraints.

structure is subject to the loading

$$\{\bar{\mathbf{P}}\} = \sum_{k=1}^{K} v_k \{\hat{\mathbf{P}}_k\} \tag{2.35}$$

and fulfills compatibility, subject to the initial relative displacements $\{\bar{\mathbf{\Delta}}^*\}$ given by equation (2.22) or (2.23) for active stress constraints. For the case of linear stress constraints of the type shown in equation (2.8), the necessary conditions of optimality for the considered class of problems are given in Fig. 2.3.

### 2.3.3 Stress constraints, several displacement constraints and several load conditions

We assume that for each load condition $l = 1, \dots, L$ we have $K_l$ displacement conditions: $k = 1, \dots, K_l$. Then equation (2.2) or (2.28) is replaced by

$$\{\hat{\mathbf{F}}_{kl}\}^{\mathrm{T}}[\mathbf{D}]\{\mathbf{F}_l\} + \{\hat{\mathbf{F}}_{kl}\}^{\mathrm{T}}\{\mathbf{\Delta}_l^*\} - t_{kl} \leqslant 0 \quad (k = 1, \dots, K_l; l = 1, \dots, L) \tag{2.36}$$

Moreover, the stress constraints in equations (2.3) and (2.8) become, respectively,

$$\theta_{hr}^e(\{\mathbf{n}_{hl}^e\}) - \sigma_{hr}^e = \theta_{hr}^e([\mathbf{\omega}_h^e]\{\mathbf{ff}_l^e\} + \{\mathbf{n}_{ohl}^e\}) - \sigma_{hr}^e \leqslant 0$$
$$(h = 1, \dots, H; r = 1, \dots, R_h^e; l = 1, \dots, L; e = 1, \dots, E) \tag{2.37}$$

and

$$\{\mathbf{s}_{hr}^e\}^{\mathrm{T}}\{\mathbf{n}_{hl}^e\} - \sigma_{hr}^e = \{\mathbf{s}_{hr}^e\}^{\mathrm{T}}([\mathbf{\omega}_h^e]\{\mathbf{ff}_l^e\} + \{\mathbf{n}_{ohl}^e\}) - \sigma_{hr}^e \leqslant 0$$
$$(h = 1, \dots, H; r = 1, \dots, R_h^e; l = 1, \dots, L; e = 1, \dots, E) \tag{2.38}$$

Finally, equations (2.6) and (2.7) are replaced by

$$\{\mathbf{P}_l\} - [\mathbf{B}]\{\mathbf{F}_l\} = \mathbf{0} \quad (l = 1, \dots, L) \tag{2.39}$$

$$\{\hat{\mathbf{P}}_{kl}\} - [\mathbf{B}]\{\hat{\mathbf{F}}_{kl}\} = \mathbf{0} \quad (k = 1, \dots, K_l; l = 1, \dots, L) \tag{2.40}$$

Then the optimality condition (2.19) becomes

$$\frac{\partial w^e}{\partial x_i^e} + \sum_{k=1}^{K_l} \sum_{l=1}^{L} v_{kl} \{\hat{\mathbf{ff}}_{kl}^e\}^{\mathrm{T}} \left( \left[ \frac{\partial \mathbf{d}^e}{\partial x_i^e} \right] \{\mathbf{ff}_l^e\} + \left\{ \frac{\partial \mathbf{\delta}_l^{*e}}{\partial x_i^e} \right\} \right)$$
$$+ \sum_{h=1}^{H} \sum_{r=1}^{R} \sum_{l=1}^{L} \lambda_{hrl}^e \{\mathbf{\omega}_h^e\}^{\mathrm{T}} \frac{\partial \theta_{hr}^e(\{\mathbf{n}_{hl}^e\})}{\partial x_i^e} - \beta_i^e \!\downarrow + \beta_i^e \!\uparrow = 0 \tag{2.41}$$

The generalized form of equations (2.20) and (2.24) can be represented as

$$\sum_{k=1}^{K_l} v_{kl}[\mathbf{D}]\{\hat{\mathbf{F}}_{kl}\} + \bar{\mathbf{\Delta}}_l^* - [\mathbf{B}]^{\mathrm{T}}\{\bar{\mathbf{L}}_k\} = \mathbf{0} \quad (l = 1, \dots, L) \tag{2.42}$$

$$v_{kl}([\mathbf{D}]\{\mathbf{F}_l\} + \{\mathbf{\Delta}_l^*\}) - [\mathbf{B}]^{\mathrm{T}}\{\mathbf{L}_{kl}\} = \mathbf{0} \quad (k = 1, \dots, K_l; l = 1, \dots, L) \tag{2.43}$$

Introducing

$$\frac{1}{v_{kl}}\{\mathbf{L}_{kl}\} = \{\mathbf{U}_l\} \quad (l = 1, \dots, L) \tag{2.44}$$

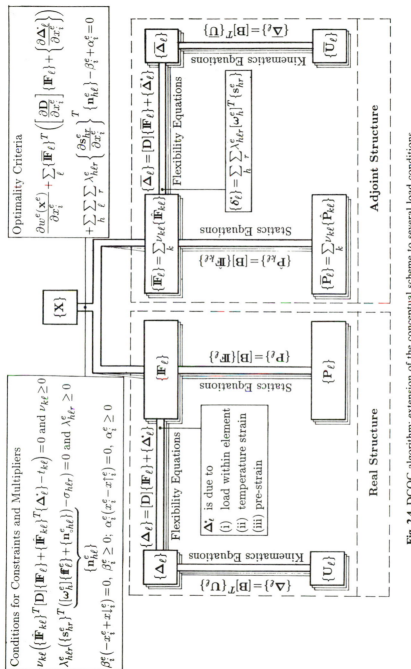

**Fig. 2.4** DCOC algorithm: extension of the conceptual scheme to several load conditions.

equation (2.43) with any $k$ value implies compatibility of the relative displacements and the nodal displacements for the loading conditions $l = 1, \ldots, L$.

Moreover, if we interpret the Lagrange multipliers $\{\bar{\mathbf{L}}_l\}$ in equation (2.42) as the adjoint nodal displacements for the $l$th load condition

$$\{\bar{\mathbf{L}}_l\} = \{\bar{\mathbf{U}}_l\} \tag{2.45}$$

then by equation (2.42) the adjoint structure must satisfy the following static and kinematic conditions for various load conditions.

- The adjoint nodal forces $\{\bar{\mathbb{F}}_l\}$ must equilibrate the adjoint load which is the factored combination of the virtual loads

$$\{\bar{\mathbf{P}}_l\} = \sum_{k=1}^{K_l} v_{kl} \{\hat{\mathbf{P}}_{kl}\} \quad (l = 1, \ldots, L) \tag{2.46}$$

- The adjoint nodal displacements $\{\bar{\mathbf{U}}_l\}$ and relative displacements $\{\bar{\mathbf{\Delta}}_l\}$ must satisfy compatibility subject to the initial relative displacements (prestrains)

$$\{\mathbf{\Delta}_l^*\} = [\bar{\boldsymbol{\delta}}_l^{*1}, \ldots, \bar{\boldsymbol{\delta}}_l^{*e}, \ldots, \bar{\boldsymbol{\delta}}_l^{*E}]^{\mathrm{T}} \tag{2.47}$$

with

$$\{\bar{\boldsymbol{\delta}}_l^{*e}\} = \sum_{h=1}^{H^e} \sum_{r=1}^{R_h^e} v_{hrl} [\boldsymbol{\omega}_h^e]^{\mathrm{T}} \frac{\partial \theta_{hr}^e}{\partial \mathbf{n}_{hl}^e} \quad (l = 1, \ldots, L) \tag{2.48}$$

Necessary conditions of optimality for the considered class of problems are summarized in Fig. 2.4.

### 2.3.4 Extended version of DCOC with allowance for support settlements, cost of supports and passive control

This extended version of DCOC covers the following design considerations:

- elastic supports,
- given support settlements,
- contact problems,
- allowance for the cost of reactions,
- variable loads,
- variable prestrains,
- variable support settlements,
- allowance for the cost of variable loads,
- allowance for the cost of variable prestrains, and
- allowance for the cost of variable support settlements.

The last six design conditions concern means of passive control and its cost. The considered problem can be stated as

$$\min_{\mathbf{x}, \tilde{\mathbf{\Delta}}^*, \check{\mathbf{\Delta}}^*, \tilde{\mathbf{P}}} \Phi = \sum_e w^e(\mathbf{x}^e) + \sum_n \check{\Omega}^n + \sum_e \chi^e(\tilde{\boldsymbol{\delta}}^{*e}) + \sum_n \vartheta^n(\check{\boldsymbol{\delta}}^{*n}) + \sum_j \psi_j(\tilde{P}_j) \tag{2.49}$$

where $\Phi$ is the objective function (e.g. weight), $\mathbf{x} = [\mathbf{x}^1, \ldots, \mathbf{x}^e, \ldots, \mathbf{x}^E]^{\mathrm{T}}$ the design variables, $\{\tilde{\mathbf{\Delta}}^*\} = [\tilde{\boldsymbol{\delta}}^{*1}, \ldots, \tilde{\boldsymbol{\delta}}^{*e}, \ldots, \tilde{\boldsymbol{\delta}}^{*E}]^{\mathrm{T}}$ the variable initial relative displacements (due to variable prestrain), $\{\check{\mathbf{\Delta}}^*\} = [\check{\boldsymbol{\delta}}^{*1}, \ldots, \check{\boldsymbol{\delta}}^{*n}, \ldots, \check{\boldsymbol{\delta}}^{*N}]^{\mathrm{T}}$ the variable support settlements, $\{\tilde{\mathbf{P}}\} = [\tilde{P}_1, \ldots, \tilde{P}_j, \ldots, \tilde{P}_J]^{\mathrm{T}}$ the variable loads, $w^e$ the objective function (e.g. weight) for the element $e$, $\check{\Omega}^n = \max_l \Omega^n(\mathbf{r}_l^n)$ is the highest cost requirement (out of all load conditions) for the reaction element (support) $n$, $\Omega^n(\ )$ is the support cost function for the reaction element $n$, $\{\mathbf{r}^n\} = \lfloor r_1^n, \ldots, r_q^n, \ldots, r_{Q^n}^n \rfloor^T$ the reactions in the support $n$, $\chi^e$ the cost of variable initial relative displacements (variable prestrain) in element $e$, $\{\tilde{\boldsymbol{\delta}}^{*e}\} = \lfloor \tilde{\delta}_1^{*e}, \ldots, \tilde{\delta}_m^{*e}, \ldots, \tilde{\delta}_{M^e}^{*e} \rfloor^T$ the adjustable initial relative displacements in element $e$, $\vartheta^n$ the cost of variable support settlements in support $n$, $\{\tilde{\boldsymbol{\delta}}^{*n}\} = \lfloor \tilde{\underline{\delta}}_1^{*n}, \ldots, \tilde{\underline{\delta}}_q^{*n}, \ldots, \tilde{\underline{\delta}}_{Q^n}^{*n} \rfloor^T$ the variable settlements at support $n$ and $\psi_j$ the cost of variable load at the degree of freedom $j$. The above minimization problem is subject to the following constraints.

The displacement constraints are

$$
\left\{ \begin{matrix} \hat{\mathbb{F}}_{kl} \\ \hat{\mathbf{R}}_{kl} \end{matrix} \right\}^T \begin{bmatrix} \mathbf{D} & \\ & \underline{\mathbf{D}} \end{bmatrix} \left\{ \begin{matrix} \mathbb{F}_l \\ \mathbf{R}_l \end{matrix} \right\} + \left\{ \begin{matrix} \hat{\mathbb{F}}_{kl} \\ \hat{\mathbf{R}}_{kl} \end{matrix} \right\}^T \left( \left\{ \begin{matrix} \boldsymbol{\Delta}_l^* \\ \underline{\boldsymbol{\Delta}}_l^* \end{matrix} \right\} + \left\{ \begin{matrix} \tilde{\boldsymbol{\Delta}}^* \\ \tilde{\underline{\boldsymbol{\Delta}}}^* \end{matrix} \right\} \right) - t_{kl} \leqslant 0
$$

$$ (k = 1, \ldots, K_l; l = 1, \ldots, L) \tag{2.50} $$

where the subscripts $k$ and $l$, respectively, identify a displacement constraint and a load condition, $\{\hat{\mathbb{F}}_{kl}\}$ and $\{\hat{\mathbf{R}}_{kl}\}$, respectively, are the virtual nodal forces (flexibility notation) and virtual reactions for the displacement constraint $k$ and load condition $l$, $\{\mathbb{F}_l\}$ and $\{\mathbf{R}_l\}$ are, respectively, real nodal forces and real reactions for the load condition $l$, $[\mathbf{D}]$ is the flexibility matrix and $[\underline{\mathbf{D}}]$ the reaction element (support) flexibility matrix, $\{\boldsymbol{\Delta}_l^*\}$ and $\{\underline{\boldsymbol{\Delta}}_l^*\}$ are, respectively, given initial relative displacements and support settlements for the load condition $l$, $\{\tilde{\boldsymbol{\Delta}}^*\}$ and $\{\tilde{\underline{\boldsymbol{\Delta}}}^*\}$ are variable initial relative displacements and support settlements and $t_{kl}$ is the prescribed limit in the displacement constraint $k$ for the load condition $l$. Since here we are dealing with passive control, the variable initial relative displacements and variable support settlements are the same for all load conditions.

The stress conditions are as in equation (2.37) or (2.38).

The side constraints are as in equations (2.4) and (2.5).

The equilibrium condition for the real forces is

$$
\{\mathbf{P}_l\} + \left\{ \begin{matrix} \mathbf{0} \\ \tilde{\mathbf{P}} \end{matrix} \right\} - [\mathbf{B}] \left\{ \begin{matrix} \mathbb{F}_l \\ \mathbf{R}_l \end{matrix} \right\} = \mathbf{0} \quad (l = 1, \ldots, L) \tag{2.51}
$$

where $\{\mathbf{P}_l\}$ are the real loads for the load condition $l$, $\{\tilde{\mathbf{P}}\}$ are the variable loads (for some of the degrees of freedom, but the same for all load conditions), $\{\mathbf{0}\}$ denotes zero variable load for some of the degrees of freedom, $[\mathbf{B}]$ is the statics matrix, $\{\mathbb{F}_l\}$ are the real nodal forces (flexibility formulation) for the load condition $l$ and $\{\mathbf{R}_l\}$ are the real reactions for the load condition $l$.

The equilibrium condition for the virtual forces is

$$
\{\hat{\mathbf{P}}_{kl}\} - [\mathbf{B}] \left\{ \begin{matrix} \hat{\mathbb{F}}_{kl} \\ \hat{\mathbf{R}}_{kl} \end{matrix} \right\} = \mathbf{0} \quad (k = 1, \ldots, K_l; l = 1, \ldots, L) \tag{2.52}
$$

where $\{\hat{\mathbf{P}}_{kl}\}$ are the virtual loads (e.g. a unit dummy load for a displacement constraint), $\{\hat{\mathbb{F}}_{kl}\}$ the virtual nodal forces and $\{\hat{\mathbf{R}}_{kl}\}$ the virtual reactions, all three for the displacement constraint $k$ under the load condition $l$.

The support cost inequality is

$$
\Omega^n(\mathbf{r}_l^n) - \check{\Omega}^n \leqslant 0 \quad (l = 1, \ldots, L; n = 1, \ldots, N) \tag{2.53}
$$

where $\check{\Omega}^n$ is the maximum cost requirement for the support $n$. It is assumed that each support element is governed by one design parameter. Alternatively, a reaction element may have several design variables and stress constraints. In that case, it can be treated the same way in the DCOC procedure as an ordinary element.

The Lagrange multipliers for the constraints (2.50), (2.37), (2.4), (2.5), (2.51), (2.52) and (2.53) will be denoted, respectively, by

$$v_{kl}, \lambda^e_{hrl}, \beta^e_i\downarrow, \beta^e_i\uparrow, \{\bar{\mathbf{L}}_l\}, \{\mathbf{L}_{kl}\}, \rho^n_l \tag{2.54}$$

For the above problem, we have the following Kuhn–Tucker conditions for the variation of the variables stated below. For the design variables $x^e_i$,

$$\frac{\partial w^e}{\partial x^e_i} + \sum_k \sum_l v_{kl} \{\hat{\mathbf{ff}}_{kl}\}^{\mathrm{T}} \left( \left[ \frac{\partial \mathbf{d}^e}{\partial x_i} \right] \{\mathbf{ff}^e_l\} + \left\{ \frac{\partial \boldsymbol{\delta}^{*e}_l}{\partial x^e_i} \right\} \right)$$

$$+ \sum_h \sum_r \sum_l \lambda^e_{hrl} \frac{\partial \theta^e_{hr} \{\mathbf{n}^e_{hl}\}}{\partial x^e_i} - \beta^e_i\downarrow + \beta^e_i\uparrow = 0 \quad (i = 1, \ldots, I^e; e = 1, \ldots, E) \tag{2.55}$$

For the real nodal forces $\{\mathbb{F}_l\}$ and real reactions $\{\mathbf{R}_l\}$,

$$\sum_k v_{kl} \begin{bmatrix} \mathbf{D} & \\ & \underline{\mathbf{D}} \end{bmatrix} \begin{Bmatrix} \hat{\mathbb{F}}_{kl} \\ \hat{\mathbf{R}}_{kl} \end{Bmatrix} - \begin{Bmatrix} \bar{\Delta}^*_l \\ \underline{\bar{\Delta}}^*_l \end{Bmatrix} + [\mathbf{B}]^{\mathrm{T}}\{\bar{\mathbf{L}}_l\} = 0 \quad (l = 1, \ldots, L) \tag{2.56}$$

where

$$\{\bar{\Delta}^*_l\} = \lfloor \bar{\boldsymbol{\delta}}^{*1}_l, \ldots, \bar{\boldsymbol{\delta}}^{*e}_l, \ldots, \bar{\boldsymbol{\delta}}^{*E}_l \rfloor^{\mathrm{T}} \tag{2.57}$$

$$\{\underline{\bar{\Delta}}^*_l\} = \lfloor \underline{\bar{\boldsymbol{\delta}}}^{*1}_l, \ldots, \underline{\bar{\boldsymbol{\delta}}}^{*n}_l, \ldots, \underline{\bar{\boldsymbol{\delta}}}^{*N}_l \rfloor^{\mathrm{T}} \tag{2.58}$$

$$\bar{\boldsymbol{\delta}}^{*e}_l = \sum_h \sum_r \lambda^e_{hrl} \left\{ \frac{\partial \theta^e_{hr}}{\partial \mathbf{n}^e_{hl}} \right\} \tag{2.59}$$

$$\underline{\bar{\delta}}^{*q}_{ql} = \rho^n_l \frac{\partial \Omega^n}{\partial r^n_{ql}}, \quad \sum_l \rho^n_l = 1 \tag{2.60}$$

For the virtual nodal forces $\{\hat{\mathbb{F}}_{kl}\}$ and virtual reactions $\{\hat{\mathbb{R}}_{kl}\}$,

$$v_{kl} \left( \begin{bmatrix} \mathbf{D} & \\ & \underline{\mathbf{D}} \end{bmatrix} \begin{Bmatrix} \mathbb{F}_l \\ \mathbf{R}_l \end{Bmatrix} + \begin{Bmatrix} \Delta^*_l \\ \underline{\Delta}^*_l \end{Bmatrix} + \begin{Bmatrix} \tilde{\Delta}^* \\ \underline{\tilde{\Delta}}^* \end{Bmatrix} \right) + [\mathbf{B}]^{\mathrm{T}}\{\mathbf{L}_{kl}\} = 0$$

$$(k = 1, \ldots, K; l = 1, \ldots, L) \tag{2.61}$$

Since the expression in parentheses after $v_{kl}$ in equation (2.61) represents the real relative nodal displacements, and $[\mathbf{B}]^{\mathrm{T}}$ is the kinematics matrix, we can see that

$$\frac{1}{v_{kl}}\{\mathbf{L}_{kl}\} = \{\mathbf{U}_l\} \tag{2.62}$$

where $\{\mathbf{U}_l\}$ denotes the real nodal displacements. Then it follows from equation (2.61) that kinematic admissibility, which was not included in the original formulation above, is automatically satisfied as an optimality condition.

Moreover, we interpret the Lagrange multiplier $\{\bar{\mathbf{L}}_l\}$ in (2.56) as the adjoint nodal displacement,

$$\{\bar{\mathbf{L}}_l\} = \{\bar{\mathbf{U}}_l\} \tag{2.63}$$

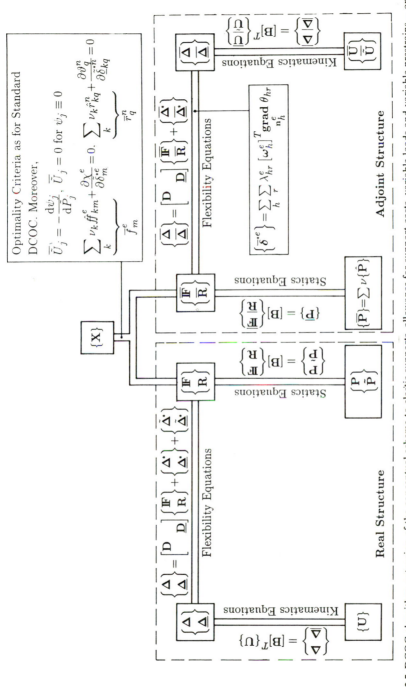

**Fig. 2.5** DCOC algorithm: extension of the conceptual scheme to elastic supports, allowance for support costs, variable loads and variable prestrains – one loading condition.

and the corresponding relative displacements ($\{\bar{\Delta}\}$ at system level and $\{\bar{\delta}^e\}$ at element level) as adjoint relative displacements. It can be seen from equation (2.56) that the adjoint system has the same flexibility matrix as the real structure, but is subject to initial relative displacements $\lfloor \Delta_i^*, \underline{\Delta}_i^* \rfloor^T$ owing to active stress constraints.

Further Kuhn–Tucker conditions for the remaining variables are as follows. For the variable loads $\tilde{P}_j$,

$$\sum_l \bar{U}_{jl} = -\frac{\mathrm{d}\psi_j}{\mathrm{d}\tilde{P}_j} \tag{2.64}$$

For the variable relative displacements (prestrains) $\{\tilde{\delta}^{*e}\}$,

$$-\sum_k \sum_l v_{kl} \hat{f\!f}^e_{klm} = \frac{\partial \chi^e}{\partial \tilde{\delta}^{*e}_m} \tag{2.65}$$

For the variable support settlements $\{\tilde{\underline{\delta}}^{*n}\}$,

$$-\sum_k \sum_l v_{kl} \hat{f}^n_{klq} = \frac{\partial \vartheta^n}{\partial \tilde{\underline{\delta}}^{*n}_q} \tag{2.66}$$

Moreover, all Lagrange multipliers for inequality constraints in equation (2.54) are nonnegative and can be positive only if the corresponding constraint is satisfied as an equality.

The necessary conditions of optimality for the above extended DCOC algorithm are summarized in Fig. 2.5 for one load condition.

In the case of contact problems between the elastic structure to be optimized and some rigid system, the contact constraint consists of two parts. If the displacement is smaller than the distance to the rigid system, then the contact constraint is inactive. If the displacement without the contact constraint would be greater than the above distance, then a given support settlement type constraint becomes active at the considered node.

For problems with given prestrain, lack of fit, temperature changes and support settlements, the appropriate value of initial relative displacements $\delta^{*e}$ or initial support displacements $\underline{\delta}^{*e}$ must be included in the real structure but the adjoint structure is not affected. In the case of adjustable loads, prestrains and support settlements, relations (2.64)–(2.66) must be taken into consideration for the adjoint structure.

### 2.3.5 Natural frequency and system stability constraints

DCOC was extended recently by O. Sigmund to dynamic constraints. The corresponding optimality criteria will be presented in chapter 10 of this book, where their applications in topology optimization are also discussed.

## 2.4 COMPUTATIONAL IMPLEMENTATION OF DCOC

As was mentioned in section 2.3, the flexibility method is suitable for deriving optimality criteria for DCOC, but in actual computations it is more convenient and more economical to use the stiffness method.

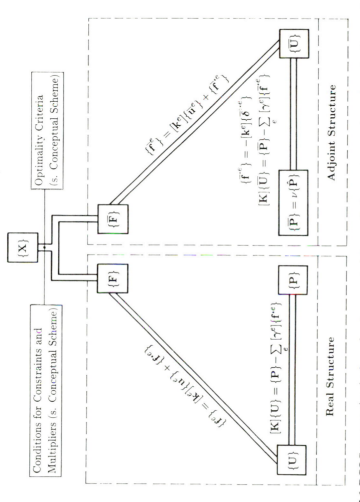

**Fig. 2.6** DCOC computational scheme for stress constraints, one displacement constraint and one loading condition.

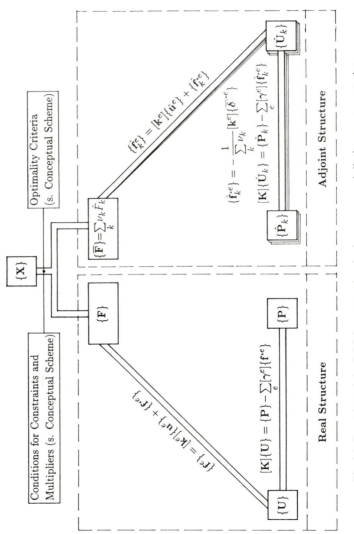

**Fig. 2.7** Extension of the DCOC computational scheme to several displacement constraints.

For the simple case of stress constraints, one displacement constraint and one loading condition, the computational scheme based on the stiffness method is shown in Fig. 2.6. For the adjoint system the initial relative displacements in the conceptual scheme (Fig. 2.2) are first extended to the stiffness formulation $\{\bar{\boldsymbol{\delta}}^{\mathrm{o}*e}\}$ and then converted via the element stiffness matrices $[\mathbf{k}^e]$ into fixed end forces $\{\bar{\mathbf{f}}^{*e}\}$. As mentioned before, the decomposed stiffness matrix for the real structure can also be used for the adjoint structure. For one displacement constraint and one design variable per element, the Lagrange multiplier $v$ can be updated explicitly from the displacement constraint (section 2.4.1(c)). The update of the design parameters is based on the optimality criteria given in Fig. 2.2.

Figure 2.7 shows the extension of the computational scheme to several displacement constraints. Although conceptually the adjoint structure is subject to the loads $\{\bar{\mathbf{P}}\} = \sum_k v_k \hat{\mathbf{P}}_k$, it is more convenient for the calculation of the Lagrange multipliers $v_k$ to analyze the structure for each virtual load $\hat{\mathbf{P}}_k$. The relative initial displacements $\{\hat{\boldsymbol{\delta}}^{*e}\}$ due to active stress constraints, or the corresponding fixed end forces $\{\hat{\mathbf{f}}_k^{*e}\}$, can be distributed arbitrarily to force and displacement systems associated with each virtual load $\hat{\mathbf{P}}_k$. One such distribution is given in Fig. 2.7. For further details see section 2.4.4.

## 2.4.1 Computational procedure for problems with one load condition, one displacement constraint, one design variable per element and flexibility matrices/stress constraints that are linear in the reciprocal variable $1/x^e$ and in the stress resultants

We consider a subclass of problems with

- one load condition and one displacement constraint,
- one design variable $x^e$ per element ($I^e = 1$ for $e = 1, \ldots, E$),
- one stress constraint ($(\theta^e(\mathbf{n})^e - \sigma^e \leqslant 0$) per element ($H^e = 1$ and $R_h^e = 1$ for $e = 1, \ldots, E$),
- a linear element objective function

$$w^e(x^e) = c^e x^e \qquad (2.67)$$

where $c^e$ is a given constant,

- element flexibility matrices of the type

$$[\mathbf{d}^e] = [\mathbf{d}_\mathrm{o}^e]/x^e, \qquad (2.68)$$

where $[\mathbf{d}_\mathrm{o}^e]$ is a given matrix,

- stress constraints that in equations (2.3) and (2.8), respectively, can be expressed as

$$\theta^e(\{\mathbf{n}^e\}) - \sigma^e = \frac{\theta_\mathrm{o}^e(\{\mathbf{n}^e\})}{x^e} - \sigma^e \leqslant 0 \qquad (2.69)$$

and

$$\{\mathbf{s}^e\}^\mathrm{T}\{\mathbf{n}^e\} - \sigma^e = \frac{\{\mathbf{s}_\mathrm{o}^e\}^\mathrm{T}}{x^e}\{\mathbf{n}^e\} - \sigma^e \leqslant 0 \qquad (2.70)$$

where $\theta_\mathrm{o}^e$ and $\{\mathbf{s}_\mathrm{o}^e\}$ are independent of the design variable $x^e$, and
- only lower side constraints.

For simplicity, the subscripts representing $i = 1$, $h = 1$ and $r = 1$ are omitted in equations (2.67)–(2.70) ($x_1^e \rightarrow x^e$, $\theta_{11}^e \rightarrow \theta^e$, $\mathbf{n}_1^e \rightarrow \mathbf{n}^e$).

For the considered class of problems, each element belongs to one of the following sets termed 'regions':

  $R_\sigma$  the stress constraint is active
  $R_s$  the lower side constraint is active
  $R_{\sigma s}$  both the stress constraint and the lower side constraint are active
  $R_d$  none of the local (stress or side) constraints is active; the element is controlled by the displacement constraint

A flowchart for the main steps of the DCOC algorithm is shown in Fig. 2.8. Some explanatory comments on various steps are given below.

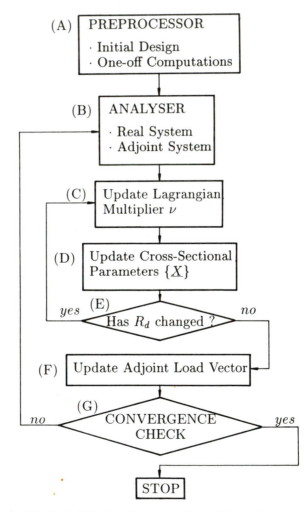

**Fig. 2.8** Flowchart for the DCOC algorithm: one design variable and one stress constraint per element, and one displacement constraint.

## (a) Pre-processor

The input includes the definition of the problem, including types of elements, geometrical and material properties, loading, permissible stresses and displacements as well as initial values of the design variables. One-off computations include calculations of certain geometrical and mechanical properties of the elements, bandwidth of the stiffness matrix, etc.

## (b) Analysis of the real and adjoint systems

As can be seen from Fig. 2.6, the stiffness equation for the real structure is of the form

$$[\mathbf{K}]\{\mathbf{U}\} = \{\mathbf{P}\} - \sum_e [\gamma^e]\{\mathbf{f}^{*e}\} \tag{2.71}$$

where $[\gamma^e]$ is a transformation matrix from element level to system level and $\{\mathbf{f}^{*e}\}$ are 'fixed end forces' due to loads on the interior of an element or prestrains or temperature strains. Using an assembly technique known as the direct stiffness method, the stiffness matrix at the system level can be generated by using the relation

$$[\mathbf{K}] = \sum_{e=1}^{E} [\gamma^e]^{\mathrm{T}}[\mathbf{k}^e][\gamma^e] \tag{2.72}$$

It can also be seen from Fig. 2.6 that the nodal forces can be calculated from the relations

$$\{\mathbf{f}^e\} = [\mathbf{k}^e]\{\mathbf{u}^e\} + \{\mathbf{f}^{*e}\} \tag{2.73}$$

at the element level.

The calculation of the fixed-end forces $\{\mathbf{f}^{*e}\}$ for the real structure is a well-known operation. However, the fixed-end forces $\{\bar{\mathbf{f}}^{*e}\}$ for the adjoint system must be calculated from the fictitious relative displacement $(\bar{\delta}^{*e}\}$ given by equation (2.22) or (2.33):

$$\{\bar{\mathbf{f}}^{*e}\} = - [\mathbf{k}^e]\{\bar{\delta}^{\circ *e}\} \tag{2.74}$$

where $\{\bar{\delta}^{\circ *e}\}$ is the extended version of $\{\bar{\delta}^{*e}\}$ for the stiffness formulation which has a greater number of components than the relative displacement vector for the flexibility formulation (zero components are added for supported degrees of freedom in the flexibility formulation).

The analysis of the adjoint system involves only the forward and backward substitution of the adjoint load vector (Fig. 2.6)

$$\{\bar{\mathbf{P}}\} = \sum_e [\gamma^e]\{\bar{\mathbf{f}}^{*e}\} \tag{2.75}$$

using the decomposed stiffness matrix, which is already available from the analysis of the real system. Since in the DCOC algorithm the analysis of the adjoint system, in effect, replaces sensitivity analysis of other methods, this is a computationally very efficient feature of DCOC.

In the analysis of the adjoint system during the first iteration, the adjoint load vector consists of the virtual loads corresponding to the displacement constraints. During subsequent iterations, the adjoint load vector is the sum of the virtual load vector and of the equivalent nodal loads (reversed fixed end forces) caused by the adjoint initial relative displacements due to active stress constraints (equation (2.74)).

### (c) Updating the Lagrange multiplier $v$

For elements within the set $R_d$ the Lagrange multipliers $\lambda_{hr}^e$, $\beta_i^e\downarrow$ and $\beta_i^e\uparrow$ take on a zero value and hence equation (2.19) with equations (2.67) and (2.68) reduces to

$$\frac{\partial w^e}{\partial x^e} + v\{\hat{\mathbf{f}}^e\}^T\left[\frac{\partial \mathbf{d}^e}{\partial x^e}\right]\{\mathbf{f}^e\} = c^e - v\{\hat{\mathbf{f}}^e\}^T[\mathbf{d}_o^e]\{\mathbf{f}^e\}/(x^e)^2 = 0 \qquad (2.76)$$

or

$$x^e = v(\{\hat{\mathbf{f}}^e\}^T[\mathbf{d}_o^e]\{\mathbf{f}^e\}/c^e)^{1/2} \qquad (2.77)$$

Then the displacement constraint can be expressed as

$$t = \sum_{e=1}^{E}(\{\hat{\mathbf{f}}^e\}^T[\mathbf{d}^e]\{\mathbf{f}^e\}) = \sum_{e\in R_d}\frac{\{\hat{\mathbf{f}}^e\}[\mathbf{d}_o^e]\{\hat{\mathbf{f}}^e\}}{(v\{\hat{\mathbf{f}}^e\}^T[\mathbf{d}_o^e]\{\mathbf{f}^e\}/c^e)^{1/2}} + \sum_{e\notin R_d}(\{\hat{\mathbf{f}}^e\}^T[\mathbf{d}^e]\{\mathbf{f}^e\}) \qquad (2.78)$$

implying

$$v^{1/2} = \frac{\sum_{e\in R_d}(c^e\{\hat{\mathbf{f}}^e\}[\mathbf{d}_o^e]\{\mathbf{f}^e\})^{1/2}}{t - \sum_{e\notin R_d}(\{\hat{\mathbf{f}}^e\}^T[\mathbf{d}^e]\{\mathbf{f}^e\})} \qquad (2.79)$$

The values of $c^e$ and $[\mathbf{d}_o^e]$ are given for some simple elements in Table 2.3, where $\rho$ is the specific weight of the structural material, $E$ is Young's modulus, $\mu$ is Poisson's ratio and other symbols are defined in Fig. 2.9. The superscript $e$ is omitted for simplicity after the symbols $h$, $L$, $\rho$, $b$, $E$, $d$ and $\mu$.

### (d) Updating the cross-sectional variables $\{X\}$

This is done on the basis of equation (2.77) for elements of the $R_d$ region. For elements contained in $R_\sigma$, $R_s$ or $R_{\sigma s}$ regions, the stress or side constraint determines the value of $x^e$.

### (e) Recycling if $R_d$ has changed

The reason for this step is as follows. If the displacement-controlled region has changed, then equation (2.79) would give an incorrect estimate of $v$ and hence steps (c) and (d) must be repeated.

### (f) Updating the adjoint load vector

The equivalent element nodal loads for $e\in R_\sigma$ are the reversed fixed end forces calculated from equations (2.74) and (2.22) or (2.23). The Lagrange multipliers $\lambda_{11}^e \to \lambda^e$ for elements of $R_\sigma$ are determined from equation (2.19) with $H_e = 1$,

**Table 2.3** Design variable, weight constant and constant part of the flexibility matrix for some simple elements

| Type of element | $x^e$ | $c^e$ | $[\mathbf{d}^e_o]$ |
|---|---|---|---|
| Truss elements (Fig. 2.9(a)) | Cross-sectional area | $L\rho$ | $\dfrac{L}{E}$ |
| Beam element, rectangular, cross section, variable width (Fig. 2.9(b)) | Width | $hL\rho$ | $\begin{bmatrix} \dfrac{4L^3}{Eh^3} & \dfrac{6L^2}{Eh^3} \\[2ex] \dfrac{6L^2}{Eh^3} & \dfrac{12L}{Eh^3} \end{bmatrix}$ |
| Frame element, rectangular cross section, variable width (Fig. 2.9(c)) | Width | $hL\rho$ | $\begin{bmatrix} \dfrac{4L^3}{Eh^3} & 0 & \dfrac{6L^2}{Eh^3} \\[2ex] 0 & \dfrac{L}{Eh} & 0 \\[2ex] \dfrac{6L^2}{Eh^3} & 0 & \dfrac{12L}{Eh^3} \end{bmatrix}$ |
| Triangular, constant-strain membrane element (Fig. 2.9(d)) | Thickness | $\dfrac{bh\rho}{2}$ | $\dfrac{2}{Ebh}\begin{bmatrix} b^2 & bd & -\mu hd \\ bd & 2(1+\mu)h^2+d^2 & -\mu hd \\ -\mu hd & -\mu hd & h^2 \end{bmatrix}$ |

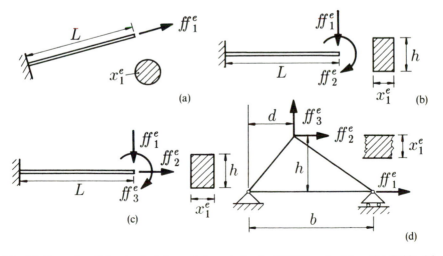

**Fig. 2.9** Examples of the simplest class of elements for the DCOC method (see also Tables 2.3 and 2.4).

$R_h^e = 1$, $\beta_i^e \downarrow = 0$ and $\beta_i^e \uparrow = 0$ (see also equations (2.67)–(2.70)):

$$\lambda^e = \frac{(x^e)^2 c^e - v\{\overline{\mathbf{ff}}^e\}^{\mathrm{T}}[\mathbf{d}_0^e]\{\mathbf{ff}^e\}}{\theta_0^e(\{\mathbf{n}^e\})} \tag{2.80}$$

or

$$\lambda^e = \frac{(x^e)^2 c^e - v\{\widehat{\mathbf{ff}}^e\}[\mathbf{d}_0^e]\{\mathbf{ff}^e\}}{\{\mathbf{s}_0\}^{\mathrm{T}}\{\mathbf{n}^e\}} \tag{2.81}$$

For the element types considered in Table 2.3 and Fig. 2.9, the function $\theta_0^e(\{\mathbf{n}^e\})$ and the vector $\{\mathbf{s}_0\}$ are given in Table 2.4. It can be seen from equation (2.8) that, in equations (2.80) and (2.81), $\{\mathbf{n}^e\}$ is given by $[\boldsymbol{\omega}^e]\{\mathbf{ff}^e\}$.

In Table 2.4, the stress resultants $n_1^e$, $n_2^e$ and $n_3^e$ correspond to the nodal forces $ff_1^e$, $ff_2^e$ and $ff_3^e$ in Fig. 2.9, but at some given cross section of the element $e$. For the first three cases, we have two transformation vectors in the last column, which correspond to tensile and compressive permissible stresses $(\sigma_T^e, \sigma_C^e)$, respectively. This represents two stress constraints per element, out of which only one at a time can be active.

### (g) The convergence criterion

The simplest criterion is

$$\frac{|\Phi_{\text{new}} - \Phi_{\text{old}}|}{\Phi_{\text{new}}} \leqslant T \tag{2.82}$$

where $T$ is a given tolerance value. One may also use the following additional convergence tests:

$$\frac{|x_{i,\text{new}}^e - x_{i,\text{old}}^e|}{x_{i,\text{new}}^e} \leqslant T_1 \quad (i = 1, ..., I^e; \quad e = 1, ..., E) \tag{2.82a}$$

**Table 2.4** The function $\theta_0^e(\{\mathbf{n}^e\})$ and vector $\{\mathbf{s}_0^e\}$ in the stress condition for the considered class of problems

| Type of element | Type of stress | $\theta_0^e(\{\mathbf{n}^e\})$ | $\{\mathbf{s}_0^e\}$ |
|---|---|---|---|
| Truss elements (Fig. 2.9a) | Axial | $\|n_1^e\|$ | $\{1\}$ or $\{-1\}$ |
| Beam element, rectangular cross section, variable width (Fig. 2.9(b)) | Axial, bottom of cross section | $\dfrac{6\|n_2^e\|}{h^2}$ | $\left\{\begin{matrix} 0 \\ 6/h^2 \end{matrix}\right\}$ or $\left\{\begin{matrix} 0 \\ -6/h^2 \end{matrix}\right\}$ |
| Frame element, rectangular cross section, variable width (Fig. 2.9(c)) | Axial, bottom of cross section | $\dfrac{\|n_2^e\|}{h} + \dfrac{6\|n_3^e\|}{h^2}$ | $\left\{\begin{matrix} 0 \\ 1/h \\ 6/h^2 \end{matrix}\right\}$ or $\left\{\begin{matrix} 0 \\ -1/h \\ -6/h^2 \end{matrix}\right\}$ |
| Triangular, constant-strain membrane element (Fig. 2.9(d)) | Mises stress, equation (2.17) | $(n_1^e)^2 + (n_2^e)^2 - n_1^e n_2^e + 3(n_3^e)^2$ | Not applicable owing to nonlinearity |

In most test examples in section 2.5, very stringent tolerance values (e.g. $T = 10^{-8}$) will be used with a view to demonstrating the accuracy of this method. In practical applications a tolerance value of $T = 10^{-3}$ or $T = 10^{-4}$ is usually sufficient.

### 2.4.2 A simple application illustrating the computational procedure for the class of problems in section 2.4.1: trusses

The design variable, element objective function, element flexibility matrix and stress constraints for a truss element were already given in equations (2.67)–(2.70), together with Tables 2.3 and 2.4. Denoting the two ends of the truss element by A and B, respectively, we have the element force vector and displacement vector in the stiffness formulation as

$$\{\mathbf{f}^e\} = \lfloor f_A^e, f_B^e \rfloor^T, \quad \{\mathbf{u}^e\} = \lfloor u_A^e, u_B^e \rfloor^T \tag{2.83}$$

and the force–displacement relation

$$\{\mathbf{f}^e\} = [\mathbf{k}^e]\{\mathbf{u}^e\} \tag{2.84}$$

with the element stiffness matrix

$$[\mathbf{k}^e] = \frac{E^e x^e}{L^e} \begin{bmatrix} 1 & -1 \\ -1 & 1 \end{bmatrix} \tag{2.85}$$

in which $x^e$ is the cross-sectional area (Table 2.3) and other symbols were defined earlier. The element–system coordinate transformation matrix for a space truss can be represented as

$$[\boldsymbol{\gamma}^e] = \begin{bmatrix} \cos \alpha^e & \cos \beta^e & \cos \zeta^e & 0 & 0 & 0 \\ 0 & 0 & 0 & \cos \alpha^e & \cos \beta^e & \cos \zeta^e \end{bmatrix} \tag{2.86}$$

where $\alpha^e$, $\beta^e$ and $\zeta^e$ are the angles between the longitudinal axis of the element $e$ and the global (system) coordinate axes. In the flexibility formulation, the end A of the truss element has a rigid support and end B can move freely. Then the displacement–force relation contains only scalar quantities:

$$\delta^e = d^e f\!\!f^e \tag{2.87}$$

where $\delta^e$ and $f\!\!f^e$ refer to the relative displacement (element elongation = displacement at end B) and axial force at the end B, respectively. Moreover, the flexibility matrix reduces to

$$d^e = \frac{L^e}{E^e x^e} \tag{2.88}$$

as in Table 2.3 with equation (2.68). The element objective function $w^e$ is also given by Table 2.3 with equation (2.67) as

$$w^e(x^e) = c^e x^e = L\rho^e x^e \tag{2.89}$$

If (i) we express the stress constraint using the function $\theta^e(\{\mathbf{n}^e\})$ and (ii) the permissible stresses in tension and compression have the same value ($\sigma^e$), then we have (Table 2.4)

$$\theta^e(\{\mathbf{n}^e\}) - \sigma^e = |n^e|/x^e - \sigma^e \leqslant 0 \tag{2.90}$$

If (i) we express the stress constraint using a transformation vector $\{s^e\}$ and (ii) we have different permissible stresses $(\sigma^e_T, \sigma^e_C)$ in tension and compression, then we have (Table 2.4)

$$\{s^e\}^T\{n^e\} - \sigma^e_T = n^e/x^e - \sigma^e_T \leqslant 0 \tag{2.91}$$

$$-\{s^e\}^T\{n^e\} - \sigma^e_C = -n^e/x^e - \sigma^e_C \leqslant 0 \tag{2.92}$$

Since for trusses $ff^e = n^e$ or $[\omega] = 1$, by equation (2.81) we have the following value for the Lagrange multiplier of elements with an active stress constraint $(e \in R_\sigma)$ and equal permissible stresses $(\sigma)$ in tension and compression:

$$\lambda^e = \frac{\rho^e(x^e)^2 - v\, ff^e\, \widehat{ff}^e/E^e}{|ff^e|} L^e \tag{2.93}$$

or

$$\lambda^e = \frac{L^e}{\sigma^e}\left(\rho^e x^e - \frac{v\, \widehat{ff}^e ff^e}{x^e E^e}\right) \tag{2.94}$$

The adjoint initial relative displacement by equations (2.22) and (2.90) with $n^e = ff^e$ becomes

$$\bar{\delta}^{*e} = \lambda^e\, \mathrm{sgn}\, ff^e/x^e \tag{2.95}$$

The adjoint initial relative displacements, which are extended for the stiffness method, are given by (Fig. 2.6)

$$\{\bar{\delta}^{o*e}\} = \begin{Bmatrix} \delta^{*e} \\ 0 \end{Bmatrix} \tag{2.96}$$

and the corresponding fixed end forces by

$$\{\bar{f}^{**e}\} = -[k^e]\{\delta^{o*e}\} \tag{2.97}$$

where $[k^e]$ is given by equation (2.85). For the updating of the Lagrange multiplier $v$, equation (2.79) implies with Table 2.3

$$v^{1/2} = \frac{\displaystyle\sum_{e \in R_d} L^e(\rho^e\{\widehat{ff}^e\, ff^e/E^e)^{1/2}}{t - \displaystyle\sum_{e \notin R_d} (L^e\, \widehat{ff}^e ff^e/E^e x^e)} \tag{2.98}$$

In equation (2.98), the value of $x^e$ for $e \in R_\sigma$ is based on the latest information using the real forces from the analysis step and equation (2.101): for $e \in R_\sigma$

$$L^e\, \widehat{ff}^e\, ff^e/(E^e x^e) = L^e\, \widehat{ff}^e\, ff^e/(E^e|ff^e|/\sigma^e) = L^e\, \widehat{ff}^e \sigma^e\, \mathrm{sgn}\, ff^e/E^e \tag{2.98a}$$

For $e \in R_s$ we have $x^e = x^e\!\downarrow$.

In step (D) of the flow-chart in Fig. 2.8, we have the following re-sizing formula:

$$(x^e)^2 = \max\{(x^e_d)^2, (x_\sigma)^2, (x^e\!\downarrow)^2\} \tag{2.99}$$

where by equation (2.77) and Table 2.3

$$(x^e_d)^2 = v\, \frac{\overline{ff}^e\, ff^e}{E^e\rho} \tag{2.100}$$

and by equation (2.90)

$$x_\sigma^e = \frac{|ff^e|}{\sigma^e} \qquad (2.101)$$

### 2.4.3 Certain computational difficulties and methods for removing them

(a) Truss element with active stress and side constraints ($e \in R_{\sigma s}$)

If both the stress and the side constraints are active for a truss element, then by equations (2.19), (2.67)–(2.70) and Tables 2.3 and 2.4 we have the relation for the Lagrange multiplier $\lambda^e$ (assuming that $v$ is given):

$$c^e - v\widehat{ff}^e ff^e L^e / E^e (x^e)^2 - \lambda^e |ff^e|/(x^e)^2 - \beta^e \downarrow = 0 \qquad (2.102)$$

which cannot be solved uniquely because there are two unknowns ($\lambda^e$ and $\beta^e \downarrow$) in one equation.

(b) No displacement constraint is active for a truss

We recall that, in the derivation of the DCOC algorithm in section 2.3.1, compatibility of the displacements was not included in the formulation and it turned out to be an optimality criterion, provided that a displacement condition was active. If only stress constraints are active, then the optimality criteria derived above are not valid and the solution algorithm must be modified.

(c) One method for avoiding the above difficulties: upgrading stress constraints to global constraints

If either of the difficulties outlined under (a) or (b) is encountered then, respectively,

• the stress constraint for all elements with $e \in R_{\sigma s}$, or
• any one stress constraint

can be upgraded into a global (displacement) constraint. This removes the difficulties outlined above, but the introduction of an extra global constraint also requires additional computer time.

Considering a truss element, for example, a stress constraint can be replaced by a constraint on the relative displacement between the two ends of the considered truss element (i.e. on the elongation of the element). Using the corresponding virtual loads $\{\widehat{\mathbf{P}}_k\} \to \{\widehat{\mathbf{P}}\}$ shown in Fig. 2.10, this additional displacement constraint becomes

$$\sum_{e=1}^{E} \frac{\widehat{ff}_k^e ff^e L^e}{x^e E^e} - t_k \leqslant 0 \qquad (2.103)$$

with

$$t_k = \frac{L^e \sigma^e}{E^e} \qquad (2.104)$$

where $e$ refers to the element for which the stress constraint is upgraded into a displacement constraint.

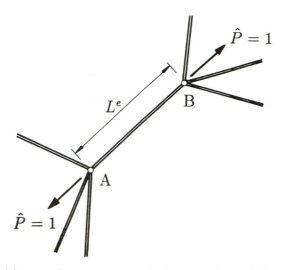

**Fig. 2.10** Method for upgrading a stress constraint for trusses into a displacement constraint.

**(d) Another possibility: forcing uniqueness of the adjoint strains by corner-rounding at constraint intersections using KS functions**

The nonuniqueness of the multiplier $\lambda^e$ in equation (2.102) can easily be understood by considering the cross-sectional area–member force relation $x^e(ff^e)$ for this problem (thick line in Fig. 2.11(a)). The above relation is based on the inequalities

$$x^e \geqslant x^e\downarrow, \quad x^e \geqslant ff^e/\sigma^e, \quad x^e \geqslant -ff^e/\sigma^e \qquad (2.105)$$

with at least one of these three constraints satisfied as an equality.

If both the stress and the side constraints are active, for example at the point R in Fig. 2.11(a), then the subgradient of $x^e(ff^e)$ is clearly nonunique, ranging from zero to $1/\sigma_e$. This causes, in turn, the nonuniqueness of the corresponding Lagrange multiplier $\lambda_i$.

The above difficulty can be avoided by replacing the exact $x^e(ff^e)$ relation by a smooth envelope function (SEF).

Applications of Kreisselmeier–Steinhauser functions in optimization problems were discussed recently (Chang, 1992; Sobieszczanski-Sobieski, 1992). On the basis of equation (2) in the latter reference, we have the following KS functions for the three linear segments in Fig. 2.11(a): for $|ff^e| \geqslant \sigma^e x^e\downarrow$,

$$\text{KS} = |ff^e|/\sigma^e + \frac{1}{\eta}\ln\left[1 + e^{-2\eta|ff^e|/\sigma^e} + e^{\eta(x^e\downarrow - |ff^e|/\sigma^e)}\right] \qquad (2.106)$$

for $|ff^e| \leqslant \sigma^e x^e\downarrow$,

$$\text{KS} = x^e\downarrow + \frac{1}{\eta}\ln\left[1 + e^{\eta(ff^e/\sigma^e - x^e\downarrow)} + e^{\eta(-ff^e/\sigma^e - x^e\downarrow)}\right] \qquad (2.107)$$

where $\eta$ is a given constant.

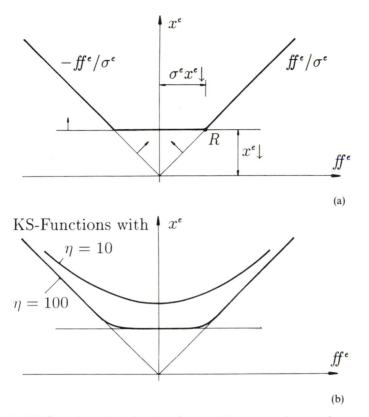

Fig. 2.11 Smooth envelope functions for several local constraints per element.

The above KS functions for $\eta = 10$ and $\eta = 100$ are shown in Fig. 2.11(b). For $\eta = 1000$, the graph of the KS function would be on this scale indistinguishable from the three straight line segments in Fig. 2.11(a). The expressions in both equations (2.106) and (2.107) represent the same function for all values of $ff^e$, but these separate expressions are necessary to avoid an overflow in the computer due to too large values of the exponentials.

Using the SEF in equations (2.106) and (2.107) or in Fig. 2.11(b), we can reformulate the Lagrangian function for the problem in equations (2.1)–(2.7) in the following form (see also the restrictions in equations (2.67)–(2.70)):

$$\min \Phi = \sum_{e=1}^{E} c^e x^e + v \sum_{e=1}^{E} \frac{\hat{ff}^e \, ff^e L^e}{E^e x^e} + \sum_{e=1}^{E} \lambda^e [\text{KS}(ff^e) - x^e]$$

$$+ \{\bar{\mathbf{L}}\}^{\mathrm{T}} (\{\mathbf{P}\} - [\mathbf{B}]\{\mathbb{F}\}) + \{\mathbf{L}\} (\{\hat{\mathbf{P}}\})^{\mathrm{T}} - [\mathbf{B}]\{\hat{\mathbb{F}}\}) \qquad (2.108)$$

Then we have the following Kuhn–Tucker conditions.

• *Variation of the design variables:* $x^e$

$$c^e - \frac{v \hat{ff}^e \, ff^e L^e}{E^e (x^e)^2} = \lambda^e \qquad (2.109)$$

- *Variation of the real forces:* $ff^e$

$$\bar{\delta}^{*e} = \frac{v\widehat{ff}^e L^e}{E^e x^e} + \lambda^e \frac{\partial(KS)}{\partial ff^e} \tag{2.110}$$

where $\partial(KS)/\partial ff^e$ is given by the following expressions: for $|ff^e| \geqslant \sigma^e x^e \downarrow$,

$$\frac{\partial(KS)}{\partial ff^e} = \frac{\operatorname{sgn} ff^e}{\sigma^e} \left[ 1 - \frac{2e^{-2\eta|ff^e|/\sigma^e} + e^{\eta(x^e\downarrow - |ff^e|/\sigma^e)}}{1 + e^{-2\eta|ff^e|/\sigma^e} + e^{\eta(x^e\downarrow - |ff^e|/\sigma^e)}} \right]$$

$$= \frac{\operatorname{sgn} ff^e}{\sigma^e} \left[ \frac{1 - e^{-2\eta|ff^e|/\sigma}}{1 + e^{-2\eta|ff^e|/\sigma^e} + e^{\eta(x^e\downarrow - |ff^e|/\sigma^e)}} \right] \tag{2.111}$$

for $|ff^e| \leqslant \sigma^e x^e \downarrow$,

$$\frac{\partial(KS)}{\partial ff^e} = \frac{1}{\sigma^e} \frac{e^{\eta(ff^e/\sigma^e - x^e\downarrow)} - e^{\eta(-ff^e/\sigma^e - x^e\downarrow)}}{1 + e^{\eta(ff^e/\sigma^e - x^e\downarrow)} + e^{\eta(-ff^e/\sigma^e - x^e\downarrow)}} \tag{2.112}$$

and then a modified form of equations (2.19) and (2.22) represents the optimality criteria and adjoint initial displacement.

In using the above SEF, a lower value of $\eta$ (Fig. 2.11(b)) should be used initially to avoid computational instabilities. The above value should then be progressively increased. Moreover, move limits are to be progressively decreased during later iterations, which slows down the convergence significantly. Using the above procedure in a test example involving a ten-bar truss and KS functions, a four-digit agreement with the exact solution was achieved with an $\eta$ value of 1000.

### 2.4.4 Computational procedure for problems with several displacement constraints

As mentioned at the beginning of section 2.4, the adjoint load for several displacement constraints (Fig. 2.3) is $\{\bar{\mathbf{P}}\} = \sum_{k=1}^{K} v_k \{\hat{\mathbf{P}}_k\}$ together with relative initial displacements $\{\bar{\delta}^{*e}\}$ given by equations (2.21)–(2.23) for elements with active stress constraints. Since for inactive displacement constraints $v_k = 0$, we can replace the above relation by

$$\{\bar{\mathbf{P}}\} = \sum_{k=1}^{K_A} v_k \{\hat{\mathbf{P}}_k\} \tag{2.113}$$

where $K_A$ is the number of active displacement constraints.

Zhou (1992) has proposed the following algorithm for calculating the Lagrange multipliers $v_k$, which is presented here in a somewhat modified form. First a structural analysis is carried out for each virtual load system $\{\hat{\mathbf{P}}_k\}$ with a fixed fraction

$$\{\hat{\bar{\delta}}_k^{*e}\} = \{\bar{\delta}^{*e}\} / \left( \sum_{k=1}^{K_A} v_k \right) \tag{2.114}$$

of the initial relative displacement $\{\bar{\delta}^{*e}\}$ allocated to each such analysis. The corresponding fixed end forces are given by (Fig. 2.7)

$$\{\bar{\mathbf{f}}_k^{*e}\} = -[\mathbf{k}^e]\{\hat{\boldsymbol{\delta}}_k^{*e}\} = -\frac{1}{\displaystyle\sum_{k=1}^{K_A} v_k}[\mathbf{k}^e]\{\bar{\boldsymbol{\delta}}^{\circ *e}\} \qquad (2.115)$$

where the superscript 0 simply means that these initial displacements have been extended from flexibility formulation to stiffness formulation by adding some zeros to these vectors. The analysis is carried out by means of the stiffness method (Fig. 2.7):

$$[\mathbf{K}]\{\hat{\mathbf{U}}_k\} = \{\hat{\mathbf{P}}_k\} - \sum_{e=1}^{E}[\boldsymbol{\gamma}^e]\{\hat{\mathbf{f}}_k^{*e}\} \qquad (2.116)$$

where $[\boldsymbol{\gamma}^e]$ is the transformation matrix from local to global coordinates. The internal nodal forces $\{\hat{\mathbf{f}}_k^e\}$ for the above virtual systems can then be calculated at the element level (Fig. 2.7):

$$\{\hat{\mathbf{f}}_k^e\} = [\mathbf{k}^e]\{\hat{\mathbf{u}}_k^e\} + \{\hat{\mathbf{f}}_k^{*e}\} \qquad (2.117)$$

On the basis of this analysis, the correct adjoint internal forces

$$\{\bar{\mathbf{f}}^e\} = \sum_{k=1}^{K_A} v_k\{\hat{\mathbf{f}}_k^e\} \qquad (2.118)$$

are calculated. These can be used in the re-design formula based on the optimality criterion (2.19) with the extension (2.31), also shown at the top of Fig. 2.3. It can be easily seen that the sum of the initial displacements in equation (2.114), after multiplication by $v_k$ in equation (2.117) adds up to the correct value

$$\sum_{k=1}^{K_A} v_k\{\hat{\boldsymbol{\delta}}_k^{*e}\} = \sum_{k=1}^{K_A} v_k\left(\frac{1}{\displaystyle\sum_{k=1}^{K_A} v_k}\{\bar{\boldsymbol{\delta}}^{*e}\}\right) = \{\bar{\boldsymbol{\delta}}^{*e}\} \qquad (2.119)$$

In Zhou's algorithm, the $K_A$ Lagrange multipliers $v_k$ $(k = 1, \ldots, K_A)$ for the displacement constraints are calculated from work equations representing active displacement constraints (equation (2.28) with an equality sign). In the above equations, the flexibility matrix $[\mathbf{D}]$ and the initial relative displacements (if any) $\{\boldsymbol{\Delta}^*\}$ depend on the design variables $\{\mathbf{X}\}$ and the latter depend, via the optimality criteria (relation (2.19) with the extension in relation (2.31)) on $v_k$. Thus the $K_A$ work equations (2.28) depend implicitly on the $K_A$ Lagrange multipliers $v_k$ $(k = 1, \ldots, K_A)$. Zhou's algorithm uses the approximation that the real and virtual forces $(\{\mathbf{F}\}, \{\hat{\mathbf{F}}_k\})$ are based on the prior analysis and are not supposed to change as a result of variations of $\{\mathbf{v}\} = \lfloor v_1, \ldots, v_k \rfloor^T$ and $\{\mathbf{X}\}$. The Lagrange multipliers can be determined within the above approximation by Newton's method, using the recurrence relation

$$\{\mathbf{v}\}_{t+1} = \{\mathbf{v}\}_t - [\mathbf{J}]_t^{-1}\{\boldsymbol{\varphi}\}_t \qquad (2.120)$$

where $\{\boldsymbol{\varphi}\} = \lfloor \varphi_1, \ldots, \varphi_{K_A} \rfloor^T$ are the LHSs of the active displacement constraints in equation (2.28), $[\mathbf{J}]$ is the Jacobian matrix containing derivatives of $\{\boldsymbol{\varphi}\}$ with respect to $\{\mathbf{v}\}$ and $t$ is the Newton iteration number. The apparent disadvantage of this algorithm is $K_A$ analyses for the virtual loads $\{\hat{\mathbf{P}}_k\}$ instead of one analysis for the adjoint load $\{\bar{\mathbf{P}}\} = \sum_{k=1}^{K_A} v_k\{\hat{\mathbf{P}}\}$, but this is more than compensated by the efficient calculation of the Lagrange multipliers $v_k$, which does not involve any further structural analyses.

### 2.4.5 Several design variables and several stress constraints per element

In the general formulation of DCOC, we have $I^e$ design variables per element $(\{\mathbf{x}^e\} = [x_1, \ldots, x_i, \ldots, x_{I^e}]^T)$ and $R_h^e$ stress conditions at each of the cross sections $h = 1, \ldots, H^e$. Assuming that the Lagrange multipliers for the displacement constraints $(v_k)$ have already been determined (see step (C) in Fig. 2.8) for the update at the element level, the unknowns for a given element $e$ are the $I^e$ design variables and the Lagrange multipliers, say $\lambda_r^e$ $(r = 1, \ldots, R_A^e)$, for the active local (i.e. stress or side) constraints where $R_A^e$ is the number of such active constraints. The number of equations available is also $I^e + R_A^e$, consisting of the $I^e$ optimality criteria (2.19) and of the $R_A^e$ stress constraint satisfied as an equality.

In the above general case one should be able to determine all unknowns required for the update operation. A difficulty arises when the number of active local (stress and side) constraints exceeds the number of design variables $(R_A^e > I^e)$ because this means that some of the $R_A^e$ active constraints with $I^e$ unknowns are not independent and they introduce $R_A^e$ unknown Lagrange multipliers, whilst determining only $I^e$-design variables. This means that the number of independent equations is $2I_e^e$ and the number of variables is $R_A^e + I^e$, with $R_A^e > I^e$, resulting in nonuniqueness of the solution. A special case of this problem was discussed in sections 2.4.3(a), 2.4.3(c) and 2.4.3(d), in which a truss element had active stress and side constraints $(I^e = 1, R_A^e = 2)$. The above difficulty can be removed by upgrading the stress constraints involved into a displacement constraint (section 2.4.3(c)) or by introducing smooth envelope functions (SEFs, section 2.4.3(d)).

## 2.5 TEST EXAMPLES SOLVED BY THE DCOC METHOD

All computational results reported in this section were obtained on an HP-9000 workstation with double precision (FORTRAN 77). The dual method used for verification of the results is based on quadratic approximation of the objective function and linear approximation of the behavioral constraints in terms of reciprocal variables (Fleury, 1979; Zhou, 1989).

### 2.5.1 An elementary example: three-bar truss

The aim of this example is mainly educational: owing to the extreme simplicity of the problem, the principles involved are not obscured by computational complexities and the solution can be obtained even by hand calculations. By working his/her way laboriously over the steps of the DCOC procedure, the reader can achieve a better intuitive insight into this method.

We consider the truss shown in Fig. 2.12(a), in which the three bars may take on different cross-sectional areas $(\{\mathbf{X}\} = \lfloor x^{(1)}, x^{(2)}, x^{(3)} \rfloor^T)$. The truss is subject to a horizontal load of $P_1 = P$ and a vertical load of $P_2 = kP$ at its free joint. The

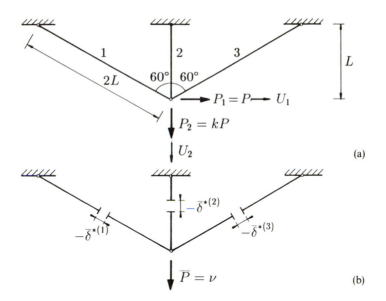

(a)

(b)

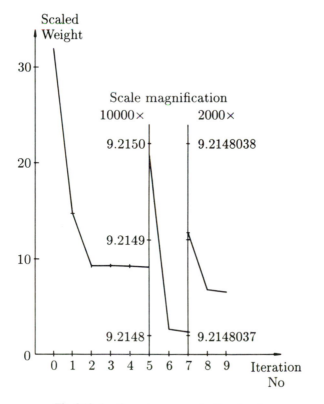

(c)

Fig. 2.12 An elementary example: three-bar truss.

structural weight is to be minimized, subject to a single displacement constraint $U_2 \leqslant t$ and constraints on all stresses, with $\sigma_C = c\sigma_T$ where $\sigma_C$ and $\sigma_T$ are the permissible stresses in compression and tension, respectively, for all members ($e = 1, 2, 3$). Young's modulus ($E$) and the specific weight $\rho$ are the same for all three bars. The adjoint load $\bar{P} = v$ for the displacement constraint is shown in Fig. 2.12(b). Superscripts representing element numbers are given in parentheses, in order to indicate that they are not exponents. For simplicity, we shall use the stiffness notation for bar forces ($f^e$).

(a) Analysis equations for the real truss

The nodal displacement–bar elongation–bar force relations can be easily determined by projecting the nodal displacements onto the member directions (e.g. Haftka *et al.* (1990), who used the variable linking $x^{(1)} = x^{(3)}$):

$$\frac{\sqrt{3}U_1}{2} + \frac{U_2}{2} = \delta^{(1)} = \frac{2f^{(1)}L}{x^{(1)}E} \tag{2.121}$$

$$U_2 = \delta^{(2)} = \frac{f^{(2)}L}{x^{(2)}E} \tag{2.122}$$

$$-\frac{\sqrt{3}U_1}{2} + \frac{U_2}{2} = \delta^{(3)} = \frac{2f^{(3)}L}{x^{(3)}E} \tag{2.123}$$

From equations (2.121)–(2.123) we can express the forces and stresses in terms of the displacements $\{U\} = \lfloor U_1, U_2 \rfloor^{\mathrm{T}}$:

$$f^{(1)} = \frac{\sqrt{3}U_1 + U_2}{4L} Ex^{(1)}, \qquad \sigma^{(1)} = \frac{\sqrt{3}U_1 + U_2}{4L} E \tag{2.124}$$

$$f^{(2)} = \frac{U_2}{L} Ex^{(2)}, \qquad \sigma^{(2)} = \frac{U_2}{L} E \tag{2.125}$$

$$f^{(3)} = \frac{-\sqrt{3}U_1 + U_2}{4L} Ex^{(3)}, \qquad \sigma^{(3)} = \frac{-\sqrt{3}U_1 + U_2}{4L} E \tag{2.126}$$

Equilibrium in the horizontal direction implies

$$P_1 = P = \sqrt{3}(f^{(1)} - f^{(3)})/2$$
$$= E[3(x^{(1)} + x^{(3)})U_1 + \sqrt{3}(x^{(1)} - x^{(3)})U_2]/8L \tag{2.127}$$

and by equilibrium in the vertical direction we have

$$P_2 = kP = (f^{(1)} + f^{(3)})/2 + f^{(2)}$$
$$= E[(x^{(1)} - x^{(3)})\sqrt{3}U_1 + (x^{(1)} + x^{(3)} + 8x^{(2)})U_2]/8L \tag{2.128}$$

Introducing the notation

$$q = 2x^{(2)}(x^{(1)} + x^{(3)}) + x^{(1)}x^{(3)}, \quad t = x^{(1)} + x^{(3)} + 8x^{(2)} \tag{2.129}$$

the nodal displacements by equations (2.127) and (2.128) become

$$U_2 = (2PL/E)[k(x^{(1)} + x^{(3)}) - (x^{(1)} - x^{(3)})/\sqrt{3}]/q \tag{2.130}$$

$$U_1 = (2PL/3E)[t - \sqrt{3}k(x^{(1)} - x^{(3)})]/q \qquad (2.131)$$

Then by equations (2.124)–(2.126) the bar forces are given by

$$f^{(1)} = Px^{(1)}[2kx^{(3)} + (t + x^{(3)} - x^{(1)})/\sqrt{3}]/2q \qquad (2.132)$$

$$f^{(2)} = 2Px^{(2)}[k(x^{(1)} + x^{(3)}) - (x^{(1)} - x^{(3)})/\sqrt{3}]/q \qquad (2.133)$$

$$f^{(3)} = Px^{(3)}[2kx^{(1)} + (x^{(3)} - x^{(1)} - t)/\sqrt{3}]/2q \qquad (2.134)$$

It can be easily checked that equations (2.130) and (2.131) reduce to known relations (Haftka, Gürdal and Kamat, 1990, p. 184) for $x^{(1)} = x^{(3)}$ and $k = 8$.

(b) Analysis equations for the adjoint truss

We consider now the adjoint truss in Fig. 2.12(b), with the load $\bar{P} = v$ (i.e. $v$ times the virtual load $\hat{P} = 1$) and initial relative displacements caused by the active stress constraints (equations (2.93)–(2.95)). The adjoint bar elongation $\bar{\delta}^e$–nodal displacement $U_j$–bar force $\bar{f}^e$ relations can be derived easily from compatibility:

$$\bar{\delta}^{(1)} = \frac{\bar{U}_1 \sqrt{3}}{2} + \frac{\bar{U}_2}{2} = \bar{\delta}^{*(1)} + \frac{\bar{f}^{(1)} 2L}{x^{(1)} E} \qquad (2.135)$$

$$\bar{\delta}^{(2)} = \bar{U}_2 = \bar{\delta}^{*(2)} + \frac{\bar{f}^{(2)} L}{x^{(2)} E} \qquad (2.136)$$

$$\bar{\delta}^{(3)} = -\frac{\bar{U}_1 \sqrt{3}}{2} + \frac{\bar{U}_2}{2} = \bar{\delta}^{*(3)} + \frac{\bar{f}^{(3)} 2L}{x^{(3)} E} \qquad (2.137)$$

Expressing the adjoint bar forces $\bar{f}^e$ from equations (2.135)–(2.137), we have

$$\bar{f}^{(1)} = Ex^{(1)} \frac{\sqrt{3}\bar{U}_1 + \bar{U}_2 - 2\bar{\delta}^{*(1)}}{4L} \qquad (2.138)$$

$$\bar{f}^{(2)} = Ex^{(2)} \frac{\bar{U}_2 - \bar{\delta}^{*(2)}}{L} \qquad (2.139)$$

$$\bar{f}^{(3)} = Ex^{(3)} \frac{-\sqrt{3}\bar{U}_1 + \bar{U}_2 - 2\bar{\delta}^{*(3)}}{4L} \qquad (2.140)$$

Equilibrium in the horizontal direction implies

$$0 = \bar{f}^{(1)} - \bar{f}^{(3)} \qquad (2.141)$$

or

$$0 = x^{(1)}(\sqrt{3}\bar{U}_1 + \bar{U}_2 - \bar{\delta}^{*(1)}) - x^{(3)}(-\sqrt{3}\bar{U}_1 + \ddot{U}_2 - 2\bar{\delta}^{*(3)}) \qquad (2.142)$$

Moreover, by equilibrium in the vertical direction and equation (2.141), we have

$$v = \frac{\bar{f}^{(1)} + \bar{f}^{(3)}}{2} + \bar{f}^{(2)} = \bar{f}^{(1)} + \bar{f}^{(2)} \qquad (2.143)$$

or

$$4Lv/E = x^{(1)}(\sqrt{3}\bar{U}_1 + \bar{U}_2 - 2\bar{\delta}^{*(1)}) + 4x^{(2)}(\bar{U}_2 - \bar{\delta}^{*(2)}) \qquad (2.144)$$

Introducing the notation

$$q = x^{(1)}x^{(3)} + 2x^{(2)}(x^{(1)} + x^{(3)}), \quad r = \bar{\delta}^{*(1)}x^{(1)} - \bar{\delta}^{*(3)}x^{(3)}$$

$$s = 2Lv/E + \bar{\delta}^{*(1)}x^{(1)} + 2\bar{\delta}^{*(2)}x^{(2)} \tag{2.145}$$

the nodal displacements become

$$\bar{U}_2 = [(x^{(1)} + x^{(3)})s - x^{(1)}r]/q \tag{2.146}$$

$$\bar{U}_1 = -[(x^{(1)} - x^{(3)})s - (x^{(1)} + 4x^{(2)})r]/\sqrt{3}q \tag{2.147}$$

and the bar forces are given by

$$\bar{f}^{(1)} = (Ex^{(1)}/2L)[(x^{(3)}s + 2x^{(2)}r)/q - \bar{\delta}^{*(1)}] \tag{2.148}$$

$$\bar{f}^{(2)} = (Ex^{(2)}/L)\{[(x^{(1)} + x^{(3)})s - x^{(1)}r]/q - \bar{\delta}^{*(2)}\} \tag{2.149}$$

$$\bar{f}^{(3)} = (Ex^{(3)}/2L)\{[(x^{(1)}s - (x^{(1)} + 2x^{(2)})r]/q - \bar{\delta}^{*(3)}\} \tag{2.150}$$

## (c) Initial design and first iteration

We consider the problem in Fig. 2.12(a) in a normalized form with $P = t = L = E = \rho = 1, P_2 = 8, P_1 = 1$ (i.e. $k = 8$) and $\sigma_T = \sigma_C = 1.3 = \sigma$ (i.e. $c = 1$). The initial design

$$x^{(1)} = x^{(2)} = x^{(3)} = 6.4 \tag{2.151}$$

has the following features:

- all members have the same cross-sectional area, and
- they have been scaled to give exactly the prescribed displacement.

Then by equations (2.130)–(2.134) we have

$$U_1 = 0.208\,333\cdots, \quad U_2 = 1.0, \quad f^{(1)} = 2.177\,350, \quad f^{(2)} = 6.4, \quad f^{(3)} = 1.022\,650 \tag{2.152}$$

As mentioned in section 2.4.1(b), the analysis of the adjoint truss in the first iteration may be based on the virtual load without initial relative displacements for active stress constraints because at the start of the first iteration the value of the Langrange multipliers is not yet known. By equations (2.148)–(2.150), we have for an initial adjoint load of $v = 1$ and $\bar{\delta}^{*1} = \bar{\delta}^{*2} = \bar{\delta}^{*3} = 0$

$$\bar{f}^{(1)} = 0.2, \quad \bar{f}^{(2)} = 0.8, \quad \bar{f}^{(3)} = 0.2 \tag{2.153}$$

At this stage it cannot be decided on the basis of equations (2.99)–(2.101) which members belong to the regions $R_d$, $R_\sigma$ and $R_s$, but it is clear that a truss element with $\bar{f}^e f^e \leqslant 0$ cannot be part of the displacement controlled set $R_d$. For this reason, we can assume for the first iteration that

$$e \in R_d \quad (\text{if } \bar{f}^e f^e > 0), \quad e \notin R_d \quad (\text{if } \bar{f}^e f^e \leqslant 0) \tag{2.154}$$

giving for this case

$$1 \in R_d, \quad 2 \in R_d, \quad 3 \in R_d \tag{2.155}$$

The values of $c^e$ and $[d_o]$ by Table 2.3 take on the values $c^e = L\rho = 1$ and

$[d_o] = L/E = 1$. Hence equations (2.98) and (2.98a) with (2.152), (2.153) and (2.155) yield

$$\sqrt{v} = 2\sqrt{2.177\,350 \times 0.2} + \sqrt{6.4 \times 0.8} + 2\sqrt{1.022\,650 \times 0.2} = 4.487\,045,$$

$$v = 20.133\,569 \tag{2.156}$$

The comparison values in equations (2.100) and (2.101) then become

$$(x_d^{(1)})^2 = v\bar{f}^{(1)}f^{(1)} = 8.767\,565, \quad (x_d^{(2)})^2 = 103.083\,873, \quad (x_d^{(3)})^2 = 4.117\,919$$

$$(x_\sigma^{(1)})^2 = \left(\frac{f^{(1)}}{1.3}\right)^2 = 2.805\,238, \quad (x_d^{(2)})^2 = 24.236\,686, \quad (x_d^{(3)})^2 = 0.618\,824 \tag{2.157}$$

Adopting a prescribed minimum cross-sectional area of $x\downarrow = 0.01$, equations (2.156) and (2.157) confirm the regions assumed in equation (2.155). It follows then from equations (2.156) and (2.157) that the design variables at the end of the first iteration are

$$x^{(1)} = 2.961\,007, \quad x^{(2)} = 10.153\,023, \quad x^{(3)} = 2.029\,266 \tag{2.158}$$

The initial design had the weight

$$\Phi_{\text{initial}} = 6.4(1 + 2 \times 2) = 32 \tag{2.159}$$

and the unscaled truss weight after the first iteration has been reduced to

$$\Phi(1) = x^{(2)} + 2(x^{(1)} + x^{(3)}) = 20.133\,569 \tag{2.160}$$

The initial relative displacements $\bar{\delta}^{*e}$ for the adjoint truss in the second iteration take on a zero value because no stress constraints are active at the end of the first iteration.

## (d) Second iteration

Using the $x^e$ values in equations (2.158), (2.130)–(2.134) yields the following displacements, forces and stresses for the real truss:

$$U_1 = 0.455\,269, \quad U_2 = 0.733\,813, \quad f^{(1)} = 1.126\,932,$$
$$f^{(2)} = 7.450\,419, \quad f^{(3)} = -0.027\,769 \quad \sigma^{(1)} = 0.380\,591,$$
$$\sigma^{(2)} = 0.733\,813, \quad \sigma^{(3)} = -0.013\,684 \tag{2.161}$$

It can be seen from relations (2.161) that for a feasible solution with at least one active constraint, we can multiply the $x^e$ values in equation (2.158) by 0.733813, thereby obtaining the following scaled designs:

$$x^{(1)} = 2.172\,825, \quad x^{(2)} = 7.450\,420, \quad x^{(3)} = 1.489\,102 \tag{2.162}$$

The corresponding scaled weight value becomes

$$\Phi(1)_{\text{scaled}} = 0.733\,813\,\Phi(1) = 14.774\,274 \tag{2.163}$$

Substituting the design variables $f^e$ from equation (2.162) into equations (2.148)–(2.150) with a virtual load of $v = 1$ and $\bar{\delta}^{*e} = 0$ (for all $e$), we obtain the virtual forces $\hat{f}^e = \bar{f}^e/v$:

$$\hat{f}^{(1)} = 0.055\,977, \quad \hat{f}^{(2)} = 0.944\,023, \quad \hat{f}^{(3)} = 0.055\,977 \tag{2.164}$$

Using the assumption in equation (2.154), we have the following regions for the calculation of $v$:

$$1 \in R_d, \quad 2 \in R_d, \quad 3 \in R_\sigma \tag{2.165}$$

Then by equations (2.98), (2.98a), (2.161) and (2.164) we have

$$\sqrt{v} = \frac{2\sqrt{1.126\,932 \times 0.055\,977} + \sqrt{7.450\,419 \times 0.944\,023}}{1 + 0.055\,977 \times 1.3 \times 2} = 2.753\,612,$$

$$v = 7.582\,384 \tag{2.166}$$

The comparison values under equations (2.100) and (2.101) then become

$$(x_d^{(1)})^2 = 0.478\,316, \quad (x_d^{(2)})^2 = 53.329\,675, \quad (x_d^{(3)})^2 < 0$$
$$(x_\sigma^{(1)})^2 = 0.751\,465, \quad (x_\sigma^{(2)})^2 = 32.845\,410, \quad (x_\sigma^{(3)})^2 = 0.000\,456, \tag{2.167}$$

which indicates the following optimal regions:

$$1 \in R_\sigma, \quad 2 \in R_d, \quad 3 \in R_\sigma \tag{2.168}$$

The corresponding Lagrange multiplier $v$ as given by equations (2.98) and (2.98a) becomes

$$\sqrt{v} = \frac{\sqrt{7.450\,419 \times 0.944\,023}}{1 - (0.055\,977 - 0.055\,977)1.3 \times 2} = \sqrt{7.450\,419 \times 0.944\,023} = 2.652\,050,$$

$$v = 7.033\,367 \tag{2.169}$$

The following comparison values are given by this new Lagrange multiplier:

$$(x_d^{(1)})^2 = 0.443\,681, \quad (x_d^{(2)}) = 49.468\,247, \quad x_d^3 < 0 \tag{2.170}$$

Then equation (2.99) with equations (2.167) and (2.170) implies

$$x^{(1)} = 0.866\,871, \quad x^{(2)} = 7.033\,367, \quad x^{(3)} = 0.021\,361 \tag{2.171}$$

The corresponding unscaled weight becomes

$$\Phi(2) = x^{(2)} + 2(x^{(1)} + x^{(3)}) = 8.809\,831 \tag{2.172}$$

By equations (2.94) and (2.95) the initial relative displacements corresponding to active stress constraints are given by

$$\bar{\delta}^{*e} = \frac{L^e}{1.3}\left[1 - \frac{v\hat{f}^e f^e}{(x^e)^2}\right] \operatorname{sgn} f^e \tag{2.173}$$

implying

$$\bar{\delta}^{*(1)} = \frac{2}{1.3}\left(1 - \frac{7.033\,367 \times 0.055\,977 \times 1.126\,932}{0.866\,871^2}\right) = 0.630\,122$$

$$\bar{\delta}^{*(2)} = \frac{1}{1.3}\left(1 - \frac{7.033\,367 \times 0.944\,023 \times 7.450\,419}{7.033\,367^2}\right) = 0$$

$$\bar{\delta}^{*(3)} = -\frac{2}{1.3}\left(1 + \frac{7.033\,367 \times 0.055\,977 \times 0.027\,769}{0.021\,361^2}\right) = -38.400\,269 \tag{2.174}$$

Since $2 \in R_d$, a zero relative initial displacement was to be expected for $e = 2$.

(e) Third iteration

Using the $x^e$ values in equation (2.171), by equations (2.130)–(2.134) we have for the real truss

$$U_1 = 2.420\,910, \quad U_2 = 1.057\,729, \quad f^{(1)} = 1.137\,957,$$
$$f^{(2)} = 7.439\,393, \quad f^{(3)} = -0.016\,744 \quad \sigma^{(1)} = 1.312\,718,$$
$$\sigma^{(2)} = 1.057\,729, \quad \sigma^{(3)} = -0.783\,867 \tag{2.175}$$

Since the constraint violation for the displacement constraint is $1.057\,729$, the scaled design variables and scaled weight become

$$x^{(1)} = 0.916\,915, \quad x^{(2)} = 7.439\,396, \quad x^{(3)} = 0.022\,594,$$
$$\Phi(2)_{\text{scaled}} = 1.057\,729\,\Phi(2) = 9.318\,413 \tag{2.176}$$

The analysis of the adjoint truss is based on equations (2.148)–(2.150), with the load $v$ from equation (2.169), the initial displacements from equation (2.174) and the scaled design variables from equation (2.176):

$$\bar{f}^{(1)} = 0.426\,220, \quad \bar{f}^{(2)} = 6.607\,147, \quad \bar{f}^{(3)} = 0.426\,220 \tag{2.177}$$

The next step is the calculation of the Lagrange multiplier from equation (2.98), in which the virtual forces $\hat{f}^e$ can be obtained by dividing the adjoint forces $\bar{f}^e$ in equation (2.177) by the old Lagrange multiplier $v = 7.033\,367$. The regions in equation (2.168) are assumed to be still valid. As in equation (2.169), the denominator in equation (2.98) becomes unity:

$$\sqrt{v} = \sqrt{7.439\,393 \times 6.607\,147/7.033\,367} = 2.643\,590, \quad v = 6.988\,568 \tag{2.178}$$

The test on the basis of equations (2.99)–(2.101) confirms the regions in equation (2.168).

We shall not continue the above procedure by hand calculations beyond this point, but show in Table 2.5 the output from a standard DCOC program, which carries out the same operations but uses a general-purpose stiffness method for the analysis of the real and adjoint trusses. We can see that there is an almost complete agreement with the iterations calculated by hand. A small difference in the last digit in some values is due to the fact that the hand calculations were

**Table 2.5** Iteration history of the three-bar truss example

| Iteration number | Unscaled weight | Scaled weight | Scaled design variables | | |
|---|---|---|---|---|---|
| | | | $x^{(1)}$ | $x^{(2)}$ | $x^{(3)}$ |
| 0 | 32 | 32 | 6.4 | 6.4 | 6.4 |
| 1 | 20.133 568 | 14.774 273 | 2.172 827 | 7.450 418 | 1.489 102 |
| 2 | 8.809 826 | 9.318 408 | 0.916 915 | 7.439 393 | 0.022 593 |
| 3 | 8.765 021 | 9.321 990 | 0.930 976 | 7.432 643 | 0.013 698 |
| 4 | 9.012 294 | 9.260 719 | 0.904 816 | 7.430 536 | 0.010 276 |
| 5 | 9.214 932 | 9.214 988 | 0.882 170 | 7.430 648 | 0.010 000 |
| ⋮ | ⋮ | ⋮ | ⋮ | ⋮ | ⋮ |
| 9 | 9.214 804 | 9.214 804 | 0.882 077 | 7.430 649 | 0.010 000 |

rounded up for six decimal digits whereas the computer program used double precision with a much higher accuracy.

The iteration history of the computer solution is also shown in Fig. 2.12(c). We can see that the convergence is monotonic, apart from the third iteration which shows a slight increase in the scaled weight. This is because initial relative displacements due to active stress constraints appear in that iteration for the first time in the iteration history. However, in subsequent iterations the convergence is incredibly fast, which can be seen from iteration 4 onwards in Fig. 2.12(c), in spite of scale magnifications by 10 000 and 2000 at iterations 5 and 7, respectively.

In iteration 9, the structural weight changed only from 9.214 803 724 2 to 9.214 803 723 9, thereby satisfying the convergence criterion in equation (2.82) with an extremely stringent tolerance value of $T = 10^{-10}$. Naturally, for practical applications such a high accuracy is not necessary. It can be seen from Table 2.5 that already after two iterations the weight differs only by 1% from the optimal weight.

### 2.5.2 Ten-bar truss with stress constraints and one displacement constraint

The ten-bar truss shown in Fig. 2.13 is a well-known example (Haftka, Gürdal and Kamat, 1990). A modified version of the ten-bar truss is considered herein, insofar as the displacement constraint is $|U_{V6}| \leqslant 5.0$ in, where $U_{V6}$ is the vertical displacement at node 6. The purpose of this modification is to set a problem, for which the optimal solution contains an $R_\sigma$ region but no $R_{\sigma s}$ region. The material properties are as follows:

$$E^e = 10^7 \, \text{lbf in}^{-2}, \quad \rho^e = 0.1 \, \text{lb in}^{-3}, \quad \sigma_a^e = 25\,000 \, \text{lbf in}^{-2},$$
$$x^e \!\downarrow = 0.1 \, \text{in}^2 \quad \text{(for } e = 1,\dots,10) \tag{2.179}$$

Since the purpose of this example is to verify the validity and accuracy of the DCOC method, a convergence tolerance value of $T = 10^{-12}$ was employed for

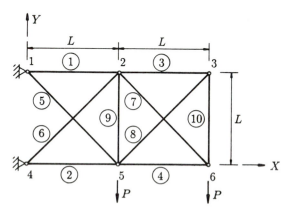

$$L = 360 \, in, P = 10^5 \, \ell b$$

**Fig. 2.13** Ten-bar truss. $L = 360$ in; $P = 10^5$ lbf.

**Table 2.6** Results for the ten-bar truss in section 2.5.2

| e | Cross-sectional area (in²) | | |
|---|---|---|---|
| | DCOC | Dual | DOC–FSD |
| 1 | 12.161 173 957 | 12.161 173 956 | 12.126 576 172 |
| 2 | 8.707 029 023 | 8.707 029 026 | 8.827 450 732 |
| 3 | 0.100 000 000 | 0.100 000 000 | 0.100 000 000 |
| 4 | 6.040 579 884 | 6.040 579 884 | 6.046 585 281 |
| 5 | 5.560 164 853 | 5.560 164 853 | 5.564 322 434 |
| 6 | 8.573 640 198 | 8.573 640 196 | 8.497 882 192 |
| 7 | 8.542 669 996 | 8.542 669 996 | 8.551 162 911 |
| 8 | 0.100 000 000 | 0.100 000 000 | 0.100 000 000 |
| 9 | 0.100 000 000 | 0.100 000 000 | 0.100 000 000 |
| 10 | 0.100 000 000 | 0.100 000 000 | 0.100 000 000 |
| Number of analyses | 24 | 18 | 25 |
| Weight (lb) | 2139.104 979 978 1 | 2139.104 979 977 9 | 2139.197 925 706 7 |

the DCOC method as well as for the DOC–FSD and dual methods used for comparisons. The results are given in Table 2.6, which shows that the optimal weight obtained by the DCOC method has 13 significant digits agreement with that of the dual method, and the design variables obtained by these two methods show an at least 9 significant digits agreement. As expected, the DOC–FSD method does not yield the same solution as the other two methods, although it is very close to them.

### 2.5.3 Clamped beam with rectangular cross-section of variable width

The beam shown in Fig. 2.14(a) has a rectangular cross-section with a constant depth and a variable width. Since the beam and loads are symmetric, the half beam shown in Fig. 2.14(b) can be considered. Normalized parameters are as follows:

$$\rho = 1, \quad E = 12, \quad d = 1, \quad q = 1, \quad t = 1, \quad a = 1 \qquad (2.180)$$

where $\rho$ is the specific weight, $E$ Young's modulus, $d$ the constant beam depth, $q$ the uniformly distributed load, $t$ the permissible displacement prescribed at midspan and $a$ the half beam span. The flexural and shear stress constraints are represented by $x^e \leqslant k_1 |M^e_{max}|$ and $x^e \leqslant k_2 |F^e_{Y\,max}|$, where $M^e_{max}$ and $F^e_{Y\,max}$ are, respectively, the maximum moment and shear force in the $e$th element, and $k_1 = 0.23$ and $k_2 = 0.03$ are given constants.

The half beam is discretized into prismatic beam elements. Two models consisting of 100 and 1000 elements are considered. For the 100 element model, a convergence tolerance of $T = 10^{-8}$ was used (equation (2.82)) for all computations concerned. The optimal weight and CPU times for the DCOC, DOC–FSD and dual methods are given in Table 2.7. The CPU times given for 'analysis' include those for the analysis of the adjoint system (DCOC), analysis for the virtual load

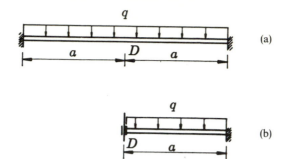

**Fig. 2.14** Clamped beam example.

**Table 2.7** Results for the clamped beam example

|  | 100 element model | | |
|---|---|---|---|
|  | DCOC | DOC–FSD | Dual |
| Optimal weight | 0.064 202 389 | 0.064 203 018 | 0.064 202 389 |
| Number of analyses | 29 | 28 | 9 |
| CPU times (s) |  |  |  |
|    Optimization | 7.84 | 3.63 | 535.28 |
|    Analysis | 116.95 | 112.70 | 117.41 |
|    Total | 124.79 | 116.33 | 706.69 |
|  | 1000 element model | | |
|  | DCOC | DOC–FSD | |
| Optimal weight | 0.063 996 543 | 0.063 999 993 | |
| Number of Analyses | 18 | 20 | |
| CPU times (s) |  |  | |
|    Optimization | 57.35 | 34.98 | |
|    Analysis | 928.59 | 1126.41 | |
|    Total | 985.84 | 1161.39 | |

system (DOC–FSD) and sensitivity analysis (dual method). The optimum weight obtained by the DCOC method has an eight significant digits agreement with that of the dual method, and the optimal design variables for these two methods have shown an at least five significant digits agreement (Zhou, 1992). Again, the solution of the DOC–FSD method results in a higher weight than the other two methods. The CPU time needed for the optimization phase of the dual method is much more than that for the DCOC method, since 76 stress constraints and 1 deflection constraint are active at the optimum. It can be seen that a much larger number of analyses were needed for the DCOC and DOC–FSD methods. This is because, in those two methods, the resizing rule for the members with active stress constraints is based on forces given by the prior computation step

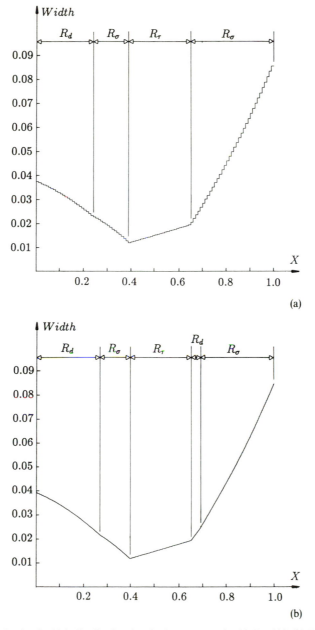

**Fig. 2.15** Optimal width distribution for the beam example: (a) $E = 100$; (b) $E = 1000$.

and the dependence of the forces on design variables is not considered. For this reason, the treatment of stress constraints in the resizing procedure of the DCOC and DOC–FSD methods can be regarded as a zero-order approximation. On the other hand, the dual method is based on first-order approximation which achieves a faster convergence. Table 2.7 shows that about the same analysis time was needed for the dual method with only 9 iterations as for the DCOC method (with 29 iterations). This is because the sensitivity analysis needed by the dual method is very expensive. The optimal width distributions obtained by the DCOC method for both models with 100 and 1000 elements are shown in Fig. 2.15.

For the 1000 element model, for which a convergence tolerance of $T = 10^{-6}$

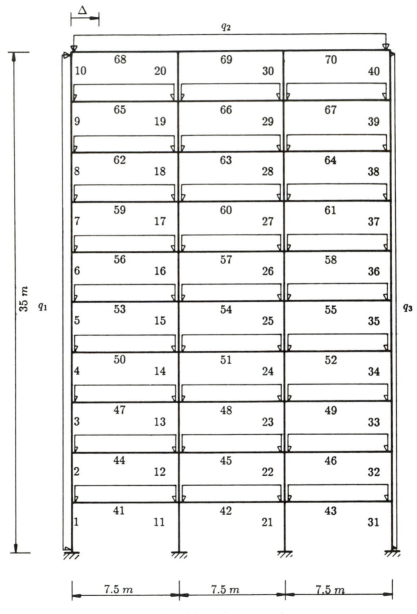

**Fig. 2.16** Multistory frame example.

was used, the results of the DCOC and DOC–FSD methods are also given in Table 2.7. The displacement constraint was active in the solutions by both methods and 730 stress constraints were active in the solution by the DCOC method. Table 2.7 shows that for both DCOC and DOC–FSD methods the CPU time needed for the optimization phase is a fraction of that needed for the analysis phase. The 1000 element model was not solved by the dual method because, as a result of over 700 active constraints, the computer time would have been prohibitively high.

### 2.5.4 Ten-storey, three-bay frame

Figure 2.16 shows the geometry and loading for the considered frame, which has 70 elements with rectangular cross sections of variable width. The material properties are as follows:

$$E = 2.1 \times 10^7 \, \text{kN m}^{-2}, \quad \sigma_a = 3.0 \times 10^5 \, \text{kN m}^{-2}, \quad \tau_a = 5.0 \times 10^4 \, \text{kN m}^{-2} \quad (2.181)$$

where $E$ is Young's modulus, $\sigma_a$ the permissible flexural stress and $\tau_a$ the permissible shear stress. A horizontal displacement at the left top corner of the frame is constrained to a value of $U_H \leqslant 0.3$ m. The given constant depth is $d = 0.3$ m for all elements. The values of the distributed loads in Fig. 2.16 are:

$$q_1 = 10.0 \, \text{kN m}^{-1}, \quad q_2 = 20.0 \, \text{kN m}^{-1}, \quad q_3 = 5.0 \, \text{kN m}^{-1} \quad (2.182)$$

With the initial design variables of $x^e = 0.6$ m for all elements, a stress constraint violation of 47.6% occurs. A convergence tolerance value of $T = 10^{-4}$ was used. The results of the DCOC method are compared with those of the DOC–FSD and dual methods in Table 2.8. The reason for small constraint violations in the results of the DCOC and DOC–FSD methods is the zero-order treatment of stress constraints discussed in section 2.5.3. The optimal width distribution is shown in Fig. 2.17. For detailed results the reader is referred to Zhou's dissertation (Zhou, 1992).

### 2.5.5 Ten-bar truss with stress constraints and two displacement constraints

The ten-bar truss considered in section 2.5.2 is used here again as a test example for problems with multiple displacement constraints. The displacement con-

**Table 2.8** Results for ten-story and three-bay frame

|                          | DCOC  | DOC–FSD | Dual   |
|--------------------------|-------|---------|--------|
| Optimal volume ($m^3$)   | 53.51 | 53.57   | 53.50  |
| Constraint violation     | 0.19% | 0.74%   | 0.00%  |
| Number of analyses       | 13    | 8       | 12     |
| CPU times (s)            |       |         |        |
|   Optimization | 2.63  | 1.20    | 108.55 |
|   Analysis     | 49.61 | 34.67   | 197.20 |
|   Total        | 52.24 | 35.87   | 305.75 |

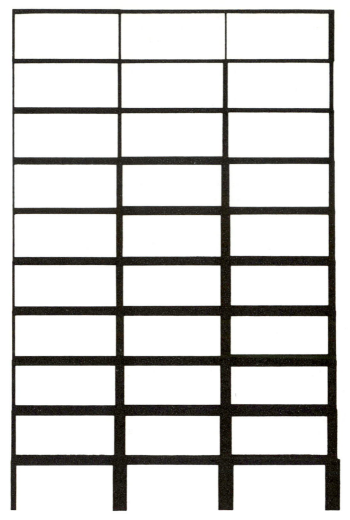

**Fig. 2.17** Optimal width distribution in the frame example.

straints are $|U_{H6}| \leqslant 1.0$ in and $|U_{V6}| \leqslant 5.0$ in, where $U_{H6}$ and $U_{V6}$ are, respectively, horizontal and vertical displacement of node 6. The other conditions are the same as described in section 2.5.2. It was shown by the dual method that both displacement constraints and the stress constraint at the 5th bar are active in the optimal design. The results from the DCOC, dual and DOC–FSD methods are indicated in Table 2.9, for which a convergence tolerance value of $T = 10^{-12}$ was used.

### 2.5.6 Ten-bar truss with active stress and side constraints for one member

The technique of upgrading stress constraints into global constraints is illustrated in this subsection by another version of the ten-bar truss considered in section

**Table 2.9** Results for the ten-bar truss in section 2.5.5

| | Cross-sectional area (in$^2$) | | |
|---|---|---|---|
| $e$ | DCOC | Dual | DOC–FSD |
| 1 | 10.827 889 1 | 10.827 889 1 | 10.836 227 6 |
| 2 | 12.295 024 3 | 12.295 024 3 | 12.331 033 0 |
| 3 | 0.100 000 0 | 0.100 000 0 | 0.100 000 0 |
| 4 | 8.602 843 0 | 8.602 843 0 | 8.569 172 0 |
| 5 | 5.641 706 0 | 5.641 706 0 | 5.643 366 7 |
| 6 | 7.619 254 7 | 7.619 254 7 | 7.567 562 9 |
| 7 | 7.605 251 3 | 7.605 251 3 | 7.648 485 4 |
| 8 | 0.100 000 0 | 0.100 000 0 | 0.100 000 0 |
| 9 | 0.100 000 0 | 0.100 000 0 | 0.100 000 0 |
| 10 | 0.100 000 0 | 0.100 000 0 | 0.100 000 0 |
| Number of analyses | 23 | 19 | 29 |
| Weight (lb) | 2220.352 475 375 | 2220.352 475 375 | 2220.390 773 364 |

2.5.2. The single displacement constraint is changed to $|U_{V6}| \leqslant 4.0$ in where $U_{V6}$ represents the vertical displacement of node 6. Other design conditions are exactly the same as those given in section 2.5.2. As a result of this modification, the optimal solution contains both $R_\sigma$ and $R_{\sigma s}$ type regions. It was found by using the dual method that the displacement constraint as well as the stress constraints for the 5th and 9th elements are active and the cross-sectional area of the 9th bar takes the lower side value. Therefore, the DCOC method including the technique of upgrading stress constraints in the $R_{\sigma s}$ region must be applied.

A convergence tolerance value of $T = 10^{-10}$ is used. The optimum results of the DCOC, DOC–FSD and dual methods are given in Table 2.10, which shows

**Table 2.10** Results for the ten-bar truss in section 2.5.6

| | Cross-sectional area (in$^2$) | | |
|---|---|---|---|
| $e$ | DCOC | Dual | DOC–FSD |
| 1 | 14.973 877 3 | 14.973 877 3 | 15.550 021 6 |
| 2 | 11.717 729 4 | 11.717 729 4 | 11.222 496 6 |
| 3 | 0.100 000 0 | 0.100 000 0 | 0.100 000 0 |
| 4 | 7.700 225 6 | 7.700 225 6 | 7.717 639 3 |
| 5 | 5.531 657 0 | 5.531 657 0 | 5.083 493 4 |
| 6 | 10.188 680 2 | 10.188 680 2 | 10.966 889 5 |
| 7 | 10.889 763 5 | 10.889 763 5 | 10.914 390 2 |
| 8 | 0.100 000 0 | 0.100 000 0 | 0.100 000 0 |
| 9 | 0.100 000 0 | 0.100 000 0 | 0.416 827 5 |
| 10 | 0.100 000 0 | 0.100 000 0 | 0.100 000 0 |
| Number of analyses | 18 | 16 | 191 |
| Weight (lb) | 2608.762 283 67 | 2608.762 283 67 | 2641.764 762 99 |

that the optimum design obtained by the DCOC method shows a 12 digit agreement with that by the dual method. The displacement constraint, stress constraints for the 5th and 9th bars and the side constraint of the 9th bar are active in the solutions of the DCOC and dual methods. The DOC–FSD method gives a significantly different solution in which the side constraint of the 9th element is not active. A much larger number of analyses is needed by the DOC–FSD method for the considered example.

*Note*

The number of iterations used in most test examples of this chapter is relatively high owing to the unusually stringent convergence criteria used, resulting in an eight to twelve digits agreement between weight values of the DCOC and dual methods. For an accuracy required in practical problems, a much smaller number of iterations is sufficient (e.g. the multistorey frame example in section 2.5.4).

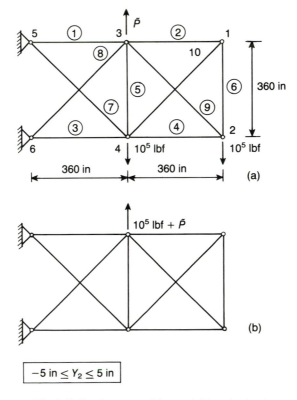

**Fig. 2.18** Ten-bar truss with a variable point load.

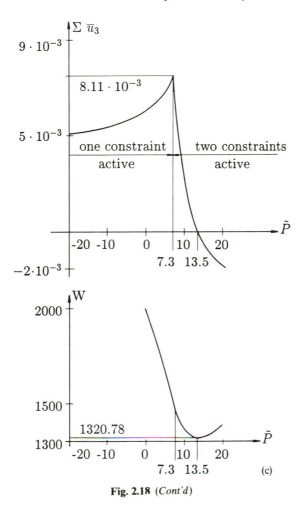

**Fig. 2.18** (*Cont'd*)

### 2.5.7 Ten-bar truss with a variable point load

In order to verify some recent extensions of DCOC introduced in section 2.3.4, we consider a ten-bar truss with the two load conditions given in Figs. 2.18(a) and 2.18(b), including a variable vertical point load $\tilde{P}$ at the joint 3. The material properties are $E^e = 10^7$ lbf in$^{-2}$, $\rho^e = 0.1$ lb in$^{-3}$ and the minimum cross-sectional area $x^e\downarrow = 0.1$ in$^2$ for all members. The vertical displacement at joint 6 is restricted in both directions to $\pm 5$ in. It is assumed that the variable load has a zero cost, $\psi \equiv 0$. Then it follows from equation (2.64) that the sum of the vertical adjoint displacements at joint 3 must take on a zero value for optimality:

$$\sum_{l=1}^{2} \bar{U}_{3vl} = 0$$

Figures 2.18(c) and 2.18(d) show, respectively, the variation of $\sum \bar{U}_3$ and the structural weight $W$ (which is optimal within a given $P$ value) in dependence of the variable load $\tilde{P}$. It can be seen that, indeed, the minimum weight corresponds to the fulfillment of the above optimality condition.

## 2.6 A BRIEF REVIEW OF THE COC ALGORITHM

### 2.6.1 Continuum formulation in structural mechanics

In sections 2.1–2.5 we used the so-called **discretized formulation**, in which the structure is divided into finite elements and then at all points of the structure stresses, strains and displacements can be derived from the displacements at certain points of the elements termed 'nodes'. In the so-called **continuum formulation**, we represent the translational and rotational displacements $\mathbf{u} = [u_1, \ldots, u_h, \ldots, u_H]$ along the centroidal axis or middle surface (the brackets { } are omitted for vectors in sections 2.6 and 2.7) of a one- or two-dimensional structural component as a function of the spatial coordinates $\xi$, i.e $\mathbf{u} = \mathbf{u}(\xi)$. Then stresses and strains at any arbitrary point of a cross section can be derived from given displacement functions $\mathbf{u}(\xi)$. It follows from the above fundamental simplification that, in the continuum approach to structural mechanics, structures may be treated as

- one-dimensional continua (e.g. bars, beams, arches, rings, frames, trusses, beam-grids or grillages, shell-grids, cable nets),
- two-dimensional continua (plates, disks, structures subject to plane stress or plane strain, shells, folded plates, etc.) or
- three-dimensional continua (stress systems for which the above simplifying idealizations are not possible).

Special cases of two-dimensional continua are grid-type or perforated structures in which the spacing of members or the microstructure is small compared with the macroscopic dimensions of the structure and hence the structure can be replaced by a continuum using a smoothing-out process or homogenization. In the case of grid-type structures, Prager and the first author (e.g. Prager and Rozvany, 1977) used terms such as 'truss-like continua' and 'grillage-like continua' for such homogenized structures.

In sections 2.6 and 2.7, we use basically Prager's notation and terminology, but some of the symbols shall be made compatible with the notation of the numerical school. Stress resultants (or, in Prager's terminology, generalized stresses) at a given cross section shall be denoted by $\boldsymbol{\sigma} = [\sigma_1, \ldots, \sigma_j, \ldots, \sigma_J]^{\mathrm{T}}$. Examples of such stress resultants are bending moments, shear forces, etc. For three-dimensional continua, $\boldsymbol{\sigma}$ denotes the local stress components ($\sigma_x$, $\tau_{xy}$, etc.). In the considered approach to structural mechanics we can use cross-sectional strains or generalized strains $\boldsymbol{\varepsilon} = [\varepsilon_1, \ldots, \varepsilon_j, \ldots, \varepsilon_J]^{\mathrm{T}}$ which refer to spatial derivatives of displacements $\mathbf{u}$ of the centroidal axis or middle surface. For example, if the deflection of a beam is given by $u_1(\xi)$, where $\xi$ is the distance along the beam, then the curvature $\varepsilon_1 = \kappa$ of the same beam is

$$\kappa = -\frac{\mathrm{d}^2 u_1}{\mathrm{d}\xi^2} \tag{2.183}$$

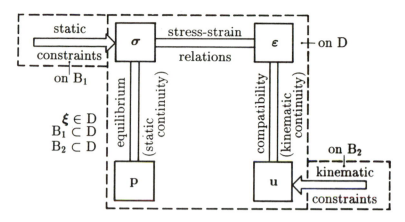

**Fig. 2.19** Fundamental relations of structural mechanics.

The term 'cross-sectional strain' is justified because $\kappa$ denotes the rate of spatial change per unit length of the angular rotation of the cross-sections (which are assumed to remain plane). The loads at a given cross-section $\xi$ may have three force and three moment components and are denoted by $\mathbf{p} = [p_1, \ldots, p_h, \ldots, p_H]^T$. The union of all admissible values of the spatial coordinates is termed the structural domain $D$, with $\xi \in D$.

The fundamental relations of structural mechanics are summarized in Fig. 2.19. On the structural domain $D$, it is necessary to satisfy static continuity (equilibrium) conditions $(\mathbf{p}, \boldsymbol{\sigma})$, kinematic continuity (compatibility) conditions $(\mathbf{u}, \boldsymbol{\varepsilon})$ and generalized stress–strain relations $(\boldsymbol{\sigma}, \boldsymbol{\varepsilon})$. On certain subsets of the structural domain, $B_1 \subset D$, $B_2 \subset D$, respectively, static and kinematic constraints (or 'boundary conditions') must be fulfilled. As a simple illustrative example, we consider a Bernoulli cantilever beam. In that case, the generalized stress is the bending moment $\sigma_1 \to M$, the generalized strain is the curvature $\varepsilon_1 \to \kappa$, the load is the transverse load $p_1 \to p$ and the displacement the transverse displacement $u_1 \to u$. The fundamental relations listed above are

$$\mathrm{d}^2 M / \mathrm{d}\xi^2 = -p, \quad \mathrm{d}^2 u / \mathrm{d}\xi^2 = -\kappa, \quad EI\kappa = M \tag{2.184}$$

where $E$ is Young's modulus and $I$ is the moment of inertia. The boundary conditions at the fixed and free ends $(B_2, B_1)$ are, respectively,

$$u = \mathrm{d}u / \mathrm{d}\xi = 0, \quad M = \mathrm{d}M / \mathrm{d}\xi = 0 \tag{2.185}$$

if the free end has no concentrated load or moment acting on it.

In the remaining sections it is assumed that the cross-sectional geometry is partially defined so that a finite number of design variables $\mathbf{x} = [x_1 \cdots x_i \cdots x_I]^T$ fully determine the cross section. Examples of such design variables, which are functions of the spatial coordinates $\xi$, are given in Fig. 2.20.

The specific cost (cost per unit length, area or volume) $\psi$ at a cross section $\xi$ depends on the design variables

$$\psi = \psi[\mathbf{x}(\xi)] \tag{2.186}$$

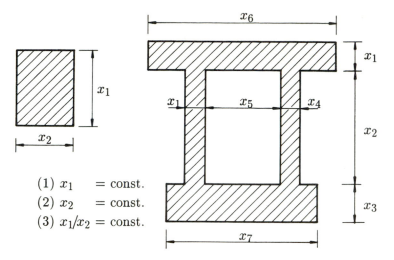

**Fig. 2.20** Examples of design variables representing cross-sectional dimensions.

The total cost or objective functional is then given by

$$\Phi = \int_D \psi[\mathbf{x}(\xi)]\,d\xi \tag{2.187}$$

Considering a beam of rectangular cross section having width $x_1$ and depth $x_2$ (Fig. 2.20(a)), the objective functional being the total beam weight, the specific cost function becomes

$$\psi = \rho x_1 x_2 \tag{2.188}$$

where $\rho$ is the specific weight of the beam material.

The generalized stress–strain relations for linearly elastic structures can be represented as

$$\varepsilon = [\mathbf{d}]\sigma, \tag{2.189}$$

where $[\mathbf{d}]$ denotes the generalized flexibility matrix.

Considering, for example, a plate in plane stress, the strain–stress relation in equation (2.189) takes the form

$$\begin{Bmatrix} \varepsilon_x \\ \varepsilon_y \\ \gamma_{xy} \end{Bmatrix} = \frac{1}{x_1 E} \begin{bmatrix} 1 & -\mu & 0 \\ -\mu & 1 & 0 \\ 0 & 0 & 2(1+\mu) \end{bmatrix} \begin{Bmatrix} N_x \\ N_y \\ N_{xy} \end{Bmatrix} \tag{2.190}$$

where $N_x$, $N_y$ and $N_{xy}$ are the stress resultants for the $x$ and $y$ directions, $\varepsilon_x$, $\varepsilon_y$ and $\gamma_{xy}$ are the usual strain components (which can also refer to an entire cross section), $x_1$ is the design variable representing the plate thickness and $\mu$ is Poisson's ratio.

## 2.6.2 Historical background

As mentioned already in section 2.2, discretized optimality criteria (DOC) methods were developed by a group of aerospace scientists (Berke, 1970; Gellatly and Berke, 1971; Venkayya, Khot and Berke, 1973; Berke and Khot, 1974; for reviews see Berke and Khot, 1987, 1988; Khot and Berke, 1984; Haftka, Gürdal and Kamat, 1990), whereas continuum-type optimality criteria (COC) methods were introduced mostly by the anaytical school of structural optimization around W. Prager. The fundamental ideas behind the COC approach first appeared in the literature, in the context of least-weight trusses, around the turn of the century in a paper by the versatile Australian inventor A.G.M. Michell (Michell, 1904), whose interests ranged from machine design (crankless engine, thrust bearing) to hydraulics and lubrication as well as structural mechanics. Michell's optimality condition for least-weight trusses with a single load condition and a constraint on the stresses consisted of the following requirement on strains $\varepsilon$

$$\varepsilon^{K} = c \operatorname{sgn} N^{S} \text{ (for } N > 0), \quad |\varepsilon^{K}| \leqslant c \text{ (for } N = 0) \tag{2.191}$$

where the superscripts K and S, respectively, denote kinematic and static admissibility, $c$ is a given constant and $N$ is the axial force along a line element of the available space. An interesting feature of the above optimality criteria, which will have a particular significance in layout optimization (Chapter 10), is that the strains are also restricted along vanishing members (of zero cross-sectional area).

Another important early development was the uniform energy dissipation principle by Drucker and Shield (1956), which states that for a simple class of cost functions expressed in terms of the generalized stresses $\boldsymbol{\sigma}$ and having the property $\psi(\lambda\boldsymbol{\sigma}) = \lambda\psi(\boldsymbol{\sigma})$, a plastically designed structure takes on a minimum cost $\psi$ if

$$\text{on } D, \varepsilon\boldsymbol{\sigma}/\psi = \text{constant} \tag{2.192}$$

Considering the optimal plastic design of beams and frames with segment-wise constant cross sections, Foulkes (1954) obtained the following optimality condition:

$$\left( \sum_{D_i} |\theta| \right) \Big/ L_i = \text{constant} \tag{2.193}$$

where $D_i = (i = 1, \dots, n)$ are segments having the length $L_i$ and $\theta$ denotes plastic hinge rotations.

For beams and frames of freely varying cross section, with a specific cost function $\psi = c|M|$, where $c$ is a given constant and $M$ is the bending moment, Heyman (1959) introduced the optimality condition

$$\kappa^{K} = c \operatorname{sgn} M^{S} \text{ (for } M > 0), \quad |\kappa^{K}| \leqslant c \text{ (for } M = 0) \tag{2.194}$$

where $c$ is a given constant, the superscripts S and K denote static and kinematic admissibility and $\kappa$ are curvature rates. A comparison of equations (2.191) and (2.194) shows that Heyman, in effect, extended Michell's optimality criteria from trusses to frames.

A general optimality condition for the plastic design of structures was introduced by Prager and Shield (1967), who extended a more restricted optimality criterion by Marcal and Prager (1964). The same optimality condition can be derived from earlier criteria by Mroz (1963). In the Prager–Shield condition, the specific cost is expressed in terms of the generalized stresses $\psi(\sigma)$,

$$\bar{\varepsilon}^{K} = \text{grad } \psi(\sigma^S) \qquad (2.195)$$

where $\bar{\varepsilon}$ denotes a fictitious strain field now termed as adjoint strain field, but is also one possible velocity field at the collapse of the plastically designed structure. It was shown by the first author (e.g. Rozvany, 1989, p. 42) that for slope discontinuities of $\psi(\sigma)$, the gradient in equation (2.195) is replaced by the subgradient, representing a convex combination of the gradient values for the adjacent stress regimes.

The first continuum-type optimality criterion for a given elastic deflection was derived for beams of varying width by Barnett (1961) in the form

$$\psi(\xi) = c(M\bar{M})^{1/2} \qquad (2.196)$$

where $\psi$ is the cross-sectional area, $c$ is a constant, $M$ is the bending moment for the external load and $\bar{M}$ is the bending moment caused by a virtual load (e.g. unit dummy load). The same optimality condition was expressed in terms of constant mutual complementary energy by Shield and Prager (1970) and Huang (1971). For sandwich beams with a given compliance (total external work) and for sandwich columns of given Euler buckling load, the following optimality condition was derived by Prager and Taylor (1968):

$$|\kappa^K| = \text{constant} \qquad (2.197)$$

Optimality criteria for eigenvalue problems have been used extensively by Olhoff (1981).

A comprehensive treatment of optimality criteria for a variety of design conditions in both elastic and plastic design is given in a recent book by the first author (Rozvany, 1989). The same text includes a large number of analytically solved illustrative examples and a bibliography with some 1200 references. An original contribution in the above book is a general treatment of linearly elastic structures with stess and displacement constraints, for which continuum-type optimality criteria (COC) are reproduced here in Fig. 2.21. It will be seen that the conceptual schemes of DCOC in Figs. 2.2–2.5 represent discretized equivalents of the COC approach. The optimality criteria in Fig. 2.21 are based on stress constraints of the type

$$S_r[\mathbf{x}(\xi), \sigma(\xi)] \leqslant 0 \text{ (for all } \xi \in D) \qquad (2.198)$$

and displacement constraints

$$\int_D \bar{\sigma}_k^T[\mathbf{d}]\sigma \, d\xi \leqslant t_k \qquad (2.199)$$

which are expressed in terms of mutual work involving the real stresses $\sigma(\xi)$ and the virtual generalized stresses $\bar{\sigma}_k(\xi)$ equilibrating the virtual loads $\bar{\mathbf{P}}_k(\xi)$ associated with the $k$th displacement condition. The virtual load consists of a unit (dummy) load for a constrained deflection at one point and of the weighting factors if the constraint limits a weighted combination of several displacements.

As exemplified by Fig. 2.21, an intrinsic feature of the COC formulation is that the optimality criteria are expressed in terms of a fictitious structure termed

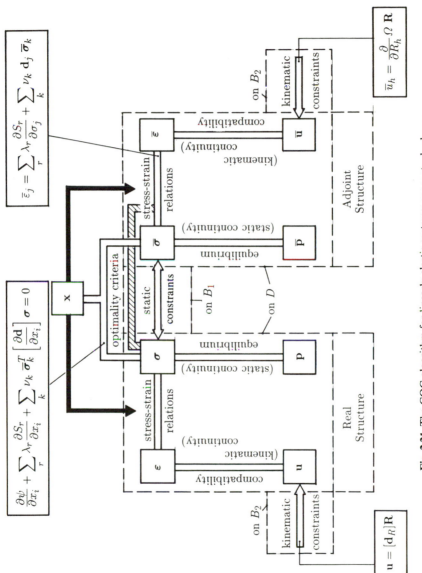

**Fig. 2.21** The COC algorithm for linearly elastic systems: conceptual scheme.

the adjoint structure (Rozvany and Zhou, 1993). The equilibrium and compatibility conditions for the real and adjoint structures are usually identical, as is one part $(\bar{\varepsilon}_j = \sum_k v_k \mathbf{d}_j \bar{\sigma}_k)$ of the generalized strain–stress relations, where $\mathbf{d}_j$ denotes the $j$th row of the generalized flexibility matrix $[\mathbf{d}]$. However, if stress constraints are active for the considered cross section $v$, then an additional strain component, which depends on the real generalized stresses $[\sum_r \lambda_r (\partial S_r / \partial \sigma_j)]$ also influences the adjoint strains.

The kinematic boundary conditions are the same for both real and adjoint structures if the supports are rigid and costless. In the case of elastic supports, however, the real support displacements are

$$\mathbf{u} = [\mathbf{d}_R] \mathbf{R} \tag{2.200}$$

where $[\mathbf{d}_R]$ is the support flexibility matrix and $\mathbf{R} = (R_1, \ldots, R_h, \ldots, R_H)$ are the reactions. Moreover, in the case of allowance for support costs the adjoint support displacements become

$$u_h = \frac{\partial}{\partial R_h} \Omega(\mathbf{R}) \tag{2.201}$$

where $\Omega(\mathbf{R})$ is the support cost function. Finally, the design variables (cross-sectional dimensions) $\mathbf{x}$ are related to the real and adjoint generalized stresses through the optimality criteria.

### 2.6.3 Derivation of the optimality criteria in Fig. 2.21

Using the Lagrangians $\bar{\theta}, \theta_k (k = 1, \ldots, K), v^k (k = 1, \ldots, K)$ and $\lambda_r(\xi) (r = 1, \ldots, R)$, and the nonnegative slack functions $\alpha_r(\xi) (r = 1, \ldots, R)$ and slack variables $\beta_k (k = 1, \ldots, K)$, the augmented (Lagrangian) functional for the considered problem can be stated as

$$\begin{aligned}
\Phi^L = &\int_D \psi[\mathbf{x}(\xi)] \, \mathrm{d}\xi + \sum_k v_k \left( \int_D \bar{\sigma}_k^T [\mathbf{d}] \sigma \, \mathrm{d}\xi - t_k + \beta_k \right) \\
&+ \int_D \lambda_r(\xi) \{ S_r[\mathbf{x}(\xi), \sigma(\xi)] + \alpha_r(\xi) \} \, \mathrm{d}\xi + \bar{\theta} \left[ \int_D \mathbf{p}^T \bar{\mathbf{u}}^K \, \mathrm{d}\xi - \int_D \sigma^T \bar{\varepsilon}^K \, \mathrm{d}\xi \right] \\
&+ \sum_k \theta_k \left[ \int_D \bar{\mathbf{p}}_k^T \mathbf{u}_k^K \, \mathrm{d}\xi - \int_D \bar{\sigma}_k^T \varepsilon_k^K \, \mathrm{d}\xi \right]
\end{aligned} \tag{2.202}$$

In equation (2.202), statical admissibility of the real loads and generalized stresses $(\mathbf{p}, \sigma)$ and of the adjoint loads and stresses $(\bar{\mathbf{p}}, \bar{\sigma})$ is expressed via the principle of virtual displacements, in which at this stage $(\bar{\mathbf{u}}^K, \bar{\varepsilon}^K)$ and $(\mathbf{u}_k^K, \varepsilon_k^K)$ represent any kinematically admissible small displacement system which is not necessarily related to the real and adjoint loads $(\mathbf{p}, \bar{\mathbf{p}})$. This is a relaxed form of the considered optimization problem, because in the original problem the real strains and displacements $(\varepsilon, \mathbf{u})$ must also fulfill kinematic admissibility.

The problem in equation (2.202) is a so-called variational problem with unknown functions which are to be optimized. Necessary conditions of optimality are provided by the Euler–Lagrange equations, which for variation of the design variables $x_i (i = 1, \ldots, I)$ and real generalized stresses $\sigma_j (j = 1, \ldots, J)$ yield, respectively, the optimality criteria and adjoint stress–strain relations at the top of Fig. 2.21. It can be seen from the latter that, although in displacement calculations the virtual forces and stresses need to satisfy only statical admissibility, in the COC algorithm they must also fulfill kinematic admissibility via the above adjoint stress–strain relations. Finally, Euler–Lagrange equations for variation of $\bar{\sigma}_{kj}$ take the form

$$\theta_k \varepsilon_{kj}^K = v_k \mathbf{d}_j \boldsymbol{\sigma} \qquad (2.203)$$

where $\mathbf{d}_j$ is the $j$th row of the generalized flexibility matrix $[\mathbf{d}]$. This means that $\varepsilon_{kj}^K$, which until now represented any kinematically admissible strain field, by means of this optimality condition has become equal to elastic strains (factored by $v_k / \theta_k$) caused by the statically admissible real stress field $\boldsymbol{\sigma}$.

This shows that kinematic admissibility, which was relaxed in the original formulation in equation (2.202), is now satisfied through the optimality criterion in equation (2.203).

In most practical problems, it is not possible to find a closed-form analytical solution satisfying simultaneously all the relations shown in Fig. 2.21. It is therefore necessary to adopt an iterative procedure consisting of the same main steps as DCOC:

• analysis of the real and adjoint structures;
• updating the design variables using optimality criteria.

## 2.7 APPLICATIONS OF THE ITERATIVE COC ALGORITHM

### 2.7.1 Classes of problems solved by COC

The COC algorithm was tested on a number of test examples (e.g. Rozvany *et al.*, 1989; Rozvany, Zhou and Gollub, 1990; Zhou and Rozvany, 1993) which included the following types of problems:

• prescribed lower and upper bounds on the cross-sectional dimensions,
• segment-wise constant cross sections,
• allowance for the cost of supports,
• allowance for selfweight,
• nonlinear and nonseparable specific cost and flexibility functions,
• two-dimensional structural systems (plates).

For example in the case of allowance for selfweight, it is sufficient to multiply the RHS of the adjoint stress–strain relation (top right corner of Fig. 2.21) by $1 + \bar{u}$, a surprisingly simple extension.

In this section we shall illustrate the iterative COC algorithm with only one test example.

### 2.7.2 Beam of independently variable width $x_1$ and depth $x_2$ with bending in both horizontal and vertical directions

(a) Optimality criteria

For the above class of problems we have

$$\boldsymbol{\sigma}=(M_1,M_2),\quad \bar{\boldsymbol{\sigma}}=(\bar{M}_1,\bar{M}_2),\quad [\mathbf{d}]=\begin{bmatrix} \dfrac{1}{rx_1^3x_2} & 0 \\ 0 & \dfrac{1}{rx_1x_2^3} \end{bmatrix},\quad \psi=cx_1x_2,\quad r=E/12$$

(2.204)

where $M_1$ and $M_2$ are the real bending moments in the horizontal and vertical directions, $\bar{M}_1$ and $\bar{M}_2$ are the corresponding adjoint moments, $[\mathbf{d}]$ is the specific generalized matrix, $c=\rho$ is the specific weight and $E$ is Young's modulus for the beam material. We consider a single deflection constraint which limits the sum of the horizontal and vertical displacements

$$t \geqslant \int_D \left( \frac{M_1\bar{M}_1}{rx_1^3x_2} + \frac{M_2\bar{M}_2}{rx_1x_2^3} \right) d\xi$$

(2.205)

and the cross-sectional dimensions are constrained from below

$$x_1 \geqslant x_1\!\downarrow,\quad x_2 \geqslant x_2\!\downarrow$$

(2.206)

For the above problem, the following optimality conditions have been derived from both the general formulas in Fig. 2.21 (with $\lambda_r=0$ for all $r$ and $v_1 \to v$) and by a variational derivation (Rozvany, Zhou and Gollub, 1990).

● Case A: $x_1 > x_1\!\downarrow,\ x_2 > x_2\!\downarrow$.

$$x_1^6 = 4v\frac{(M_1\bar{M}_1)^2}{M_2\bar{M}_2},\quad x_2^6 = 4v\frac{(M_2\bar{M}_2)^2}{M_1\bar{M}_1}$$

(2.207)

The conditions for the validity of case A and equation (2.207) are

$$M_2\bar{M}_2 > 0,\quad M_1\bar{M}_1 > 0,\quad x_1^6\!\downarrow < 4v\frac{(M_1\bar{M}_1)^2}{M_2\bar{M}_2},\quad x_2^6\!\downarrow < \frac{(M_2\bar{M}_2)^2}{M_1\bar{M}_1}$$

(2.208)

● Cases B and C: $x_i = x_i\!\downarrow,\ x_k > x_k\!\downarrow$ ($i=1,\ k=2$ and $i=2,\ k=1$, respectively).

$$x_k^2 = \frac{vM_i\bar{M}_i}{2x_i^4\!\downarrow} + \left[ \left( \frac{vM_i\bar{M}_i}{2x_i^4\!\downarrow} \right)^2 + \frac{3vM_k\bar{M}_k}{x_i^2\!\downarrow} \right]^{1/2}$$

(2.209)

The conditions for the validity of case B or C and equation (2.209) are

$$x_i^4\!\downarrow x_k^4 > v(3M_i\bar{M}_ix_k^2 + M_k\bar{M}_kx_{ia}^2),\quad M_k\bar{M}_k > 0$$

(2.210)

● Case D: $x_1 = x_1\!\downarrow,\ x_2 = x_2\!\downarrow$. We have the following conditions for this case:

$$x_1^4\!\downarrow x_2^4 \geqslant v(3M_2\bar{M}_2x_1^2\!\downarrow + M_1\bar{M}_1x_2^2\!\downarrow),\quad x_1^4\!\downarrow x_2^4\!\downarrow \geqslant v(M_2\bar{M}_2x_1^2\!\downarrow + 3M_1\bar{M}_1x_2^2\!\downarrow)$$

(2.211)

The calculation of the Lagrangian from the deflection condition in equation (2.205) requires a Newton–Raphson procedure.

(b) Solutions by the iterative COC and SQP methods

Considering a clamped beam having a span of 2.0 and a uniformly distributed load of unit itensity in the vertical direction and a central unit point load in the horizontal direction (Fig. 2.22), a deflection value of $\Delta = 3$ in equation (2.205) and $c = r = 1$ were adopted.

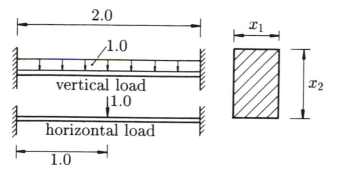

Fig. 2.22 Test example for the COC algorithm: beam of variable width and depth.

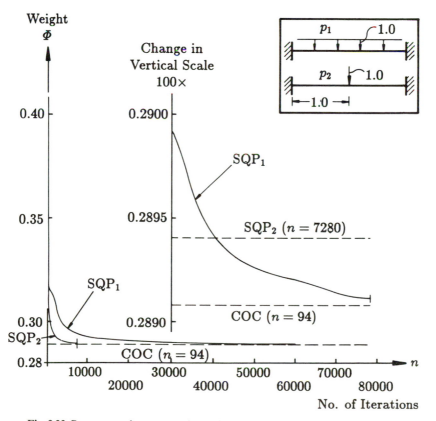

Fig. 2.23 Beam example: a comparison of iteration histories by various methods.

**Table 2.11** A comparison of weight values obtained by the SQP and COC methods after convergence and percentage differences

|         | Weight $\Phi$ | $\Delta\%$ | Number of iterations | CPU time (s) |
|---------|---------------|------------|----------------------|--------------|
| COC     | 0.289 081     | 0          | 94                   | 57           |
| $SQP_1$ | 0.289 114     | 0.011      | 77564                | 1 507 451 (17.45 days) |
| $SQP_2$ | 0.289 408     | 0.113      | 7280                 | 176 746 (2.05 days) |

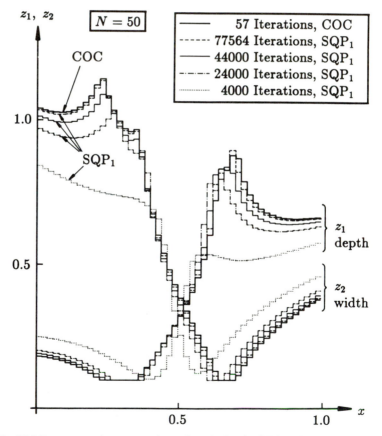

**Fig. 2.24** Beam example: a comparison of results by the COC and SQP algorithms.

Using the COC method with only 50 elements and 100 variables (for a comparison with the SQP method), and a tolerance criterion of $(\Phi_{new} - \Phi_{old})/\Phi_{new} \leqslant 10^{-8}$, a weight value of $\Phi = 0.289\,081$ was obtained after 94 iterations, with a total CPU time (analysis plus optimization) of 57 s. On repeating the COC procedure with 10 000 elements and 20 000 variables, the above convergence criterion was reached after 103 iterations with a CPU time of 12 096 s and a weight of $\Phi = 0.288\,782$.

The same problem with 50 elements and 100 variables was also computed using the SQP method with reciprocal variables. The above convergence criterion was satisfied after 77 564 iterations, requiring a CPU time of 1 507 451 s (17.45 days). In Fig. 2.23, the iteration history of this calculation is compared with the result given by the COC method after convergence.

Since the convergence was extremely slow, a modified procedure was used in which the coupling effect was relaxed through multiplying the mixed second derivatives by various factors. After testing several such factors, a value of 0.2 was adopted, which resulted in a much faster convergence (SQP$_2$ in Fig. 2.23).

The SQP$_2$ procedure reached the convergence criterion after 7280 iterations, requiring a CPU time of 'only' 176 746 s (2.05 days). However, because of small instabilities in the convergence, satisfaction of the convergence criterion was somewhat accidental giving a fairly large error in comparison to the COC and SQP$_1$ results (enlarged right-hand part of Fig. 2.22). The results of the COC, as well as the unmodified and modified SQP methods, are compared in Table 2.11. The width and depth distributions given by the two SQP methods after various numbers of iterations are compared with those obtained by the COC method in Fig. 2.24. It can be concluded from these diagrams and Table 2.11 that the SQP method fully confirms the results by the COC algorithm. However, the more accurate SQP method (SQP$_1$) requires almost 30 000 times as much time as the COC procedure.

## 2.8 CONCLUDING REMARKS

A new discretized optimality criteria method (DCOC) was discussed in detail in this chapter, together with a brief review of the corresponding continuum-type optimality criteria (COC) method. It was shown that DCOC increases our optimization capability by several orders of magnitude if a large number of stress constraints are active, which is usual in structural design, and hence this method should become an indispensable tool in the optimization of very large structural systems. Applications of the DCOC and COC algorithms in topology optimization will be discussed in Chapter 10.

## ACKNOWLEDGEMENTS

The authors are greatly indebted to the Deutsche Forschungsgemeinschaft for financial support (Projects Ro 744/1-1 and Ro 744/1-2), to Sabine Liebermann for processing the text, preparing some diagrams and checking algebraically part of the equations, to Peter Moche for the remaining diagrams, and to Susann Rozvany for editing and correcting the text.

## APPENDIX 2A   NOTATION

### 2A.1   Notation – discretized systems

| | |
|---|---|
| $A$ | cross-sectional area |
| $b$ | given width of cross section |
| $[\mathbf{B}], [\mathbf{B}]^{\mathrm{T}}$ | statics and kinematics matrices |
| $c^e$ | cost constant for element $e$ |
| $[\mathbf{D}], [\mathbf{d}^e]$ | flexibility matrix |
| $[\underline{\mathbf{D}}], [\underline{\mathbf{d}}^n]$ | reaction element (support) flexibility matrix |
| $e = 1, \dots, E$ | superscript identifying elements |
| $[\mathbf{d}_o^e]$ | constant part of element flexibility matrix |
| $\{\mathbf{F}\}, \{\mathbf{f}^e\}$ | nodal forces (stiffness method) |
| $\{\mathbb{F}\}, \{\mathbf{ff}^e\}$ | nodal forces (flexibility method) |
| $\{\hat{\mathbb{F}}\}, \{\hat{\mathbf{ff}}^e\}$ | virtual nodal forces (flexibility method) |
| $\{\bar{\mathbb{F}}\}, \{\bar{\mathbf{ff}}^e\}$ | adjoint nodal forces (flexibility method) |
| $\{\mathbf{F}^*\}, (\mathbf{f}^{*e}\}$ | fixed end forces |
| $\{\bar{\mathbf{F}}^*\}, \{\bar{\mathbf{f}}^{*e}\}$ | adjoint fixed end forces |
| $h$ | given depth of cross section |
| $h = 1, \dots, H^e$ | subscript identifying cross sections along element $e$ |
| $i = 1, \dots, I^e$ | subscript identifying design variables for element $e$ |
| $I$ | moment of inertia |
| $j = 1, \dots, J$ | subscript identifying degrees of freedom (joint displacement and loads) |
| $k = 1, \dots, K$ | subscript identifying displacement constraints |
| $k = 1, \dots, K_l$ | subscript identifying displacement constraints (for the $l$th load condition) |
| $[\mathbf{K}], [\mathbf{k}^e]$ | stiffness matrix |
| $L$ | length of member |
| $\{\bar{\mathbf{L}}_l\}, \{\mathbf{L}_{kl}\}$ | Lagrange multipliers |
| $l = 1, \dots, L$ | subscript identifying load conditions |
| $m = 1, \dots, M^e$ | subscript identifying nodal forces and relative displacements at nodes of element $e$ |
| $n = 1, \dots, N$ | superscript identifying reaction elements (supports) |
| $\{\mathbf{n}\}$ | stress resultants at an arbitrary cross section |
| $\{\mathbf{n}_h^e\}$ | stress resultants at cross section $h$ of element $e$ |
| $\{\mathbf{n}_{oh}^e\}$ | stress resultants at cross section $h$ of element $e$ caused by the load acting on the interior of element $e$ |
| $\{\mathbf{P}\}$ | real loads |
| $\{\hat{\mathbf{P}}\}$ | virtual loads |
| $\{\bar{\mathbf{P}}\}$ | adjoint loads |
| $\{\tilde{\mathbf{P}}\}$ | variable loads |
| $q = 1, \dots, Q^n$ | subscript identifying reaction components and displacements at support $n$ |
| $Q$ | section modulus |

| | |
|---|---|
| $r = 1, \ldots, R_h^e$ | subscript identifying a type and/or location of a stress or stress condition at a cross section $h$ of element $e$ |
| $R_\sigma$ | set of elements for which a stressed constraint is active |
| $R_s$ | set of elements for which a side constraint is active |
| $R_{\sigma s}$ | set of elements for which both a stress constraint and a side constraint are active |
| $R_d$ | set of elements controlled by a deflection constraint |
| $\{\mathbf{R}\}, \{\mathbf{r}^n\}$ | real reactions |
| $\{\bar{\mathbf{R}}\}, \{\bar{\mathbf{r}}^n\}$ | adjoint reactions |
| $\{\mathbf{s}_{hr}^e\}$ | transformation vector relating stress resultants $\{\mathbf{n}_h^e\}$ to stresses $\sigma_{hr}$ of type and/or location $r$ at cross section $h$ |
| $\{\mathbf{s}_0^e\}$ | constant part of stress transformation vector |
| $t, t_k, t_{kl}$ | specified value of deflections |
| $T, T_1$ | tolerance values in convergence criteria |
| $\{\mathbf{U}\}$ | real nodal displacements |
| $\{\bar{\mathbf{U}}\}$ | adjoint nodal displacements |
| $\{\tilde{\bar{\mathbf{U}}}\}$ | adjoint nodal displacements corresponding to variable loads |
| $v = 1, \ldots, V_h^e$ | subscript identifying stress resultants at the cross section $h$ |
| $W(\mathbf{X}), w^e(\mathbf{x})$ | objective functions associated with elements |
| $\{\mathbf{X}\}, \{\mathbf{x}^e\}$ | design variables |
| $\{\mathbf{X}{\downarrow}\}, \{\mathbf{x}{\downarrow}^e\}$ | lower limit on design variables |
| $\{\mathbf{X}{\uparrow}\}, \{\mathbf{x}{\uparrow}^e\}$ | upper limit on design variables |
| $\alpha$ | coefficient of thermal expansion |
| $\beta_i^e{\downarrow}, \beta_i^e{\uparrow}, \nu_{kl},$ $\lambda_{hrl}^e, \rho_e^n$ | Lagrange multipliers |
| $[\boldsymbol{\Gamma}], \{\boldsymbol{\gamma}^e\}$ | transformation matrix from element level to system level |
| $\{\boldsymbol{\Delta}\}, \{\boldsymbol{\delta}^e\}$ | real relative displacements |
| $\{\bar{\boldsymbol{\Delta}}\}, \{\bar{\boldsymbol{\delta}}^e\}$ | adjoint relative displacements |
| $\{\boldsymbol{\Delta}^*\}, \{\boldsymbol{\delta}^{*e}\}$ | real initial relative displacements |
| $\{\bar{\boldsymbol{\Delta}}^*\}, \{\bar{\boldsymbol{\delta}}^{*e}\}$ | adjoint initial relative displacements |
| $\{\boldsymbol{\Delta}^{\circ *}\}, \{\boldsymbol{\delta}^{\circ *e}\}$ | real initial relative displacements extended for stiffness method |
| $\{\bar{\boldsymbol{\Delta}}^{\circ *}\}, \{\bar{\boldsymbol{\delta}}^{\circ *e}\}$ | adjoint initial relative displacements extended for stiffness method |
| $\{\underline{\boldsymbol{\Delta}}\}, \{\underline{\boldsymbol{\delta}}^n\}$ | displacements of a reaction element (support) $n$ |
| $\{\underline{\boldsymbol{\Delta}}^*\}, \{\underline{\boldsymbol{\delta}}^{*n}\}$ | prescribed initial displacements of a reaction element $n$ (support settlements) |
| $\{\tilde{\boldsymbol{\Delta}}^*\}, \{\tilde{\boldsymbol{\delta}}^{*e}\}$ | variable initial relative displacements (due to prestrain) |
| $\{\tilde{\underline{\boldsymbol{\Delta}}}^*\}, \{\tilde{\underline{\boldsymbol{\delta}}}^{*n}\}$ | variable settlements at support $n$ |
| $\{\bar{\underline{\boldsymbol{\Delta}}}\}, \{\bar{\underline{\boldsymbol{\delta}}}^n\}$ | displacements at supports of adjoint structure |
| $\{\bar{\underline{\boldsymbol{\Delta}}}^*\}, \{\bar{\underline{\boldsymbol{\delta}}}^{*n}\}$ | initial displacements at supports of adjoint structure |
| $\eta$ | constant in Kreisselmeier–Steinhauser functions |
| $\theta_{hr}^e(\mathbf{n}_h^e)$ | stress conditions of type $r$ (function of stress resultants $\{\mathbf{n}_h^e\}$) at a cross section $h$ of element $e$ |
| $\theta_o^e$ | constant part of stress condition |
| $\vartheta^n(\tilde{\underline{\boldsymbol{\delta}}}^{*n})$ | cost of variable support settlement |
| $\mu$ | Poisson's ratio |

| | |
|---|---|
| $\rho^e$ | specific weight of structural material of element $e$ |
| $\sigma^e_{hr}$ | local stress of type and/or location $r$ at cross-section $h$ of element $e$ |
| $\tau$ | magnitude of temperature change |
| $\Phi$ | objective function |
| $\chi^e(\tilde{\delta}^{*e})$ | cost of variable initial relative displacement |
| $\psi_j(\tilde{P}_j)$ | cost of variable load $\tilde{P}_j$ |
| $[\omega^e_h]$ | matrix relating nodal forces $\{\mathbf{ff}^e\}$ and stress resultants $\{\mathbf{n}^e_h\}$ at cross section $h$ of element $e$ |
| $\Omega^n(r^n)$ | reaction cost function for the reaction element (support) $n$ |

Brackets used for matrices

| | |
|---|---|
| $\{\ \}$ | column matrix (vector) |
| $\lfloor\ \rfloor$ | row matrix |
| $[\ ]$ | rectangular or square matrix |

Explanatory notes on the notation for discretized systems

Bold upper-case letters represent vectors and matrices at system level and bold lower-case letters vectors and matrices at element level. Symbols in italics denote scalar quantities. $\mathbb{F}$ and $\mathbf{ff}$ refer to nodal forces in flexibility formulation whilst $\mathbf{F}$ and $\mathbf{f}$ denote nodal forces in stiffness formulation. An asterisk * signifies initial relative displacements or fixed end forces. A tilde ~ indicates quantities that are variable by passive control (variable prestrains or loads). Symbols with an underline _ refer to reaction elements (supports). An overline ‾ indicates an adjoint quantity and a hat ^ a quantity associated with a virtual load.

## 2A.2    Notation – structural continua

| | |
|---|---|
| $B_1$ | subset of the structural domain $D$, on which static boundary conditions are given |
| $B_2$ | subset of the structural domain, on which kinematic boundary conditions are given |
| $D$ | structural domain $(\xi \in D)$ |
| $[\mathbf{d}]$ | generalized flexibility matrix |
| $h = 1,\ldots,H$ | subscript identifying load components $p_h$ and displacement components $u_h$ at a given cross section |
| $i = 1,\ldots,I$ | subscript identifying design variables $x_i$ for a given cross section |
| $j = 1,\ldots,J$ | subscript identifying stress resultants $\sigma_j$ and generalized strains $\varepsilon_j$ at a given cross section |
| $k = 1,\ldots,K$ | subscript identifying displacement constraints |
| $\mathbf{p}(\xi)$ | loads at a given cross section |
| $\bar{\mathbf{p}}(\xi)$ | adjoint loads |
| $r = 1,\ldots,R$ | subscripts identifying stress constraints |
| $S_r(\mathbf{x},\boldsymbol{\sigma}) \leqslant 0$ | stress constraints |
| $\mathbf{u}(\xi)$ | displacements at a given cross section |

| $\bar{\mathbf{u}}(\xi)$ | adjoint displacements |
| $\mathbf{x}(\xi)$ | design variables at a given cross section |
| $\varepsilon(\xi)$ | cross-sectional strains (generalized strains, e.g. beam curvature) |
| $\bar{\varepsilon}(\xi)$ | adjoint generalized strains |
| $\lambda_r(\xi)$ | Lagrange multiplier for stress constraints |
| $\mu$ | Poisson's ratio |
| $\nu_k$ | Lagrange multiplier for displacement constraints |
| $\xi \in D$ | spatial coordinates |
| $\sigma$ | stress resultant (generalized stress) |
| $\bar{\sigma}$ | adjoint generalized stress |
| $\Phi = \int_D \psi[\mathbf{x}(\xi)]\,d\xi$ | total cost, objective functional |
| $\psi[\mathbf{x}(\xi)]$ | specific cost at a given cross section |

## REFERENCES

Barnett, R.L. (1961) Minimum-weight design of beam for deflection. *Journal of Engineering Mechanics ASCE* **87**, 75–109.

Berke, L. (1970) An efficient approach in the minimum weight design of deflection limited structures. *AFFDL-TM-70-FDTR*.

Berke, L. and Khot, N.S. (1974) Use of optimality criteria methods for large scale systems. *AGARD LS-70*.

Berke, L. and Khot, N.S. (1987) Structural optimization using optimality criteria, in *Computer Aided Optimal Design: Structural and Mechanical Systems* (ed. C.A. Mota Soares), Springer, Berlin, pp. 271–312.

Berke, L. and Khot, N.S. (1988) Performance characteristics of optimality criteria methods, in *Structural Optimization* (eds. G.I.N. Rozvany and B.L. Karihaloo). Proceedings of the IUTAM Symposium, Melbourne, 1988, Kluwer, Dordrecht, pp. 39–46.

Chang, K.J. (1992) Optimality criteria methods using K–S functions. *Structural Optimization*, **4**, 213–17.

Drucker, D.C. and Shield, R.T. (1956) Design for minimum weight. Proceedings of the 9th International Congress on Applied Mechanics, Brussels, 1956, Vol. 5, pp. 212–22.

Fleury, C. (1979) Structural weight optimization by dual methods of convex programming. *International Journal of Numerical Methods in Engineering*, **14**, 1761–83.

Foulkes, J. (1954) The minimum-weight design of structural frames. *Proceedings of the Royal Society*, **223**, 482–94.

Gellatly, R.A. and Berke, L. (1971) Optimal structural design. *AFFDL-TR-70-165*.

Haftka, R., Gürdal, Z. and Kamat, M.P. (1990) *Elements of Structural Optimization*, Kluwer, Dordrecht.

Heyman, J. (1959) On the absolute minimum weight design of framed structures. *Quarterly Journal of Mechanics and Applied Mathematics*, **12**, 314–24.

Huang, N.-C. (1971) On the principle of stationary mutual complementary energy and its application to structural design. *Zeitschrift für angewandte Mathematik und Physik*, **22**, 608–20.

Khot, N.S. and Berke, L. (1984) Structural optimization using optimality criteria methods, in *New Directions in Optimum Structural Design* (eds. E. Atrek, R.H. Gallagher, K.M. Ragsdell and O.C. Zienkiewicz). Wiley, Chichester, pp. 47–74.

Marcal, P.V. and Prager, W. (1964) A method of optimal plastic design. *Journal de Mécanique*, **3**, 509–30.

Michell, A.G.M. (1904) The limits of economy of material in frame-structures. *Philosophical Magazine*, **8**, 589–97.

Mroz, Z. (1963) Limit analysis of plastic structures subject to boundary variations. *Archiwum Mechaniki Stosowanej*, **15**, 63–76.

Olhoff, N. (1981) Optimization of columns against buckling. Optimization of transversely vibrating beams and rotating shafts, in *Optimization of Distributed Parameter Structures* (eds. E.J. Haug and J. Cea). Proceedings of NATO ASI, Iowa, 1980, Sijthoff and Noordhoff, Alphen aan der Rijn, pp. 152–76, 177–99.

Prager, W. and Rozvany, G.I.N. (1977) Optimization of structural geometry, in *Dynamical Systems* (eds. A.R. Bednarek and L. Cesari). Academic Press, New York, pp. 265–93.

Prager, W. and Shield, R.T. (1967) A general theory of optimal plastic design. *Journal of Applied Mechanics*, **34**, 184–6.

Prager, W. and Taylor, J. (1968) Problems of optimal structural design. *Journal of Applied Mechanics*, **5**, 102–6.

Rozvany, G.I.N. (1989) *Structural Design via Optimality Criteria*, Kluwer, Dordrecht.

Rozvany, G.I.N. and Zhou, M. (1991) The COC algorithm, part I: cross-section optimization or sizing (presented 2nd World Congr. on Computational Mechanics, Stuttgart, 1990). *Computer Methods in Applied Mechanics and Engineering*, **89**, 281–308.

Rozvany, G.I.N. and Zhou, M. (1993) Continuum-based optimality criteria (COC) methods: an introduction, in *Optimization of Large Structural Systems*, (ed. G.I.N. Rozvany). Proceedings of NATO/DFG ASI, Berchtesgaden, 1991, Kluwer, Dordrecht, pp. 1–26.

Rozvany, G.I.N., Zhou, M. and Gollub, W. (1990) Continuum-type optimality criteria methods for large finite element systems with a displacement constraint – part II. *Structural Optimization*, **2**, 77–104.

Rozvany, G.I.N., Zhou, M., Rotthaus, M., Gollub, W. and Spengemann, F. (1989) Continuum-type optimality criteria methods for large systems with a displacement constraint – part I. *Structural Optimization*, **1**, 47–72.

Shield, R.T. and Prager, W. (1970) Optimal structural design for given deflection. *Zeitschrift für angewandte Mathematik und Physik*, **21**, 513–23.

Sobieszczanski-Sobieski, J. (1992) A technique for locating function roots and for satisfying equality constraints in optimization. *Structural Optimization*, **4**, 241–3.

Venkayya, V.B., Khot, N.S. and Berke, L. (1973) Application of optimality criteria approaches to automated design of large practical structures. *AGARD-CP-123*, pp. 3.1–3.9.

Zhou, M. (1989) Geometrical optimization of trusses by a two-level approximation concept. *Structural Optimization*, **1**, 235–40.

Zhou, M. (1992) A new discretized optimality criteria method in structural optimization. Doctoral Dissertation, DVI-Verlag, Düsseldorf.

Zhou, M. and Rozvany, G.I.N. (1992) A new discretized optimality criteria method in structural optimization. Proceedings of 33rd AIAA/ASME/ASCE/AHS/ASC Structures, Structural Dynamics and Materials Conference, Dallas, TX, 1992, AIAA, Washington, DC, pp. 3106–20.

Zhou, M. and Rozvany, G.I.N. (1993) Iterative COC methods – parts I and II, in *Optimization of Large Structural Systems* (ed. G.I.N. Rozvany). Proceedings of NATO/DFG ASI, Berchtesgaden, 1991, Kluwer, Dordrecht, pp. 27–75.

# 3

---

# Model reduction and verification

---

PANOS Y. PAPALAMBROS

## 3.1 INTRODUCTION

A successful optimization study depends strongly on the appropriateness of the underlying mathematical model used for performing optimization. Indeed in most cases the model will determine whether a solution may or may not be achieved. Numerical algorithms are very sensitive to the precise mathematical form in which the model is cast. Moreover, preliminary models resulting from early modeling efforts may be poorly bounded so that either no finite optimum exists or an optimum is located by artificially forcing activity of constraints that have little or no engineering meaning. Even when a model is properly bounded the solution may be obtained, some times trivially, by forcing activity of as many constraints as the number of degrees of freedom in the model, thus making further optimization unnecessary. This tends to happen in preliminary model formulations with objective and constraint functions depending monotonically on most of the design variables.

Proper modeling requires experience and good understanding of the numerical algorithms that will be used for computing solutions. There are also some simple and rigorous principles that can be applied to exploit monotonicity properties of the model. This chapter reviews some of the basic ideas for proper model formulation and illustrates by example how they may be applied during an optimization study. Modeling considerations prior to embarking on numerical computations are examined first. Next, an extended example illustrates how a model may be reduced and verified systematically. Finally some automation efforts for such modeling considerations are discussed.

## 3.2 MODELING CONSIDERATIONS PRIOR TO COMPUTATION

At some point during the design process, it becomes possible to represent design decisions by a precise mathematical optimization statement. Usually a nonlinear programming (NLP) model formulation, such as

$$
\begin{aligned}
&\text{minimize} && f(x) && x \in X \\
&\text{subject to} && h_j(x) = 0 && j = 1, 2, \ldots, m_1 \\
&&& g_j(x) \leqslant 0 && j = m_1 + 1, m_1 + 2, \ldots, m_1 + m_2
\end{aligned}
\tag{3.1}
$$

is adequate. In equation (3.1), $f$, $h$ and $g$ are the scalar objective, equality and inequality constraints respectively, $x$ is the vector of design variables $x_1, x_2, \ldots, x_n$, and the set $X$ is a subset of the $n$-dimensional real space $R^n$, included in the statement to indicate that there may exist other restrictions on the variables that are not explicit in the constraints, for example, integer values. (Note that boldface characters indicate vector quantities.) Here the objective is shown as a single one, since multiobjective formulations are usually transformed to a single objective. The above mathematical statement will be referred to as the **design optimization model** cast into the so-called **negative null form**, a standard form where the objective is to minimize, while equality and inequality constraints are written with zero and less than or equal to zero, respectively, placed at the right-hand side of the equation.

### 3.2.1 Explicit vs implicit models

The design model implicitly assumes the existence of an **analysis model** used to evaluate the functions $f, h, g$. For example, the analysis models may be finite element analyses of structural behavior, system simulation, or curve-fitted experimental data. Analysis models may be explicit or implicit. In an **explicit** model, the model function forms are explicitly represented by mathematical, usually algebraic, functions; in an **implicit** model, the model functions are usually represented by a procedure or subroutine, for example, an iterative numerical solution of a set of differential equations.

It is generally preferable to deal with explicit models since then it is easier to make assertions about optimality, for example, by detecting mathematical properties such as monotonicity or convexity (a property that guarantees a unique minimum); also, it is often possible to manipulate the model in order to extract additional information or properties, or to recast it in a more convenient form, as we will see below. From a practical viewpoint, the models in many important classes of design optimization problems are implicit, which poses a substantially increased burden on ascertaining reliability of the optimization results. In practice one seldom seeks a mathematical optimum. Rather, one aims only at so-called **satisficing** designs (Wilde, 1978), namely, design solutions with a value of the objective function within a known acceptable distance from the (unknown) true optimal value, and frequently only at designs improved by a certain percentage over current ones. Thus the disadvantage of implicit models is significantly tempered.

From an operational viewpoint, it is useful to distinguish between two phases in the optimization process. At the point where the optimal design model has been precisely formulated, the **modeling phase** of the optimization procedure is concluded. What follows constitutes the **solution phase** of the procedure. For most realistic models, the solution phase will include an iterative procedure that should converge to the optimum. An iteration begins by selecting an initial design corresponding to a starting value for the design variables and executing the calculation represented by the analysis model(s), so the values for objective and constraint functions are obtained. The cost of subsequent calculations for evaluating the next iterant design increases with the size of the problem, as measured by the number of variables and constraints.

When an implicit analysis model is computationally expensive, the iterative nature of numerical solution strategies makes the use of such a model unattractive. In a preliminary design effort it may be possible to use a simplified explicit model instead of the more accurate implicit one. The process could be similar to the design of physical experiments, such as Taguchi methods. These methods use a carefully planned efficient set of experiments to discover how design variables relate to each other. In the present context the experiments are performed computationally using the 'full' model and the results are curve fitted. After an initial optimization study has been successfully completed with the simplified model, a few iterations using the full model and the identified optimum as a starting point may complete the study.

### 3.2.2 Design parameters vs variables

The optimization model for any given design problem is more accurately described as

$$\begin{aligned}
&\text{minimize} && f(x, p) && x \in X, p \in P \\
&\text{subject to} && h_j(x, p) = 0 && j = 1, 2, \ldots, m_1 \\
& && g_j(x, p) \leqslant 0 && j = m_1 + 1, m_1 + 2, \ldots, m_1 + m_2
\end{aligned} \qquad (3.2)$$

where $p$ is a vector of parameters, $p = (p_1, p_2, \ldots, p_q)$, and $P$ is a subset of $R^q$. The quantities $p_1, p_2, \ldots, p_q$ are the **design parameters** and they are considered fixed to specific values during the optimization process. Variables and parameters together give a complete description of the design, and the constraint functions expressed as equalities and inequalities in terms of variables and parameters give a complete description of the **design space**, i.e. the space that contains all acceptable solutions. The selection of variables and parameters is generally a modeling decision that depends on the position of the designer in the hierarchy of decision-making and on expediency for actually solving eventually the underlying mathematical problem.

When posing the design problem one must consider the position one occupies in the decision-making hierarchy. The higher is the position the fewer are the quantities that one considers as fixed; hence more quantities will be considered as variables than as parameters. A more subtle consideration in classifying parameters and variables comes from the relative ease of solution that results from this classification. For example, material selection is a typical key decision in design. In an optimization model, one material will correspond to several quantities in the model, e.g. density, yield strength, modulus of elasticity. Treating the material as a variable will require inclusion of additional relations in the model that link these quantities, such relations usually given at best in some tabular form. In subsequent numerical processing table look-up or curve-fitting will be then required. Table look-up should not be used as it may introduce discontinuities and nondifferentiabilities that could defy many NLP algorithms. Even with curve-fitting the problem will still contain discrete variables corresponding to the different materials, so continuous NLP algorithms cannot be used and the problem becomes considerably more difficult. Thus the best choice is to treat material as a parameter and to find the optimal solutions for different materials separately. The strategy will be appropriate as long as the number of available materials is not very large.

Even when there is no other good reason to treat a variable as a parameter, one may wish to do so just to make the problem simpler to solve. Generally, the smaller the number of design variables is, the easier is the problem to solve. A variable may appear in complicated nonlinear expressions and fixing it as a parameter may make the problem substantially easier, perhaps linear. Finally, there may be engineering evidence that some variables have little influence on the problem, so fixing them as parameters may be a good idea, at least in early studies.

### 3.2.3 Constraint activity

The concept of constraint activity is very important in design optimization. Loosely speaking, an active constraint corresponds to a 'critical' design requirement, i.e. one whose presence determines where the optimum will be. In traditional structural design, 'fully stressed' designs would correspond to finding minimum weight designs with all stress constraints active. Formally, a constraint (equality or inequality) is called (globally) **active** if and only if removing it from the model changes the set of (globally) optimal solutions; it is **inactive** otherwise. An optimal solution is called **constraint bound** if and only if there are exactly as many active constraints as there are design variables. A constraint is called **tight** if and only if it is satisfied as a strict equality at the optimum. For continuous variables, active constraints will also be tight. However, not all tight constraints are active. A feasible point is called **constraint bound** if and only if at this point there are as many tight constraints as there are design variables. As mentioned earlier, in a constraint-bound optimum one needs only to find the active constraints. Thus, constraint-bound points are defined by solving a system of $n$ equations in $n$ unknowns, these equations constituting a **working (active) set**. Of course, not all such solutions are constraint-bound optima. For some sufficiency criteria see Papalambros (1988a). If the point is optimal, then the working set is the (true) active set. The term 'active' is sometimes used for local minima, but then all relevant properties must be considered valid within an appropriate neighborhood or region. A more comprehensive discussion of activity is given in Pomrehn and Papalambros (1992). Identified active inequalities are indicated with the symbol $\leqq$ instead of $\leq$.

An active constraint is a direct contributor to optimality and corresponds either to an equality constraint (such as equilibrium) or to a critical design requirement (such as setting the maximum stress equal to the strength of the material used). When a subset of the original constraints is selected for further processing, these constraints together with the objective function constitute a **submodel** of the original problem. The submodel contains only those constraints that are known to be active or that have been judiciously selected as having a high likelihood to become active. An **active set strategy** is a set of rules which are used to decide which constraints may be active at the optimum. An **active set** (at a given point) is a set that contains only those constraints that are considered as being definitely active at the current iteration, which usually means that they are satisfied as equalities (at that point). An **extended active set** is a set that contains both the currently active constraints and additional ones that are considered candidates for activity in one or more subsequent iterations. Use of such sets is often necessary in order to control computational costs associated with having a small number of variables but a very large number of inequality constraints (only few of which can possibly be active).

Clearly, at a given point the active set is a subset of the extended one. The rules in an active set strategy will in general apply to both an active set and an extended one. In the mathematical programming literature the term 'active set strategy' applies only to the active set, while in the structural optimization literature the term usually applies to the extended active set.

The main idea here is to emphasize that knowing the active constraints as early as possible may be very advantageous both in locating and in verifying an optimal solution.

### 3.2.4 Monotonicity principles

In many preliminary design models, usually because of simplification of trade-offs, the objective and constraint functions are **monotonic** with respect to the design variables. A continuous differentiable function $f(x)$ is increasing (decreasing) with respect to (wrt) a design variable $x_i$, if $\partial f/\partial x_i > 0$ ($\partial f/\partial x_i < 0$). We say that $f$ is coordinate-wise monotonic wrt $x_i$, or that $x_i$ is a monotonic variable (in $f$). This concept can be extended to other functions but this is not necessary here. We will also assume that all design variables are strictly positive. This is usually the case in design problems; if not, a simple shift in the datum can make the variables positive, e.g. using an absolute temperature scale rather than Farenheit or Celsius. Monotonicity analysis is based on two simple principles (see Papalambros and Wilde (1988) for further details).

> *First monotonicity principle (MP1)* In a well-constrained objective function every increasing (decreasing) variable is bounded below (above) by at least one active constraint.
>
> *Second monotonicity principle (MP2)* Every monotonic variable not occurring in a well-constrained objective function is either irrelevant and can be deleted from the problem together with all constraints in which it occurs, or is relevant and bounded by two active constraints, one from above and one from below.

We will now illustrate how these principles can be used to identify a model that is not well constrained or even to obtain an optimal solution in simple cases. The modeling example of a linear actuator will later demonstrate the approach in a more complicated situation.

Functional relations in a model usually must be rearranged in order to cast them into a standard form, such as equation (3.1). In general, a given relation may be 'equivalent' to several standard forms, some of which may appear monotonic and some not. Therefore, monotonicity should be examined in the original formulation of the problem and, if algebraic manipulations are employed, care should be taken not to disguise monotonic properties. However, once a relation has identified monotonicities, these are invariant in the sense that any standard form resulting in identified monotonicities will have exactly the same types of monotonicities. The following implicit function theorem specifically stated for monotone functions is useful in studying how monotonicity properties may be inherited when equalities are eliminated through variable substitution.

*Implicit function theorem*    Let $X_i \subseteq R$, $i = 1, \dots, n$, be $n$ subsets (finite or infinite) of $R$ and let $X = \{x \mid x = (x_1, \dots, x_n)$, $x_i \in X_i$, $i = 1, \dots, n\}$. Let $F: X \rightarrow R$ be a function coordinate-wise monotonic on $X$. Then, for each $s$ in the range of $F$ and for each $i = 1, \dots, n$, there exists a function $\phi(i, s; x_i')$ of the variable vector $x_i' = (x_1, \dots, x_{i-1}, x_{i+1}, \dots, x_n)$ such that $\phi(i, s; x_i')$ is (coordinate-wise) monotonic wrt $x_i$. Furthermore, for $1 \leqslant j \leqslant n$ and $i \neq j$, if $F$ is monotonic in the same (opposite) sense wrt $x_i$ and $x_j$, then $\phi(i, s; x_i')$ is decreasing (increasing) wrt $x_j$.

Monotonicity principles can be applied directly when the model contains only inequalities, i.e.

$$
\begin{aligned}
\text{minimize} \quad & f(x) & & x \in X \\
\text{subject to} \quad & g_j(x) \leqslant 0 & & j = 1, 2, \dots, m
\end{aligned}
\tag{3.3}
$$

When active constraints (equalities or inequalities) have been identified, a new round of monotonicity analysis can be performed if the active constraints are eliminated through variable substitution. In the new reduced model the exact functions may not be known but the relevant monotonicities may be implicitly defined using the above theorem.

Before proceeding with some examples, it is interesting to recall that in the presence of differentiability and the usual constraint qualifications the Karush–Kuhn–Tucker (KKT) necessary optimality conditions for the model in equation (3.3) are

$$
\nabla f + \mu^T \nabla g = 0, \, g \leqslant 0
$$

$$
\mu^T g = 0, \, \mu \geqslant 0
\tag{3.4}
$$

where $\nabla g$ is the Jacobian matrix of the vector function $g(x)$. For this case, the two monotonicity principles can be derived trivially from equation (3.4). The principles are more general necessary conditions than the KKT conditions, as they apply globally and do not assume continuity or differentiability. On the other hand, they assume (global or regional) monotonicity, a property that could be considered more restrictive. Their utility is based on the argument that many early design optimization models do have extensive monotonicity properties (see for example, Papalambros and Wilde (1988)). The principles frequently allow a drastic reduction of the number of cases needed to be examined under the complementary slackness conditions of equation (3.4). This can be seen in some simple examples (Papalambros and Li, 1983). Note that when an implicit function is used, a positive (negative) superscript sign for a variable indicates increasing (decreasing) function wrt that variable.

---

## EXAMPLE 3.1

Consider the problem

$$
\begin{aligned}
\text{minimize} \quad & f = (x_1 - 3)^2 + (x_2 - 3)^2 \\
\text{subject to} \quad & g_1 = -x_1 \leqslant 0 \\
& g_2 = -x_2 \leqslant 0 \\
& g_3 = x_1 + x_2 - 4 \leqslant 0
\end{aligned}
\tag{3.5}
$$

There are eight possible cases dictated by the complementary slackness conditions. The objective is nonmonotonic but we observe that if $x_i \geq 3, i = 1, 2$, there is no solution since $g_3$ is violated. Thus $x_i < 3$ for at least one $i, i = 1, 2$, and the objective is decreasing wrt at least one $i$, which then makes $g_3$ always active by the first monotonicity principle (MP1). The problem has now one degree of freedom and the solution can be obtained quickly using constrained derivatives (Wilde and Beightler, 1967). If $f_1$ is the reduced objective resulting from elimination of $x_1$ and $g_3$ we have

$$\mathrm{d}f_1/\mathrm{d}x_2 = (\partial f/\partial x_1)(\partial \phi/\partial x_2) + \partial f/\partial x_2 = 0 \tag{3.6}$$

where $\phi(x_2) \equiv x_1 = 4 - x_2$. Evaluation of equation (3.6) gives $x_1^* = 2, x_2^* = 2, f_o^* = 2$.

---

## EXAMPLE 3.2

Consider the problem

$$\begin{aligned} \text{minimize} \quad & f = (1/2)(x_1^2 + x_2^2 + x_3^2) - (x_1 + x_2 + x_3) \\ \text{subject to} \quad & g_1 = x_1 + x_2 + x_3 - 1 \leq 0 \\ & g_2 = 4x_1 + 2x_2 - 7/3 \leq 0 \\ & x_i \geq 0, i = 1, 2, 3 \end{aligned} \tag{3.7}$$

From the KKT conditions there are 32 cases to be examined. The objective appears nonmonotonic, but in fact $\partial f/\partial x_i = x_i - 1 \leq 0$ because $g_1$ implies $x_i \leq 1, i = 1, 2, 3$. Hence, by MP1 wrt $x_3$ we have two cases: if $x_3 = 1$ then $x_2 = x_1 = 0$ ($g_1$ is tight) and $f = -1/2$; if $x_3 < 1$, then $g_1$ is active. Defining $x_3 = 1 - x_1 - x_2 = \phi_1(x_1^-, x_2^-)$ we can calculate the constrained derivative of the reduced objective $f_1$ wrt one of the remaining variables, say $x_1$.

$$\partial f_1/\partial x_1 = (\partial f/\partial x_3)(\partial \phi_1/\partial x_1) + \partial f/\partial x_1 = (1/2)(4x_1 + 2x_2 - 2) \tag{3.8}$$

Now we observe the following: (a) if $4x_1 + 2x_2 < 2$, then $g_2$ is inactive with $f_1(x_1^-, x_2^-)$ and no solution exists; (b) if $4x_1 + 2x_2 = 2$, then $g_2$ is inactive and the remaining one-degree-of-freedom problem has the solution $x_1 = x_2 = x_3 = \frac{1}{3}$, $f = -\frac{5}{6}$; (c) if $4x_1 + 2x_2 > 2$, then the problem becomes $\{\min f_1(x_1^+, x_2)$, subject to $g_2(x_1^+, x_2^+) \leq 0$ with $x_1 \geq 0, x_2 \geq 0, x_1 + x_2 \leq 1\}$ and the solution is $x_1 = 0$, $x_2 > 1$ which is not permitted. The conclusion is a global optimum at $x_i = \frac{1}{3}, i = 1, 2, 3$.

---

## EXAMPLE 3.3

Consider the problem

$$\begin{aligned} \text{maximize} \quad & f = 0.0201 \, d^4 \, wn^2 \\ \text{subject to} \quad & g_1 = d^2 w - 675 \leq 0 \\ & g_2 = d - 36 \leq 0 \\ & g_3 = n - 125 \leq 0 \\ & g_4 = n^2 d^2 - (0.419)(10^7) \leq 0 \\ & d, n, w \geq 0 \end{aligned} \tag{3.9}$$

This problem involves maximizing the stored energy in a flywheel and is reduced to the above-mentioned form after some manipulations (Siddall, 1972). Constraint $g_1$ is active by MP1 wrt $w$. Note that the standard negative null form is obtained by using the objective $\{\text{minimize} - f\}$. Elimination of $w$ from the objective gives $f = (0.0201)(675)d^2n^2$. Clearly $g_4$ will be active except if $g_2, g_3$ combined impose a stricter bound on $d^2n^2$. For the given numbers this is not the case, so there are infinite solutions:

$$\max f^* = (5.68)(10^7), \ w_* = 675d_*^{-2}, \ n_* = 2047d_*^{-1}$$

$$16.4 \leqslant d_* \leqslant 36 \tag{3.10}$$

## EXAMPLE 3.4

Consider now the problem

$$
\begin{aligned}
\text{maximize} \quad & f = x_1^{-2} + x_2^{-2} + x_3^{-2} \\
\text{subject to} \quad & g_1 = 1 - x_1 - x_2 - x_3 \leqslant 0 \\
& g_2 = x_1^2 + x_2^2 - 2 \leqslant 0 \\
& g_3 = 2 - x_1 x_2 x_3 \leqslant 0 \\
& x_i \geqslant 0, i = 1, 2, 3
\end{aligned}
\tag{3.11}
$$

By MP1 constraint $g_2$ must be active providing upper bounds on $x_1$ and $x_2$. However, $x_3$ is unbounded from above since $f, g_1$, and $g_3$ are all decreasing wrt $x_3$. The problem has no solution unless the objective and/or constraint functions are appropriately modified by remodeling. Any nonredundant equality constraint would also serve. The usual practice in such cases is to add a simple inequality constraint, e.g. $x_3 \leqslant a, a > 0$. However, this constraint will be always active by construction, which means that the optimum is in fact determined by this artificial bound. In real problems, such information must be consciously considered because the optimum is essentially arbitrarily fixed by the modeler. We will explore this further in the linear actuator example.

### 3.2.5 Modeling decisions

It is good practice as one embarks on an optimization study to make a checklist of all the decisions that are required for developing a model. As an example, Table 3.1 shows what this checklist might look like for a structural optimization problem using finite elements for the analysis model (Papalambros, 1988b). This list is certainly not complete and it may change depending on the problem. With experience, the need for such lists may diminish. Yet even experienced users will benefit by going through a detailed itemization of all the decisions they make as they go. It is not uncommon for some overlooked decision to have a strong bearing on the optimization results, particularly when something goes wrong. Furthermore, such decisions can be increasingly automated with the use of artificial intelligence techniques; see, for example, Thomas (1990). This is a particularly attractive approach for developing quickly and accurately optimization

**Table 3.1** Some decisions related to modeling (finite element based models)

General
    model cost level (what is an 'expensive' model)
        size of model
            number of elements, constraints, and variables
        complexity of geometry
        preparation time for FEM model
    identification of problem type
        shape or size variables
    problem decomposition (reduction of dimensionality)
        substructuring
        separability of objective
        multilevel model
        variable linking
Objective
    single
    multiple
        transformation to single objective
            weight factors
Constraints
    selection of 'appropriate' constraints
    selection of parametric bounds
        hard and soft bounds
    constraint interaction, compatibility rules
    consistency (non-empty feasible domain)
Design variables
    selection
    transformation
    continuous or discrete values
        maximum number of discrete variable values
        maximum number of coordinate variables in an element
    starting point values (initial design)
Design parameters
    selection
    values for current application
    range of values
Submodel
    active set strategy (many decisions, part of them model related)
Program parameters related to model
    maximum number of nets
    maximum number of nodes
    maximum number of active elements
    maximum number of element variables
    maximum number of composite stacks
    maximum number of master variables
    maximum number of layers in stack
Scaling
    functions
    variables
Finite differencing
    step size for material angle variables
    step size for coordinate variables
    step size for thickness variables

models within a specific domain, for example, optimal trusses or plate girders in bridge construction (Adeli and Balasubramanyam, 1988a; Adeli and Mak, 1988). For further discussion on the use of expert systems in structural design see Adeli and Balasubramanyam (1988b).

## 3.3 MODELING EXAMPLE: A LINEAR ACTUATOR

In this extensive example we will go through various modeling steps using several of the ideas mentioned earlier and some new ones. The procedure followed is typical for models with relatively small number of variables and explicit functions, but can be used in other situations to advantage.

### 3.3.1 Model setup

The example presented here is part of a larger classroom project dealing with the optimal design of a drive screw linear actuator (Alexander and Rycenga, 1990). The drive screw is part of a power module assembly that converts rotary to linear motion, so that a given load is linearly oscillated at a specified rate. Such a device is used is some household appliances. The assembly consists of an electric motor, drive gear, pinion (driven) gear, drive screw, load-carrying nut, and a chassis providing bearing surfaces and support. The present example addresses only the design of the drive screw, schematically shown in Fig. 3.1.

The objective function in the design model is to minimize product cost consisting of material and manufacturing costs. Machining costs for a metal drive screw or injection molding costs for a plastic one are considered fixed for relatively small changes in the design, hence only material cost is taken as the objective to minimize, namely

$$f_o = (C_m \pi/r)(d_1^2 L_1 + d_2^2 L_2 + d_3^2 L_3) \qquad (3.12)$$

Here $C_m$ is the material cost ($\$ \text{lb}^{-1}$), $d_1, d_2$ and $d_3$ are the diameters of gear–drive screw interface, threaded and bearing surface segments, respectively, $L_1, L_2$ and $L_3$ being the respective segment lengths.

There are operational, assembly, and packaging constraints. For strength against bending during assembly we set

$$Mc_1/I \leqslant \sigma_{\text{all}} \qquad (3.13)$$

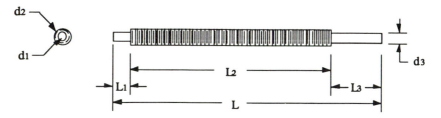

**Fig. 3.1** Schematic of drive screw design.

where the bending moment $M = F_a L/2$, $F_a$ being the force required to snap the drive screw into the chassis during assembly, and $L$ being the total length of the component:

$$L = L_1 + L_2 + L_3 \qquad (3.14)$$

Furthermore, $c_1 = d_2/2$, $I = \pi d_2^4/64$ is the moment of inertia, and $\sigma_{all}$ is the maximum allowable bending stress for a given material.

During operation, a constraint against fatigue failure in shear must be imposed:

$$K T c_3/J \leqslant \tau_{all} \qquad (3.15)$$

Here, $K$ is a stress concentration factor, $T$ is the applied torque, $C_3 = d_1/2$, $J = \pi d_1^4/32$ is the polar moment of inertia, and $\tau_{all}$ is the maximum allowable shear stress. The torque is computed from the equation

$$T = T_m C_2 N_S/N_m \qquad (3.16)$$

where $T_m$ is the motor torque, $c_2 = 1/16 \, (\text{lb oz}^{-1})$ is a conversion factor, and $N_s$ and $N_m$ are the number of teeth on the screw (driven) and motor (drive) gear, respectively.

To meet the specified linear cycle rate of oscillation, a speed constraint is imposed:

$$c_4 N_m S_m/N_s N_T \leqslant S \qquad (3.17)$$

where $c_4 = 60^{-1}$ (number of threads $\text{rev}^{-1}$)$(\text{min s}^{-1})$ is a conversion factor, $S_m$ is the motor speed ($\text{rev min}^{-1}$), $N_T$ is the number of threads per inch, and $S$ is the specified linear cycle rate (in $\text{s}^{-1}$).

In order for the screw to operate in a drive mode the following constraint must be satisfied (Juvinall, 1983).

$$\frac{W d_2}{2} \frac{\pi f d_2 + N_T^{-1} \cos \alpha_n}{\pi d_2 \cos \alpha_n - N_T^{-1} f} \leqslant T \qquad (3.18)$$

Here $W$ is the drive screw load, $f$ is the friction coefficient, $N_T^{-1}$ is the lead of screw threads, and $\alpha_n$ is the thread angle measured in the normal plane. There is also an upper bound on the number of threads per inch imposed by mass production considerations,

$$N_T \leqslant 24 \qquad (3.19)$$

From gear design considerations, particularly avoidance of interference, limits on the numbers of gear teeth are imposed:

$$N_m \geqslant 8 \qquad N_s \leqslant 52 \qquad (3.20)$$

Finally, there are some packaging and geometric considerations that impose constraints:

$$8.75 \leqslant L_1 + L_2 + L_3 \leqslant 10.0 \qquad (3.21)$$

$$7.023 \leqslant L_2 \leqslant 7.523 \qquad (3.22)$$

$$1.1525 \leqslant L_3 \leqslant 1.6525 \qquad (3.23)$$

$$d_1 \leqslant d_2, d_3 \leqslant d_2, d_2 \leqslant 0.625 \qquad (3.24)$$

Note that several assumptions were invoked in the model above: manufacturing costs remain fixed; a high volume production is planned; Standard Unified Threads are used; the assembly force for the drive screw is concentrated at the midpoint; frictional forces are considered only between threads and load nut, and all others are assumed negligible.

### 3.3.2 Model validity constraints

During the early stages of developing a mathematical optimization model many assumptions are made in order to obtain reasonably simple expressions for the objective and constraint functions. One must always check whether subsequent results from optimization conform to these assumptions, lest they are violated. This would indicate that the model used is inappropriate for the optimal design obtained and the optimization results are at least suspect and possibly erroneous. The remedy is usually a more accurate, probably also more complicated, mathematical model of the phenomenon under question. For example, equation (3.13) is valid only if the length/diameter ratio is more than 10.

### 3.3.3 Material choice as a parameter

A final observation on the initial drive screw model is that a significant trade-off exists on the choice of material: stainless steel vs plastic. A steel screw will have higher strength and be smaller in size but will require secondary processing, such as rolling of threads and finishing of bearing surfaces. A plastic screw would be made by injection molding in a one-step process that is cheaper but more material would be used because of lower strength, plastic having a higher cost per pound than steel. Specialty plastics with high strength would be even more expensive and less moldable. Thus the choice of material must be based on a model that contains more information than the current one. The constant term representing manufacturing costs should be included in the objective. Indeed a more accurate cost objective should include capital investment costs for manufacturing.

Nevertheless, substantial insight can be gained from the present model if we include material as a parameter; in fact each material is represented by four parameters $C_m$, $\sigma_{all}$, $\tau_{all}$, $f$. In the model analysis that follows we keep these parameters in the model with their symbols, rather than giving numerical values insofar as possible. The goal is to derive as many additional results as possible independently of the material used. This will substantially facilitate a post-optimal parametric study on the material. It would be much more difficult to treat material as a variable, because then we would have four additional variables with discrete values and implicitly linked, perhaps through a table of material properties. As mentioned earlier, this would destroy the continuity assumed in nonlinear programming formulations.

### 3.3.4 Standard null form

The model is now summarized in the negative null form, all parameters represented by their numerical values (Table 3.2), except for material parameters.

**Table 3.2** List of parameters (material values for stainless steel)

| | |
|---|---|
| $C_m$ | material cost (\$/lb) |
| $f$ | friction coefficient (0.35 for steel on plastic) |
| $F_a$ | force required to snap drive screw into chassis during assembly (6 lbf) |
| $K$ | stress concentration fator (3) |
| $L_1$ | length of gear/drive screw interface segment (0.405 in) |
| $S$ | linear cycle rate (0.0583 in s$^{-1}$) |
| $S_m$ | motor speed (300 rpm) |
| $T_m$ | motor torque (2 in oz) |
| $W$ | drive screw load (3 lb) |
| $a_n$ | thread angle in normal plane (60°) |
| $\sigma_{all}$ | maximum allowable bending stress (20 000 lbf in$^{-2}$) |
| $\tau_{all}$ | maximum allowable shear stress (22 000 lbf in$^{-2}$) |

All 'intermediate' variables defined through equalities are eliminated together with the associated equality constraints by direct substitution. This should be always done when possible, in order to arrive at a model with only inequality constraints, thus facilitating subsequent monotonicity analysis. This is a model reduction step, since the number of design variables is reduced. Note, however, that this may not be always a model simplification step, as the resulting expressions may become more complex with undetermined monotonicities. Some judgement must be exercised here. Occasionally, 'directing' an equality may be useful (see further below) in avoiding direct elimination. In the model below the intermediate variables $M, L, T, I, c_1, c_3$ and $J$ together with the corresponding defining equalities have been eliminated. The variables are $d_1, d_2, d_3, L_2, L_3, N_m$, $N_S, N_T$.

**Model 1**

minimize $f_0 = (C_m \pi/4)(0.405 d_1^2 + L_2 d_2^2 + L_3 d_3^2)$
subject to

$g_1 = 38.88 + 96 L_2 + 96 L_3 - \pi \sigma_{all} d_2^3 \leqslant 0$

$g_2 = 6 N_s / N_m - \pi \tau_{all} d_1^3 \leqslant 0$

$g_3 = 8.345 - L_2 - L_3 \leqslant 0 \qquad g_4 = -9.595 + L_2 + L_3 \leqslant 0$

$g_5 = L_2 - 7.523 \leqslant 0 \qquad g_6 = 7.023 - L_2 \leqslant 0$

$g_7 = L_3 - 1.6525 \leqslant 0 \qquad g_8 = 1.1525 - L_3 \leqslant 0$

$g_9 = d_2 - 0.625 \leqslant 0 \qquad g_{10} = d_3 - d_2 \leqslant 0$

$g_{11} = d_1 - d_2 \leqslant 0 \qquad g_{12} = 5 N_m / N_S - 0.0583 N_T \leqslant 0$

$$g_{13} = 1.5 d_2 \frac{f \pi d_2 + 0.5 N_T^{-1}}{0.5 \pi d_2 - f N_T^{-1}} - 0.125 N_S / N_m \leqslant 0$$

$g_{14} = N_T - 24 \leqslant 0 \qquad g_{15} = 8 - N_m \leqslant 0$

$g_{16} = N_S - 52 \leqslant 0 \qquad\qquad\qquad\qquad\qquad\qquad (3.25)$

As there are no equality constraints, there are eight degrees of freedom corresponding to the eight design variables. Three of these variables, $N_m, N_T$ and $N_S$, must take integer values, so this problem is in fact a mixed continuous–integer variable nonlinear programming problem and standard numerical NLP methods will not work. We will see later how this is dealt with in the particular example.

### 3.3.5 Feasibility checking

Before embarking on analyzing the model, it is a good idea to check that the feasible domain is not empty, i.e. there exists at least one proven feasible point. From a mathematical viewpoint this may be a hard problem (possibly as hard as the optimization itself), but from an engineering viewpoint past experience can be a guide. In the drive screw example, an existing design using stainless steel has the following values: $d_1 = 0.1875$ in, $d_2 = 0.3125$ in, $d_3 = 0.2443$ in, $L_2 = 7.273$ in, $L_3 = 1.4025$ in, $N_m = 8, N_S = 48, N_T = 18$. The design is feasible with an objective function value of $f_0 = 0.635C_m$. Now we can proceed knowing that an optimization attempt is possible.

### 3.3.6 Monotonicity

Looking at model 1 one notes that most constraints were written in a form that requires no divisions. This is always advisable, since in subsequent numerical processing a denominator may become zero and cause an abrupt termination by overflow error. This can happen even if the imposed constraints exclude the relevant variable values, because many numerical algorithms will temporarily operate in the infeasible domain. In model 1, constraint $g_{13}$ has not been rewritten yet because of concern that this might obscure its monotonicity wrt $d_2$ and $N_T$. Let us examine this more carefully. Assuming a strictly positive denominator of the first term in $g_{13}$ we multiply both sides by it and collect terms:

$$g_{13}: 1.5f\pi d_2^2 + 0.75N_T^{-1}d_2 - 0.0625\pi(N_S/N_m)d_2 + 0.125f(N_S/N_m)N_T^{-1} \leqslant 0 \tag{3.26}$$

Clearly $g_{13}$ decreases wrt $N_T$ but is nonmonotonic wrt $d_2$. In fact,

$$\partial g_3/\partial d_2 = 3\pi f d_2 + 0.75N_T^{-1} - 0.0625\pi N_S/N_m \tag{3.27}$$

which can be positive or negative depending on the variable values. All remaining functions in the constraints have obvious monotonicities.

### 3.3.7 Variable transformation

The model can be further simplified by a variable transformation. We observe that the two variables $N_S, N_m$ appear together as a ratio everywhere except in the simple bounds $g_{15}, g_{16}$. We can define a new variable $R$,

$$R = N_m/N_S \tag{3.28}$$

which indeed is the reduction ratio of the gear drive, and eliminate variable $N_m$

using $N_m = RN_S$. The new model, including the reformulated constraint $g_{13}$, is as follows.

**Model 2**

$$\min f_0 = (C_m\pi/4)(0.405d_1^2 + L_2 d_2^2 + L_3 d_3^2)$$

subject to

$$g_1 = 38.88 + 96L_2 + 96L_3 - \pi\sigma_{all}d_2^3 \leqslant 0$$

$$g_2 = 6 - \pi\tau_{all}Rd_1^3 \leqslant 0$$

$$g_3 = 8.345 - L_2 - L_3 \leqslant 0 \qquad g_4 = -9.595 + L_2 + L_3 \leqslant 0$$

$$g_5 = L_2 - 7.523 \leqslant 0 \qquad g_6 = 7.023 - L_2 \leqslant 0$$

$$g_7 = L_3 - 1.6525 \leqslant 0 \qquad g_8 = 1.1525 - L_3 \leqslant 0$$

$$g_9 = d_2 - 0.625 \leqslant 0 \qquad g_{10} = d_3 - d_2 \leqslant 0$$

$$g_{11} = d_1 - d_2 \leqslant 0 \qquad g_{12} = 5R - 0.0583 N_T \leqslant 0$$

$$g_{13} = 1.5\pi f R d_2^2 + 0.75 R N_T^{-1} d_2 - 0.0625\pi d_2 + 0.125 f N_T^{-1} \leqslant 0$$

$$g_{14} = N_T - 24 \leqslant 0 \qquad g_{15} = 8 - RN_S \leqslant 0$$

$$g_{16} = N_S - 52 \leqslant 0 \tag{3.29}$$

Note that the requirement of integer values for $N_m$ is now converted to one of rational values for $R$.

### 3.3.8 Repairing a model

Model 2 is now used to perform the first cycle of monotonicity analysis. The monotonicity table is a convenient tool to do this: Table 3.3. The columns are the design variables and the rows are the objective and constraint functions, the entries in the table being the monotonicities of each function with respect to each variable. Positive (negative) sign indicates increasing (decreasing) function, U indicates undetermined or unknown monotonicity. An empty entry indicates that the function does not depend on the respective variable, so the table acts also as an incidence table. (Items in parentheses will be explained in the next subsection.)

Monotonicity principles can be quickly applied by inspection using the monotonicity table. Looking at Table 3.3, by MP1 wrt $d_3$ we see that model 2 is not well constrained because no lower bound exists for $d_3$. Note that $d_3 > 0$ is not an appropriate bound because of the strict inequality. If the model were treated numerically as is, no convergence would occur if the algorithm was successful or an erroneous result would be found if the algorithm was led astray. Examining the engineering meaning of this model deficiency we see that an adequate thrust surface must be provided to keep the shaft from wearing into the bearing support, so we accept the simple remedy of adding a new constraint

$$g_{17} = 0.1875 - d_3 \leqslant 0 \tag{3.30}$$

Poor boundedness is cause for concern, so the above deficiency triggers also

**Table 3.3** Monotonicity table for model 2 (with model repairs in parentheses)

| Functions | Variables | | | | | | | |
|---|---|---|---|---|---|---|---|---|
| | $d_1$ | $d_2$ | $d_3$ | $L_2$ | $L_3$ | $R$ | $N_S$ | $N_T$ |
| $f_0$ | + | + | + | + | + | | | |
| $g_1$ | | − | | + | + | | | |
| $g_2$ | − | | | | | − | | |
| $g_3$ | | | | − | − | | | |
| $g_4$ | | | | + | + | | | |
| $g_5$ | | | | + | | | | |
| $g_6$ | | | | − | | | | |
| $g_7$ | | | | | + | | | |
| $g_8$ | | | | | − | | | |
| $g_9$ | | + | | | | | | |
| $(g_{10})$ | | − | + | | | | | |
| $g_{11}$ | + | − | | | | | | |
| $g_{12}$ | | | | | | + | | − |
| $g_{13}$ | | U | | | | + | | − |
| $g_{14}$ | | | | | | | | + |
| $g_{15}$ | | | | | | − | − | |
| $g_{16}$ | | | | | | | + | |
| $(g_{17})$ | | − | | | | | | |
| $(g_{18})$ | | − | + | | | | | − |

a closer examination of constraint $g_{10}$, the only other one containing $g_3$: $d_3 \leqslant d_2$. On reflection, constraint activity for $g_{10}$ should not be allowed, i.e. $d_2 > d_3$. Examining the geometry of the screw more closely a new constraint is discovered:

$$g_{18}: d_2 \geqslant d_3 + 1.2990/N_T \tag{3.31}$$

### 3.3.9 Optimality rules

At this point we decide to add constraints $g_{17}(d_3^-)$ and $g_{18}(d_2^-, d_3^+, N_T^-)$ to the model and delete $g_{10}$ as redundant. These changes are shown in parentheses in Table 3.2 and applying monotonicity analysis to this model we now obtain the following results, which represent necessary rules for optimality.

(R1) By MP1 wrt $d_1$, $g_2$ is active.

(R2) By MP1 wrt $d_2$, at least one constraint from the set $\{g_1, g_{11}, g_{13}\}$ is active.

(R3) By MP1 wrt $L_2$, at least one constraint from the set $\{g_3, g_6\}$ is active.

(R4) By MP1 wrt $L_3$, at least one constraint from the set $\{g_3, g_8\}$ is active.

(R5) By MP2 wrt $R$, either all constraints $g_2, g_{12}, g_{13}$ and $g_{15}$ are inactive or at least one from each of the sets $\{g_2, g_{15}\}$, $\{g_{12}, g_{13}\}$ is active.

(R6) By MP2 wrt $N_S$, either $g_{15}$ and $g_{16}$ are both active or they are both inactive.

(R7) By MP2 wrt $R$, if $g_2$ is active then at least one of $\{g_{12}, g_{13}\}$ is active. Then, by MP2 wrt $N_T$ and (R1), $g_{14}$ is active.

The original eight degrees of freedom have now been reduced to five, because of the identified active constraints:

$$g_2 : \pi \tau_{\text{all}} R d_1^3 = 6$$

$$g_{14} : N_{\text{T}} = 24$$

$$g_{17} : d_3 = 0.1875 \tag{3.32}$$

The remaining rules above give only conditional activity and in order to identify a single constraint as active dominance arguments are required. We will proceed with these later below. Thus, because of (R1), (R5) must be modified as

(R5′) By MP2 wrt $R$, at least one constraint from the set $\{g_{12}, g_{13}\}$ is active.

This interaction was also used in deriving (R7).

### 3.3.10 Active constraint elimination

The three active constraints, equations (3.32), are used to eliminate three variables from model 2, namely $d_1, d_3$ and $N_{\text{T}}$. There are two reasons for this. One is that monotonicity analysis on the new reduced model may reveal additional activity requirements. Another is that dominance arguments will be simpler in the reduced model. Which variables to eliminate is a judicious choice, based on what may be algebraically simpler and what may be a desirable form of the reduced model. The new model is as follows.

**Model 3**

minimize $f_0 = 0.25 \pi C_{\text{m}} [0.405 (6/\pi \tau_{\text{all}} R)^{2/3} + L_2 d_2^2 + L_3 (0.1875)^2]$

subject to

$$g_1 = 38.88 + 96 L_2 + 96 L_3 - \pi \sigma_{\text{all}} d_2^3 \leqslant 0$$

$$g_3 = 8.345 - L_2 - L_3 \leqslant 0 \qquad g_4 = -9.595 + L_2 + L_3 \leqslant 0$$

$$g_5 = L_2 - 7.523 \leqslant 0 \qquad g_6 = 7.023 - L_2 \leqslant 0$$

$$g_7 = L_3 - 1.6525 \leqslant 0 \qquad g_8 = 1.1525 - L_3 \leqslant 0$$

$$g_9 = d_2 - 0.625 \leqslant 0 \qquad g_{11} = (6/\pi \tau_{\text{all}} R)^{1/3} - d_2 \leqslant 0$$

$$g_{12} = R - 0.2798 \leqslant 0$$

$$g_{13} = 1.5 \pi f R d_2^2 + 0.0313 R d_2 - 0.0625 \pi d_2 + 0.0052 f \leqslant 0$$

$$g_{15} = 8 - R N_{\text{s}} \leqslant 0 \qquad g_{16} = N_{\text{s}} - 52 \leqslant 0$$

$$g_{18} = 0.2416 - d_2 \leqslant 0 \tag{3.33}$$

The monotonicity table for model 3 is shown in Table 3.4. No new results are obtained from this table. So dominance arguments must be sought in order to clarify the previously stated conditional activities and to obtain further model reduction.

**Table 3.4** Monotonicity table for model 3

| Functions | Variables | | | | |
|---|---|---|---|---|---|
| | $d_2$ | $L_2$ | $L_3$ | $R$ | $N_S$ |
| $f_0$ | + | + | + | − | |
| $g_1$ | − | + | + | | |
| $g_3$ | | − | − | | |
| $g_4$ | | + | + | | |
| $g_5$ | | + | | | |
| $g_6$ | | − | | | |
| $g_7$ | | | + | | |
| $g_8$ | | | − | | |
| $g_9$ | + | | | | |
| $g_{11}$ | − | | | − | |
| $g_{12}$ | | | | + | |
| $g_{13}$ | U | | | + | |
| $g_{15}$ | | | | − | − |
| $g_{16}$ | | | | | + |
| $g_{18}$ | − | | | | |

### 3.3.11 Activity map and dominance

Consider constraints $g_3$, $g_6$ and $g_8$ and the derived rules (R3) and (R4). An **activity map**, as shown in Fig. 3.2, can assist in dominance analysis. In this map all possible activity combinations for the three constraints are examined. Only constraint numbers are shown for simplicity, the overbar on a number indicating an inactive constraint while a plain number indicates an active one. The crosshatched areas indicate combinations that are not possible. In Fig. 3.2, three combinations are excluded since they violate rules (R3) and/or (R4), as indicated. The combination 368 is excluded by the maximal activity principle (Papalambros and Wilde, 1988), which basically says that the number of active constraints cannot exceed the number of variables in them. Finally, if $g_3$ is inactive then both $g_6$ and $g_8$ must be active giving $L_2 = 7.023$, $L_3 = 1.1525$, and $L_2 + L_3 = 8.1755$, which violates $g_3$. Hence this case is infeasible and excluded. From Fig. 3.2, it is now obvious that $g_3$ must be active irrespective of the activity of $g_6$ and $g_8$. This also makes $g_4$ and at least three constraints from the set $\{g_5, g_6, g_7, g_8\}$ be inactive. The conditional inactivity can be resolved easily as $L_2, L_3$ appear only in a small part of the model.

Consider eliminating $L_2 (= 8.345 - L_3)$ from the model. The submodel containing $L_2$ and $L_3$ becomes now

$$\text{minimize } f_0 = 0.25\pi C_m \{0.405(6/\pi\tau_{all}R)^{2/3} + 8.345d_2^2 + L_3[(0.1875)^2 - d_2^2]\}$$

$$g_1 = 38.88 + 96(8.345) - \pi\sigma_{all}d_2^3 \leqslant 0$$

$$g_5 = 0.822 - L_3 \leqslant 0 \qquad g_6 = L_3 - 1.322 \leqslant 0$$

$$g_7 = L_3 - 1.6525 \leqslant 0 \qquad g_8 = 1.1525 - L_3 \leqslant 0 \qquad (3.34)$$

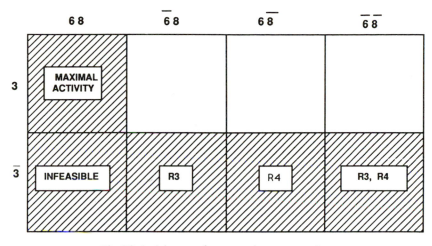

**Fig. 3.2** Activity map for constraints $g_3$, $g_6$ and $g_8$.

In the objective, the monotonicity wrt $L_3$ is determined by the sign of the quantity

$$(0.1875)^2 - d_2^2 \tag{3.35}$$

which is negative since $d_2 \geqslant 0.2416$ from $g_{18}$. Hence, the objective is decreasing wrt $L_3$ and an upper bound is required by MP1. Constraint $g_6$ is obviously the dominant and active one, so

$$L_{3*} = 1.322, L_{2*} = 7.023 \tag{3.36}$$

and $g_5, g_7, g_8$ are inactive.

The above results lead to yet another further reduced model with only three degrees of freedom.

**Model 4**

minimize $f_0 = 0.25\pi C_m [0.405(6/\pi\tau_{all} R)^{2/3} + 7.023 d_2^2 + 1.322(0.1875)^2]$

subject to

$$g_1 = 840 - \pi\sigma_{all} d_2^3 \leqslant 0 \qquad\qquad g_9 = d_2 - 0.625 \leqslant 0$$

$$g_{11} = (6/\pi\tau_{all} R)^{1/3} - d_2 \leqslant 0 \qquad g_{12} = R - 0.2798 \leqslant 0$$

$$g_{13} = 1.5\pi f R d_2^2 + 0.0313 R d_2 - 0.0625\pi d_2 + 0.0052 f \leqslant 0$$

$$g_{15} = 8 - R N_S \leqslant 0 \qquad\qquad g_{16} = N_S - 52 \leqslant 0$$

$$g_{18} = 0.2416 - d_2 \leqslant 0 \tag{3.37}$$

The monotonicity table for this model is shown in Table 3.5. Rules (R2), (R5′) and (R6) are still the only results derived from monotonicity principles.

At this point, the unknown monotonicity of $g_{13}$ wrt $d_2$ prevents us from continuing the reduction process. Indeed,

$$\partial g_{13}/\partial d_2 = 3\pi f R d_2 + 0.0313 R - 0.0625\pi \tag{3.38}$$

**Table 3.5** Monotonicity table for model 4

|  | Variables | | |
| --- | --- | --- | --- |
| Functions | $d_2$ | $R$ | $N_S$ |
| $f_0$ | + | − | |
| $g_1$ | − | | |
| $g_9$ | + | | |
| $g_{11}$ | − | − | |
| $g_{12}$ | | + | |
| $g_{13}$ | U | + | |
| $g_{15}$ | − | | − |
| $g_{16}$ | | | + |
| $g_{18}$ | − | | |

and for $g_{13}$ to be increasing wrt $d_2$ we would need to have (for $f = 0.34$)

$$d_2 \geqslant 0.0595 R^{-1} - 0.0033 \triangleq D_2(R^-) \tag{3.39}$$

for the intire feasible range of $d_2, R$. We note that

$$\max D_2(R^-) = 0.0595 R_{\min}^{-1} - 0.0033 = 0.0595 N_{S\max}/8 - 0.0033$$
$$= (0.0595)(52)/8 - 0.0033 = 0.3983 \tag{3.40}$$

This last number falls within the known feasible range of $d_2, 0.2416 \leqslant d_2 \leqslant 0.625$, so $g_{13}$ appears really nonmonotonic in the feasible domain.

### 3.3.12 Parametric models and case decomposition

The idea of **regional monotonicity** could be used here, i.e. try to identify in what interval of values of $d_2$ is $g_{13}$ monotonic and examine each case separately, comparing the results at the end. This would be unnecessarily complicated here. Instead, a simple problem decomposition can be applied: case A with $g_{13}$ inactive, and case B with $g_{13}$ active. We can examine these two cases separately and compare the results.

Before we proceed with the cases, it is instructive to recast model 4 in a simplified parametric form, by introducing the parameters $K_0', K_i, i = 0, \ldots, 4$, and rearranging (Table 3.6). The revised model is as follows. Note that all parameters $K_i$ relate to material properties.

**Model 5**

minimize $f_0 = K_0 R^{2/3} + K_0'[7.023 d_2^2 + 1.322(0.1875)^2]$

subject to

$$g_1 = K_1 - d_2 \leqslant 0 \qquad g_9 = d_2 - 0.625 \leqslant 0$$
$$g_{11} = K_2 R^{1/3} - d_2 \leqslant 0 \qquad g_{12} = R - 0.2798 \leqslant 0$$
$$g_{13} = K_3 R d_2^2 + 0.0313 R d_2 - 0.1963 d_2 + K_4 \leqslant 0$$

**Table 3.6** Parameter definitions for model 5

$$K'_0 = 0.25\pi C_m$$
$$K_0 = K'_0 (0.405)(6/\pi\tau_{all})^{2/3}$$
$$K_1 = (840/\pi\sigma_{all})^{1/3}$$
$$K_2 = (6/\pi\tau_{all})^{1/3}$$
$$K_3 = 1.5\pi f$$
$$K_4 = 0.0052 f$$

$$g_{15} = 8 - RN_S \leqslant 0 \qquad\qquad g_{16} = N_S - 52 \leqslant 0$$

$$g_{18} = 0.2416 - d_2 \leqslant 0 \tag{3.41}$$

Consider now case A with $g_{13}$ inactive. Then $g_{12}$ is active from rule (R5') and $R_* = 0.2798$. From rule (R2) we now have

$$d_{2*} = \max(K_1, K_2 R_*^{1/3}, 0.2416) \tag{3.42}$$

and from rule (R5) we have

$$8/R_* \leqslant N_{S*} \leqslant 52 \tag{3.43}$$

Note that one degree of freedom remains at the optimum, as $N_m, N_S$ can be selected to give the largest rational number not exceeding 0.2798 and satisfying the range in equation (3.43). Any solution thus obtained must be checked that it satisfies the remaining inactive constraints. This **feasibility** check is frequently overlooked in the application of monotonicity analysis, leading to erroneous conclusions.

Next, consider case B with $g_{13}$ active. Rules (R2) and (R5') are satisfied and no new activity results can be obtained. Note that now, locally, $g_{13}$ must be decreasing wrt $d_2$, i.e. $g_{13}(d_2^-, R^+) \leqslant 0$, while $f_0(d_2^+, R^-)$. An implicit solution of $g_{13}$ gives $d_2 = \varphi_{13}(R^+)$ and substitution in the objective gives $f_0(d_2^+, R^-) = f_0[\varphi_{13}^+(R^+), R^-] = f_0(R)$ with $R$ having unclear monotonicity. There are two degrees of freedom left, but a one-dimensional search in $d_2$ would suffice if $g_{13}$ is solved explicitly for $R$ and the objective is expressed as a function of $d_2$ only. Constraints $g_{15}$ and $g_{16}$ could be replaced by

$$R \geqslant 8N_S^{-1} \geqslant 8/52 = 0.1538 \tag{3.44}$$

There is a lingering concern, though, regarding the physical meaning of $g_{13}$ being active. Essentially, friction forces and applied forces on an equivalent inclined plane would be equal and motion would be impending. This would not represent a stable design for the lead screw, albeit possibly an optimal one. The designer must then examine the implications on the appropriateness of the model and/or the parameter values selected. Also, satisfaction of model validity for the beam stress formula (3.13) should be checked. The results obtained numerically using SQP code NLPQL (Schittkowski, 1984) with $C_m = 10.0$ and continuous values for the integer variables indicate constraints $\{g_2, g_3, g_6, g_{14}, g_{17}\}$ and $\{g_{12}, g_{18}\}$ are active. The activity of the first five was the one discovered *a priori* by analysis. Constraints $\{g_1, g_4, g_5, g_7, g_8, g_9, g_{10}, g_{11}, g_{13}, g_{15}, g_{16}\}$ are inactive.

### 3.3.13  Directing an equality

It is interesting to note that $g_{18}$ is active in the final solution. Indeed, according to Spotts (1985) the required relation is an equality:

$$h_{10}: d_3 = d_2 - 1.2990/N_T \qquad (3.45)$$

rather than the inequality (3.31). Instead of using $g_{18}$ in the model we could have used $h_{10}$, which would prevent direct application of MP1.

However, an equality constraint can be viewed as an active inequality that has been 'properly directed' (Papalambros and Wilde, 1988). One way to determine such a direction is to use the monotonicity principles. Ignoring equation (3.30) for the moment, and with equation (3.45) replacing $g_{10}$, we see that by MP1 wrt $d_3$ a lower bound is required. Then $h_{10}$ can provide this bound if directed as $d_3 \geqq d_2 - 1.2990 N^{-1}$ or $d_2 - 1.2990 N_T^{-1} - d_3 \leqq 0$.

Not much inference is possible when the directed $h_{10}$ is used in the model instead of $g_{18}$. One could proceed by assuming $h_{10}$ inactive and include $g_{17}$ which would then be active. This would take us essentially through the same steps as before, checking $h_{10}$ for violation in the final solution. Interestingly, numerical results obtained for such a scenario using NLPQL indicate $h_{10}$ is satisfied in the final solution with a zero multiplier value.


## 3.4  MODELING AUTOMATION

Traditional numerical solutions rely almost exclusively on iterative use of local information, such as values of functions, gradients or composition of locally active sets. Knowledge accumulated during such an iterative process may be used heuristically to speed up the solution process, for example, as discussed in Arora and Baenzinger (1986) and Adeli and Balasubramanyam (1988a, b). Analytical techniques aiming at active constraint identification, such as monotonicity analysis, use primarily **global** information, which is true for all points in the design space. A production system using rules for global information processing was first introduced in Li and Papalambros (1985a). Combined qualitative and quantitative reasoning for optimization in an AI environment has also been discussed in Agogino and Almgren (1987), Hammond and Johnson (1987, 1988), Hansen, Jaumard and Lu (1988a, b), and Watton and Rinderle (1991). The need for developing new types of algorithms using both global and local knowledge was explored in Li (1985) and Li and Papalambros (1988). Artificial intelligence (AI) techniques, including symbolic manipulations, have been proposed as modeling tools; see, for example, Li and Papalambros (1985b) and Choy and Agogino (1986).

One possibility offered by the use of AI techniques in modeling is to generate automatically global information about constraint activity and model boundedness, which can then be given directly to a global or local–global strategy, as described in the above references. This will remove, at least in part, the difficulty that relatively unskilled users may encounter in discovering rigorous global knowledge. Moreover, it eliminates tedious procedures prone to errors, when manually performed. The system PRIMA (production system for implicit elimination in monotonicity analysis) was implemented in the OPS5/OPS87 production system development tool (Forgy, 1981).

In PRIMA (Rao and Papalambros, 1987) the monotonicity table forms the very basic abstract data structure in the problem domain. Rules based on the monotonicity principles are applied to obtain **facts** about the design mode. These facts, called **state predicates**, being derived from necessary conditions, must be satisfied by every optimum solution, called a **state**. As these facts are derived, the system automatically generates LISP-like macros in the form of predicate functions, each of which should evaluate to true at every optimal design or state. In the presence of monotonicity, additional facts can be derived after one or more active constraints have been implicitly eliminated from the model (together with one or more corresponding variables). The result is again a new derived monotonicity table that may generate new state predicates. This process continues until all active constraints are eliminated, or no useful monotonicities remain.

The primary task, then, is to search an 'implicit elimination tree' (Fig. 3.3), where (i) visiting a node, i.e. monotonicity table, consists of applying mathematically rigorous necessary conditions implemented in the form of **rules,** (ii) the tree arcs (connecting one node to another) correspond to the process of implicit elimination, and (iii) the choice of **pivots** used in elimination (specific variable and constraint that will be eliminated) corresponds to selective expansion of nodes and is based on heuristic rules. This classification of rigorous rules (at

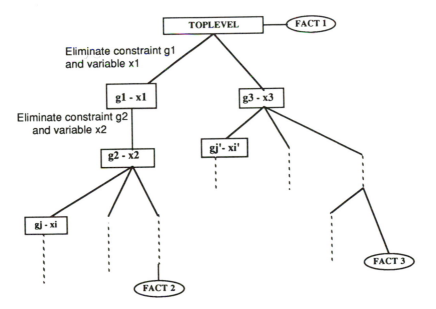

**Fig. 3.3** Generic implicit elimination tree in PRIMA. The nodes $g_1 - x_1$ etc. are reduced monotonicity tables obtained by implicit elimination. Facts 1, 2, 3 etc. could typically be obtained at any node in this tree.

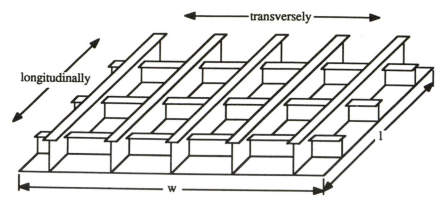

**Fig. 3.4** Ship bottom structure.

the nodes) and heuristic rules (at the arcs) can be considered a characteristic of this problem domain.

The model of the desired automation procedure is as follows. A tree data structure, called the **implicit elimination tree**, Fig. 3.3, is built in the space of feasible designs. Each node is a monotonicity table containing all the boundedness and constraint activity information that characterizes it. The root of the tree is the given top-level monotonicity table. The arcs of the tree structure correspond to a pivot used during the implicit elimination process. The root node has a degree of freedom (dof) greater than or equal to unity, which might decrease if more constraints are identified as active using MP1 or MP2. The terminal nodes along a depth-first search path correspond to the following cases: (i) the dof has been reduced to zero (i.e. the solution is constraint bound in a non-empty feasible domain); (ii) all active constraints have been eliminated in that branch; (iii) the heuristics do not provide a pivot, the particular path is deemed unprofitable, and backtracking must be performed. Full or partial traversal of this implicit elimination tree yields state predicates embodying global information.

The heuristic rules used in PRIMA are discussed in detail in Papalambros (1988). The main function of the rules is to guide the search through an implicit elimination tree in the most profitable generate-and-test manner, i.e. to make good pivot selections and to control backtracking. The use of an AI approach here offers the usual advantages of having a flexible representation of heuristic knowledge and an efficient data structure for repeated list processing.

An example application is repeated here from Rao and Papalambros (1991) (a revised version of Rao and Papalambros (1987)). The objective is to minimize the production cost of a ship bottom structure modeled as a structural grillage (Fig. 3.4).

The model, based on Winkle and Baird (1985), is stated as follows.

**Mathematical model for ship bottom structure**

minimize $VC = MW + k\,MH$
subject to

$h_1$:    $SM_{freqd}$          $= (4.74 c_{freqd} h_{freqd} s_{floor} l_{floor}^2)/100^3$
$h_2$:    $c_{freqd}$          $= 0.9$

$$
\begin{aligned}
h_3\colon\quad & s_{\text{floor}} && = s_{\text{frame}} \\
h_4\colon\quad & A_{\text{kr}} && = w_{\text{kr}}t_{\text{kr}} \\
h_5\colon\quad & A_{\text{tp}} && = w_{\text{tp}}t_{\text{tp}} \\
h_6\colon\quad & l_{\text{floor}} && = B - 0.750 \\
h_7\colon\quad & \text{MH} && = f_1(l,w,\text{etc.}) \\
h_8\colon\quad & \text{MW} && = w_{\text{ms}}f_2(l,w,\text{etc.}) \\
h_9\colon\quad & n_{\text{floor}} && = l/s_{\text{floor}} \\
h_{10}\colon\quad & n_{\text{keelson}} && = w/s_{\text{keelson}} - 1 \\
h_{11}\colon\quad & \text{SM}_{\text{floor}} && = f_3(l,w,\text{etc.})
\end{aligned}
$$

$$
\begin{aligned}
g_1\colon\quad & t_{\text{keel}} && \geqslant t_{\text{bmin}} + 0.0015 \\
g_2\colon\quad & t_{\text{cg}} && \geqslant (0.063L + 5)/1000 \\
g_3\colon\quad & A_{\text{tp}} && \geqslant (0.168L^{3/2} - 8)/10000 \\
g_4\colon\quad & s_{\text{keelson}} && \leqslant 2.13 \\
g_5\colon\quad & h_{\text{cg}} && \geqslant h_{\text{keelson}} \\
g_6\colon\quad & h_{\text{keelson}} && \geqslant h_{\text{floor}} \\
g_7\colon\quad & t_{\text{keelson}} && \geqslant (0.063L + 4)/1000 \\
g_8\colon\quad & A_{\text{kr}} && \geqslant (0.038L^{3/2} + 17)/10000 \\
g_9\colon\quad & h_{\text{freqd}} && \geqslant d \\
g_{10}\colon\quad & h_{\text{freqd}} && \geqslant 0.66D \\
g_{11}\colon\quad & w_{\text{fr}}(h_{\text{floor}} + t_{\text{fr}}) && \geqslant w_{\text{fr}}0.0625\,l_{\text{floor}} \\
g_{12}\colon\quad & h_{\text{floor}} && \geqslant 0.0625 l_{\text{floor}} \\
g_{13}\colon\quad & w_{\text{fr}}t_{\text{floor}} && \geqslant w_{\text{fr}}[(h_{\text{floor}} + t_{\text{fr}})/100 + 0.003] \\
g_{14}\colon\quad & t_{\text{floor}} && \geqslant h_{\text{floor}}/100 + 0.003 \\
g_{15}\colon\quad & t_{\text{floor}} && \leqslant 0.0115 \\
g_{16}\colon\quad & \text{SM}_{\text{floor}} && \geqslant \text{SM}_{\text{freqd}} \\
g_{17}\colon\quad & s_{\text{frame}} && \leqslant (2.08L + 438)/1000 \\
g_{18}\colon\quad & D_{\text{s}} && \geqslant D \\
g_{19}\colon\quad & t_{\text{bottom}} && \geqslant t_{\text{bmin}} \\
g_{20}\colon\quad & t_{\text{bmin}} && \geqslant (s_{\text{frame}}/519)[(L - 19.8)\max(d/D_{\text{s}}, 0.65)]^{1/2} + 0.0025 \\
g_{21}\colon\quad & t_{\text{bmin}} && \geqslant (s_{\text{frame}})(L + 45.73)/(25L + 6082) \\
g_{22}\colon\quad & w_{\text{keel}} && \geqslant 0.750
\end{aligned}
$$

(3.46)

Functions $f_1, f_2$, and $f_3$ are defined using man–hour estimates, total volume of structural elements, and plate–stiffener combination section moduli respectively. The problem has 11 equality constraints and 20 degrees of freedom. The design variables are shown in Table 3.7.

Equality constraints are explicitly eliminated and a top-level monotonicity table is obtained where PRIMA deduces six different facts. The first monotonicity principle is repeatedly applied eight times and constraints $g_1, g_3, g_4, g_5, g_8, g_{22}$ are identified as being active. These six active constraints and the variables appearing in each of them provide several possibilities for selecting implicit elimination pivots, and the heuristic rules are now used. Starting with the top-level monotonicity table, the sequence of pivots $\{g_1-x_8, g_3-x_{20}, g_8-x_{19}, g_5-x_2, g_{22}-x_{18}, g_4-x_7\}$ leads to the table shown in Fig. 3.5, where the following additional facts or rules are obtained.

Fact 1: At least one of $g_{16}, g_{13}$ and $g_{11}$ is active.
Fact 2: At least one of $g_{21}$ and $g_{20}$ is active.
Fact 3: At least one of $g_{16}, g_{14}$ and $g_{13}$ is active.
Fact 4: At least one of $g_{16}$ and $g_{19}$ is active.
Fact 5: At least one of $g_{21}, g_{20}, g_{17}$ and $g_{16}$ is active.

**Table 3.7** Variables in the ship structure design problem

| | | |
|---|---|---|
| $x_1$ | $A_{kr}$ | Cross-sectional area of keelson rider plate |
| $x_2$ | $A_{tp}$ | Horizontal top plate area at midship |
| $x_3$ | $c_{freqd}$ | A constant, dependent upon ship type |
| $x_4$ | $D_s$ | Scantling depth |
| $x_5$ | $h_{cg}$ | Height of center girder |
| $x_6$ | $h_{floor}$ | Height of vertical floor plate |
| $x_7$ | $h_{freqd}$ | Height for calculating floor section modulus |
| $x_8$ | $h_{keelson}$ | Height of vertical keelson plate |
| $x_9$ | $l_{floor}$ | Length of floor |
| $x_{10}$ | MH | Manhours of labor |
| $x_{11}$ | MW | Material weight |
| $x_{12}$ | $n_{floor}$ | Number of floors |
| $x_{13}$ | $n_{keelson}$ | Number of keelsons |
| $x_{14}$ | $s_{floor}$ | Floor spacing |
| $x_{15}$ | $s_{frame}$ | Frame spacing |
| $x_{16}$ | $s_{keelson}$ | Keelson spacing |
| $x_{17}$ | $SM_{floor}$ | Floor section modulus |
| $x_{18}$ | $SM_{freqd}$ | Floor required section modulus |
| $x_{19}$ | $t_{bmin}$ | Minimum bottom shell thickness |
| $x_{20}$ | $t_{bottom}$ | Bottom shell thickness |
| $x_{21}$ | $t_{cg}$ | Center girder thickness at midships |
| $x_{22}$ | $t_{floor}$ | Floor thickness |
| $x_{23}$ | $t_{fr}$ | Floor rider plate thickness |
| $x_{24}$ | $t_{keel}$ | Keel plate thickness |
| $x_{25}$ | $t_{keelson}$ | Keelson thickness |
| $x_{26}$ | $t_{kr}$ | Keelson rider plate thickness |
| $x_{27}$ | $t_{tp}$ | Top plate thickness |
| $x_{28}$ | $w_{fr}$ | Width of floor rider plate |
| $x_{29}$ | $w_{keel}$ | Width of keel plate |
| $x_{30}$ | $w_{kr}$ | Keelson rider plate width |
| $x_{31}$ | $w_{tp}$ | Width of top plate |

These not very obvious facts are not sufficient for identifying an optimum, but they can disqualify any existing solution that does not satisfy them.

Another use of AI technology in optimal design modeling is the selection and assembly of appropriate optimization models from a database of available configurations within a design class. For example, in the design of truss structures an expert system can be used to determine the appropriate truss configuration, the associated parameter values, such as various bounds, and to invoke the mathematical model for subsequent numerical optimization (Adeli and Bala-subramanyam, 1988a, b).

A generic model for developing a knowledge-based methodology to aid design optimization formulations was proposed by Balachandran and Gero (1987) and further articulated by Balachandran (1991). Many other authors have commented on the general complementary nature of mathematical optimization and knowledge-based systems. The common idea is the obvious one: if a precise analytical model can be generated with resonable effort, then a decision model based on mathematical optimization is preferable. In the absence of analytical

MONOTONICITY TABLE toplevel-->g1--x8-->g3--x20-->g8--x19-->g5--x2-->g22--x18-->g4--x7

| | x1 | x3 | x4 | x5 | x6 | x9 | x10 | x11 | x12 | x13 | x14 | x15 | x16 | x17 |
|---|---|---|---|---|---|---|---|---|---|---|---|---|---|---|
| obj | U | U | U | U | − | + | | + | + | + | U | U | U | + |
| g2 | | + | − | | | | | | | | | | | |
| g6 | | | | − | | | | | | | | | | |
| g7 | | | | − | | | | | | | − | | | |
| g9 | | | | | | | − | | | | | | | |
| g10 | | − | | | | | | | | | | | | − |
| g11 | | − | | | | | | | + | | | | | − |
| g12 | | + | | | | | | − | | + | | | | |
| g13 | | + | | | | | | − | | | | | | |
| g14 | | | | | | | | + | | | | | | |
| g15 | | U | | | U | U | | U | U | | | | | U |
| g16 | | | | + | + | | | | | | | | | |
| g17 | − | | | | | | | | | | | | | |
| g18 | | | | | | − | | | | + | | | | |
| g19 | − | | | + | + | | | | | − | | | | |
| g20 | | | | + | + | | | | | − | | | | |

THE FOLLOWING FACT IS DERIVED W.R.T. VAR x17
AT LEAST ONE OF THE FOLLOWING CONSTRAINTS IS ACTIVE- g16--g13--g11

THE FOLLOWING FACT IS DERIVED W.R.T. VAR x13
AT LEAST ONE OF THE FOLLOWING CONSTRAINTS IS ACTIVE- g21--g20

THE FOLLOWING FACT IS DERIVED W.R.T. VAR x11
AT LEAST ONE OF THE FOLLOWING CONSTRAINTS IS ACTIVE- g16--g14--g13

THE FOLLOWING FACT IS DERIVED W.R.T. VAR x9
AT LEAST ONE OF THE FOLLOWING CONSTRAINTS IS ACTIVE- g19--g16

THE FOLLOWING FACT IS DERIVED W.R.T. VAR x6
AT LEAST ONE OF THE FOLLOWING CONSTRAINTS IS ACTIVE- g21--g20--g17--g16

**Fig. 3.5** Final PRIMA results for ship bottom structure problem.

models codification of domain knowledge may be possible in the form of rules or other AI data structures, which can be then manipulated to make useful deductions.

Of special interest here is also work by Schittkowski on an intelligent modeling and mathematical programming package (Schittkowski, 1985), and the available commercial package GAMS (Brooke, Kendrick and Meeraus, 1988). In Schittkowski's system a set of rules assists the user in selecting algorithms and parameters within algorithms, based on an evaluation of the type of model that needs to be solved. The rules are a result of numerical analysis knowledge and experimentation with actual problems. Each rule has a level of certainty or veracity probability associated with it. These are propagated and indicated in the deductions produced by the rules. The system keeps a record of all major decisions made and attendant results every time a new problem is being solved. This information is used to modify or sharpen the rules and their associated veracity probabilities. Thus a learning function is supported. GAMS (General Algebraic Modeling System) is a high level language that integrates features of a relational database in formulating explicit (algebraic) mathematical programming models. Originally it was conceived to facilitate easy data entry and modification of very large linear programming models. Current extensions can handle nonlinear and binary (zero–one) linear programming models. The major advantage of a GAMS environment in the present context is the ability to generate quickly modifications of models by adding, deleting or changing the expressions of constraints without worrying about FORTRAN reprogramming. Such functions for optimization computations are now beginning to appear also in general-purpose packages such as MATHEMATICA (Wolfram, 1988) and MATLAB (Grace, 1991).

## 3.5 CONCLUDING REMARKS

When an optimization study is undertaken, a careful model analysis will save a lot of subsequent effort in trying to obtain good and reliable results from numerical computations. An initial model must be properly cast into the standard form for numerical processing and possible unboundedness should be checked. A clear understanding of the physical meaning of each constraint is necessary, as well as an exploration of how constraints may interact with each other. Model validity limitations must be always checked to ascertain that they are indeed satisfied by the optimal values of the design variables. Although we have demonstrated these ideas on relatively small explicit models, such ideas are important also when implicit models are used. Proper modeling is somewhat of an art but the tactics involved are grounded frequently on a rigorous understanding of how optimization methods work. Many of these ideas will continue to become increasingly automated.

## ACKNOWLEDGEMENTS

Thanks are due to the University of Michigan graduate students B. Alexander and B. Rycenga for posing the linear actuator problem and R.P. Krishnamachari for helping with the final form of the manuscript.

# REFERENCES

Adeli, H. and Balasubramanyam, K.V. (1988a) A knowledge-based system for design of bridge trusses. *Journal of Computing in Civil Engineering*, **2** (1), 1–20.

Adeli, H. and Balasubramanyam, K.V. (1988b) *Expert Systems for Structural Design – A New Generation*, Prentice-Hall, Englewood Cliffs, NJ.

Adeli, H. and Mak, K.Y. (1988) Architecture of a coupled expert system for optimum design of plate girder bridges. *Engineering Applications of AI*, **1**, 277–85.

Agogino, A.M. and Almgren, A.S. (1987) Techniques for integrating qualitative reasoning and symbolic computation in engineering optimization. *Engineering Optimization*, **12** (2), 117–35.

Alexander, B. and Rycenga, B. (1990) Optimization of a linear actuator. *Rep. UM-MEAM-555-91*, Department of Mechanical Engineering and Applied Mechanics, The University of Michigan, Ann Arbor, MI.

Arora, J.S. and Baenziger, G. (1986) Uses of artificial intelligence in design optimization. *Computer Methods in Applied Mechanics and Engineering*, **54**, 303–23.

Balachandran, M. (1991) *Knowledge Based Optimum Design*, Computational Mechanics Publications, Southampton.

Balachandran, M. and Gero, J.S. (1987) A knowledge-based approach to mathematical design, modeling, and optimization. *Engineering Optimization*, **12**, 91–116.

Brooke, A., Kendrick, D. and Meeraus, A. (1988) *GAMS – A User's Guide*, The Scientific Press, Redwood City, CA.

Choy, J.K. and Agogino, A.M. (1986) SYMON: automated symbolic monotonicity analysis for qualitative design optimization, in *Proc. ASME 1986 Int. Comp. in Eng. Conf.*, Chicago, IL, pp. 207–12.

Forgy, C.L. (1981) *OPS5 User's Manual*, Department of Computer Science, Carnegie Mellon University, Pittsburgh, PA.

Grace, A. (1991) *Optimization Toolbox for Use with MATLAB*, The Math Works Inc.

Hammond, C.R. and Johnson, G.E. (1987) A general approach to constrained optimal design based on symbolic mathematics, in *Advances in Design Automation – 1987* (ed. S.S. Rao), DE. Vol. 10-1, ASME, New York, pp. 31–40.

Hammond, C.R. and Johnson, G.E. (1988) The method of alternate formulations, an automated strategy for optimal design. *Journal of Mechanisms, Transmissions, and Automation in Design*, **110**, 459–63.

Hansen, P., Jaumard, B. and Lu, S.H. (1988a) A framework for algorithms in globally optimal design. *Journal of Mechanisms, Transmissions, and Automation in Design*, **111** (3), 353–60.

Hansen, P., Jaumard, B. and Lu, S.H. (1988b) An automated procedure for globally optimal design. *Journal of Mechanisms, Transmissions, and Automation in Design*, **111** (3), 361–7.

Juvinall, R.C. (1983) *Fundamentals of Machine Components*, Wiley, New York.

Li, H.L. (1985) Design optimization strategies with global and local knowledge. PhD Dissertation, Department of Mechanical Engineering and Applied Mechanics, University of Michigan, Ann Arbor, MI.

Li, H.L. and Papalambros, P. (1985a) A production system for use of global optimization knowledge. *Journal of Mechanisms, Transmissions, and Automation in Design*, **107** (2), 277–84.

Li, H.L. and Papalambros, P. (1985b) REDUCE applications in design optimization, in CAD/CAM Robotics and Auto, Int. Conf. Proc., Tucson, A.

Li, H.L. and Papalambros, P. (1988) A combined local–global active set strategy for nonlinear design optimization. *Journal of Mechanisms, Transmissions, and Automation in Design*, **110** (4), 464–471.

Papalambros, P. (1988a) Remarks on sufficiency of constraint-bound solutions in optimal design, in *Advances in Design Automation – 1988* (ed. S.S. Rao), ASME, New York. Also (1993) *Journal of Mechanical Design*, **115** (3), 374–9.

Papalambros, P. (1988b) Enhancements in design optimization problem solving: a knowledge-based approach. *Rep. TKHS-88.79*, SAAB-Scania aircraft Division, Linköping.

Papalambros, P. and Li, H.L. (1983) Notes on the operational utility of monotonicity in optimization. *Journal of Mechanisms, Transmissions, and Automation in Design*, **105**(2), 174–81.

Papalambros, P.Y. and Wilde, D.J. (1988) *Principles of Optimal Design*, Cambridge University Press, New York.

Pomrehn, L.P. and Papalambros, P.Y. (1992) Constraint activity revisited: application to global and discrete design optimization, in *Advances in Design Automation* (ed. D.A. Hoeltzel), DE-Vol.44-1, ASME, New York, pp. 223–9.

Rao, J.R. and Papalambros, P. (1987) Implementation of semi-heuristic reasoning for boundedness analysis of design optimization models, in *Advances in Design Automation – 1987* (ed. S.S. Rao), ASME, New York.

Rao, J.R. and Papalambros, P. (1991) PRIMA: a production-based implicit elimination system for monotonicity analysis of optimal design models. *Journal of Mechanical Design*, **113** (4), 408–15.

Schittkowski, K. (1984) NLPQL: a FORTRAN subroutine solving constrained nonlinear programming problems. *Tech. Rep.*, Institute für Informatik, University of Stuttgart.

Schittkowski, K. (1985) EMP: an expert system for mathematical programming. *Mathematical Institute Rep.*, University of Bayreuth.

Siddall, J.N. (1972) *Analytical Decision-Making in Engineering Design*, Prentice-Hall, Englewood Cliffs, NJ.

Spotts, M.F. (1985) *Design of Machine Elements*, Prentice-Hall, Englewood Cliffs, NJ.

Thomas, R. (1990) Development of an intelligent front-end for design optimization programs. Diploma Thesis, *Rep. LiTH-IKP-R-604*, Department of Mechanical Engineering, University of Linköping.

Watton, J.D. and Rinderle, J.R. (1991) Symbolic design optimization: a computer aided method to increase monotonicity through variable reformulation. *Rep. 24-56-91*, Engineering Design Research Center, Carnegie Mellon University, Pittsburgh, PA.

Wilde, D.J. (1978) *Globally Optimal Design*, Wiley–Interscience, New York.

Wilde, D.J. and Beightler, C.S. (1967) *Foundations of Optimization*, Prentice-Hall, Englewood Cliffs, NJ.

Winkle, I.E. and Baird, D. (1985) Towards more effective structural design through synthesis and optimization of relative fabrication costs. Proceedings of the Joint Meetings of the Royal Institute of Naval Architecture (RINA) and the Institute of Engineers and Shipbuilders, Glasgow.

Wolfram, S. (1988) *Mathematica: A System for Doing Matematics by Computer*, Addition-Wesley, Redwood City, CA.

# 4

# Generalized geometric programming and structural optimization

SCOTT A. BURNS and NARBEY KHACHATURIAN

## 4.1 INTRODUCTION

This chapter contains a brief presentation of some of the new developments in optimal structural design which have proven to be effective tools in design of structures, and which have the potential of introducing major changes in our approach to optimization of structures. Specifically, our attention will be directed to the generalized geometric programming (GGP) method and its application to problems with both inequality and equality constraints. This represents typical problems in optimization of structures.

The GGP method represents a significant development in mathematical programming. It is an extension of classical geometric programming, which originally concerned problems with positive coefficients on all terms only. In its generalized form, it reduces a nonlinear problem to a stable sequence of linear problems, and permits the coefficients to be either positive or negative. The method provides a general and convenient approach for nonlinear optimization problems (Avriel, Dembo and Passy, 1975). The method has been applied successfully to many areas of engineering; several dozen engineering applications of geometric programming are reported in a 1978 survey paper (Rijckaert and Martens, 1978), and more recently, geometric programming algorithms have been developed for optimization of space structures and nonprismatic plate girders (Adeli and Kamal, 1986; Abuyounes and Adeli, 1986; Adeli and Chompooming, 1989). With recent extensions to include equality constraints in the problem formulation (Burns, 1987b), GGP becomes particularly well suited to the optimization of structures (Burns, 1985).

A second development is the integrated approach to structural optimization, in which a mathematical programming problem is posed in the mixed space of design variables (such as dimension) and behavior variables (such as forces and displacements). This integrated formulation contains both equality and inequality constraints. The equality constraints interrelate the structural behavior variables and perform the dual function of a structural analysis and an implicit sensitivity analysis; the inequality constraints establish the limits of safety and serviceability of the structure.

In the conventional approach of structural optimization by mathematical programming, the solution process is carried out in several iterations of analysis and design. The analysis phase is concerned with the equality constraints assuring that the equilibrium, compatibility, and constitutive equations are satisfied. This phase is also concerned with determining sensitivity information, such as changes in stress with respect to changes in geometric proportion. This information is passed to the design phase, which is concerned with adjusting the design variables so that an objective function is optimized while inequality constraints on behavior remain satisfied. The nature of the two-phase approach is such that a restricted amount of information is passed between the analysis and design phases. The integrated formulation eliminates this bottleneck, yielding significant improvements in performance, as discussed in subsequent sections.

The integrated formulation has appeared in the literature periodically. In early work, Fox and Schmit applied the penalty function and conjugate gradient methods to solve the integrated formulation of several trusses, giving rise to nonlinear programs with 30–78 design variables (Schmit and Fox, 1965; Fox and Schmit, 1966). Later, Fuchs (1982) developed an integrated formulation for optimization of trusses using three different sets of analysis constraints within the integrated formulation. Haftka (1985) applied a preconditioned conjugate gradient method to solve the integrated formulation of a 72-bar space truss (120 design variables), and Newton's method to optimize a damage tolerant wing-box structure (45 design variables). Soeiro and Hoit (1987) have applied the integrated formulation to optimize a piecewise prismatic cantilever beam and a one-story one-bay frame. Chibani, Burns and Khachaturian (1989) apply decomposition techniques to the problem of member sizing of truss structures under alternative loading conditions using the integrated formulation.

So far, the research effort in structural optimization has been predominantly in the solution of problems in which the geometry is given and the optimal proportioning of the members is sought. Shape optimization has many applications although they are generally of a higher level of complexity than the problem of member sizing. This chapter extends the integrated formulation to the problem of shape optimization of continuum structures modeled by finite elements.

In the discussions that follow, the GGP method will be reviewed briefly for application to problems with both equality and inequality constraints. Two examples will be presented to demonstrate the application of the GGP method to different problems in structural optimization. Then an integrated formulation will be developed for shape optimization of continuum structures and an example will be presented.

## 4.2 GENERALIZED GEOMETRIC PROGRAMMING

This section presents a mathematical programming technique to solve large-scale, nonlinear, mixed equality–inequality constrained problems. It is based on a method called generalized geometric programming (GGP) (Avriel, Dembo and Passy, 1975), with a modification to include equality constraints (Burns, 1987b).

Consider the following three algebraic constructions: monomial,

$$u(x) = c \prod_{j=1}^{n} x_j^{a_j}$$

for example

$$u = 3.2x_1^2 x_2^{1.5} x_3^{-0.5}$$

posynomial,

$$p(x) = \sum_{i=1}^{t} u_i(x)$$

for example

$$p = 3.2x_1^2 x_2^{1.5} x_3^{-0.5} + 5x_2^{-1}$$

polynomial,

$$g(x) = p^+(x) - p^-(x)$$

for example

$$g = 4x_1^2 x_2^2 - 3.2x_1^2 x_2^{1.5} x_3^{-0.5} - 5x_2^{-1}$$

All monomial coefficients, $c$, are required to be positive, but the exponents, $a$, are unrestricted in sign. The variables, $x$, are assumed to be strictly positive in value. Note that a posynomial is simply a polynomial with all coefficients positive.

GGP solves mathematical programs that are posed in polynomial form, as defined above. It operates on the posynomial parts of the polynomial objective and constraint functions using a process called 'condensation', as described in the next section.

### 4.2.1 Condensation

Condensation is the process of approximating a posynomial function with a monomial function (Duffin, 1970; Avriel and Williams, 1970; Pascual and Ben-Israel, 1970; Passy, 1971). It is based on the weighted arithmetic–geometric (A–G) mean inequality (Hardy, Littlewood and Polya, 1959; Beckenbach and Bellman, 1961),

$$\sum_i u_i \geqslant \prod_i (u_i/\delta_i)^{\delta_i}$$

where $u_i$ are positive values, the $\delta_i$ are positive weights, and $\sum \delta_i = 1$. For example, if $u = \{1, 2, 3\}$ and $\delta = \{1/3, 1/3, 1/3\}$, then the sum of these three numbers, 6, is greater than the weighted product of the numbers, 5.451. If the $u_i$ in this inequality are chosen to represent monomial functions instead of numbers, then the left-hand side of this inequality becomes a posynomial. The right-hand side is a product of weighted monomials, which is itself a monomial. It is easily shown that when all of the $u_i/\delta_i$ are equal, the inequality holds as an equality.

When applied to a posynomial, the A–G inequality converts the posynomial into an approximating monomial. The monomial produced is dependent on the selection of weights, which can be any set of positive values that sum to unity. One very useful choice is to set the weights equal to the fraction that each monomial term of the posynomial contributes to the total value of the posynomial, when evaluated at some operating point, $\bar{x}$:

$$\delta_i = \frac{u_i(\bar{x})}{p(\bar{x})} \qquad (4.1)$$

It can be seen that all $u_i/\delta_i$ are equal when $u$ is evaluated at the operating point.

Thus the approximating monomial is equal in value to the posynomial when evluated at $\bar{x}$.

The process of condensing a posynomial to a monomial may be represented symbolically as

$$C[p(x), \bar{x}] = \prod_{i=1}^{t} [u_i(x)/\delta_i]^{\delta_i}$$

where the $\delta_i$ are defined by equation (4.1). Consider the following example.
Condense $2x_1^2 x_2 + 4x_1^{-1}$ at $\bar{x} = (2, 1)$.

$$p(\bar{x}) = 2(2)^2(1) + 4(2)^{-1} = 10$$

$$\delta_1 = 2(2)^2(1)/10 = 0.8$$

$$\delta_2 = 4(2)^{-1}/10 = 0.2$$

$$\prod_{i=1}^{t} [u_i(x)/\delta_i]^{\delta_i} = \left(\frac{2x_1^2 x_2}{0.8}\right)^{0.8} \left(\frac{4x_1^{-1}}{0.2}\right)^{0.2} = 3.79 x_1^{1.4} x_2^{0.8}$$

Thus

$$C[2x_1^2 x_2 + 4x_1^{-1}, \bar{x} = (2, 1)] = 3.79 x_1^{1.4} x_2^{0.8}$$

### 4.2.2 Problem statement

GGP is concerned with minimizing a polynomial objective function subject to polynomial inequality constraints:

$$
\begin{aligned}
\text{minimize} \quad & g_0(x) \\
\text{subject to} \quad & g_k(x) \leqslant 0 \qquad k = 1, 2, \ldots, m \\
& 0 < x_j^{\text{LB}} \leqslant x_j \qquad j = 1, 2, \ldots, n
\end{aligned}
$$

where $x_j^{\text{LB}}$ are positive lower bounds on the variables, and the polynomial functions, $g$, are differences of posynomial functions. The GGP method first introduces a new variable so that the objective function becomes linear. The old nonlinear objective function is absorbed as an additional constraint:

$$
\begin{aligned}
\text{minimize} \quad & x_0 \\
\text{subject to} \quad & g_0(x) \leqslant x_0 \\
& g_k(x) \leqslant 0 \qquad k = 1, 2, \ldots, m \\
& 0 < x_j^{\text{LB}} \leqslant x_j \qquad j = 1, 2, \ldots, n
\end{aligned}
$$

Next, each polynomial is separated into its positive and negative parts, giving differences of pairs of posynomials:

$$
\begin{aligned}
\text{minimize} \quad & x_0 \\
\text{subject to} \quad & p_0^+(x) - p_0^-(x) \leqslant x_0 \\
& p_k^+(x) - p_k^-(x) \leqslant 0 \qquad k = 1, 2, \ldots, m \\
& 0 < x_j^{\text{LB}} \leqslant x_j \qquad j = 1, 2, \ldots, n
\end{aligned}
$$

Finally, all of the negative terms are brought to the right-hand side of the inequalities and then divided through to yield a quotient form:

*Quotient form*

minimize $\quad x_0$

subject to $\quad \dfrac{p_0^+(x)}{p_0^-(x) + x_0} \leqslant 1$

$\qquad\qquad \dfrac{p_k^+(x)}{p_k^-(x)} \leqslant 1 \qquad k = 1, 2, \ldots, m$

$\qquad\qquad 0 < x_j^{LB} \leqslant x_j \qquad j = 1, 2, \ldots, n \qquad\qquad (4.2)$

Note that because the variables are positive, there is no danger of dividing by zero when the negative terms are divided into the positive terms. The problem in quotient form has the same feasible region and solutions as the original GGP problem.

From this point on, the following problem will be used to demonstrate the operation of the GGP method:

$$\text{minimize} \quad x_2$$
$$\text{subject to} \quad x_1^2 - 10x_1 - 2x_2 + 29 \leqslant 0$$
$$- x_1^2 + 8x_1 + 4x_2 - 40 \leqslant 0$$
$$- x_1^2 + 12x_1 - x_2 - 32 \leqslant 0$$
$$x_1 \geqslant 1, x_2 \geqslant 1 \qquad\qquad (4.3)$$

The feasible region of this problem is the shaded area in Fig. 4.1. Here, $x_1$ and $x_2$ correspond to the horizontal and vertical axes, respectively. The objective function is chosen to be linear from the start to facilitate a two-dimensional

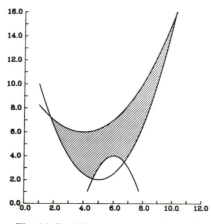

**Fig. 4.1** Feasible region of problem (4.3).

graphical representation. This problem in quotient form is

$$\text{minimize} \quad x_2$$

$$\text{subject to} \quad \frac{x_1^2 + 29}{10x_1 + 2x_2} \leqslant 1$$

$$\frac{8x_1 + 4x_2}{x_1^2 + 40} \leqslant 1$$

$$\frac{12x_1}{x_1^2 + x_2 + 32} \leqslant 1$$

$$x_1 \geqslant 1, x_2 \geqslant 1$$

### 4.2.3 Condensation of the denominators

Solution of the GGP problem is a two-level iterative process. The first level consists of condensing the denominators of each of the constraints at the operating point. This results in a posynomial divided by an approximating monomial. The monomial can be divided into the posynomial in the numerator, yielding an approximating posynomial that is always greater than or equal to the parent quotient form:

$$\frac{p^+(x)}{C[p^-(x), \bar{x}]} \geqslant \frac{p_k^+(x)}{p_k^-(x)}$$

Since this inequality relationship holds for all positive values of $x$, the feasible side of the approximating posynomial constraint is a subset of the feasible side of the parent quotient constraint.

Now consider the example problem. If the operating point is selected to be $\bar{x} = (3, 5)$, the approximating posynomial constraint corresponding to the first quotient constraint is

$$p^-(\bar{x}) = 10(3) + 2(5) = 10$$

$$\delta_1 = 10(3)/40 = 0.75$$

$$\delta_2 = 2(5)/40 = 0.25$$

$$\frac{p^+(x)}{C[p^-(x), \bar{x} = (3, 5)]} = \frac{x_1^2 + 29}{(10/0.75)^{0.75}(2/0.25)^{0.25} x_1^{0.75} x_2^{0.25}} \leqslant 1$$

$$= 0.0852 x_1^{1.25} x_2^{-0.25} + 2.47 x_1^{-0.75} x_2^{-0.25} \leqslant 1$$

The feasible region of the approximating posynomial is the darker shaded region in Fig. 4.2(a). Similarly, the denominators of the other two constraints are condensed to give

$$\frac{8x_1 + 4x_2}{C[x_1^2 + 40, \bar{x} = (3, 5)]} = 0.244 x_1^{0.633} x_2^{-0.25} + 0.122 x_1^{-0.367} x_2 \leqslant 1$$

$$\frac{12x_1}{C[x_1^2 + x_2 + 32, \bar{x} = (3, 5)]} = 0.478 x_1^{0.609} x_2^{-0.109} \leqslant 1$$

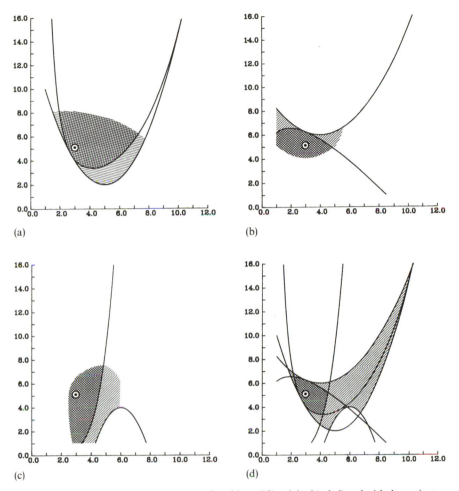

**Fig. 4.2** (a) Feasible region of constraint 1 of problem (4.3), original (ruled) and with denominator condensed at $\bar{x} = (3, 5)$ (cross-hatched). (b) Feasible region of constraint 2 of problem (4.3), original (ruled) and with denominator condensed at $\bar{x} = (3, 5)$ (cross-hatched). (c) Feasible region of constraint 3 of problem (4.3), original (ruled) and with denominator condensed at $\bar{x} = (3, 5)$ (cross-hatched). (d) Feasible region of all constraints of problem (4.3), original (ruled) and with denominators condensed at $\bar{x} = (3, 5)$ (cross-hatched).

The corresponding feasible regions are shown in Figs. 4.2(b) and 4.2(c), respectively. The feasible region resulting from the intersection of these three approximations is shown in Fig. 4.2(d). Notice that it is a subset of the original feasible region.

It is interesting to observe how the posynomial subregion changes with respect to different operating points. Figure 4.3 shows four different posynomial sub-regions corresponding to four different operating points. Notice that when the operating point falls within the feasible region of the parent problem, it also falls within the feasible region of the posynomial subregion. Also note that when the operating point falls outside of the parent feasible region, a well-defined posynomial

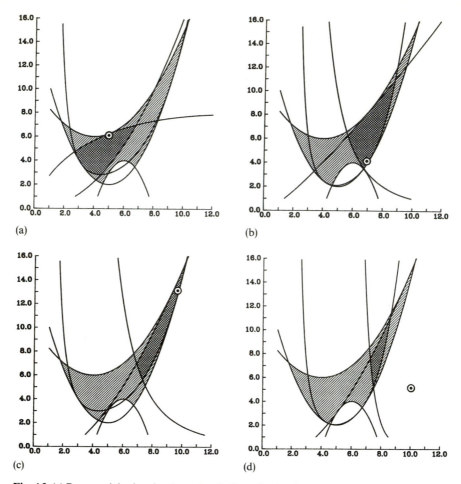

**Fig. 4.3** (a) Posynomial subregion (cross-hatched) production by condensation of denominators at $\bar{x} = (5, 6)$ for problem (4.3). (b) Posynomial subregion (cross-hatched) produced by condensation of denominators at $\bar{x} = (7, 4)$ for problem (4.3). (c) Posynomial subregion (cross-hatched) produced by condensation of denominators at $\bar{x} = (9.7, 13)$ for problem (4.3). (d) Posynomial subregion (cross-hatched) produced by condensation of denominators at $\bar{x} = (10, 5)$ for problem (4.3).

subregion still exists within the parent feasible region. Although this latter case will not always be true (certain operating points outside of the parent feasible region will produce a set of posynomial approximations with the null set as their intersection), it indicates that in some cases an infeasible operating point can be used as a starting point in the GGP process. Regardless, an operating point that satisfies the parent constraints will always produce a posynomial subregion that contains the operating point and that is completely contained within the parent feasible region.

The condensation of denominators gives rise to the following posynomial subproblem:

*Posynomial subproblem*

minimize  $x_0$

subject to  $\dfrac{p_0^+(x)}{C[p_0^-(x) + x_0, \bar{x}]} \geqslant 1$

$\dfrac{p_k^+(x)}{C[p_k^-(x), \bar{x}]} \geqslant 1 \qquad k = 1, 2, \ldots, m$

$0 < x_j^{\mathrm{LB}} \leqslant x_j \qquad j = 1, 2, \ldots, n \qquad\qquad (4.4)$

This subproblem is contained completely within the parent problem (4.2). The solution of this problem is, therefore, feasible to the parent problem.

## 4.2.4 The outer loop

In the previous section, it was demonstrated how a problem in quotient form is converted to a posynomial-constrained problem by condensing the denominators at some operating point. The outer loop of the two-level GGP process consists of the formation of a series of such posynomial subproblems, each condensed at the solution of the previous posynomial problem.

The outer loop is initiated by choosing a starting point for the variables. It is assumed that the initial operating point is feasible to the GGP problem. This starting point is used to generate a posynomial subproblem of the form (4.4). This subproblem is solved by the inner loop process described in the next section.

Once the solution to the posynomial subproblem has been obtained, it is used as the new operating point and a new posynomial subproblem is generated from it. This subproblem is solved and the process is repeated until convergence to a solution has been achieved. This process is shown graphically in Fig. 4.4. Aviel and Williams (1970) have shown that this process will converge to a point satisfying the Kuhn–Tucker necessary conditions for optimality if mild regularity conditions are satisfied.

All that remains now is the solution of the sequence of posynomial subproblems generated by the outer loop. This takes place in the inner loop, and it is where the bulk of the computation occurs. One may question the merit of solving a series of posynomial problems instead of a single polynomial problem. The advantage of doing so lies in the structure of the posynomial subproblem. The parent problem is generally nonconvex and may possess several local minima. The posynomial subproblem has only one solution because it is convex when transformed by a logarithmic change of variables (Duffin, Peterson and Zener, 1967). This convexity gives rise to very efficient solution strategies for the posynomial subproblem.

## 4.2.5 The inner loop

This section presents a linear-programming-based solution technique for the posynomial subproblem generated by the outer loop of the GGP solution process. It is an adaptation of the Kelly cutting plane method for convex problems (Kelly, 1960).

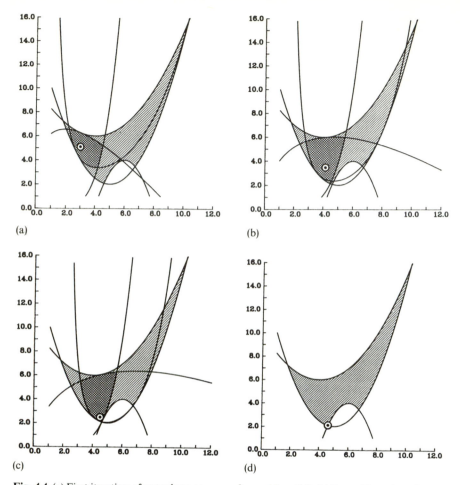

**Fig. 4.4** (a) First iteration of outer loop sequence for problem (4.3). (b) Second iteration of outer loop sequence for problem (4.3). (c) Third iteration of outer loop sequence for problem (4.3). (d) Fourth iteration of outer loop sequence and final solution for problem (4.3).

The inner loop uses a second operating point, $\hat{x}$, that is initially set equal to $\bar{x}$. The first step of the inner loop is to condense each posynomial constraint of the subproblem at $\hat{x}$. Recall that the denominators of the parent quotient constraints have already been condensed at $\bar{x}$ to arrive at the posynomial constraints. This second round of condensations results in a set of monomial approximations to the posynomial constraints. Observing the direction of the inequality sign in the A–G inequality, it is apparent that the feasible side of the monomial approximation completely contains the feasible region of the posynomial constraint. Thus, the intersection of this set of monomial approximations results in a feasible region that completely contains the posynomial subproblem. Since the logarithm of a monomial function is a linear function, the logarithm of the monomial problem gives rise to the following linear program (LP), which is an exterior approximation of the posynomial subproblem:

*Linear (log monomial) program*

minimize   $x_0$

subject to   $\ln\left(C\left[\dfrac{p_0^+(x)}{C[p_0^-(x)+x_0,\bar{x}]},\hat{x}\right]\right)\leqslant 0$

$\ln\left(C\left[\dfrac{p_k^+(x)}{C[p_k^-(x),\bar{x}]},\hat{x}\right]\right)\leqslant 0$        $k=1,2,\ldots,m$

$0<x_j^{\mathrm{LB}}\leqslant x_j$        $j=1,2,\ldots,n$        (4.5)

This linear program is put into standard form once the variables are transformed into log space according to

$$z_j=\ln\left(\frac{x_j}{x_j^{\mathrm{LB}}}\right)\quad j=1,2,\ldots,n \qquad (4.6)$$

Now the lower bounds on all $z$ variables are zero.

This LP problem is solved, and the solution is assigned to be the new $\hat{x}$. Since the solution of an LP always lies on the boundary of the LP feasible region, $\hat{x}$ will generally not be feasible to the posynomial subproblem. Therefore, the most violated of the posynomial constraints is condensed to a monomial, this time at the new $\hat{x}$. It is then transformed into log space and appended to the previous LP. The solution of this new LP becomes the new $\hat{x}$ and another check is made to determine whether all posynomial constraints are satisfied (to within some prespecified tolerance). If not, the most violated posynomial is recondensed at the new $\hat{x}$ and appended to the growing LP. This process is repeated until an acceptable solution to the posynomial subproblem is reached.

To clarify this inner loop process, consider again the example problem (4.3). As a result of the first round of denominator condensations, the posynomial subproblem, which was shown in Fig. 4.4(a), is produced:

minimize   $x_2$
subject to   $0.0852x_1^{1.25}x_2^{-0.25}+2.47x_1^{-0.75}x_2^{-0.25}\leqslant 1$
$0.244x_1^{0.633}x_2^{-0.25}+0.122x_1^{-0.367}x_2\leqslant 1$
$0.478x_1^{0.609}x_2^{-0.109}\leqslant 1$
$x_1\geqslant 1, x_2\geqslant 1$

Condensation of each of these posynomial constraints at the operating point $\hat{x}=(3,5)$ yields

minimize   $x_2$
subject to   $1.92x_1^{-0.276}x_2^{-0.25}\leqslant 1$
$0.355x_1^{0.178}x_2^{0.455}\leqslant 1$
$0.478x_1^{0.609}x_2^{-0.109}\leqslant 1$
$x_1\geqslant 1, x_2\geqslant 1$

The feasible region produced by the intersection of these three monomials is shown in Fig. 4.5. The feasible region of the posynomial problem is also shown in this figure. Note that it is a subset of the monomial feasible region. Also note that the third posynomial is already in monomial form, and is unchanged by the condensation process.

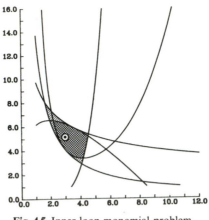

**Fig. 4.5** Inner loop monomial problem.

After making the logarithmic change of variables in equation (4.6), the following LP is produced:

$$\text{minimize} \quad z_2$$
$$\text{subject to} \quad 0.276z_1 + 0.250z_2 \geqslant 0.652$$
$$0.178z_1 + 0.455z_2 \leqslant 1.04$$
$$0.609z_1 - 0.109z_2 \leqslant 0.738$$
$$z_1 \geqslant 0, z_2 \geqslant 0$$

The solution of this LP, after transforming out of log space, is $(x_1, x_2) = (4.07, 2.91)$. The posynomial constraints are each evaluated at this point, and it is determined that the first constraint is the most violated of the three:

$$0.0852(4.07)^{1.25}(2.91)^{-0.25} + 2.47(4.07)^{-0.75}(2.91)^{-0.25} = 1.04$$

$$0.244(4.07)^{0.633}(2.91)^{-0.25} + 0.122(4.07)^{-0.367}(2.91) = 0.805$$

$$0.478(4.07)^{0.609}(2.91)^{-0.109} = 1.00$$

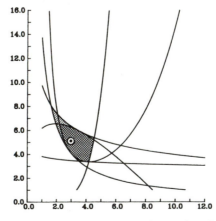

**Fig. 4.6** Monomial problem with first cutting plane.

Recondensing the first posynomial constraint at $\hat{x} = (4.07, 2.91)$, and transforming to log space, gives the cutting plane $0.0221z_1 + 0.250z_2 \leqslant 0.336$. The monomial version of this cutting plane is shown in Fig. 4.6. Note that it 'cuts off' a portion of the feasible region that is infeasible to the posynomial problem. Solving the appended LP gives the new solution $(x_1, x_2) = (4.18, 3.36)$. This satisfies the posynomial constraints sufficiently well and therefore, the inner loop procedure has been completed. To continue with this problem, one would return to the outer loop and generate a new posynomial problem (that shown in Fig. 4.4(b)) by condensing the denominators of the parent quotient constraints at the point $\bar{x} = (4.18, 3.36)$.

The inner loop would be very inefficient if the LP were to be solved from scratch for each additional cutting plane. Fortunately, this is not necessary because the dual formulation of the LP has special properties that result in significant computational savings.

The LP (4.5) may be expressed in a more standard form as

$$
\begin{array}{ll}
\text{minimize} & z_0 \\
\text{subject to} & -\mathbf{Az} \geqslant \mathbf{b} \\
& \mathbf{Iz} \geqslant \mathbf{0} \\
& -\mathbf{Iz} \geqslant -\mathbf{z}^{UB} \\
& -\mathbf{A}^{CP}\mathbf{z} \geqslant \mathbf{b}^{CP}
\end{array}
\tag{4.7}
$$

where $\mathbf{A}$ is the exponent matrix of the monomial problem, $\mathbf{b}$ is the vector of logarithms of the coefficients of the monomials, $\mathbf{I}$ is the identity matrix, $\mathbf{z}^{UB}$ is a vector of upper bounds on the variables transformed into log space, and $\mathbf{A}^{CP}$ and $\mathbf{b}^{CP}$ are similar to $\mathbf{A}$ and $\mathbf{b}$, but relate to the cutting planes. For every cutting plane added to the problem, an additional row is added to the LP. Since the new cutting plane excludes the previous LP solution from the new feasible region, feasibility must be regained before the new LP can be solved.

Recall from the duality theory of linear programming that associated with an LP of the form

$$
\begin{array}{ll}
\text{minimize} & \mathbf{cz} \\
\text{subject to} & \mathbf{Az} \geqslant \mathbf{b} \\
& \mathbf{z} \geqslant \mathbf{0}
\end{array}
$$

there is a dual formulation of the form

$$
\begin{array}{ll}
\text{maximize} & \mathbf{by} \\
\text{subject to} & \mathbf{A}^T\mathbf{y} \leqslant \mathbf{c} \\
& \mathbf{y} \geqslant \mathbf{0}
\end{array}
$$

Thus, the dual formulation of the LP (4.7) is

$$
\text{maximize} \quad [b \quad 0 \quad -z_j^{UB} \quad b_i^{CP}]\{y\}
$$

$$
\text{subject to} \quad [-\mathbf{A}^T \quad \mathbf{I} \quad -\mathbf{I} \quad -(\mathbf{A}^{CP})^T]\{y\} \leqslant \begin{pmatrix} 1 \\ 0 \\ \vdots \\ 0 \end{pmatrix}
$$

The dual variables, *y*, are associated with each constraint and cutting plane. Note that a basic feasible solution is always available for this problem, namely the zero vector. No 'phase I' procedure is necessary to initiate the LP solution process. Also, since the dual problem solves from the infeasible side inward, each cutting plane makes the previous LP solution feasible to the dual problem, and the new LP solution can be obtained after only a few additional pivot operations. Therefore, the inner loop solution process consists of the solution of an initial LP and a few additional pivot operations as cutting planes are added, until the posynomial constraints are satisfied by the LP solution.

### 4.2.6 Equality constraints

The GGP procedure outlined in the previous sections has concerned only inequality-constrained problems. The method can be extended to include equality constraints with minor changes to the algorithm. The equality constraints are first put in quotient form and the denominators are condensed to make posynomial equality constraints. Then, in the inner loop, the equality constraints are again condensed to form monomial constraints. These monomial equality constraints are transformed into log space and added to the linear program. Up to this point, the only difference between the treatment of the equality and inequality constraints is that, in the dual LP, the equality constraints appear as free variables instead of as variables bounded below by zero.

Once the LP solution has been found, only the posynomial inequalities are checked to determine which one is most violated. In other words, the equality constraints appear only once in the LP and do not produce cutting planes. It has been shown that this modification to the GGP process is equivalent to a Newton–Raphson solution of the equality constraints in log space conducted simultaneously with the cutting plane treatment of the inequality constraints (Burns, 1987b). More recent developments have shown that this treatment of the equality constraints has special invariance properties not shared by the Newton–Raphson method that give rise to enhanced performance. Further details are available in Burns and Locascio (1991). Figure 4.7 presents a flowchart of the modified equality–inequality constrained GGP algorithm.

To demonstrate the GGP method discussed in this section, the following two examples are presented.

---

### EXAMPLE 4.1 SYMMETRICAL ONE-STORY, ONE-BAY FRAME

Figure 4.8 shows a symmetrical one-story, one-bay frame subjected to a uniformly distributed load of $0.6\,\text{klbf ft}^{-1}$. Each member is prismatic and the columns have the same section properties. The bases of the columns are assumed to be free to rotate. For convenience, the shape of all members will be assumed to be square.

The allowable stress in the beam is given as $1.8\,\text{klbf in}^{-2}$ and the allowable stress in the columns is given as $1.6\,\text{klbf in}^{-2}$.

It is intended to determine the dimensions of the beam and the column for each local minimum. The objective function is the volume of the structure.

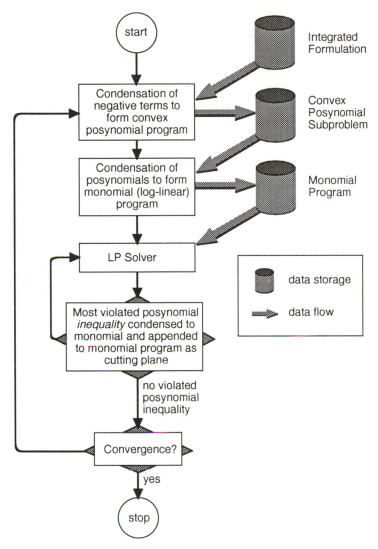

**Fig. 4.7** Generalized geometric programming flowchart.

The assumption that the cross section of each member is square is made for convenience. With this simplification, it becomes unnecessary to assume a depth for each section, making comparisons more direct. However, the method can be used for any other shape in the nondiscrete formulation of the problem. In particular, for I-sections of given depth, the section of the smallest acceptable web thickness with the highest flexural efficiency can be selected.

In this simple example the analysis is carried out in a general form in order to obtain relationships between the forces, dimensions and section properties of the structure.

Assuming $I_b/I_c = m$, the analysis of the frame by the slope–deflection method

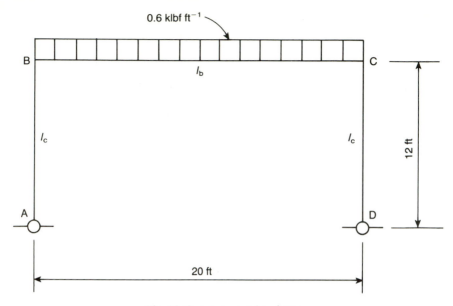

**Fig. 4.8** One-story, one-bay frame.

will yield the following forces:

$$M_{BA} = 60/(3.0 + 1.2m)\,\text{ft klbf}$$

$$M_{BC} = -\,60/(3.0 + 1.2m)\,\text{ft klbf}$$

$$M_{CB} = 60/(3.0 + 1.2m)\,\text{ft klbf}$$

$$M_{CD} = -\,60/(3.0 + 1.2m)\,\text{ft klbf}$$

$$R_H\,(\text{horizontal reaction}) = 5/(3.0 + 1.2m)\,\text{klbf}$$

$$R_V\,(\text{vertical reaction}) = (0.6 \times 20)/2 = 6\,\text{klbf}$$

**Formulation of the problem**

Assume the dimensions of the cross sections of the column and the beam to be $x_1$ and $x_2$ respectively:

$$m = I_b/I_c = x_2^4/x_1^4$$

The optimization problem can be described as follows:

$$\text{minimize} \quad V = 24x_1^2 + 20x_2^2$$

$$\text{subject to}$$

1. stress at the top of the column should be less than $1.60\,\text{klbf in}^{-2}$:

$$6/x_1^2 + 4320x_1^4/(3x_1^4 + 1.2x_2^4)x_1^3 \leqslant 1.6\,\text{klbf in}^{-2}$$

2. stress at the end of the beam should be less than $1.80\,\text{klbf in}^{-2}$:

$$5x_1^4/(3x_1^4 + 1.2x_2^4)x_2^2 + 4320x_1^4/(3x_1^4 + 1.2x_2^4)x_2^3 \leqslant 1.8\,\text{klbf in}^{-2}$$

3. stress at the midspan of the beam should be less than $1.80\,\text{klbf in}^{-2}$:

$$5x_1^4/(3x_1^4 + 1.2x_2^4)x_2^2 + 2160/x_2^3 - 4320x_1^4/(3x_1^4 + 1.2x_2^4)x_2^3 \leqslant 1.8\,\text{klbf in}^{-2}$$

For this simple problem there are only inequality constraints, since the equations relating the forces to the loads and relative stiffnesses have been conveniently included in the constraint expressions.

Solution of the above optimization problem by geometric programming results in two solutions corresponding to the following two local minima:

local minimum 1

$$x_1 = 2.654\,\text{in}$$
$$x_2 = 10.604\,\text{in}$$
$$V = 2417.95\,\text{in}^2\,\text{ft}$$

local minimum 2

$$x_1 = 8.933\,\text{in}$$
$$x_2 = 8.480\,\text{in}$$
$$V = 3353.37\,\text{in}^2\,\text{ft}$$

Figure 4.9 shows the two local minima.

Local minimum 1 is also the global minimum. However, it is clear that the columns are subjected mainly to axial forces. The moment at the top of the column, or at the end of the beam, is small and as a result the beam acts almost as a simply supported beam resting on piers subjected to axial compressive force.

Specifically, the following are the forces and stresses for local minimum 1:

$$H \text{ (horizontal reaction)} = 0.0162\,\text{klbf}$$

$$M_{BA} \text{ (moment, top of column)} = 0.1944\,\text{ft klbf}$$

$$M_M \text{ (moment, midspan)} = 29.81\,\text{ft klbf}$$

$$f_{BA} \text{ (stress, top of column)} = 1.600\,\text{klbf in}^{-2}$$

$$f_M \text{ (stress, midspan)} = 1.800\,\text{klbf in}^{-2}$$

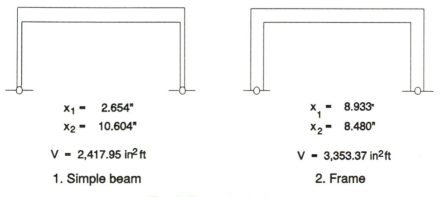

$$x_1 = 2.654''$$
$$x_2 = 10.604''$$

$$V = 2,417.95\,\text{in}^2\,\text{ft}$$

1. Simple beam

$$x_1 = 8.933''$$
$$x_2 = 8.480''$$

$$V = 3,353.37\,\text{in}^2\,\text{ft}$$

2. Frame

**Fig. 4.9** The two local minima.

The solution would suggest that if the frame were subjected only to a vertical uniformly distributed load, a better design would be a simply-supported beam resting on two piers. Local minimum 1 is fully stressed.

Local minimum 2 corresponds to a structure that is heavier than that in local mimimum 1 by about 39%. However, in this case the structure behaves as a frame and the column is subjected to a large moment. Since the column is relatively long, to make the structure behave as a frame requires a large increase in the size of the column which results in an increased weight.

Specifically, the following are the forces and stresses for local minimum 2:

$$H \text{ (horizontal reaction)} = 1.258 \text{ klbf}$$

$$M_{BA} \text{ (moment, top of column)} = 15.10 \text{ ft klbf}$$

$$M_M \text{ (moment, midspan)} = 14.90 \text{ ft klbf}$$

$$f_{BA} \text{ (stress, top of column)} = 1.600 \text{ klbf in}^{-2}$$

$$f_M \text{ (stress, midspan)} = 1.777 \text{ klbf in}^{-2}$$

In this example since the column is long the local minimum 1 corresponds to a lighter structure. However, if the column were very short in relation to the girder, local minimum 2 would become the global minimum. In this case local minimum 2 is not fully stressed, since the computed stress in the girder is somewhat below the allowable values.

---

### EXAMPLE 4.2  TWO-SPAN NONPRISMATIC CONTINUOUS BEAM

Figure 4.10 shows a two-span nonprismatic continuous beam subjected to a uniformly distributed fixed load of $1.8 \text{ klbf ft}^{-1}$. The beam is fabricated in three sections that are to be spliced in the field at the points of inflection of the uniformly distributed load. Section B–B shows the cross section of the beam in the negative moment region and Section A–A shows the cross section in the positive moment region. The width of flange, which is given as 8 in, and the $50 \text{ in} \times \frac{9}{16}$ in web are the same for the entire structure.

Assume that the uniformly distributed load includes the weight of the beam, and that the increased weight in the negative moment region need not be considered.

The flexural stress in the structure cannot exceed $20 \text{ klbf in}^2$.

Calculate $\alpha$, and the flange thicknesses $t_1$ and $t_2$ such that the volume of the structure is minimum.

This problem is constrained so that the points of splice where the section properties change fall at the points of inflection of the uniformly distributed load. This constraint is introduced to reduce the level of stress at the point of splice in order to increase the fatigue life of the structure. Although this constraint results in reversal of stress at the point splice, the stress level is so low that allowable range stress is not exceeded. It should be pointed out, however, that this constraint does not contribute to the optimal design of a continuous beam. It results in an increase in the weight of the structure since it tends to increase the value of $\alpha$. The moment at the interior support can be expressed in the following

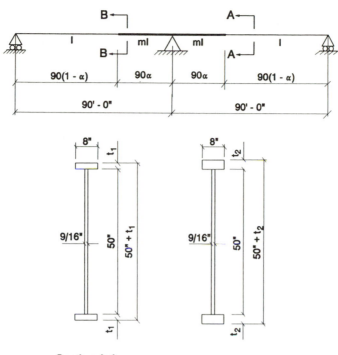

**Fig. 4.10** Two-span continuous beam.

form:

$$M = CwL^2/8$$

where

$$C = [m - (3\alpha^4 - 8\alpha^3 + 6\alpha^2)(m-1)]/[1 + (1-\alpha)^3(m-1)]$$

and $m$ is the ratio of the moment of inertia of the section in the negative moment region to that of the positive moment region. Figure 4.11 shows the variation of $C$ with $m$ and $\alpha$.

For the splice to be at the point of inflection, the following expressions must hold:

$$m = (\alpha^4 - 4\alpha^3 + 6\alpha^2)/(1-\alpha)^4$$

and

$$M_2 \text{ (bending moment at interior support)} = 4\alpha wL^2/8$$

The following are the section properties and forces: cross-sectional area of the beam in the positive moment region (in$^2$),

$$A_1 = 8(50 + 2t_1) - 50(8 - 9/16) = 16t_1 + 225/8$$

moment of inertia of the beam in the positive moment region (in$^4$),

$$I_1 = 8(50 + 2t_1)^3/12 - (50)^3(8 - 9/16)/12$$
$$= 16(25 + t_1)^3/3 - 77\,474$$

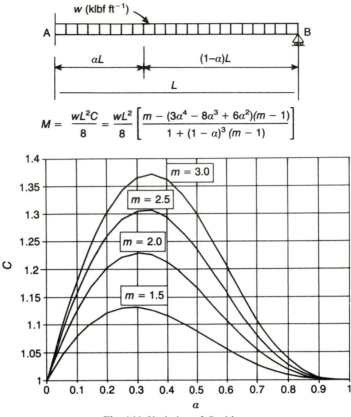

$$M = \frac{wL^2 C}{8} = \frac{wL^2}{8}\left[\frac{m-(3\alpha^4-8\alpha^3+6\alpha^2)(m-1)}{1+(1-\alpha)^3(m-1)}\right]$$

**Fig. 4.11** Variation of $C$ with $\alpha$.

distance from the neutral axis to the extreme fiber in the positive moment region (in),

$$c_1 = 25 + t_1$$

maximum positive moment in each span (in klbf),

$$M_1 = w[90(1-\alpha)]^2/8 = (12 \times 1.8 \times 8100)(1-\alpha)^2/8 = 21\,870(1-\alpha)^2$$

cross-sectional area of the beam in the negative moment region (in$^2$),

$$A_2 = 8(50 + 2t_2) - 50(8 - 9/16) = 16t_2 + 225/8$$

moment of inertia of the beam in the negative moment region (in$^4$),

$$I_2 = 8(50 + 2t_2)^3/12 - (50)^3(8 - 9/16)/12$$
$$= 16(25 + t_2)^3/3 - 77\,474$$

distance from the neutral axis to the extreme fiber in the negative moment region (in),

$$c_2 = 25 + t_2$$

maximum negative moment at the interior support (in klbf)

$$M_2 = wL^2(4\alpha)/8 = (12 \times 1.8)(90)^2(4\alpha)/8 = 21\,870(4\alpha)$$

The objective function is taken as $U$ which is a measure of the volume of the structure, as follows:

$$\text{minimize } U = A_1(1 - \alpha) + A_2(\alpha)$$
$$= (16t_1 + 225/8)(1 - \alpha) + (16t_2 + 225/8)\alpha$$
$$= 16t_1 + 16\alpha(t_2 - t_1) + 28.125$$

The inequality constraints are

1. the stress due to the maximum positive moment should be less than $20\,\text{klbf in}^{-2}$:

$$M_1 c_1/I_1 \leqslant 20\,\text{klbf in}^{-2}$$

or

$$M_1 c_1 - 20 I_1 \leqslant 0$$
$$21\,870(1 - \alpha)^2(25 + t_1) - 20[16(25 + t_1)^3/3 - 77\,474] \leqslant 0$$

and

$$(1 - \alpha)^2(25 + t_1) - (25 + t_1)^3/205 + 70.85 \leqslant 0$$

2. and stress due to the maximum negative moment should be less than $20\,\text{klbf in}^{-2}$:

$$M_2 c_2/I_2 \leqslant 20\,\text{klbf in}^{-2}$$
$$M_2 c_2 - 20 I_2 \leqslant 0$$
$$21\,870(4\alpha)(25 + t_2) - 20[16(25 + t_2)^3/3 - 77\,474] \leqslant 0$$

and

$$(4\alpha)^2(25 + t_2) - (25 + t_2)^3/205 + 70.85 \leqslant 0$$

The equality constraint is, since the point of splice is at the point of inflection, the following equation must hold:

$$m = (\alpha^4 - 4\alpha^3 + 6\alpha^2)/(1 - \alpha)^4$$

or

$$[(25 + t_2)^3 - 14\,526](1 - \alpha)^4 - [(25 + t_1)^3 - 14\,526][(1 - \alpha)^4 - 4(1 - \alpha) + 3] = 0$$

The optimization problem can now be stated as

$$\text{minimize } \quad 16t_1 + 16\alpha(t_2 - t_1) + 28.125$$
$$\text{subject to}$$
$$(1 - \alpha)^2(25 + t_1) - (25 + t_1)^3/205 + 70.85 \leqslant 0$$
$$(4\alpha)(25 + t_2) - (25 + t_2)^3/205 + 70.85 \leqslant 0$$
$$[(25 + t_2)^3 - 14\,526](1 - \alpha)^4 - [(25 + t_1)^3 - 14\,526]$$
$$\times [(1 - \alpha)^4 - 4(1 - \alpha) + 3] = 0$$

It should be noted that the first two constraints are inequalities and the third constraint is an equation.

Solution of the above problem by geometric programming will yield the following answers:

$$\alpha = 0.351$$
$$t_1 = 0.718\,\text{in}$$
$$t_2 = 3.303\,\text{in}$$
$$U = 54.13\,\text{in}^2$$

and
$$V = 54.13 \times 90 \times 12 = 58\,460\,\text{in}^3$$
$$M_1 = 9211.7\,\text{in klbf}$$
$$I_1 = 13\,248\,\text{in}^4$$
$$c_1 = 25.718\,\text{in}$$

$s_1$ (maximum stress in the positive moment region)
$$= (9211.7 \times 25.718)/(13\,248)\,\text{klbf in}^{-2}$$
$$= 17.88\,\text{klbf in}^{-2} < 20.00\,\text{klbf in}^{-2}$$
$$M_2 = 30\,705\,\text{in klbf}$$
$$I_2 = 43\,445\,\text{in}^4$$
$$c_2 = 28.303\,\text{in}$$

$s_2$ (stress at the interior support)
$$= (30\,705.5 \times 28.303)/43\,445 = 20.00\,\text{klbf in}^{-2}$$

It should be noted that in this case the maximum stress in the region of positive moment is less than the allowable stress. However, the maximum stress at the first interior support is equal to the allowable stress. The optimum design is not fully stressed.

## 4.3 APPLICATION TO SHAPE OPTIMIZATION

In this section, the integrated formulation is applied to the shape optimization of continuum structures modeled by finite elements. The stiffness matrix relating nodal forces to nodal displacements is expressed explicitly as a set of nonlinear equality constraints. The stiffness equations are nonlinear because they explicitly include a representation of the shape of the structure, as defined by the element nodal coordinates. The variables in this formulation are chosen to be the nodal coordinates of each element, the displacements of each node, and the stresses in each element. This selection of variables permits the stiffness matrix equations to be expressed in a convenient explicit form.

One advantage of this integrated formulation is that an explicit sensitivity analysis is not required, which eliminates the two-phase nature of typical shape optimization methods. The problem is related as one large mathematical program in the mixed space of shape variables and structural behavior variables, related to one another through a large set of equality and inequality constraints, such as the nonlinear stiffness equations. The sensitivity information, that is the changes in structural behavior with respect to changes in shape, are implicit in the constraint functions and need not be explicitly calculated.

The integrated formulation requires that very large mathematical programs be solved with each iteration of the method. Because of this, extreme care must be taken in implementing a solution technique to minimize the computational effort required for a solution. The GGP method is extremely efficient for solving large-scale nonlinear problems because of the special properties of condensation (Burns and Locascio, 1991), and because of its reliance on linear programming for the bulk of the computational effort.

In the following sections, the integrated shape optimization formulation is developed and examples are presented. Then, issues relating to practical computer-based implementation are addressed.

### 4.3.1 Integrated formulation for shape optimization

This formulation considers the class of continuum structures modeled by two-dimensional finite elements. The objective is to determine the shape of the boundary of the structure which minimizes some quantity, such as total volume, while satisfying side constraints on stress, displacement, shape, etc. throughout the structure.

The specific finite element considered in this formulation is the six-noded linear strain triangle (LST) element. This element has a quadratically varying displacement field and linearly varying stress and strain fields over the face of the element. Isoparametric elements are not considered because they require numerical integration and an explicit analytical representation of the inverse of the Jacobian matrix is cumbersome (Babu and Pinder, 1984). The formulation is easily extendable, however, to all nonisoparametric rectangular and triangular families of elements, including three-dimensional finite elements. The particular choice of the LST element is based on its ability to model flexure well, without an excessive number of elements, and because its triangular shape provides a good fit to a wide variety of shapes. This element has been found to work well for shape optimization (Pedersen, 1981).

The objective function is chosen to be volume, although any algebraic function of the design variables may be used instead. The shape variables are taken as the $x, y$ coordinates of all element corner nodes in the finite element mesh. This allows the generality of a piecewise linear boundary shape. The coordinates of the interior corner nodes are tied to the coordinates of the boundary corner nodes by equality constraints. This avoids the problems of insensitivity of the objective function to interior node movement. These equality constraints are also chosen to maintain a well-proportioned mesh as the shape of the boundary changes.

Behavioral design variables consist of the displacement components at each node for each load case and the stress components at each corner of each element for each load case. It will be shown that this selection of design variables leads to a convenient formulation.

Constraints may be placed on stress, displacement, shape, or any algebraic functions of the design variables: nodal coordinates, nodal displacements, and element stresses. Alternate load cases are considered in the formulation. All load cases are independent; combinations of loads must be specified as additional independent load cases.

### (a) Development of the stiffness equality constraints

In this section, the explicit stiffness equality constraints are derived. They relate nodal displacements to externally applied nodal forces, utilizing an intermediate

162     *Geometric programming and structural optimization*

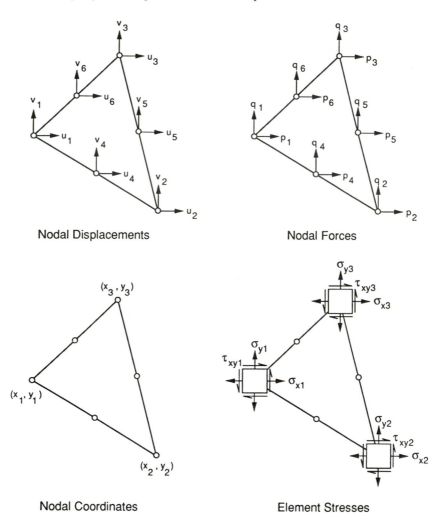

**Fig. 4.12** Linear strain triangle finite element.

set of variables to simplify the formulation. These constraints may be viewed as being equivalent to individual rows of the stiffness matrix, but with an explicit representation of the nonlinearities that are introduced by the nodal coordinates of the finite elements being treated as variables.

The assumed quadratic displacement component fields within the LST element are uniquely determined by the two nodal displacement components of the three corner nodes and three midside nodes. These twelve degrees of freedom are shown in Fig. 4.12. The displacement component fields and the twelve nodal displacements are related by shape functions, $\mathbf{N}$:

$$\begin{Bmatrix} u \\ v \end{Bmatrix} = [\mathbf{N}]\{u_1\,u_2\,u_3\,u_4\,u_5\,u_6\,v_1\,v_2\,v_3\,v_4\,v_5\,v_6\}^{\mathrm{T}} \tag{4.8}$$

where

$$[\mathbf{N}] = \begin{bmatrix} f_1 f_2 f_3 f_4 f_5 f_6 & 0 & 0 & 0 & 0 & 0 & 0 \\ 0 & 0 & 0 & 0 & 0 & 0 & f_1 f_2 f_3 f_4 f_5 f_6 \end{bmatrix}$$

and

$$f_1 = L_1(2L_1 - 1), f_2 = L_2(2L_2 - 1), f_3 = L_3(2L_3 - 1),$$
$$f_4 = 4L_1L_2, f_5 = 4L_2L_3, f_6 = 4L_3L_1$$

$$\begin{Bmatrix} L_1 \\ L_2 \\ L_3 \end{Bmatrix} = \frac{1}{2A} \begin{bmatrix} x_2y_3 - x_3y_2 & y_2 - y_3 & x_3 - x_2 \\ x_3y_1 - x_1y_3 & y_3 - y_1 & x_1 - x_3 \\ x_1y_2 - x_2y_1 & y_1 - y_2 & x_2 - x_1 \end{bmatrix} \begin{Bmatrix} 1 \\ x \\ y \end{Bmatrix}$$

In these equations, $u$ and $v$ are the displacement components at an arbitrary location $(x, y)$ within the element. The nodal displacement components in the $x$ and $y$ directions are $u_i$ and $v_i$ respectively, for $i = 1, 2, ..., 6$. The $x$ coordinates and $y$ coordinates of the corner nodes are $x_i$ and $y_i$ for $i = 1, 2$, and 3. The symbol $A$ represents the area of the element. Note that matrix $\mathbf{N}$ has elements that are nonlinear functions of the nodal coordinates.

The strains within the element are determined using the usual strain-displacement relationships, $\varepsilon_x = \partial u / \partial x, \varepsilon_y = \partial v / \partial y$, and $\gamma_{xy} = \partial u / \partial y + \partial v / \partial x$, which results in the strain–displacement matrix, $\mathbf{B}$:

$$\begin{Bmatrix} \varepsilon_x \\ \varepsilon_y \\ \gamma_{xy} \end{Bmatrix} = [\mathbf{B}] \{ u_1 \, u_2 \, u_3 \, u_4 \, u_5 \, u_6 \, v_1 \, v_2 \, v_3 \, v_4 \, v_5 \, v_6 \}^T$$

where

$$[\mathbf{B}] = \begin{bmatrix} \partial / \partial x & 0 \\ 0 & \partial / \partial y \\ \partial / \partial y & \partial / \partial x \end{bmatrix} [\mathbf{N}]$$

Here, $\varepsilon$ and $\gamma$ are the components of strain within the element.

Stresses within the element are related to strains by the stress–strain matrix, $\mathbf{D}$:

$$\begin{Bmatrix} \sigma_x \\ \sigma_y \\ \tau_{xy} \end{Bmatrix} = [\mathbf{D}] \begin{Bmatrix} \varepsilon_x \\ \varepsilon_y \\ \gamma_{xy} \end{Bmatrix}$$

where

$$[\mathbf{D}] = \frac{E}{1 - v^2} \begin{bmatrix} 1 & v & 0 \\ v & 1 & 0 \\ 0 & 0 & (1 - v)/2 \end{bmatrix}$$

Here, $\sigma$ and $\tau$ are the components of stress throughout the element, $E$ is the modulus of elasticity, and $v$ is Poisson's ratio. Because the stresses vary linearly throughout the LST finite element, nine stress component quantities are sufficient to uniquely define the state of stress throughout the element. These nine quantities are shown in Fig. 4.12.

The element stiffness matrix is produced by a volume integral over a product of matrices $\mathbf{B}$ and $\mathbf{D}$ (Zienkiewicz, 1977). The element stiffness matrix relates nodal displacements to internal nodal forces, $p_i$ and $q_i$:

$$\{p_1 \cdots p_6 q_1 \cdots q_6\}^T = \left[ \iiint_V [\mathbf{B}]^T [\mathbf{D}][\mathbf{B}] \, dV \right] \{u_1 \cdots u_6 v_1 \cdots v_6\}^T$$

If the shape of the structure were not permitted to vary, then the nodal coordinates would be constant, yielding linear equilibrium equations as rows of the stiffness matrix. In this case, however, the nodal coordinates are permitted to vary and **B** becomes a matrix of nonlinear functions of the nodal coordinates. Much of the algebraic complexity of expressing the rows of the stiffness matrix in explicit form may be circumvented by introducing an additional set of variables, the stress components at each corner of each element. This allows the element stiffness equations to be decomposed into two sets of equations:

$$\{p_1 p_2 p_3 p_4 p_5 p_6 q_1 q_2 q_3 q_4 q_5 q_6\}^T = \iiint_V [\mathbf{B}]^T \left\{ \begin{array}{c} \sigma_x \\ \sigma_y \\ \tau_{xy} \end{array} \right\} dV$$

and

$$\left\{ \begin{array}{c} \sigma_x \\ \sigma_y \\ \tau_{xy} \end{array} \right\} = [\mathbf{D}][\mathbf{B}] \{u_1 \, u_2 \, u_3 \, u_4 \, u_5 \, u_5 \, u_6 \, v_1 \, v_2 \, v_3 \, v_4 \, v_5 \, v_6\}^T$$

For example, the explicit form of the $x$ component of nodal force at node $1, p_1$, is

$$p_1 = \frac{t}{6}(-y_2\sigma_{x1} + y_3\sigma_{x1} - x_3\tau_{xy1} + x_2\tau_{xy1})$$

and the explicit form of the $x$ component of normal stress at node $1, \sigma_{x1}$, is

$$\sigma_{x1} = \frac{E}{2A(1-v^2)}(3y_2u_1 - 3y_3u_1 - y_3u_2 + y_1u_2 - y_1u_3 + y_2u_3 + 4y_3u_4 - 4y_1u_4$$

$$+ 4y_1u_6 - 4y_2u_6 + 3vx_3v_1 - 3vx_2v_1 - vx_1v_2 + vx_3v_2 - vx_2v_3$$
$$+ vx_1v_3 + 4vx_1v_4 - 4vx_3v_4 + 4vx_2v_6 - 4vx_1v_6)$$

The entire set of explicit stiffness expressions for the LST element has been compiled in Burns (1987a).

Equilibrium is maintained by summing all internal forces acting on each node and setting the sum equal to the externally applied, work-equivalent nodal loads, $F_x$ and $F_y$:

$$F_x = \sum_{inc} p \quad \text{and} \quad F_y = \sum_{inc} q$$

Here, 'inc' represents a summation over all elements incident to the node. The summation is equivalent to the process of assembly of the global stiffness matrix.

### (b) Stress limits

The introduction of stress components as intermediate variables not only simplifies the stiffness equations, but also provides a convenient means of constraining the stress magnitudes throughout the structure. Principal stresses are easily computed from the stress components,

$$\sigma_{1,2} = \frac{\sigma_x + \sigma_y}{2} \pm \left[ \left( \frac{\sigma_x - \sigma_y}{2} \right)^2 + \tau_{xy}^2 \right]^{1/2}$$

and inequality constraints which limit the principal stresses throughout the structure can be imposed:

$$\left| \frac{\sigma_x + \sigma_y}{2} \pm \left[ \left( \frac{\sigma_x - \sigma_y}{2} \right)^2 + \tau_{xy}^2 \right]^{1/2} \right| \leqslant \sigma_{\text{allowable}}$$

This is equivalent to the maximum normal stress failure theory. Other failure theories, such as the maximum distortion energy failure theory, could be implemented instead, when appropriate.

## (c) Problem formulation

The final step in formulating the integrated shape optimization problem is to relate nodal coordinates to element areas. This is done for two reasons. First, the objective function formulation is greatly simplified, and, second, the shape function expressions (4.8) are also greatly simplified when element area is treated as a single quantity. Combining all of the previous equality and inequality constraints, the final integrated formulation is

minimize

$$\sum_i A_i t_i \qquad\qquad\qquad\text{total volume (}t\text{ is element thickness)}$$

subject to

$$2A = x_3 y_1 - x_2 y_1 + x_1 y_2 - x_3 y_2 + x_2 y_3 - x_1 y_3 \qquad\qquad\text{element area}$$

$$\begin{Bmatrix} \sigma_x \\ \sigma_y \\ \tau_{xy} \end{Bmatrix} = [\mathbf{D}][\mathbf{B}] \{ u_1\, u_2\, u_3\, u_4\, u_5\, u_6\, v_1\, v_2\, v_3\, v_4\, v_5\, v_6 \}^{\mathrm{T}} \qquad\text{stress–displacement}$$

$$\{ p_1\, p_2\, p_3\, p_4\, p_5\, p_6\, q_1\, q_2\, q_3\, q_4\, q_5\, q_6 \}^{\mathrm{T}} = \iiint_V [\mathbf{B}]^{\mathrm{T}} \begin{Bmatrix} \sigma_x \\ \sigma_y \\ \tau_{xy} \end{Bmatrix} \mathrm{d}V \qquad\text{force–stress}$$

$$F_x = \sum_{\text{inc}} p \quad\text{and}\quad F_y = \sum_{\text{inc}} q \qquad\qquad\qquad\text{equilibrium}$$

$$\left| \frac{\sigma_x + \sigma_y}{2} \pm \left[ \left( \frac{\sigma_x - \sigma_y}{2} \right)^2 + \tau_{xy}^2 \right]^{1/2} \right| \leqslant \sigma_{\text{allowable}} \qquad\text{stress limits}$$

$$f(x, y) = 0 \qquad\qquad\qquad\qquad\text{coordinate linking}$$

$$u_{\text{min}} \leqslant u \leqslant u_{\text{max}}$$
$$v_{\text{min}} \leqslant v \leqslant v_{\text{max}} \qquad\qquad\text{support conditions and displacement limits}$$

$$x_{\text{min}} \leqslant x \leqslant x_{\text{max}}$$
$$y_{\text{min}} \leqslant y \leqslant y_{\text{max}} \qquad\qquad\text{envelopes on geometry and move limits}$$

The 'coordinate linking' constraints in the above formulation serve several purposes. The most important is the control of the location of interior nodes. A piecewise linear boundary shape is determined by the locations of the boundary nodes only. The interior nodes must be related to these boundary nodes so that a well-proportioned finite element mesh is maintained as the shape of the structure evolves during the optimization process. This is accomplished through a set of constraints that specify the locations of the interior nodes relative to other nodes, or that specify an upper bound on the aspect ratio of each finite element.

Another use of the coordinate linking constraints is to regulate the shape of the boundary. For example, it may be desirable to restrict a portion of the boundary to be flat, or to limit the curvature of a portion of the boundary. Algebraic constraint functions of the nodal coordinates may be imposed to do this. Envelopes on the allowable geometry may exist, leading to inequality coordinate linking constraints.

Of course, other constraints may be added to the formulation, and they need not be restricted to the nodal coordinate variables. Any algebraic function of the stress, displacement, coordinate, and element area variables can be specified. For example, displacement-dependent stress limits could be imposed.

The support conditions on the structure are handled through the bounds on the displacement variables. A fixed node would have its displacement degrees of freedom fixed between identical upper and lower bounds. More exotic support conditions are implemented with constraint functions relating the displacements to the other variables. Constraints on maximum displacement are enforced as bounds on the corresponding displacement variables. The bounds on the nodal coordinate variables can serve as move limits if the change in shape of the structure is too rapid from one iteration to the next.

---

## EXAMPLE 4.3 VARIABLE DEPTH FIXED-ENDED BEAM

This section presents an example which illustrates the application of the integrated formulation to the shape optimization of a fixed-ended beam of variable depth under the action of a unformly distributed load. The beam has a 60 in span length, a selfweight of $150 \, \text{lb ft}^{-3}$, and a distributed load of $1 \, \text{klbf in}^{-1}$ applied along the top surface. The beam has a uniform thickness of 12 in. The mesh consists of 48 elements but symmetry is used to reduce it to 24. The $x$ coordinates of all corner nodes are fixed. The $y$ coordinates of the interior corner nodes are constrained to lie at the centroid of the corner nodes above and below them. The $y$ coordinates of the corner nodes on the top and bottom surface of the beam are unrestricted. Principal stresses are constrained between $-4 \, \text{klbf in}^{-2}$ and $4 \, \text{klbf in}^{-2}$ everywhere throughout the beam; no constraints are placed on maximum displacement. The modulus of elasticity and Poisson's ratio of this hypothetical material are specified as $4000 \, \text{klbf in}^{-2}$ and 0.15, respectively.

Figure 4.13 illustrates the iterative solution sequence using GGP to solve the integrated formulation. The solution is obtained in six iterations, with each iteration requiring an average of 1.7 min of computation on a Apple Macintosh IIfx computer.

A second solution of this problem was performed, this time with the allowable tensile stress reduced from $4 \, \text{klbf in}^{-2}$ to $3 \, \text{klbf in}^{-2}$. The allowable compressive stress was kept at $-4 \, \text{klbf in}^{-2}$. The sequence of iterations is presented in

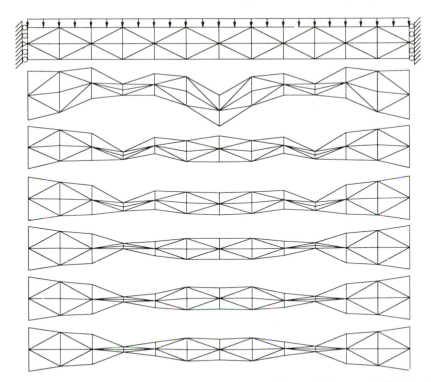

**Fig. 4.13** First six iterations of the shape optimization of a fixed-ended beam with distributed load.

Fig. 4.14. This time, the structure evolves into an arch which experiences uniform axial compression everywhere. The solution process has completely avoided the more restrictive allowable tensile stress constraints, which are not active in this solution at all.

### 4.3.2 Discussion of example problem results

It is interesting to note that the second solution above has far less volume than the first solution. Since the second problem is more highly constrained than the first, this indicates that the first solution was not a global minimum. Tightening the allowable tensile stress constraints forced the solution out of the symmetric solution and into the arch-shaped global minimum. Because the allowable tensile stress constraints are not binding in the arch solution, this indicates that the arch solution is an alternative solution to the first problem with equal tensile and compressive allowable stresses. Had the first problem been initiated with a starting configuration more closely resembling the arch-shaped solution, the solution process could have converged to the arch solution instead. A third optimal solution to the first problem is a downward sweeping arch in uniform

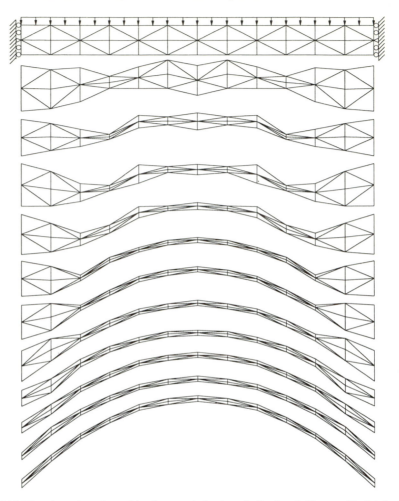

**Fig. 4.14** First eleven iterations of the shape optimization of a fixed-ended beam with distributed load and tightened allowable tensile stress constraints.

axial tension. This problem suggests that local minima are certainly possible, and probably quite prevalent, in shape optimization.

This example also raises another important issue. When the two-phase approach of shape optimization is used, it becomes necessary to define 'shape functions' which limit the range of possible shapes the structure may assume during the course of optimization. Shape functions provide a means of specifying the shape of the entire structure through the use of a relatively small number of shape parameters. This simplifies the sensitivity analysis because derivatives of behavior are calculated only with respect to the shape parameters. Unfortunately, the final shape becomes highly dependent on the particular choice of shape functions. Extreme care must be taken to ensure that the shape functions will not overly constrain the solution, possibly excluding useful shapes that are not expressible by the shape functions. With the integrated formulation, each nodal coordinate can be taken as a shape variable, if desired. This results in a general piecewise linear boundary shape. In areas of the structure where it is desirable to have less variability in shape, for example, on a surface that is desired to be flat, shape functions can easily be imposed as additional constraints in the problem formulation.

This suggests a potentially useful interactive strategy. Initially, all boundary nodes can be allowed to vary, with the exception of perhaps the location of the supports. Then as the shape optimization progresses, those areas where the shape is evolving in an undesirable way can be constrained interactively. This strategy improves the likelihood of discovering useful, yet unexpected, optimal shapes without having to build expectations into a set of imposed shape functions.

## 4.4  IMPLEMENTATION OF GGP FOR SHAPE OPTIMIZATION

Solution of the large mathematical program produced by the integrated formulation requires a great deal of computational effort and also substantial computer storage. Each finite element requires nine stress variables per load case, and each node requires two displacement variables per load case. In addition, each corner node requires two coordinate variables. Each stress variable produces an equality constraint relating it to displacement and nodal coordinate variables. There is an additional pair of equality constraints for each node that maintains equilibrium in the $x$ and $y$ directions at that node for each load case. Finally, there are two inequality constraints for each corner node, for each load case, constraining the principal stress magnitudes. The example presented in the previous section had 412 variables and 517 constraints.

Figure 4.7 illustrates that three separate problem formulations coexist during the GGP solution process – the original integrated formulation, the posynomial subproblem, and the LP. These three formulations must be represented internally in the computer, which may require a great deal of storage capacity. The bulk of the computational effort lies with the solution of the LP. Although LPs are routinely solved with thousands of variables, the computational effort is not small. Having to solve a new LP for each outer loop iteration may increase the required effort by an order of magnitude. This section addresses the issues of storage and computational effort and suggests implementation strategies for both.

### 4.4.1  Constraint emitters

This section concerns the issue of efficient use of computer memory. Up to this point, it has been assumed that the integrated formulation is supplied in explicit form. This, of course, would be impractical to do by hand for even a small shape optimization problem. It would be possible to construct, in software, a system to generate the explicit formulation from a standard structural database describing the structure to be optimized. While this is certainly a step in the right direction, it still requires that all three problem formulations shown in Fig. 4.7 be created.

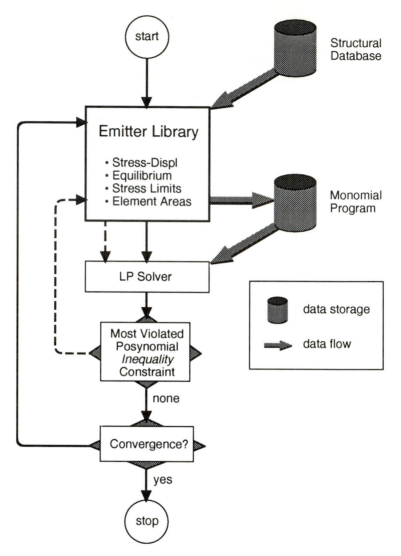

**Fig. 4.15** Constraint emitters in the solution process.

A more useful strategy makes use of what are termed 'constraint emitters'. Since the format of the constraints in the integrated formulation is known in advance, a general-purpose condensation routine is not needed. This activity is more efficiently performed by special-purpose software routines called 'constraint emitters', which generate the log–monomial linearizations directly from the structural database, given the two operating points belonging to the outer and inner loops, $\bar{x}$ and $\hat{x}$. Consequently, the only problem formulation that ever exists in memory is the LP problem. The integrated formulation and the posynomial subproblems never exist explicitly in memory. They are replaced by a much smaller structural database describing the problem being solved (nodal coordinates, element connectivity, loads, supports, etc.). Figure 4.15 shows the modified program flowchart after the emitters have been incorporated. One emitter exists for each class of constraint in the integrated formulation and together they form a library of emitter subroutines.

The constraint emitters may be viewed as producing a 'convex linearization' directly from the general structural description. This can be viewed as a form of approximate structural modeling, a strategy that has appeared frequently in shape optimization literature (Schmit and Farshi, 1974; Bennett, 1981; Kirsch, 1981; Kirsch and Hofman, 1981; Schmit, 1981; Prasad, 1983).

### 4.4.2 Preferred pivots

The second strategy concerns computational effort rather than efficient use of computer memory. Recall that for each outer loop iteration, a complete cutting plane problem is solved. The only information passed from one outer loop iteration to the next is the updated set of design variable values. A considerable saving in computational effort is possible if additional information is passed between outer loop iterations.

At the end of any given cutting plane procedure, the solution of a posynomial subproblem exists. However, in addition, an extended LP tableau that gave rise to this solution exists, including a set of columns of the LP tableau that form a square basis matrix. Recall that the dual formulation is being solved. Each column of the tableau is associated with one constraint of the original problem formulation, and each row is associated with one of the design variables. Through careful bookkeeping, it is possible to keep track of which constraints of the original problem formulation are associated with the specific columns that constitute the basis matrix.

The 'preferred pivot' strategy makes use of this bookkeeping. The set of constraints associated with the basis matrix, or the 'active set', tends to be similar from one upper-level iteration to the next, particularly toward later iterations. If, at the start of an inner loop iteration, the LP is initialized with a basis consisting of the columns derived from the active set of the previous cutting plane problem, then far less pivoting is required to solve the new LP than would otherwise be required by starting with a standard basis of slack variables. It has been demonstrated that a 50% saving in total computation can result from the preferred pivot strategy (Burns, 1985).

Some risk is associated with this strategy, however. It is possible, particularly during the early iterations when the active sets tend to be less similar, that a basis matrix formed from a previous active set may violate initial feasibility of the dual problem. In this case, the method must revert to the safe approach of starting with a basis of slack variables. A useful hybrid strategy is to give preference to the active-set constraints during pivoting operations in the early upper-level iterations instead of forcing them into the initial basis. In other words, the hybrid method would start with a basis of slack variables, but when searching for a column to enter the basis, would be restricted to the active-set constraints until no further improvement can be made, at which time the other columns would be included. During these early upper-level iterations, the active set would be monitored. When successive active sets start to become similar, then the forced-basis strategy would take over.

## 4.5 CONCLUSIONS

The generalized geometric programming method has been presented as an effective tool for optimizing nonlinear problems with equality and inequality constraints. It is demonstrated how the method reduces the optimization problem to a stable sequence of linear programs through a convex linearization scheme. Other linearization-based methods tend to experience difficulties when treating nonlinear problems. By establishing the convex subproblem prior to linearization, the GGP method avoids these difficulties.

In addition, an integrated formulation for shape optimization has been presented as an alternative to the standard two-phase procedure. The integrated approach has several notable attributes: (1) an explicit sensitivity analysis is not needed; (2) the bulk of the computational effort is in the solution of linear programs, for which highly developed and efficient computer codes are readily available; (3) shape functions are no longer necessary to make the sensitivity analysis computationally economical – useful optimal shapes are more likely to be found that might otherwise be disallowed by poorly chosen or overly restrictive shape functions.

## APPENDIX 4A CONVERSION FACTORS BETWEEN US CUSTOMARY AND SI UNITS

| To convert | To | Multiply by |
|---|---|---|
| in | m | 0.02540 |
| ft | m | 0.3048 |
| lbf | N | 4.448 |
| klbf | kN | 4.448 |
| $klbf\,ft^{-1}$ | $kN\,m^{-1}$ | 14.59 |
| $klbf\,in^{-2}$ | MPa | 6.894 |

## REFERENCES

Abuyounes, S. and Adeli, H. (1986) Optimization of steel plate girders via general geometric programming. *Journal of Structural Mechanics*, **14**(4), 501–24.

Adeli, H. and Chompooming, K. (1989) Interactive optimization of nonprismatic girders. *Computers and Structures*, **31**(4), 505–22.

Adeli, H. and Kamal, O. (1986) Efficient optimization of space trusses. *Computers and Structures*, **24**(3), 501–11.

Avriel, M. and Williams, A. (1970) Complementary geometric programming. *SIAM Journal of Applied Methemetics*, **19**(1), 125–41.

Avriel, M., Dembo, R. and Passy, U. (1975) Solution of generalized geometric programs. *International Journal of Numerical Methods in Engineering*, **9**, 149–68.

Babu, D.K. and Pinder, G.F. (1984) Analytical integration formulae for linear isoparametric finite elements. *International Journal of Numerical Methods in Engineering*, **20**, 1153–66.

Beckenbach, E. and Bellman R. (1961) *An Introduction to Inequalities*, Singer.

Bennett, J.A. (1981) Application of linear constraint approximations to frame structures.

Proceedings of the 11th ONR Naval Structural Mechanics International Symposium on Optimum Structural Design, Tucson, AZ, pp. 7-9–7-15.

Burns, S.A. (1985) Structural optimization using geometric programming and the integrated formulation. Ph.D. Thesis, University of Illinois at Urbana-Champaign.

Burns, S.A. (1987a) Simultaneous design and analysis using geometric programming and the integrated formulation. Proceedings of the Swanson Analysis Systems, Inc. ANSYS 1987 Conference and Exhibition, Newport Beach, CA.

Burns, S.A. (1987b) Generalized geometric programming with many equality constraints. *International Journal of Numerical Methods in Engineering*, **24**(4), 725–41.

Burns, S.A. and Locascio, A. (1991) A monomial-based method for solving systems of nonlinear algebraic equations. *International Journal of Numerical Methods in Engineering*, **31**(7), 1295–318.

Chibani, L., Burns, S.A. and Khachaturian, N. (1989) A move coordination method for alternative loads in structural optimization. *International Journal of Numerical Methods in Engineering*, **28**, 1041–60.

Duffin, R.J. (1970) Linearizing geometric programs. *SIAM Review*, **12**(2), 211–27.

Duffin, R.J., Peterson, E. and Zener, C. (1967) *Geometric programming*, Wiley, New York.

Fox, R.L. and Schmit, L.A. Jr (1966) Advances in the integrated approach to structural synthesis. *Journal of Spacecraft*, **3**(6), 858–66.

Fuchs, M.B. (1982) Explicit optimum design. *International Journal of Solids and Structures*, **18**(1), 13–22.

Haftka, R.T. (1985) Simultaneous analysis and design. *AIAA Journal*, **23**, 1099–103.

Hardy, G., Littlewood, J. and Polya, G. (1959) *Inequalities*, Cambridge University Press, Cambridge.

Kelly, J.E. (1960) The cutting plane method for solving convex problems. *Journal of the Society for Industrial and Appled Mathematics*, **8**(4), 703–12.

Kirsch, U. (1981) *Optimum Structural Design*, McGraw-Hill, New York, p. 18.

Kirsch, U. and Hofman, B. (1981) Approximate behavior models for optimum structural design. Proceeding of the 11th ONR Naval Structural Mechanics International Symposion on Optimum Structural Design, Tucson, AZ, pp. 7-17–7-26.

Pascual, L. and Ben-Israel, A. (1970) Constrained maximization of posynomials by geometric programming. *Journal of Optimization Theory and Applications*, **5**, 73–86.

Passy, U. (1971) Generalized weighted mean programming. *SIAM Journal of Applied Mathematics*, **20**, 763–78.

Pedersen, P. (1981) The integrated approach of FEM–SLP for solving problems of optimal design, in *Optimization of Distributed Parameter Structures*, Vol. 14 (eds E. Haug and J. Cea), Sijthoff and Noordhoff, NATO-ASI, Series E, No. 49, pp. 757–80.

Prasad, B. (1983) Explicit constraint approximation forms in structural optimization. *Computer Methods in Applied Mechanical Engineering*, **40**, 1–26.

Rijckaert, M.J. and Martens, X.M. (1978) Bibliographical note on geometric programming. *Journal of optimization Theory and Applications*, **26**(2), 325–37.

Schmit, L.A. (1981) Structural optimization, some key ideas and insights. Proceedings of the 11th ONR Naval Structural Mechanics International Symposium on Optimum Structural Design, Tucson, AZ, pp. 1–7.

Schmit, L.A. and Farshi, B. (1974) Some approximation concepts for structural synthesis. *AIAA Journal*, **12**(5), 692–9.

Schmit, L.A., Jr, and Fox, R.L. (1965) An integrated approach to structural synthesis and analysis. *AIAA Journal*, **3**, 1104–12.

Soeiro, A. and Hoit, M. (1987) Sizing optimization, in *Computer Applications in Structural Engineering* (ed. D.R. Jenkins), American Society of Civil Engineers, New York, pp. 342–56.

Zienkiewicz, O.C. (1977) *The Finite Element Method*, 3rd edn, McGraw-Hill (UK) Ltd., London, p. 104.

# 5

# Nonlinear mixed integer programming

MANOHAR P. KAMAT and LUIS MESQUITA

## 5.1 INTRODUCTION

Many variables entering into engineering design problems are restricted to assume only discrete values. There are two important reasons for this. The first one could be due to manufacturing constraints or commercial viability. Examples would be pipes, I-beams, screws etc. that are commercially available only in certain discrete sizes. The second reason is the nature of the design variables. Examples would be number of items, such as the number of stiffeners on a plate, the number of transverse bulkheads on a ship, the number of bars in a truss, the number of blades in a gas turbine engine etc. Practical engineering problems could be such that some of the design variables can take any finite continuous values while others may take certain integer values. Furthermore, if the objective function for such problems is nonlinear then the problem falls into the class of nonlinear mixed integer programming. Most optimization packages assume that the design variables can take any continuous values subject to certain design constraints while the objective function is minimized. In problems where the design variables can only take integer values, the practice is to round off these variables in the final continuous solution thereby obtaining only an approximate solution to the truly mixed integer programming problem.

Integer programming involves the use of an algorithm that forces the solution obtained to integer values. Techniques used therein involve the solution of a series of optimization subproblems. The subproblems are obtained by imposing additional constraints on the original problem. The conventional way of obtaining the integer solution by rounding off the continuous solution involves making a decision as to which design variables should be rounded off to the next lower or higher integer. Usually there is no rational way to make this decision. Secondly, the rounded-off solution may be in the infeasible region and moving from the infeasible region to the feasible region could be difficult. More importantly the true integer solution to the nonlinear mixed integer programming problem has a better chance of being globally optimal when the solution space is concave. This follows from the fact that the subproblems solved have disjoint feasible spaces.

The computational effort required to solve a continuous optimization problem followed by rounding off the integer variables is almost always far less than the effort to solve the same problem using integer programming. If only a few of the design variables are required to be integers, then it may be possible to obtain a solution by simple enumeration. However, if the number of integer variables is large (say greater than five) then the number of possible combinations is prohibitively large. For example, if we have six integer design variables each of which can take on integer values in the range from 1 to 10, the possible number of combinations is one million. In instances such as these, integer programming is the only viable option.

## 5.2 NONLINEAR MIXED INTEGER PROGRAMMING

A general optimization problem could be stated as follows:

$$\text{minimize} \quad f(\mathbf{X}), \mathbf{X} = (x_1 \, x_2 \cdots x_n)^t \tag{5.1}$$

$$\text{subject to} \quad h_i(\mathbf{X}) = 0 \qquad i = 1, 2, ..., n_e \tag{5.2}$$

$$g_j(\mathbf{X}) > 0 \qquad j = 1, 2, ..., n_g \tag{5.3}$$

The vector $\mathbf{X}$ will be referred to as the design variable vector. For a structural optimization problem, it normally represents some characteristics of structural elements such as cross-sectional diameters, areas, etc. If $f$, $h_i$, and $g_j$ are all linear functions of the design variable vector $\mathbf{X}$ the problem is a linear programming problem (LP). If any of the $f$, $h_i$ or $g_j$ are nonlinear functions of $\mathbf{X}$, a nonlinear programming problem (NLP) results. If all the design variables are such that they can assume any real values within specified bounds then one has the classical continuous variable optimization problem. In several structural design problems, however, some or all of the design variables may assume only integer or discrete values. Such problems can be formulated as nonlinear mixed integer programming problems (NMIP).

Integer programming techniques fall into two broad types: (1) search methods and (2) cutting plane methods. The first type is motivated by the fact that the integer solution can be regarded as consisting of a finite number of points. In its simplest form, search methods seek the enumeration of all such points. This would be equivalent to a simple exhaustive enumeration. However, what makes search methods more promising than simple exhaustive enumeration is that they have been developed to enumerate only a portion of all candidate solutions while automatically discarding the remaining points because they are either infeasible or inferior to points that are retained. The efficiency of the resulting search algorithm depends on the power of such a technique to discard nonpromising solution points. Thus, search methods primarily include implicit enumeration techniques and branch-and-bound techniques. The first type is mostly suited for the 'zero–one' problem, wherein the integer design variables may take on values of either zero or one. Implicit enumeration techniques are not suited for discrete or integer problems which are not of the zero–one type as posed.

Cutting plane methods were the first systematic techniques available for the solution of linear integer optimization problems. The early work of Dantzig, Fulkerson and Johnson (1954) directed the attention of researchers to the importance of solving linear integer programming problems. Markovitz and

Manne (1957) considered the more general case of discrete variables. Dantzig (1959) was the first to propose the cutting plane method for linear integer solution in a finite number of steps. The first cutting plane method that guaranteed an integer solution in a finite number of steps is due to Gomory (1958). Though this algorithm converged in a finite number of steps, it ran into difficulties caused by machine round-off. A new algorithm was developed by Gomory (1960a) to overcome the roundoff problem. In his subsequent work, Gomory (1960b) extended the above methods to cover mixed integer problems. Glover (1965) introduced a new type of cutting plane method known as the bound escalation method. Young (1971), Balas (1971) and Glover (1971) developed what is known as the 'convexity' cut method.

The above cutting plane methods have the drawback that the solution is not available until the algorithm converges. A primal algorithm was first introduced by Ben-Israel and Charnes (1962); however, the first finite primal algorithm (a problem that converges in a finite number of steps) was developed by Young (1965). Improvements on these algorithms were given by Young (1968) and Glover (1968).

Cutting plane methods cannot be used for nonlinear problems as they are based on linearity assumptions. Branch and bound algorithms do not rely on linearity and can be used to solve nonlinear integer programming problems. The branch and bound procedure does not deal directly with the integer problem. Rather, it considers a continuous problem obtained by relaxing the integer restrictions on the variables. Thus the solution space of the integer problem is only a subset of the continuous space. The prime reason for dealing with the continuous problem is that it is simpler to manipulate such a problem. Furthermore, one can draw upon the tremendous advances in the theory of continuous optimization methods. If the optimal continuous solution is integer, then it is also optimum for the integer problem. The branch and bound technique consists of two basic operations:

- *Branching*. This operation partitions the continuous solution spaces into subspaces (subproblems), which are also continuous. The purpose of partitioning is to eliminate part of the continuous space that is not feasible for the integer problem. This elimination is achieved by imposing (mutually exclusive) constraints that are necessary conditions for producing integer solutions, but in a way that no feasible integer point is eliminated. In other words, the resulting collection of subproblems defines every feasible integer point of the original problem. Because of the nature of the partitioning operations it is called branching.
- *Bounding*. Assuming the original problem is of the minimization type the optimal objective values for each subproblem created by branching set a lower bound on the objective value associated with any of its integer feasible values. This bound is essential for 'ranking' the optimum solutions of the subsets, and hence in locating the optimum integer solution.

The first known branch and bound algorithm was developed by Land and Doig (1960) as an application to the mixed and pure integer problems. Dakin (1965) modified the Land and Doig algorithm to facilitate computer implementation, and also to make it applicable to nonlinear mixed integer programming problems. Gupta (1980) and Gupta and Ravindran (1983) use numerical results to show that Dakin's algorithm is well suited to solve relatively large, nonlinear mixed integer programming problems.

In addition to the above two basic classes of methods, several researchers have devised other variations of what may be called primal and dual methods primarily with a view to solving structural optimization problems involving integer or discrete variables along with continuous variables.

Schmit and Fleury (1980) use approximate concepts and dual methods to solve structural problems involving a mix of discrete and continuous sizing type of design variables. The optimization problem is converted into a sequence of explicit approximate primal problems of separable form. These problems are solved by constructing continuous explicit dual functions, which are maximized subject to simple nonnegativity constraints. Dual methods are not guaranteed to yield an optimum solution to the discrete problem, because of the nonconvexity involved. However, encouraging results are presented by Schmit and Fleury (1980).

Imai (1979) uses the above concepts for the problem of material selection. Hua (1983) uses an implicit enumeration scheme for finding the minimum weight design of a structure with discrete member sizes. A simplifying assumption is made, namely that, if reducing the size of a member causes an originally feasible solution to become infeasible, further reducing the size will not make it feasible. Johnson (1981) uses a branch and bound algorithm to find the minimum weight design of a rigid plastic structure with discrete sizes. Linking of elements is done to reduce the number of design variables. The objective function and the constraints are linear functions of the design variables, making it possible to use linear programming techniques to obtain the solution to the continuous problem.

Gisvold and Moe (1971) use a penalty method approach to minimize the weight of a structure with two discrete variables. The objective function that is to be minimized is augmented by a term that is positive everywhere and vanishes at the discrete points.

Templeman and Yates (1983) use a segmental method for the discrete design of a minimum weight structure. The problem of finding the discrete sizes of the cross-sectional areas is converted to one of finding the continuous lengths of element segments each having one of the two possible discrete cross-sectional areas. The resulting continuous problem is solved using linear programming. Linearity considerations are then used to force the discrete variables to assume discrete values.

Olsen and Vanderplaats (1989) present a numerical method for the solution of nonlinear discrete optimization problems. The method uses approximation techniques to create subproblems suitable for linear mixed integer programming methods.

Dong, Gurdal and Griffin (1990) have used a penalty formulation to solve a nonlinear integer optimization problem. The penalty approach is used for converting a constrained optimization problem into a sequence of unconstrained optimization problems. To force the solution toward discrete values sine function penalty terms are introduced. The penalty terms add zero penalty at discrete points and nonzero penalty at nondiscrete values.

In this chapter we discuss and illustrate the application of Dakin's modification of Land and Doig's algorithm for the solution of nonlinear mixed integer programming problems. The Land and Doig algorithm is unsuitable for general nonlinear mixed integer programming problems on two counts. Firstly, it employs a tree structure that requires excessive computer storage (Little *et al.*, 1960) and secondly the branching rules are tied to the assumption of linearity (Handy, 1975).

Dakin's modification of the Land and Doig algorithm makes the branching rule independent of the linearity condition. The algorithm starts by finding a solution to the continuous problem wherein the integrality requirements are relaxed. If this solution is integer, then it is the optimal solution to the given discrete problem. If the solution is noninteger then at least one integer variable, say $x_k$, can be divided into integral and fractional parts $[b]$ and $c$ respectively as

$$b_k = [b] + c \tag{5.4}$$

where $[b]$ is integral and $0 < c < 1$. If $x_k$ has to take an integer value then either one of the following two conditions must be satisfied.

$$x_k \leqslant [b] \tag{5.5}$$

or

$$x_k \geqslant [b] + 1 \tag{5.6}$$

Constraint conditions expressed by equations (5.5) and (5.6) have nothing to do with linearity and as such can be used to solve nonlinear mixed integer programming problems. Equations (5.5) and (5.6) lead to two subproblems each satisfying one of the above conditions.

*Subproblem (1)*

$$\text{minimize} \quad f(\mathbf{X}), \mathbf{X} = (x_1 \, x_2 \cdots x_n)^t \tag{5.7}$$
$$\text{subject to} \quad h_i(\mathbf{X}) = 0 \quad i = 1, 2, \ldots, n_e \tag{5.8}$$
$$g_i(\mathbf{X}) > 0 \quad j = 1, 2, \ldots, n_g \tag{5.9}$$
$$x_k \leqslant [b] \tag{5.10}$$

*Subproblem (2)*

$$\text{minimize} \quad f(\mathbf{X}), \mathbf{X} = (x_1 \, x_2 \cdots x_n)^t \tag{5.11}$$
$$\text{subject to} \quad h_i(\mathbf{X}) = 0 \quad i = 1, 2, \ldots, n_e \tag{5.12}$$
$$g_i(\mathbf{X}) > 0 \quad j = 1, 2, \ldots, n_g \tag{5.13}$$
$$x_k \geqslant [b] + 1 \tag{5.14}$$

In the subproblems, the integrality requirement on the design variables has been removed. Furthermore, the two subproblems have removed the space $[b] < x_k < [b] + 1$ from the feasible region as this space is not allowable for an integral solution. It should also be noted that none of the integer feasible solutions has been eliminated. Each of these subproblems is then solved again as a continuous problem and the information regarding the optimal solution is stored at a node with corresponding value of the objective function, and also the lower and upper bounds on the design variables.

The foregoing procedure of branching and solving a sequence of continuous problems is continued until a feasible integer solution to one of the continuous problems is found. This integer feasible solution becomes an upper bound on the objective function. At this point, all nodes that have a value of the objective function higher than the upper bound are eliminated from further consideration, and the corresponding nodes are said to be fathomed.

The procedure of branching and fathoming is repeated for each of the unfathomed nodes. When a feasible integer solution is found, and when the value of the objective function is less than the upper bound to date, it becomes the new upper bound on the objective function. A node is fathomed (i.e. no further consideration is required) in any of the following cases.

- The continuous solution is a feasible integer solution.
- The continuous solution is infeasible.
- The optimal value (of the continuous problem) is higher than the current upper bound.

This search for the optimal solution terminates when all nodes are fathomed. The current best integer solution gives the optimal solution to the given discrete optimization problems.

This process is illustrated by the following simple example involving two variables:

$$\text{minimize} \quad f(n_1, n_2) = 4n_1^2 - 5n_1 n_2 + n_2^2 + 20 \qquad (5.15)$$

subject to the constraints

$$n_1 + n_2 \leqslant 10 \qquad (5.16)$$

$$n_1 > 0 \qquad (5.17)$$

$$n_2 > 0 \qquad (5.18)$$

where $n_1$, $n_2$ are integers.

Solving the above problem yields a continuous solution:

$$n_1 = 3.8461, \quad n_2 = 6.1538, \quad f(n_1, n_2) = -1.3014$$

Choosing the first variable $n_1$ for branching, we obtain two subproblems.

*Subproblem (1)*

$$\text{minimize} \quad f(n_1, n_2) = 4n_1^2 - 5n_1 n_2 + n_2^2 + 20$$
$$\text{subject to} \quad n_1 + n_2 \leqslant 10$$
$$n_1 > 0$$
$$n_2 > 0$$
$$n_1 \leqslant 3 \qquad (5.19)$$

*Subproblem (2)*

$$\text{minimize} \quad f(n_1, n_2) = 4n_1^2 - 5n_1 n_2 + n_2^2 + 20$$
$$\text{subject to} \quad n_1 + n_2 < 10$$
$$n_1 > 0$$
$$n_2 > 0$$
$$n_1 \geqslant 4 \qquad (5.20)$$

Conditions (5.19) and (5.20) are the additional constraints imposed to branch the solution. Solution of subproblem (1) leads to $n_1 = 3$, $n_2 = 7$, with the objective function $f(n_1, n_2) = 0$. Solution of subproblem (2) leads to $n_1 = 4$, $n_2 = 6$ with the objective function $f(n_1, n_2) = 0$. Solutions to subproblems (1) and (2) are both integer feasible solutions. In this case both integer solutions have the same value for the objective function, namely 0. The above solutions are obtained by branching the first design variable $n_1$. By branching the second design variable $n_2$, leads to two subproblems that are denoted as subproblems (3) and (4).

*Subproblem (3)*

$$\text{minimize} \quad f(n_1, n_2) = 4n_1^2 - 5n_1 n_2 + n_2^2 + 20$$
$$\text{subject to} \quad n_1 + n_2 \leqslant 10$$
$$n_1 > 0$$
$$n_2 > 0$$
$$n_2 \leqslant 6 \tag{5.21}$$

*Subproblem (4)*

$$\text{minimize} \quad f(n_1, n_2) = 4n_1^2 - 5n_1 n_2 + n_2^2 + 20$$
$$\text{subject to} \quad n_1 + n_2 \leqslant 10$$
$$n_1 > 0$$
$$n_2 > 0$$
$$n_2 \geqslant 7 \tag{5.22}$$

Relations (5.21) and (5.22) are additional constraints due to the branching. Solution of subproblem (3) leads to $n_1 = 3.75$, $n_2 = 6$, with the objective function $f(n_1, n_2) = -0.25$. Solution of subproblem (4) leads to $n_1 = 3$, $n_2 = 7$ with the objective function $f(n_1, n_2) = 0$. Solution of subproblem (4) is an integer feasible solution, whereas the solution to subproblem (3) is not. We need to investigate subproblem (3) further as the value of the objective function $(-0.25)$ is lower than the least value amongst the integer feasible solutions (the least value of the integer feasible solutions is zero).

Once again branching the variable $n_1$, we obtain two subproblems (5) and (6):

*Subproblem (5)*

$$\text{minimize} \quad f(n_1, n_2) = 4n_1^2 - 5n_1 n_2 + n_2^2 + 20$$
$$\text{subject to} \quad n_1 + n_2 \leqslant 10$$
$$n_1 > 0$$
$$n_2 > 0$$
$$n_2 \leqslant 6$$
$$n_1 \leqslant 3 \tag{5.23}$$

*Subproblem (6)*

$$\text{minimize} \quad f(n_1, n_2) = 4n_1^2 - 5n_1 n_2 + n_2^2 + 20$$
$$\text{subject to} \quad n_1 + n_2 \leqslant 10$$
$$n_1 > 0$$
$$n_2 > 0$$
$$n_2 \leqslant 6$$
$$n_1 \geqslant 4 \tag{5.24}$$

Solution of subproblem (5) leads to $n_1 = 3$, $n_2 = 6$, with the objective function $f(n_1, n_2) = 2$. Solution of subproblem (6) leads to $n_1 = 4$, $n_2 = 6$, with the objective function $f(n_1, n_2) = 0$. These are no further nodes to fathom. Integer feasible solutions are $n_1 = 4$, $n_2 = 6$, $f(n_1, n_2) = 0$ and $n_1 = 3$, $n_2 = 7$, $f(n_1, n_2) = 0$. This is very characteristic of real problems wherein multiple optimum solutions often exist. The strength of the Land–Doig algorithm lies in its ability to find all possible optimum solutions.

## 5.3 APPLICATION TO A VIBRATING STIFFENED COMPOSITE PLATE

A simply supported laminated composite rectangular plate of dimensions $L_x$ and $L_y$ is considered. The laminate is assumed to be of a symmetric construction of $n_1$ plies at 90°, $n_2$ plies at 60°, $n_3$ plies at 45°, $n_4$ plies at 30° and $n_5$ at 0°. Each ply has a uniform thickness, $t_0$, and $n_1, n_2, ..., n_5$ form the five integer variables of the problem. For the purpose of analysis, the plate is discretized using a mesh of 4 by 4, 8-noded, shear deformable, isoparametric, penalty plate bending elements (Reddy, 1979). The plate is reinforced by two stiffeners placed symmetrically with respect to the laminate midplane. Each of the two stiffeners is discretized using 8 frame elements. The cross-sectional area of the 16 elements are reduced to 8 independent unknowns through linking. The linking reduces the total number of continuous design variables from 16 to 8. For each frame element the cross-sectional area, $A_k$, is assumed to be related to its cross-sectional moment of inertia, $I_k$, by

$$I_k = \alpha(A_k)^n \tag{5.25}$$

where

$$\alpha = \frac{I_0}{A_0^n}; \quad n = 3 \tag{5.26}$$

To obtain plates of different aspect ratios, the dimension $L_y$ was set to 50 units while the dimension $L_x$ was varied from 40 to 80 units. A nonstructural mass equal to 2% of the mass of the plate was assumed as its center.

Finally, the nondimensional material constants of the plate were assumed as $E_1/E_2 = 40$, $G_{12}/E_2 = 0.6$, $G_{23}/E_2 = 0.5$, $G_{31}/E_2 = 0.6$, $v_{13} = v_{12} = 0.25$ while the shear correction factors $k_{12} = k_{22}$ were assumed to be $\frac{5}{6}$. With these preliminaries the optimization problem for the simply supported plate of Fig. 5.1 can be stated as

$$\text{maximize} \quad f(n_i, A_j) = \frac{\lambda_1}{\lambda_{01}} l_1 + \frac{\lambda_2}{\lambda_{02}} l_2 + \frac{\lambda_3}{\lambda_{03}} l_3 \tag{5.27}$$

$$i = 1, 2, ..., 5, \quad j = 1, 2, ..., 8$$

$$\text{subject to} \quad \sum n_i \leqslant 10 \tag{5.28}$$

$$\lambda_k > \lambda_{0k_1}, \ k = 1, 2, 3 \tag{5.29}$$

$$0.1 < \frac{A_j}{A_0} < 0.6 \tag{5.30}$$

$$\sum_{r=1}^{2} m_{sr} < 0.006 m_{plate} \tag{5.31}$$

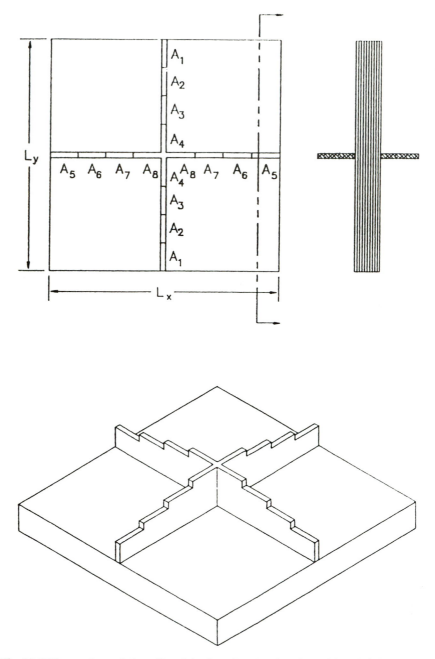

**Fig. 5.1** Different views of the stiffened laminated composite plate with a definition of the design variables linking.

where $\lambda_k$ is the $k$th eigenvalue of the plate obtained by solving the eigenvalue problem

$$\mathbf{Kq} = \mathbf{Mq} \qquad (5.32)$$

$\lambda_{0k}$ is the $k$th eigenvalue of the original unoptimized plate, $l_1$, $l_2$, $l_3$ are decision variables that take on the values of either zero or unity. Equation (5.28) restricts the total number of plies to ten or less. Equations (5.29) are the frequency separation constraints which are intended for obtaining coalescence-free designs, but as posed they do not guarantee them. Equation (5.30) specifies upper and lower bounds on the continuous design variables with $A_0$ being a reference area. Equation (5.31) ensures that the total mass of the two stiffeners is less than 0.6% of the mass of the plate. For the initial design all the elements of the two stiffeners are assumed to have the same area

$$A_s = 0.006 \frac{L_x L_y h}{L_x + L_y} \qquad (5.33)$$

$h$ being the total thickness of the plate.

In the process of applying Dakin's algorithm to the solution of the optimization problem as posed, an algorithm for the solution of the continuous optimization subproblem is necessary. Each of the continuous optimization subproblems is solved using the projected Lagrangian or variable metric method for constrained optimization (VMCON) proposed by Powell (1978). This method has been found to be ideal for small problems of the type being considered herein. It is based on a theorem which states that the optimum is a minimum of the Lagrangian function in the subspace of vectors orthogonal to the active constraint gradients. It employs a quadratic approximation to the Lagrangian in the subspace with the Hessian of the Lagrangian function being based on a BFGS (Broyden–Fletcher–Goldfarb–Shanno) update. Finally, the direction-seeking algorithm requires the solution of a quadratic programming problem (quadratic objective function and linearized constraints).

VMCON requires derivatives of the objective function and the constraints which involve the plate frequencies. It can be easily verified that (Haftka and Kamat, 1984) the derivatives of the square of the $k$th frequency, i.e. of $\lambda_k$, with respect to the $j$th design variable are given by

$$\frac{\partial \lambda_k}{\partial x_j} = \mathbf{q}_k^t \left( \frac{\partial \mathbf{K}}{\partial x_j} - \lambda_k \frac{\partial \mathbf{M}}{\partial x_j} \right) \mathbf{q}_k \qquad (5.34)$$

Calculation of the frequency derivatives using analytical formulae is computationally cheap and does not require the solution of any additional eigenvalue problems other than the one for computing the frequencies and modeshapes. In equation (5.34) $\mathbf{q}_k$ is the $k$th eigenvector, $\mathbf{K}$ is the stiffness matrix and $\mathbf{M}$ is the mass matrix. The global $\mathbf{K}$ and $\mathbf{M}$ matrices are assembled from the element matrices. To differentiate $\mathbf{K}$ and $\mathbf{M}$, it is sufficient to assemble the matrices obtained by differentiating the corresponding element matrices. The number of plies of a given fiber orientation and the stiffener areas form the two sets of design variables. The derivatives of the frame element stiffness and mass matrices with respect to the cross-sectional areas or moments of inertia are straightforward and hence are not presented here. Of interest here are the expressions for sensitivity derivatives of the stiffness and mass matrices of an individual laminated composite plate element.

Calculation of $\partial K^e/\partial x_j$ for the $e$th element requires the differentiation of matrices **A** and **D** defined by

$$(A_{ij}, D_{ij}) = \int_{-h/2}^{h/2} \sum_m \bar{Q}_{ij}^m (1, Z^2)\, dZ, \quad i, j = 1, 2, 6 \tag{5.35}$$

$$A_{ij} = k_\alpha k_\beta \int_{-h/2}^{h/2} \bar{Q}_{ij}\, dZ, \quad i, j = 4, 5 \tag{5.36}$$

$$\alpha = 6 - i,$$
$$\beta = 6 - j$$

where $\bar{Q}_{ij}^m$ are the stiffness coefficients for the $m$th ply in a standard laminate analysis. The differentiation of matrices **A** and **D** defined above will now be illustrated for a six-ply symmetric laminate with

$$Z_1 = -(n_1 + n_2 + n_3)t_0 \tag{5.37}$$
$$Z_2 = -(n_2 + n_3)t_0$$
$$Z_3 = -n_3 t_0$$
$$Z_4 = 0$$

with $t_0$ being the individual ply thickness and $h$ being the total thickness of the laminate. Thus

$$A_{ij} = 2[\bar{Q}_{ij}^1(Z_2 - Z_1) + \bar{Q}_{ij}^2(Z_3 - Z_2) + \bar{Q}_{ij}^3(Z_4 - Z_3)] \tag{5.38}$$
$$D_{ij} = \tfrac{2}{3}[\bar{Q}_{ij}^2(Z_2^3 - Z_1^3) + \bar{Q}_{ij}^2(Z_3^3 - Z_2^3) + \bar{Q}_{ij}^3(Z_4^3 - Z_3^3)] \tag{5.39}$$

From which the relations

$$\frac{\partial A_{ij}}{\partial n_k} = 2\left[ \bar{Q}_{ij}^1\left(\frac{\partial Z_2}{\partial n_k} - \frac{\partial Z_1}{\partial n_k}\right) + \bar{Q}_{ij}^2\left(\frac{\partial Z_3}{\partial n_k} - \frac{\partial Z_2}{\partial n_k}\right) + \bar{Q}_{ij}^3\left(\frac{\partial Z_4}{\partial n_k} - \frac{\partial Z_2}{\partial n_k}\right)\right] \tag{5.40}$$

and

$$\frac{\partial D_{ij}}{\partial n_k} = 2\left[ \bar{Q}_{ij}^1\left( Z_2^2\frac{\partial Z_2}{\partial n_k} - Z_1^2\frac{\partial Z_1}{\partial n_k}\right) + \bar{Q}_{ij}^2\left( Z_3^2\frac{\partial Z_3}{\partial n_k} - Z_2^2\frac{\partial Z_2}{\partial n_k}\right)\right.$$
$$\left. + \bar{Q}_{ij}^3\left( Z_4^2\frac{\partial Z_4}{\partial n_k} - Z_3^2\frac{\partial Z_3}{\partial n_k}\right)\right] \tag{5.41}$$

are obtained wherein the derivatives for $\partial Z_1/\partial n_k$ are known from equation (5.37).

To calculate $\partial M^e/\partial x_j$ for the $e$th element the expression for $p$ and $I$ defined by

$$(p, I) = \int_{-h/2}^{h/2} \sum_m (1, z^2)\rho^m\, dz \tag{5.42}$$

must be differentiated. $\rho^m$ in equation (5.42) is the material density of the $m$th layer. For a six-ply symmetric laminate

$$p = 2[(Z_2 - Z_1)\rho^1 + (Z_3 - Z_2)\rho^2 + (Z_4 - Z_3)\rho^3] \tag{5.43}$$

and

$$I = \tfrac{2}{3}[(Z_2^3 - Z_1^3)\rho^1 + (Z_3^3 - Z_2^3)\rho^2 + (Z_4^3 - Z_3^3)\rho^3]$$

Hence the derivatives of $p$ and $I$ are again given by equations (5.40) and (5.41) respectively with $\bar{Q}_{ij}^m$ being replaced by $\rho^m$.

### 5.3.1 Measures to accelerate the subproblem optimization

Depending on the number of integer design variables and their bounds, it is necessary to solve a large number of continuous subproblems. A number of steps are taken to make the solution of the subproblems computationally efficient. These steps are discussed below.

- Each of the subproblems inherits the constraints of the 'parent' problem (relations (5.7)–(5.9) or (5.11)–(5.13)) plus the additional constraints (5.10) and (5.14). While equation (5.10) lowers the upper bound constraint, equation (5.14) increases the lower bound on $x_k$ leading ultimately to coalescence of the upper and lower bounds for a given integer design variable. A tightening of the bounds on a given integer variable leads to slight reduction of the solution times for the two subproblems. The corresponding variable is then removed from the design space. This reduces the number of design variables that are removed. Reduction in the problem size results in drastic reduction in solution times for the subproblem.
- As the geometry of the structural model does not change, certain integration results computed during the initial assembly of the structure stiffness and mass matrices can be stored for later use. This reduces the time for assembly of the two matrices by a factor of about six. It should be noted that the assembly of the two matrices needs to be performed once each time the frequencies are computed. Also, during the computation of sensitivity derivatives one needs to perform the assembly for each design variable with respect to which the derivative is required.
- It is possible to make the computation of frequencies extremely efficient. For the very first time the frequencies are computed using the method of subspace iteration in conjunction with Jacobi diagonalization. For subsequent calculations of frequencies, the sensitivity derivative information can be used to predict approximate values of the new frequencies on the basis of a Taylor series expansion. The approximate frequencies are then used to shift the stiffness matrix. Inverse iteration on the shifted stiffness matrix yields the required frequencies. This method preserves the skyline structure of the stiffness and mass matrices.
- The VMCON algorithm uses the identity matrix as an initial approximation estimate to the Hessian matrix of the Lagrangian function. The solution time for the continuous subproblem depends on how close the initial estimate is to the actual Hessian matrix at the optimal point. The Hessian matrix at the end of the previous optimization subproblem serves as an excellent initial approximation to the Hessian matrix of the current subproblem. Indeed, the number of design variables may change from one subproblem to the next. Only those elements of the Hessian matrix corresponding to design variables of the current subproblem that were also present in the previous subproblem

are retained. For those new design variables which enter the design space all the elements of the columns and rows corresponding to these design variables except the diagonal elements are set to zero. The new diagonal elements are set to unity.

### 5.3.2 Discussion of results

To obtain plates of differing aspect ratios, $L_y$ is set to 50 units and $L_x$ is varied over the three values 40, 50, and 80 units. Six different designs are considered in Tables 5.1–5.5. Three of these are labeled 1(a), 1(b) and 1(c) for $l_1 = l_2 = l_3 = 0$ and the remaining three are labeled 2(a), 2(b) and 2(c) for $l_1, l_2, l_3 = 1$.

Table 5.1 compares the first three eigenvalues for the initial and optimum designs together with those for the first continuous solution for the six cases. An examination of Table 5.3 indicates that the rounded-off design is clearly inferior to the optimal design in all the cases except for case 2(c) where the optimal design is infeasible. In case 2(b) the rounded-off design not only has a smaller value for the objective function compared with the optimal design but is also infeasible. A further examination of Table 5.3 indicates that every integer feasible optimal design has an objective function that is larger than that for the continuous design for all the cases except case 2(c). This behavior is due to the fact that the design space is nonconvex and has multiple optimal solutions. The integer programming algorithm therefore greatly increases the probability of locating the global

**Table 5.1** Eigenvalues for the different designs

| Case number | $\lambda_1^*$ | $\lambda_2^*$ | $\lambda_3^*$ | Type of design |
|---|---|---|---|---|
| 1(a) | 0.404 36 | 1.6609 | 2.5336 | * |
|  | 0.616 95 | 1.7947 | 3.0845 | ** |
|  | 0.666 97 | 1.8460 | 3.9932 | *** |
| 1(b) | 0.300 77 | 0.981 90 | 2.2262 | * |
|  | 0.407 75 | 1.1855 | 2.4987 | ** |
|  | 0.417 65 | 1.5373 | 2.2236 | *** |
| 1(c) | 0.200 70 | 0.393 10 | 0.842 50 | * |
|  | 0.252 18 | 0.419 94 | 0.926 07 | ** |
|  | 0.253 91 | 0.396 39 | 0.851 92 | *** |
| 2(a) | 0.404 36 | 1.6609 | 2.5336 | * |
|  | 0.628 42 | 1.977 91 | 4.1896 | ** |
|  | 0.631 05 | 1.9910 | 4.1398 | *** |
| 2(b) | 0.300 77 | 0.981 90 | 2.2262 | * |
|  | 0.359 82 | 1.3395 | 2.222 62 | ** |
|  | 0.347 91 | 1.4392 | 2.224 65 | *** |
| 2(c) | 0.200 70 | 0.393 10 | 0.842 50 | * |
|  | 0.202 91 | 0.536 48 | 1.4154 | ** |
|  | 0.201 45 | 0.553 69 | 1.3654 | *** |

*, initial unoptimized structure; **, first continuous solution; ***, optimal design; $\lambda_i^*$, $i = 1, 2, 3$, normalized eigenvalues.

**Table 5.2** Summary of structural optimization results

| Converged design variable vector | Initial design variable vector | Case 1(a) $A_s = 0.2667$ | Case 1(b) $A_s = 0.3000$ | Case 1(c) $A_s = 0.3692$ | Case 2(a) $A_s = 0.2667$ | Case 2(b) $A_s = 0.3000$ | Case 2(c) $A_s = 0.3692$ |
|---|---|---|---|---|---|---|---|
| $n_1 (90°)$ | 2 | 1 | 1 | 2 | 1 | 3 | 1 |
| $n_2 (60°)$ | 2 | 0 | 0 | 4 | 0 | 0 | 0 |
| $n_3 (45°)$ | 2 | 1 | 2 | 2 | 1 | 2 | 3 |
| $n_4 (30°)$ | 2 | 2 | 3 | 1 | 2 | 0 | 0 |
| $n_5 (0°)$ | 2 | 6 | 4 | 1 | 6 | 5 | 6 |
| $A_1$ | $A_s$ | 0.100 | 0.600 | 0.600 | 0.100 | 0.100 | 0.600 |
| $A_2$ | $A_s$ | 0.100 | 0.100 | 0.100 | 0.100 | 0.100 | 0.100 |
| $A_3$ | $A_s$ | 0.100 | 0.100 | 0.100 | 0.100 | 0.100 | 0.100 |
| $A_4$ | $A_s$ | 0.100 | 0.100 | 0.100 | 0.100 | 0.100 | 0.100 |
| $A_5$ | $A_s$ | 0.600 | 0.600 | 0.600 | 0.600 | 0.600 | 0.600 |
| $A_6$ | $A_s$ | 0.600 | 0.600 | 0.600 | 0.600 | 0.200 | 0.100 |
| $A_7$ | $A_s$ | 0.600 | 0.200 | 0.538 | 0.100 | 0.600 | 0.600 |
| $A_8$ | $A_s$ | 0.100 | 0.100 | 0.100 | 0.600 | 0.600 | 0.538 |

**Table 5.3** Objective function comparison for the different designs

| Case number | Initial design | First continuous design | Rounded-off design | Optimal design | |
|---|---|---|---|---|---|
| 1(a) | 1.0 | 1.525 | 1.548 | 1.649 | * |
|  | 0 | 0 | 0 | 0 | ** |
| 1(b) | 1.0 | 1.355 | 1.350 | 1.388 | * |
|  | 0 | 0 | 0 | 0 | ** |
| 1(c) | 1.0 | 1.256 | 1.245 | 1.265 | * |
|  | 0 | 0 | 0 | 0 | ** |
| 2(a) | 3.0 | 4.399 | 4.393 | 4.393 | * |
|  | 0 | 0 | 0 | 0 | ** |
| 2(b) | 3.0 | 3.560 | 3.574 | 3.631 | * |
|  | 0 | 0 | 1 | 0 | ** |
| 2(c) | 3.0 | 4.055 | 4.591 | 4.033 | * |
|  | 0 | 0 | 1 | 0 | ** |

*, value of objective function to be maximized; **, number of constraints violated.

**Table 5.4** Iteration time history for problem 1(c)

| Subproblem number | Number of design variables | Time (s) | $\frac{T}{T_0} \times 100$ |
|---|---|---|---|
| 1 | 13 | $T_0 = 792$ | 100 |
| 2 | 13 | 237 | 30 |
| 3 | 12 | 60 | 7 |
| 4 | 12 | 150 | 19 |
| 5 | 11 | 39 | 5 |
| 6 | 11 | 40 | 5 |
| 7 | 10 | 34 | 4 |
| 8 | 12 | 45 | 6 |
| 9 | 12 | 81 | 10 |
| 10 | 12 | 95 | 12 |
| 11 | 12 | 82 | 10 |
| 12 | 12 | 45 | 6 |
| 13 | 12 | 81 | 10 |

optimum because it searches for an optimal solution in a number of disjoint design spaces (each of the two subproblems generated is disjoint).

Table 5.4 provides the CPU time history for the subproblems of case 1(c). It is evident that the total CPU time is less than proportional to the total number of subproblems solved. For example in Table 5.5 for case 1(c), the CPU time to solve 15 subproblems is only 4.7 times the CPU time for the first continuous

**Table 5.5** Solution statistics

| Case number | Number of continuous solutions | CPU time for first continuous solution $T_0$(s) | CPU time for optimal solution $T$(s) | $\dfrac{T}{T_0}$ |
|---|---|---|---|---|
| 1(a) | 11 | 1148 | 7332 | 6.4 |
| 1(b) | 12 | 420 | 2049 | 4.9 |
| 1(c) | 15 | 792 | 3727 | 4.7 |
| 2(a) | 6 | 1504 | 5717 | 3.8 |
| 2(b) | 17 | 5243 | 27243 | 5.2 |
| 2(c) | 7 | 7152 | 9647 | 1.4 |

solution. The additional subproblems are therefore obtained very cheaply because of the measures taken to accelerate the optimization process.

## 5.4 SOME RECENT METHODS FOR NONLINEAR MIXED INTEGER STRUCTURAL OPTIMIZATION PROBLEMS

### 5.4.1 Sequential linear discrete programming (SLDP)

This method proposed by Olson and Vanderplaats (1989) uses SLDP to solve the discrete optimization problem. The optimization problem tackled can be expressed as

minimize $f(\mathbf{X})$
subject to

$$g_j(\mathbf{X}) \leqslant 0 \qquad j = 1, \ldots, m$$
$$h_k(\mathbf{X}) = 0 \qquad k = 1, \ldots, l \qquad (5.44)$$
$$\mathbf{X} = (x_1 \cdots x_p \, x_{p+1} \cdots x_n)^t$$
$$x_i \in (d_{i1}, d_{i2}, \ldots, d_{iq}) \qquad i = 1, \ldots, p$$

where

$$x_k^l \leqslant x_k \leqslant x_k^u \qquad k = p+1, \ldots, n$$

where $p$ is the number of discrete design variables, $n$ the total number of design variables, $d_{ij}$ is the $j$th discrete value for design variable $i$, $x_k^l$ the lower bound for design variable $k$ and $x_k^u$ the upper bound for design variable $k$. The corresponding linear programming problem at $\mathbf{X}^0$ becomes

$$f(\mathbf{X}) = f(\mathbf{X}^0) + \nabla F(\mathbf{X}^0) \, \delta \mathbf{X}$$

Subject to

$$g_j(\mathbf{X}) = h_k(\mathbf{X}^0) + \nabla h_k(\mathbf{X}^0) \, \delta \mathbf{X} \leqslant 0 \quad j = 1, \ldots, m$$
$$h_k(\mathbf{X}) = h_k(\mathbf{X}^0) + \nabla h_k(\mathbf{X}^0) \, \delta \mathbf{X} = 0 \quad k = 1, \ldots, l \qquad (5.45)$$

where

$$\delta \mathbf{X} = \mathbf{X} - \mathbf{X}^0$$

$$\mathbf{X} = (x_1 \cdots x_p \, x_{p+1} \cdots x_n)^t$$

$$x_i^0 + \delta x_i \in (d_{i1}, d_{i2}, \dots, d_{iq}) \qquad i = 1, \dots, p$$

$$x_k^l \leqslant x_k^0 + \delta x_k \leqslant x_k^u \qquad k = p+1, \dots, n$$

The following construct is made to incorporate the discrete design:

$$x_i = z_{i1} d_{i1} + z_{i2} d_{i2} + \cdots + z_{iq} d_{iq} \tag{5.46}$$

with

$$z_{i1} d_{i1} + z_{i2} d_{i2} + \cdots + z_{iq} d_{iq} = 1 \tag{5.47a}$$

and

$$z_{ij} = 0 \quad \text{or} \quad 1 \qquad j = 1, \dots, q \tag{5.47b}$$

Equations (5.46) and (5.47) ensure that the design variable will have one of the discrete values $d_{iq}$. The above equations for $x_i$ can be inserted into the linear approximate problem, yielding

minimize

$$f(\mathbf{X}) \cong f(\mathbf{X}^0) + \sum_{i=1}^{p} \left[ \left( \sum_{j=1}^{q} z_{ij} d_{ij} \right) - x_k^0 \right] + \sum_{k=p+1}^{n} \frac{\partial F}{\partial x_k} (x_k - x_k^0) \tag{5.48}$$

subject to

$$g_j(\mathbf{X}^0) \cong g_j(\mathbf{X}^0) + \sum_{i=1}^{p} \frac{\partial g_j}{\partial x_i} \left[ \left( \sum_{m=1}^{q} z_{im} d_{im} \right) - x_i^0 \right] + \sum_{k=p+1}^{n} \frac{\partial g_j}{\partial x_k} (x_k - x_k^0) \leqslant 0$$

$$\sum_{l=1}^{q} z_{il} = 1 \qquad i = 1, \dots, p$$

$$z_{ij} = 0 \quad \text{or} \quad 1 \qquad \text{all } i \text{ and } j$$

$$x_k^l < x_k^0 + \sum x_k x_k^u \qquad k = p+1, \dots, n$$

To choose the initial guess $X^0$, problem (5.48) can be solved using a continuous optimization package and rounding off the solution to the discrete values. While rounding off, the gradient information can be used to decide whether to round to the higher or lower discrete value.

The continuous problem is solved using the ADS program (Olson and Vanderplaats, 1989). The LINDO program uses a branch and bound approach to solve a discrete linear optimization problem. The values of $z_{iq}$ returned from the LINDO program are converted to the design variables, using equation (5.46), and used as an initial guess for a subsequent ADS–LINDO iteration.

### 5.4.2 Penalty approach for nonlinear optimization with discrete design variables

This method proposed by Dong, Gurdal and Griffin (1990) uses a sequential unconstrained minimization technique (SUMT) whereby the constrained optimization is transformed into a sequence of unconstrained minimization problems.

To ensure that the design variables take integral values a penalty term is defined to take on a zero value at the discrete points. At nondiscrete points the penalty term assumes a nonzero value. This forces the optimal solution to be at the selected discrete points.

The SUMT technique transforms the original constrained optimization problem into a sequence of unconstrained optimization problems. The sequence of the unconstrained problems tends to the solution of the original constrained problem. Suppose the original constrained optimization problem is defined by the following relations.

$$\text{minimize} \quad f(\mathbf{X})$$
$$\text{such that} \quad g_j(\mathbf{X}) \geqslant 0, \quad j = 1, 2, \ldots, n_g \tag{5.49}$$

where $\mathbf{X} = (x_1 \, x_2 \cdots x_n)^t$ and $n$ is the total number of design variables and $n_g$ is the total number of constraints.

The constrained optimization problem is replaced by the following unconstrained minimization problem:

$$\text{minimize} \quad \Phi(\mathbf{X}, r) = f(\mathbf{X}) + r \sum_{j=1}^{n_g} y(g_j) \tag{5.50}$$

where $y(g_j)$ is a constraint violation penalty function that can be defined in a number of ways to force the solution of the unconstrained problem (5.50) towards the solution of the constrained optimization problem (5.49). The positive multiplier, $r$, in equation (5.50) controls the contribution of the constraint penalty terms. If $r = 0$, the penalty terms do not contribute to the solution of equation (5.50) at all. If the solution obtained (with $r = 0$) does not violate any constraints, then it is the solution to equation (5.49) and no further computations are necessary. In general, this will not be the case, as solution of equation (5.50) with $r = 0$ will violate a number of constraints. As the value of $r$ is increased, solution of equation (5.50) will tend toward the solution of equation (5.49).

*SUMT with discrete design variables*

$$\text{minimize} \quad f(\mathbf{X})$$
$$\text{such that} \quad g_j(\mathbf{X}) \geqslant 0, \quad j = 1, 2, \ldots, n_g$$
$$\text{where} \quad \mathbf{X} = (x_1 \, x_2 \cdots x_n)^t \tag{5.51}$$
$$x_i = (d_{i1} \, d_{i2} \cdots d_{iq})^t, \quad i = 1, 2, \ldots, n_d$$

and $n_d$ is the number of discrete design variables, $d_{ik}$ the $k$th discrete value of the $i$th design variable and $q$ the number of discrete values for each design variable.

To account for the discrete design variables, the objective function in equation (5.50) is modified to

$$\psi(\mathbf{X}, r, s) = f(\mathbf{X}) + r \sum_{j=1}^{n_g} p(g_j) + s \sum_{i=1}^{n_d} \Phi_d^i(\mathbf{X}) \tag{5.52}$$

Different forms for $\Phi_d^i(\mathbf{X})$ are possible, but Dong, Gurdal and Griffin (1990) recommend the following sine function:

$$\Phi_d^i(\mathbf{X}) = \frac{1}{2} \left\{ \sin \frac{2\pi [x_i - (1/4)(d_{ij+1} + 3d_{ij})]}{d_{ij+1} - d_{ij}} + 1 \right\} \tag{5.53}$$

where $d_{ij} < x_i < d_{ij+1}$. The proposed function $\Phi_d^i(\mathbf{X})$ penalizes only nondiscrete design variables and assures the continuity of the first derivatives of the modified pseudo-function at the discrete values of the design variables.

This method has the advantage that a discrete optimization problem can be solved using a computer program for continuous structural optimization, such as NEWSUMT (Dong, Gurdal and Griffin, 1990). Only the additional penalty term defined by the above equation needs to be specified as an addition to the objective function. The numerical examples described by Dong, Gurdal and Griffin also seem to indicate that the method is computationally very efficient in terms of the number of iterations required to converge. The drawback of the approach is that it makes the design space highly nonconvex and this increases the chances that the solution will become stuck in a local optimum. Great care would have to be exercised to ensure that the solution does not converge to a local optimum that is far from the global optimum. This can be done by carefully controlling the penalty multiplier during the optimization process.

## 5.5 CONCLUDING REMARKS

In this chapter, the authors have reviewed some of the techniques for nonlinear mixed integer programming and tried to demonstrate how they can be a viable structural optimization alternative to the conventional practice of rounding off a continuous optimum solution. The latter practice leads not only to suboptimal but often to infeasible solutions. With the recent advent of parallel distributed processing, it would appear that the branch and bound algorithm may be nicely parallelizable in which case computation times could be drastically reduced making the use of nonlinear mixed integer programming algorithms even more popular and routine.

## REFERENCES

Balas, E. (1971) Intersection cuts – a new type of cutting planes method for integer programming. *Operations Research*, **19**, 19–39.

Ben-Israel, A. and Charnes, A. (1962) On some problems in Diophantine programming. *Cahiers du Centre d'Etudes de Recherche Operationelle*, **4**, 215–80.

Dakin, R.J. (1965) A tree search algorithm for mixed integer programming problems. *Computer Journal*, **8**(3), 250–5.

Dantzig, G.B. (1959) Notes on solving linear problems in integers. *Nadal Research Logistics Quarterly*, 75–6.

Dantzig, G.B., Fulkerson, D.R. and Johnson, S.M. (1954) Solution of large scale travelling salesman problem. *Operations Research*, **2**, 393–410.

Dong, K.S., Gurdal, Z. and Griffin, O.H., Jr (1990) A penalty approach for nonlinear optimization with discrete design variables. *Engineering Optimization*, **16**, 29–42.

Gisvold, K.M. and Moe, M. (1971) A method for nonlinear mixed-integer programming and its application to design problems. *Journal of Engineering for Industry*, **94**, 353–64.

Glover, F. (1965) A bound escalation method for the solution of integer linear programs. *Cahiers du Centre d'Etudes de Recherche Operationelle*, **6** 131–68.

Glover, F. (1968) A new foundation for a simplified primal integer programming algorithm. *Operations Research*, **16**, 727–40.

Glover, F. (1971) Convexity cut and search. *Operations Research*, **21**, 123–4.

Gomory, R.E. (1958) Outline of an algorithm for integer solutions to linear problems. *Bulletin of the American Mathematical Society*, **64**, 275–8.

Gormory, R.E. (1960a) All integer programming algorithm. *RC-189*, IBM, Yorktown Heights, NY.

Gormory, R.E. (1960b) An algorithm for the mixed integer problem. *RM-2597*, Rand Corp., Santa Monica, CA.

Gupta, O.K. (1980) Branch and bound experiments in nonlinear integer programming. PhD Thesis, Purdue University, West Lafayette, IN.

Gupta, O. and Ravindran, A. (1983) Nonlinear integer programming and discrete optimization. *Journal of Mechanisms, Transmission, and Automation in Design*, **105**, 160–4.

Haftka, R.T. and Kamat, M.P. (1984) *Elements of Structural Optimization*, Martinus Nijhoff, Netherlands.

Handy, A.T. (1975) *Interger Programming*, Academic Press, New York.

Hua, H.M. (1983) Optimization of structures of discrete size elements. *Computers and Structures*, **17**(3), 327–33.

Imai, K. (1979) Structural optimization by material selection. Information Processing Center, Kajima Corp.

Johnson, R.C. (1981) Rigid plastic minimum weight plane frame design using hot rolled shapes. Proceedings of the International Symposium on Optimal Structural Design, Tucson, AZ, pp. 9-1–9-6.

Land, A.H. and Doig, A.G. (1960) An automatic method for solving discrete programming problems. *Econometrica*, **28**, 497–520.

Little, J.D.C., Murthy, K.G., Sweeney, D.W. and Kavel, C. (1960) An algorithm for the travelling salesman problem. *Operations Research*, **11**, 979–89.

Markovitz, H.M. and Manne, A.S. (1975) On the solution of discrete programming problems. *Econometrica*, **25**, 84–110.

Olson, G.R. and Vanderplaats, G.N. (1989) Methods for nonlinear optimization with discrete design variables. *AIAA Journal*, **27**(11), 1584–9.

Powell, M.J.D. (1978) A fast algorithm for nonlinearly constrained optimization calculations. Proceedings of the Biennial, 1977 Dundee Conference on Numerical Analysis, *Lecture Notes in Mathematics*, Vol. 630, Springer, Berlin, pp. 144–57.

Reddy, J.N. (1979) Simple finite elements with related continuity for nonlinear analysis of plates. Proceedings of the Third International Conference in Australia on Finite Element Methods, University of New South Wales, Sydney.

Schmit, L.A. and Fleury, C. (1980) Discrete–continuous variable structural synthesis using dual method. *AIAA Journal*, **18**(12), 1515–24.

Templeman, A.B. and Yates, D.F. (1983) A segmented method for the discrete optimum design of structures. *Engineering Optimization*, **6**, 145–55.

Young, R.D. (1965) A primal (all integer) integer programming algorithm. *Journal of Research of the National Bureau of Standards, Section B*, **69**, 213–50.

Young, R.D. (1968) A simplified primal (all integer) integer programming algorithm. *Operations Research*, **16**, 750–82.

Young, R.D. (1971) Hypercylindrically deduced cuts in 0–1 integer programming. *Operations Research*, **19**, 1393–405.

# 6

# Multicriterion structural optimization

JUHANI KOSKI

## 6.1 INTRODUCTION

Ideally structural mechanics can be viewed as a discipline which comprises the whole interactive process of analysis and design resulting in an optimal structure. After the advent of the modern computer the finite element method and optimization have gradually developed into essential parts of this process. In optimization the research has mainly been focused on the numerical solution of problems with a constantly increasing number of design variables and constraints. Usually a scalar-valued objective function, which in most cases is the weight of the structure, is optimized in the feasible set defined by the equality and inequality constraints. In practical applications, however, the weight rarely represents the only measure of the performance of a structure. In fact, several conflicting and noncommensurable criteria usually exist in real-life design problems. This situation forces the designer to look for a good compromise design by performing trade-off studies between the conflicting requirements. Consequently, he must take a decision-maker's role in an interactive design process where generally several optimization problems must be solved. Multicriterion optimization offers one flexible approach for the designer to treat this overall decision-making problem in a systematic way.

Multicriterion (multicriteria, multiobjective, Pareto, vector) optimization has recently achieved an established position also in structural design. One reason for the introduction of this approach is its natural property of allowing participation in the design process after the formulation of the optimization problem. It is generally considered that multicriterion optimization in its present sense originated towards the end of the last century when Pareto (1848–1923) presented a qualitative definition for the optimality concept in economic problems with several competing criteria (Pareto, 1896). Some other even earlier contributors have been discussed for example in Stadler (1987). A wider interest in this subject concerning the fields of optimization theory, operations research and control theory was aroused at the end of the 1960s and since then the research has been very intensive also in engineering design (Osyczka, 1984; Stadler, 1988; Eschenauer, Koski and Osyczka, 1990). Especially in structural optimization, the first applications in the English-language literature appeared in the late 1970s (Stadler,

1977; Leitmann, 1977; Stadler, 1978; Gerasimov and Repko, 1978; Koski, 1979), giving an impetus to emphasizing the decision-maker's viewpoint also in the design of load-supporting structures.

The purpose of this chapter is to introduce the basic concepts and methods used in multicriterion structural optimization. Also a brief literature review is given to describe the present research in the field. Special attention has been paid to those fundamental matters which are common to most of the published applications. These general ideas have been illustrated by some example problems where the emphasis is rather on the multicriterion view than on the numerical solution techniques. Specifically, the multicriterion problem formulation and the generation of Pareto optima are considered in these examples. The decision-making process for finding the best Pareto optimal design is also briefly discussed. On the whole, this chapter should both clarify the logical structure of the multicriterion approach and describe its diverse possibilities as well as recent advances in the field.

## 6.2 PARETO OPTIMAL DESIGN

### 6.2.1 Criteria and conflict

Any designer who applies optimization is faced by the question of which criteria are suitable for measuring the economy and performance of a structure. Such a quantity that has a tendency to improve or deteriorate is actually a criterion in nature. On the other hand, those quantities which must only satisfy some imposed requirements are not criteria but they can be treated as constraints. Most of the commonly used design quantities have a criterion nature rather than a constraint nature because in the designer's mind they usually have better or worse values. As an example of a strict constraint the structural analysis equations or any physical laws governing the system can be mentioned. They represent equality constraints whereas different official regulations and norms generally impose inequality constraints. For example such matters as space limitations, strength and manufacturing requirements are often treated as inequality constraints. One difficulty appears in choosing the allowable constraint limits which may be rather fuzzy in real-life problems. If these allowable values cannot be determined it seems reasonable to treat the quantity in question as a criterion. In general, the separation of the criteria and constraints presupposes a balance between two goals; the computational feasibility and the flexibility in the design process.

One important basic property in the multicriterion problem statement is a conflict between the criteria. Only those quantities which are competing should be treated as criteria whereas the others can be combined as a single criterion or one of them may represent the whole group. In the literature the concept of the conflict has deserved only a little attention while on the contrary the solution procedures have been studied to a great extent. In the problem formulation, however, it is useful to consider the conflict properties because it helps to create a good optimization model. For example in Cohon (1978) this topic is discussed in general terms and in Koski (1984b), where truss design is studied, the concepts of a local and global conflict have been proposed. According to the latter presentation the local conflict between two criteria can be defined as follows.

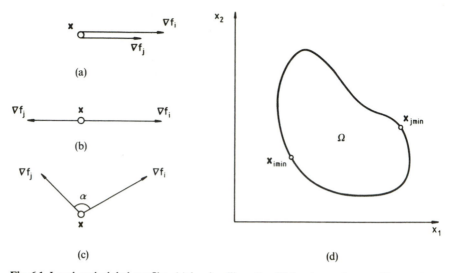

**Fig. 6.1** Local and global conflict: (a) local collinearity; (b) local complete conflict; (c) local conflict; (d) global conflict in the case of two design variables. Points $x_{i\,min}$ and $x_{j\,min}$ represent the individual minima of the criteria $f_i$ and $f_j$ in $\Omega$, respectively.

**Definition 6.1** Functions $f_i$ and $f_j$ are called **collinear** with no conflict at point **x** if there exists $c > 0$ such that $\nabla f_i(\mathbf{x}) = c\nabla f_j(\mathbf{x})$. Otherwise, the functions are called **locally conflicting** at **x**.

This definition states that any two criteria are locally conflicting at a point in the design space if their maximum improvements are achieved in different directions. The angle between the gradients can be used as a natural measure of the local conflict. Even if two criteria are locally conflicting almost everywhere in the design space they still can achieve their optimum value at the same point. Thus it seems necessary to consider separately the concept of the global conflict where also the feasible set is involved.

**Definition 6.2** Functions $f_i$ and $f_j$ are called **globally conflicting** in $\Omega$ if the optimization problems $\min_{\mathbf{x} \in \Omega} f_i(\mathbf{x})$ and $\min_{\mathbf{x} \in \Omega} f_j(\mathbf{x})$ have different solutions.

Both of these concepts have been illustrated in Fig. 6.1 where the relevant situations in the design space are shown. In structural optimization, usually the weight and any chosen displacement are both strongly locally and globally conflicting quantities. Displacements often achieve their minima at the same point but still they may be locally conflicting in that part of $\Omega$ where the best design locates.

### 6.2.2 Multicriterion problem

In formulating an optimization problem the choice of the design variables, criteria and constraints certainly represents the most important decision. The designs which will be available in the continuation are fixed at this very early stage. For example in scalar optimization the minimization of a single criterion in the feasible set usually results in one optimal solution only. Certainly numerical computations are needed to get that optimum design but, as a matter of fact, all the decisions have been made already in the problem formulation.

The multicriterion problem inherently offers a possibility to perform a systematic sensitivity analysis for the chosen criteria. As was pointed out earlier, the difference between criteria and constraints is that the designer wants to improve the value of a criterion whereas this kind of desire is not associated with the constraints. As a natural consequence of the separation of the criteria $f_i, i = 1, 2, \ldots, m$, and the constraints the following multicriterion problem is obtained:

$$\min_{\mathbf{x} \in \Omega} [f_1(\mathbf{x}) f_2(\mathbf{x}) \cdots f_m(\mathbf{x})]^T \tag{6.1}$$

Here $\mathbf{x} = [x_1 \, x_2 \cdots x_n]^T$ represents a design variable vector and $\Omega$ is the feasible set in design space $R^n$. It is defined by inequality and equality constraints in the form

$$\Omega = \{\mathbf{x} \in R^n | \mathbf{g}(\mathbf{x}) \leqslant \mathbf{0}, \mathbf{h}(\mathbf{x}) = \mathbf{0}\} \tag{6.2}$$

By using the notation $\mathbf{f}(\mathbf{x}) = [f_1(\mathbf{x}) f_2(\mathbf{x}) \cdots f_m(\mathbf{x})]^T$ for the vector objective function, which contains the $m$ conflicting and possibly noncommensurable criteria as the components, the image of the feasible set in criterion space $R^m$ is expressed as

$$\Lambda = \{\mathbf{z} \in R^m | \mathbf{z} = \mathbf{f}(\mathbf{x}), \mathbf{x} \in \Omega\}. \tag{6.3}$$

This is called the attainable (criteria) set and apparently it is more interesting for the decision maker than the feasible set. Usually there exists no unique point which would give an optimum for all $m$ criteria simultaneously. Thus the common optimality concept used in scalar optimization must be replaced by a new one, especially adapted to the multicriterion problem.

Only a partial order exists in criterion space $R^m$ and thus the concept of Pareto optimality offers the most natural solution in this context.

*Definition 6.3* A vector $\mathbf{x}^* \in \Omega$ is **Pareto optimal** for problem (6.1) if and only if there exists no $\mathbf{x} \in \Omega$ such that $f_i(\mathbf{x}) \leqslant f_i(\mathbf{x}^*)$ for $i = 1, 2, \ldots, m$ with $f_j(\mathbf{x}) < f_j(\mathbf{x}^*)$ for at least one $j$.

In words, this definition states that $\mathbf{x}^*$ is Pareto optimal if there exists no feasible vector $\mathbf{x}$ which would decrease some criterion without causing a simultaneous increase in at least one other criterion. In the literature also some other terms have been used instead of the Pareto optimality. For example words such as nondominated, noninferior, efficient, functional-efficient and EP-optimal solution have the same meaning. Here only the mathematical programming problem has been shown and applied but the corresponding control theory formulation can be found for example in Stadler (1988) and Eschenauer, Koski and Osyczka (1990).

Two different spaces $R^n$ and $R^m$, called the design and the criterion space, appear in a multicriterion problem. In order to avoid any confusion it is necessary to distinguish the optimal solutions in these separate spaces. Consequently, the vector $\mathbf{z}^* = \mathbf{f}(\mathbf{x}^*)$, which represents the image of the Pareto optimum $\mathbf{x}^*$ in the criterion space, is called the **minimal solution**. Optimality concepts in both spaces have been illustrated in Fig. 6.2 where the bicriterion case has been considered. So-called weak solutions, which are also shown in the figure, and their existence

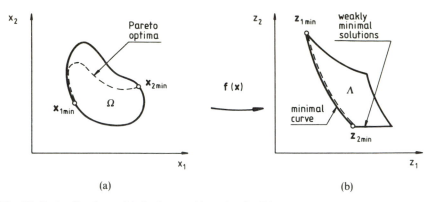

**Fig. 6.2** Optimality in multicriterion problem: (a) feasible set and Pareto optimal curve; (b) attainable set and minimal curve. In this bicriterion case both the Pareto optimal and minimal set are represented by curves. Points $x_{1\,min}$ and $x_{2\,min}$ in the design space correspond to points $z_{1\,min}$ and $z_{2\,min}$ in the criterion space.

in structural optimization have been discussed in Koski and Silvennoinen (1987). In scalar optimization, one optimal solution is usually characteristic of the problem, whereas there generally exists a set of Pareto optima as a solution to the multicriterion problem. Mathematically, problem (6.1) can be regarded as solved immediately after the Pareto optimal set has been determined. In practical applications, however, it is necessary to order this set further because only one final solution is wanted by the designer. Thus he must take a decision-maker's role and introduce his own preferences to find the best compromise solution among Pareto optima.

### 6.2.3 Plate bending problem

As an introductory example a bicriterion plate bending problem shown in Fig. 6.3 is considered. The weight of this simply supported uniformly loaded plate and the vertical displacement of the central point A should be simultaneously minimized. These two criteria have been denoted by $W$ and $\Delta$, respectively. The plate thicknesses $t_i$, $i = 1, 2, \ldots, 6$, are used as the design variables and they have been organized as zones offering a realistic design also from the manufacturing point of view. The Mindlin–Reissner thick plate element is applied in the analysis of the structure and the finite element mesh is shown in Fig. 6.3(b). The design variable zones, which correspond to certain groups of the elements, are presented in Fig. 6.3(c). By introducing constraints for the stresses and plate thicknesses the following bicriterion optimization problem is obtained:

$$\min\left[\,W(\mathbf{x})\,\Delta(\mathbf{x})\,\right]^{\mathsf{T}} \tag{6.4}$$

subject to

$$\sigma_i^{M}(\mathbf{x}) \leqslant \bar{\sigma}, \quad i = 1, 2, \ldots, 36$$

$$\underline{t} \leqslant t_i \leqslant \bar{t}, \quad i = 1, 2, \ldots, 6$$

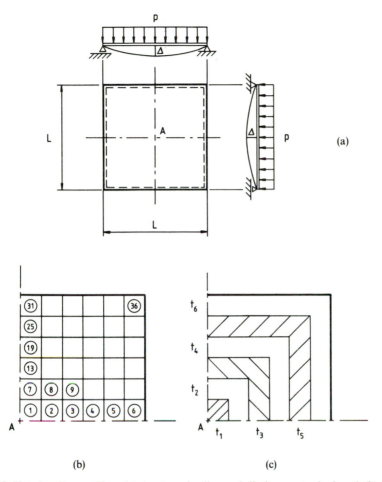

**Fig. 6.3** Plate bending problem: (a) structure, loading and displacement criterion $\Delta$; (b) finite element model; (c) design variables $t_1 \cdots t_6$ which represent the plate thicknesses. Design data for the problem: $p = 0.4 \, \text{N mm}^{-2}$, $E = 206 \times 10^3 \, \text{N mm}^{-2}$, $v = 0.3$, $\rho = 7800 \, \text{kg m}^{-3}$, $L = 600 \, \text{mm}$, $\bar{\sigma} = 140 \, \text{MPa}$, $\underline{t} = 2 \, \text{mm}$, $\bar{t} = 40 \, \text{mm}$.

Here $\mathbf{x} = [t_1 \, t_2 \cdots t_6]^T$ is the design variable vector and $\sigma_i^M$ represents the von Mises stress computed at the lower surface middle point of each element. Notations $\bar{\sigma}$, $\underline{t}$ and $\bar{t}$ are used for the allowable stress, the lower and upper limit of the design variables, respectively. The numerical design data for this problem are given in the figure legend.

Several methods discussed in the next section are available for the generation of Pareto optima for this plate problem. Ten Pareto optimal plates have been computed by using the constraint method and the sequential quadratic programming optimization algorithm. The corresponding minimal solutions and three Pareto optima, which represent the minimum weight plate, the design at point 5 and the minimum displacement structure, are shown in Fig. 6.4. Also the curve

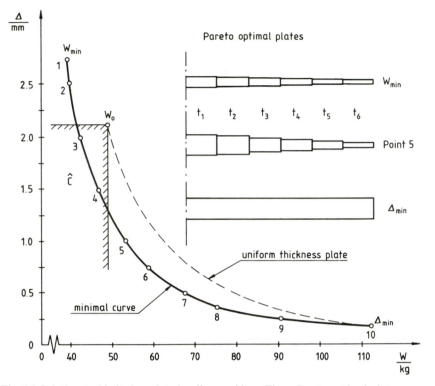

**Fig. 6.4** Solution to bicriterion plate bending problem. Three Pareto optimal plates, corresponding to minimal points 1, 5 and 10, are shown here. The curve of the uniform thickness plate has been depicted for comparison. For example, minimal points 3 and 4 located in the negative cone $\hat{C}$ are better designs than point $W_0$. The minimal curve represents the set of the minimal solutions in the bicriterion case.

**Table 6.1** Pareto optimal plate thicknesses and criteria values at the minimal points shown in Fig. 6.4 (here millimeter and kilogram units are used)

| Point | $t_1$ | $t_2$ | $t_3$ | $t_4$ | $t_5$ | $t_6$ | $W$ | $\Delta$ |
|-------|-------|-------|-------|-------|-------|-------|------|----------|
| 1 | 20.6 | 19.7 | 18.4 | 16.4 | 13.8 | 8.6 | 39.4 | 2.73 |
| 2 | 26.1 | 20.8 | 18.4 | 16.4 | 13.8 | 8.6 | 40.0 | 2.50 |
| 3 | 30.2 | 26.1 | 20.6 | 16.4 | 13.8 | 8.6 | 42.4 | 2.00 |
| 4 | 31.0 | 28.9 | 24.7 | 19.4 | 14.1 | 8.6 | 46.8 | 1.50 |
| 5 | 37.3 | 34.3 | 26.8 | 22.1 | 16.3 | 9.8 | 53.3 | 1.00 |
| 6 | 40.0 | 37.1 | 30.2 | 24.0 | 18.3 | 10.8 | 58.8 | 0.75 |
| 7 | 40.0 | 40.0 | 36.4 | 27.8 | 21.0 | 12.8 | 67.6 | 0.50 |
| 8 | 40.0 | 40.0 | 40.0 | 32.6 | 24.6 | 14.4 | 75.6 | 0.375 |
| 9 | 40.0 | 40.0 | 40.0 | 40.0 | 33.5 | 20.5 | 90.8 | 0.25 |
| 10 | 40.0 | 40.0 | 40.0 | 40.0 | 40.0 | 40.0 | 112.3 | 0.1746 |

representing the uniform thickness designs has been depicted in the criterion space for comparison. This curve is located inside the attainable set everywhere except at the minimum displacement point. Accordingly, the minimal curve gives better designs because every point except $\Delta_{\min}$ on the uniform thickness curve is dominated by a minimal curve segment, as is shown by the negative cone $\hat{C}$ located at the lightest uniform design point $W_0$. The stiffest structure and the plate which is cheapest to fabricate seem to unite in this case, but in general it is reasonable to treat the manufacturing cost and displacement $\Delta$ as separate criteria because they may be locally conflicting elsewhere. This leads to a three-criterion problem which is not considered here. All the ten Pareto optima and the corresponding minimal points are presented numerically in Table 6.1.

## 6.3 GENERATION OF PARETO OPTIMA

### 6.3.1 Linear weighting method

Different methods for generating Pareto optimal solutions to a multicriterion optimization problem have been developed. Usually their application leads to the solution of several scalar problems which include certain parameters. Typically, each parameter combination corresponds to one Pareto optimum and by varying their values it is possible to generate the Pareto optimal set or its part. In the sequel those fundamental methods, which have been applied repeatedly in the structural optimization literature, are briefly described. They are also illustrated graphically in the criterion space in order to show the reasons for their different potential to cover the Pareto optimal set.

The linear weighting method combines all the criteria into one scalar objective function by using the weighted sum of the criteria. If the weighting coefficients are denoted by $w_i, i = 1, 2, \ldots, m$, this scalar optimization problem takes the form

$$\min_{\mathbf{x} \in \Omega} \sum_{i=1}^{m} w_i f_i(\mathbf{x}) \tag{6.5}$$

where the normalization

$$\sum_{i=1}^{m} w_i = 1 \tag{6.6}$$

can be used without losing generality. By varying these weights it is now possible to generate Pareto optima for problem (6.1). The main disadvantage of this method is the fact that only in convex problems it can be guaranteed to generate the whole Pareto optimal set. According to the author's experience, such nonconvex cases where the weighting method fails to generate all Pareto optima are not typical of structural optimization. Some simple truss examples, which demonstrate that this phenomenon really exists in applications, have been reported in the literature (Koski, 1985). The geometrical interpretation of the weighting method in a bicriterion problem is shown in Fig. 6.5(a) where it corresponds to the case $p = 1$. It is interesting to notice that problem (6.5) expressed in the criterion space has the form

$$\min_{\mathbf{z} \in \Lambda} \sum_{i=1}^{m} w_i z_i \tag{6.7}$$

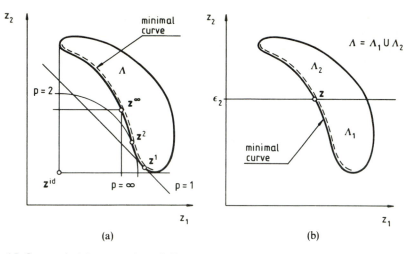

**Fig. 6.5** Geometrical interpretation of distance and constraint methods in bicriterion case: (a) linear weighting method ($p = 1$), weighted quadratic ($p = 2$) and minimax ($p = \infty$) methods illustrated in the criterion space; (b) constraint method where $f_1(\mathbf{x})$ is chosen as the scalar objective function and $f_2(\mathbf{x})$ is removed into the constraints.

where $z_i = f_i(\mathbf{x})$ for $i = 1, 2, \ldots, m$. Thus a linear objective function is minimized in the attainable set.

### 6.3.2 Minimax approach and use of distance function

The distance methods are based on the minimization of the distance between the attainable set and some chosen reference point in the criterion space. In the literature they have also been called metric, norm, and global criterion methods. The resulting scalar problem is

$$\min_{\mathbf{x} \in \Omega} d_p(\mathbf{x}) \tag{6.8}$$

where the distance function

$$d_p(\mathbf{x}) = \left\{ \sum_{i=1}^{m} w_i [f_i(\mathbf{x}) - \hat{z}_i]^p \right\}^{1/p} \tag{6.9}$$

has been widely used in structural optimization. The reference point $\hat{z} \in R^m$ may be chosen by the designer and often the so-called ideal or utopia point

$$\mathbf{z}^{\mathrm{id}} = [f_{1\,\mathrm{min}} \, f_{2\,\mathrm{min}} \cdots f_{m\,\mathrm{min}}]^{\mathrm{T}} \tag{6.10}$$

can be found in the applications. This ideal vector contains all the individual minima of the criteria in $\Omega$ as components. Thus it is necessary to solve $m$ scalar optimization problems

$$\min_{\mathbf{x} \in \Omega} f_i(\mathbf{x}), \quad i = 1, 2, \ldots, m \tag{6.11}$$

if $\mathbf{z}^{\text{id}}$ is used as a reference point $\hat{\mathbf{z}}$. The normalization given in equation (6.6) is also applicable for the weights $w_i$ here. Usually $\hat{\mathbf{z}}$ and $p$ are fixed and $w_i$ are the only parameters but also other possibilities exist (Koski and Silvennoinen, 1987). The extreme case $p = \infty$ in equation (6.9) corresponds to the weighted minimax problem

$$\min_{\mathbf{x} \in \Omega} \max_{i} [w_i f_i(\mathbf{x})], \quad i = 1, 2, \ldots, m \tag{6.12}$$

which is capable of generating all Pareto optima also in nonconvex problems. The other extreme case $p = 1$ can be interpreted as the linear weighting method if the origin is used as a reference point, i.e. $\hat{\mathbf{z}} = \mathbf{0}$. Correspondingly, the case $p = 2$ might be called as a weighted quadratic method. All these three cases have been illustrated together in Fig. 6.5(a). In practical applications, where the numerical values of the noncommensurable criteria may have huge variations with respect to each other, it is useful to normalize all the criteria before computations. One possibility is to use the formula

$$\tilde{f}_i(\mathbf{x}) = \frac{f_i(\mathbf{x}) - f_{i\min}}{f_{i\max} - f_{i\min}} \tag{6.13}$$

where all the nondimensional criteria are limited to an equal range, i.e. $\tilde{f}_i(\mathbf{x}) \in [0, 1]$, $i = 1, 2, \ldots, m$. The quadratic case $p = 2$ seems to be the most popular choice in the literature but also both of the extreme cases have been used frequently in structural design applications.

### 6.3.3 Constraint method

One possibility is to replace the original multicriterion problem by a scalar problem where one criterion $f_k$ is chosen as the objective function and all the other criteria are removed into the constraints. By introducing parameters $\varepsilon_i$ into these new constraints an additional feasible set

$$\Omega_k(\varepsilon_i) = \{\mathbf{x} \in R^n | f_i(\mathbf{x}) \leqslant \varepsilon_i, i = 1, 2, \ldots, m, i \neq k\} \tag{6.14}$$

is obtained. If the resulting feasible set is denoted by $\bar{\Omega}_k = \Omega \cap \Omega_k$, the parametrized scalar problem can be expressed as

$$\min_{\mathbf{x} \in \bar{\Omega}_k} f_k(\mathbf{x}) \tag{6.15}$$

Here each parameter combination yields a separate problem usually corresponding to one Pareto optimum. This technique, called the **constraint method**, can generate the whole Pareto optimal set also in nonconvex cases and it has been applied to some extent in structural optimization. If so-called weak solutions (Koski and Silvennoinen, 1987) shown in Fig. 6.2 exist, the constraint method can be modified to cover that case as well but then equality constraints appear and several new scalar problems must be solved for just one Pareto optimum. The geometric interpretation of the method is given in Fig. 6.5(b) where $\mathbf{z}_1$ is minimized in the set $\Lambda_1$ which is the image of the set $\bar{\Omega}_1$.

## 6.4 ANALYTIC SOLUTION FOR STATICALLY DETERMINATE TRUSSES

The necessary conditions of Pareto optimality offer one more possibility to solve the multicriterion optimization problem (6.1). In structural optimization this approach has been used mainly in problems where the formulation is based on the optimal control theory and generally analytic solutions to certain bicriterion problems have been obtained (Stadler, 1988, Leitmann, 1977). Naturally, different numerical solution techniques, corresponding to the successful optimality criteria methods in scalar optimization of structures, could be developed for multi-criterion problems. Thus far, however, only a few structural design problems (Bendsøe, Olhoff and Taylor, 1983–1984) have been solved by using directly the necessary conditions of the nonlinear programming problem (6.1). Next an application, where an analytic solution to a class of truss problems has been obtained in this way, is described.

Statically determinate and indeterminate structures are often considered separately in structural mechanics. In analyzing determinate trusses the member forces can be solved directly from the static equilibrium equations of the nodes only whereas for indeterminate structures the compatibility equations are also needed to couple the elongations of the members at each node. This property of obtaining the member forces directly is worth utilizing in optimization as well. Next a relatively general multicriterion problem is formulated for statically determinate trusses and the analytic solution presented in Koski and Silvennoinen (1982) is considered in broad outline.

The material volume and several arbitrary nodal displacements of a structure are chosen as the criteria to be minimized. Member areas $A_i$ are used as the design variables and upper limits $\bar{A}_i$ as well as lower limits $\underline{A}_i$ for them can be imposed. If the number of members is denoted by $k$, the design variable vector will be

$$\mathbf{x} = [A_1 \, A_2 \cdots A_k]^{\mathrm{T}} \tag{6.16}$$

Both material and geometrical linearity are assumed in analyzing a truss. Because all member forces $N_i$ can be solved directly from the nodal equilibrium equations, new lower limits for the design variables are obtained immediately from the stress constraints as follows:

$$\underline{A}_i = \max_j (N_i/\sigma_i^{\mathrm{a}})_j, \quad i = 1, 2, \dots, k, \quad j = 1, 2, \dots, q$$

$$\sigma_i^{\mathrm{a}} = \bar{\sigma}_i \quad \text{for tension members} \tag{6.17}$$

$$\sigma_i^{\mathrm{a}} = \underline{\sigma}_i \quad \text{for compression members}$$

where the allowable stresses $\sigma_i^{\mathrm{a}}$ are chosen for every member $i$ and $q$ is the number of loading conditions. The local instability of the compression members in the elastic range may be prevented by applying the Euler buckling constraints in the expression

$$\sigma_i \leqslant \frac{\pi^2 E I_i}{n A_i L_i^2}, \quad i = 1, 2, \dots, k \tag{6.18}$$

which states that the compressive stress $\sigma_i$ in member $i$ must be less than or equal to the Euler buckling stress divided by the safety factor $n$. On the right-hand side of this inequality $EI_i$ and $L_i$ are the bending rigidity and the length of member $i$, respectively. In order to convert the constraint into the form which has the member areas $A_i$ as the only design variables the well-known relation $I_i = cA_i^p$ is introduced. By this substitution the buckling constraints (6.18) yield another lower limit for each compression member in addition to the imposed ones and those resulting from the stress constraints (6.17). When the most severe of these lower limits is chosen for every design variable subject to compression, the feasible set

$$\Omega = \{ \mathbf{x} \in R^k | \underline{A}_i \leqslant A_i \leqslant \bar{A}_i, i = 1, 2, \ldots, k \} \tag{6.19}$$

is obtained. This region consists of a rectangular prism in the design space generated by the member areas. It is further assumed that all the lower limits of the member areas are strictly positive, i.e. $\underline{A}_i > 0$ for $i = 1, 2, \ldots, k$, as is natural in real statically determinate applications. The vector objective function consists of the material volume $V$ and $m - 1$ arbitrary nodal displacements $\Delta_i$ which all can be written in an explicit form in this determinate case. The multicriterion optimization problem for statically determinate trusses can thus be stated as

$$\min_{\mathbf{x} \in \Omega} [V(\mathbf{x}) \Delta_1(\mathbf{x}) \Delta_2(\mathbf{x}) \cdots \Delta_{m-1}(\mathbf{x})]^T \tag{6.20}$$

where

$$V = \sum_{i=1}^{k} a_i A_i, \quad a_i > 0$$

$$\Delta_j = \sum_{i=1}^{k} \frac{\alpha_i^j}{A_i}, \quad \alpha_i^j \gtrless 0, \quad j = 1, 2, \ldots, m - 1$$

This explicit problem is not convex, because displacement criteria are not convex functions of the member areas, but it can be transformed into a convex one by replacing the design variables by their inverses. This property can be utilized to prove that the necessary conditions of Pareto optimality are also sufficient for this problem. It turns out that these conditions can be solved exactly in the case $m = 2$, i.e. if the bicriterion problem

$$\min_{\mathbf{x} \in \Omega} [V(\mathbf{x}) \Delta(\mathbf{x})]^T \tag{6.21}$$

where

$$V = \sum_{i=1}^{k} a_i A_i, \quad a_i > 0, \quad \Delta = \sum_{i=1}^{k} \frac{\alpha_i}{A_i}, \quad \alpha_i \in R$$

is considered. The necessary conditions and the following mathematical proof, which is rather lengthy, is skipped and only the results are presented here. First a theorem (Koski and Silvennoinen, 1982) that gives a complete solution to the above bicriterion problem is given.

*Theorem 6.1* The set of Pareto optima for problem (6.21) consists of a connected polygonal line $l_1 \cup l_2 \cup \cdots \cup l_N$. The consecutive line segments $l_n$,

$n = 1, 2, \ldots, N$, have the parametric equations

$$
\begin{aligned}
A_i &= \bar{A}_i, && i \in I_n \\
A_j &= \underline{A}_j, && j \in J_n \\
A_s &= c_s^{-1} t, && s \in K \backslash (I_n \cup J_n)
\end{aligned}
$$

$$t_{n-1} \leqslant t \leqslant t_n$$

(6.22)

where $K = \{1, 2, \ldots, k\}$, $Q = \{i \in K | \alpha_i \leqslant 0\}$, $Q \neq K$, $N = \min \{n \in \mathbb{N} | I_{n+1} = K \backslash Q\}$ and $c_s = a_s^{1/2} \alpha_s^{-1/2}$ for $s \in K \backslash Q$. Further, $I_0 = \varnothing$, $J_0 = K$, $t_0 = \min \{c_j \underline{A}_j | j \in K \backslash Q\}$ and for $n = 1, 2, \ldots, N$

$$
\begin{aligned}
I_n &= I_{n-1} \cup \{s \in K \backslash (I_{n-1} \cup J_{n-1}) | c_s \bar{A}_s = t_{n-1}\} \\
J_n &= J_{n-1} \backslash \{j \in J_{n-1} \backslash Q | c_j \underline{A}_j = t_{n-1}\} \\
t_n &= \min \{c_s \bar{A}_s, c_j \underline{A}_j | s \in K \backslash (I_n \cup J_n), j \in J_n \backslash Q\}
\end{aligned}
$$

(6.23)

The notation $\mathbb{N} = \{1\,2\,3\ldots\}$ is used for the set of all positive integers and the set difference is denoted by $K \backslash Q = \{i \in K | i \notin Q\}$ in this theorem. It shows that the set of all Pareto optima will be a polygonal line in the design space. Index sets $I$ and $J$ change from one Pareto optimal line segment to another, whereas index set $Q$ remains constant. After solving the Pareto optimal set of problem (6.21), the corresponding minimal solutions can be found easily by substituting relations (6.22) into the expressions of $V$ and $\Delta$. It is also possible to eliminate parameter $t$, resulting in an analytic presentation of the function $V(\Delta)$. Even if the expressions (6.22) and (6.23) look complicated, their application is very easy as is illustrated by the following small example.

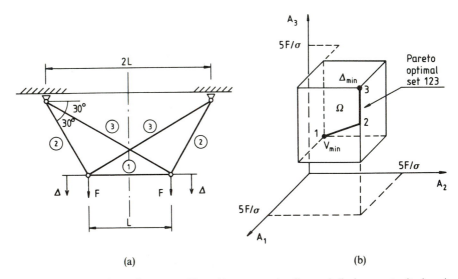

(a)                                      (b)

**Fig. 6.6** Bicriterion isostatic truss problem: (a) structure, loading and displacement criterion $\Delta$; (b) feasible set $\Omega$ in the design space and the Pareto optimal polygonal line 1–2–3. Allowable stresses are $\sigma$ in tension and $-\sigma$ in compression. Only upper limits $\bar{A} = 5F/\sigma$ have been imposed for all member areas whereas the lower limits are obtained from the stress constraints.

A five-bar truss shown in Fig. 6.6 should be optimized by using only three design variables $A_1, A_2$ and $A_3$, owing to the symmetry. The structure is subjected to one loading condition and the equal vertical displacement of both the loaded nodes is chosen as criterion $\Delta$ in problem (6.21). Stress constraints $-\sigma \leqslant \sigma_i \leqslant \sigma$ for all members and upper limits $\bar{A} = 5F/\sigma$ for the member areas are imposed in this example whereas buckling constraints are excluded. The lower limits for the member areas can be computed directly from the stress constraints because no other lower limits have been imposed. After determining the member forces the following explicit bicriterion problem is obtained:

$$\min \begin{bmatrix} (A_1 + 2A_2 + 2\sqrt{3}A_3)L \\ \left( \dfrac{3/2}{A_1} + \dfrac{3}{A_2} + \dfrac{\sqrt{3}}{A_3} \right) \dfrac{FL}{E} \end{bmatrix} \qquad (6.24)$$

subject to

$$\sqrt{3}F/\sigma \leqslant A_1 \leqslant 5F/\sigma$$
$$\sqrt{3}F/\sigma \leqslant A_2 \leqslant 5F/\sigma$$
$$F/\sigma \leqslant A_3 \leqslant 5F/\sigma$$

Next the Pareto optimal polygonal line is generated for this problem by using the scheme given in relations (6.22) and (6.23). In this case set $Q = \varnothing$ because all $\alpha_i > 0$ in the displacement criterion. Two Pareto optimal line segments, which are given as functions of parameter $t$, are obtained for this problem.

First it is practical to remove constants $L$ and $FL/E$ from the expressions of the criteria for a while. From the dimensionless coefficients $a_i$ and $\alpha_i$, $i = 1, 2, 3$, appearing in the criteria the numbers $c_1 = \sqrt{6}/3$, $c_2 = \sqrt{6}/3$ and $c_3 = \sqrt{2}$ are directly obtained. At the beginning the index sets are $I_0 = \varnothing$, $J_0 = K = \{1, 2, 3\}$ and parameter $t_0 = \sqrt{2}F/\sigma$. This Pareto optimum, where all the member areas are at their lower limits, corresponds to the minimum volume solution.

Line segment 12 lies in the interior of $\Omega$ because $I_1 = J_1 = \varnothing$ and $K\backslash(I_1 \cup J_1) = \{1, 2, 3\}$. It is given by

$$A_1 = \frac{\sqrt{6}}{2}t, \quad A_2 = \frac{\sqrt{6}}{2}t, \quad A_3 = \frac{\sqrt{2}}{2}t, \quad \sqrt{2}\frac{F}{\sigma} \leqslant t \leqslant \frac{5\sqrt{6}}{3}\frac{F}{\sigma} \qquad (6.25)$$

The second line segment 23 lies on that edge of $\Omega$ where $I_2 = \{1, 2\}$, $J_2 = \varnothing$ and $K\backslash(I_2 \cup J_2) = \{3\}$. Its expression is

$$A_1 = 5F/\sigma, \quad A_2 = 5F/\sigma, \quad A_3 = \frac{\sqrt{2}}{2}t, \quad \frac{5\sqrt{6}}{3}\frac{F}{\sigma} \leqslant t \leqslant 5\sqrt{2}\frac{F}{\sigma} \qquad (6.26)$$

These parametric equations (6.25) and (6.26) represent the Pareto optimal set of problem (6.24). This polygonal line, starting from the minimum volume solution and ending at the point where $\Delta$ achieves its minimum value, has been depicted in Fig. 6.6(b). In this specific case no Pareto optima are located on the faces of the feasible set. In large-scale problems more line segments appear but the solution procedure is quite insentive to the size of the problem.

The original multicriterion problem (6.20) is convex in the reciprocal variables and thus all Pareto optima for it can be generated by the linear weighting method.

This leads to one scalar optimization problem for each weight combination. In the present case, however, it is preferable to convert problem (6.20) into a parametric bicriterion problem. Considerable advantage is obtained in this way compared with the weighting method or any scalarization technique. First, the number of parameters is reduced by one and, secondly, each parameter combination gives a large set of Pareto optima instead of only one point. In addition, the Pareto optimal set of each bicriterion problem is known exactly and no approximate optimization procedure, possibly involving high computation costs and difficulties in convergence, is needed. The parametric bicriterion problem

$$\min_{\mathbf{x} \in \Omega} \left[ V(\mathbf{x}) \sum_{j=1}^{m-1} \lambda_j \Delta_j(\mathbf{x}) \right]^{\mathrm{T}} \tag{6.27}$$

includes a linear combination of the original displacement criteria as the second component of the vector objective function. This combined displacement criterion again has the general form

$$\sum_{j=1}^{m-1} \lambda_j \Delta_j(\mathbf{x}) = \sum_{i=1}^{k} \frac{\beta_i}{A_i}, \quad \beta_i \in R \tag{6.28}$$

which is similar to the expression of the displacement criterion $\Delta(\mathbf{x})$ in problem (6.21). Here the coefficients $\beta_i$ can be computed from the corresponding expressions of the original displacement criteria given in problem (6.20) as follows:

$$\beta_i = \lambda_1 \alpha_i^1 + \lambda_2 \alpha_i^2 + \cdots + \lambda_{m-1} \alpha_i^{m-1}, \quad i = 1, 2, \dots, k \tag{6.29}$$

For every fixed combination of parameters $\lambda_j, j = 1, 2, \dots, m-1$, it is possible to formulate problem (6.27) which can be solved completely by the given theorem. Apparently, if these parameters have the following properties

$$\sum_{i=1}^{m-1} \lambda_j = 1, \quad \lambda_j \geq 0, \quad j = 1, 2, \dots, m-1 \tag{6.30}$$

the solution set of each bicriterion problem with fixed $\lambda_j$ values represents also a Pareto optimal polygonal line of the original problem (6.20). It can be further proved that the whole Pareto optimal set of problem (6.20) is obtained in this way by going through all the possible parameter combinations.

By combining the preceding results a numerical method for generating Pareto optima for the $m$-criterion problem (6.20) may be constructed. First the original problem is converted into a bicriterion problem by choosing parameters $\lambda_j$, $i = 1, 2, \dots, m-1$, and combining the original displacement criteria into one. The given theorem can then be applied to compute the corresponding Pareto optimal polygonal line in the design space. The solution scheme uses no approximate optimization technique and thus any accuracy wanted for the results may be achieved. The free parameter $t$ is used for convenience in order to attain a clear representation form for the results and to enable an easy movement along the polygonal line in the decision-making stage. The whole Pareto optimal set of problem (6.20) can be generated as the union of the polygonal lines, each corresponding to one bicriterion problem (6.27) with fixed parameters $\lambda_j$.

The present method has the capacity of generating Pareto optima at a relatively high speed because each parameter combination gives an entire polygonal line. Moreover, the procedure can easily be coded as a computer program capable of obtaining any Pareto optimal polygonal line by a finite number of calculation steps without computational difficulties even in large problems. In order to solve a traditional scalar problem, where the material volume of an isostatic truss should be minimized subject to stress, elastic buckling and displacement constraints, an approximate optimization procedure is usually needed. This may be the optimality criteria approach or some numerical nonlinear programming algorithm. Instead of applying them, it is advantageous to remove the constrained displacements into the vector objective function in addition to the volume and to use the described method. Plenty of Pareto optimal solutions, including the wanted one, are obtained very quickly with any accuracy.

## 6.5 DECISION-MAKING PROCESS

Pareto optima represent solutions which are mathematically better than the other feasible points. Thus it is natural to concentrate merely on Pareto optima in choosing the final design called usually the best compromise solution. Comparisons between different Pareto optima can be made by introducing the designer's personal preferences based on his experience and some other factors not included in the multicriterion problem formulation (6.1). The mathematical optimization model hardly ever contains all the detailed requirements existing in practical design applications. In the multicriterion approach the decision making is concentrated on the choice of the relative importances of the competing criteria. Accordingly, this entirety consisting of the search of the best compromise solution among Pareto optima is called the multicriterion decision-making process.

Optimum structural design differs considerably from synthesis problems in other fields, such as social welfare theory, resource planning and management, where the multicriterion approach has been widely applied. Structural analysis equations are very reliable and the optimization problem is generally well defined, offering an excellent starting point for the decision maker. According to the classification especially adapted to structural optimization and proposed in Koski (1992) the multicriterion design process can be divided into three main phases. They are the problem formulation, generation of Pareto optima and decision making. The last phase can be further separated into four classes: interactive methods, choice by comparisons, procedures that use *a priori* fixed parameters and other methods. Different interactive approaches, where the designer proceeds from one Pareto optimum to a better one until a satisfactory solution is achieved, have been published. Another possibility is to generate a large collection of Pareto optima and to choose the best one by direct comparisons. The third class includes methods where some parameters have been fixed in advance. A typical example of this class is the so-called utility function methods, where some chosen scalar combination of the criteria is optimized. Also some other methods, such as goal programming and game theory approach for example, have been applied in structural optimization.

Plenty of relevant information about the system can be obtained by optimization calculations. In order to make good decisions Pareto optimal solutions and some trade-off information associated with them are usually necessary. The term trade-off has been used quite freely in the literature and it seems to take different forms depending on the context. A common feature in most trade-off analyses is that they are concerned with the advantage gained for one criterion by making concessions in another criterion. At any Pareto optimum $\mathbf{x}^*$ a trade-off number, defined symbolically by $\partial f_k / \partial f_j$, can be defined as a change of $f_k$ per unit change in $f_j$ on the minimal surface when all the other criteria remain unchanged. Naturally, these trade-off numbers are closely related to the Kuhn–Tucker multipliers of any corresponding scalar problem (6.5), (6.8), (6.12) and (6.15) at $\mathbf{x}^*$. For example in the constraint method

$$\frac{\partial f_k}{\partial f_i}(\varepsilon_j) = -u_{ki}^* \tag{6.31}$$

where $u_{ki}^*$ is the Kuhn–Tucker multiplier associated with the constraint $f_i(\mathbf{x}) \leqslant \varepsilon_i$ and $\varepsilon_j, j = 1, 2, \ldots, m, j \neq k$, represent the constraint limits corresponding to Pareto optimum $\mathbf{x}^*$. Geometrically speaking, the computation of a trade-off number at any Pareto optimum provides the designer with useful information about the tangent plane of the minimal surface in the criterion space at $\mathbf{z}^*$. If the whole minimal set is known, then complete trade-off information is directly available. In large-scale applications, however, the decision maker needs trade-off numbers at those few Pareto optima which have been computed.

## 6.6 LITERATURE REVIEW

Most of the citations used in the preceding text are the author's own contributions because they form a unified basis for this presentation. Altogether, however, they represent only a small part of the versatile production found in the field today.

According to a state-of-the-art study (Koski, 1993), about 70 articles dealing with multicriterion structural optimization have been published in English since the late 1970s. Stadler (1977, 1978) applied the control theory approach to different bicriterion problems with weight and stored energy as criteria and obtained analytical solutions for some structures, calling the results 'natural structural shapes'. Also, Leitmann (1977) used control theory in solving some bicriterion problems in viscoelasticity. Gerasimov and Repko (1978) optimized a truss and a framed plate by applying the mathematical programming approach. Koski (1979) formulated a vector optimization problem, where the material volume and several chosen nodal displacements of a truss were chosen as the criteria, and generated Pareto optima for several truss examples.

In the 1980s the number of publications increased considerably including many different academic and industrial applications. For example, such authors as Adali (1983a, b), Baier (1983), Bühlmeier (1990), Bendsøe, Olhoff and Taylor (1983), Carmichael (1980), Dhingra, Rao and Miura (1990), Diaz (1987a, b), Eschenauer and his coworkers (Eschenauer, 1983; Eschenauer, Kneppe and Stenvers, 1986; Eschenauer, Schäfer and Bernau, 1988; Eschenauer and Vietor, 1990; Eschenauer, Koski and Osyczka, 1990), Fu and Frangopol (1990), Fuchs *et al.* (1988), Hajela and Shih (1990), Jendo, Marks and Thierauf (1985), Koski (1981, 1984a), Lógó and Vásárhelyi (1988), Nafday, Corotis and Cohon (1988), Olhoff (1989), Osyczka (1981, 1984), Post (1990), Rao and his coworkers (Rao, 1984, 1987; Rao and Hati, 1986; Rao, Venkayya and Khot, 1988), Rozvany (1989),

Stadler (1977, 1978, 1987, 1988, 1989), Tseng and Lu (1990) and Yoshimura *et al.* (1984) have contributed in this field. The choice of the criteria varies from one article to another but usually physical quantities rather than economic are used. The most commonly used criterion is the weight or the material volume of a structure and the second in number is the flexibility which is usually measured by displacements or some energy expression. These two criteria are usually strongly conflicting and they offer an excellent possibility to study the multi-criterion problem with the lowest dimension ($m = 2$), which also allows the graphic presentation of the minimal curve in the criterion space. For certain specific problems the Pareto optimal set has been determined analytically but commonly some numerical optimization procedure has been applied. Interactive methods especially adapted to structural optimization are rare for the present. In addition, stochastic and fuzzy optimization problems have been treated to some extent. Industrial applications (Eschenauer, Koski and Osyczka, 1990) have increased considerably during the last decade and a multicriterion module has been added to some structural optimization software packages. In addition to different design problems the multicriterion approach has also been utilized in the plastic analysis of frames (Nafday, Corotis and Cohon, 1988a, b).

## 6.7 INDUSTRIAL APPLICATION AND ILLUSTRATIVE EXAMPLES

### 6.7.1 Shape optimization of ceramic piston crown

Different new materials and their composites have recently appeared to replace such traditional alternatives as construction steels and cast iron for example. Among these, ceramics undoubtedly deserve special attention in applications where high temperature capability, wear and corrosion resistance, and certain electrical or optical properties are needed. In combustion engines, bearings, cutting parts of tools and nozzles, just to mention some industrial products, ceramic materials have been used successfully. Most evidently the application of ceramics offers totally new solution possibilities for many traditional constructions. The economic potential associated with ceramics seems large, but severe restrictions appear if a high reliability for load-supporting components is required. Especially, shape optimization becomes an important part in the design process because the reliability is closely associated with the geometric shape of the structure. The shape optimization of ceramic components differs considerably from the design of most conventional construction materials. This is due to the fact that no clear allowable stresses exist but the strength of the ceramic component must be evaluated by the statistical approach. Furthermore, it seems impractical to replace the stress constraints by the reliability constraints because the choice of the constraint limit is extremely difficult in practical circumstances. Especially, in structural design, where both economic and human losses are involved, it is reasonable to consider the reliability as a criterion instead of a constraint function. In the following an industrial application (Koski and Silvennoinen, 1990), where the multicriterion approach is used instead of the traditional scalar problem with stress constraints, is considered.

A two-parametric Weibull distribution is used to describe the results of the strength tests for ceramics. The two material parameters, which are the mean fracture stress $\sigma$ and the Weibull modulus $m$, can both be obtained from the material tests. Also, the volume effect, which appears such that the strength decreases when the size of the specimen and the loading are proportionally increased, should be considered in the statistical approach. The mean fracture stress $\sigma$ is associated with a certain reference volume $V_0$ which usually comprises that part of the specimen where tensile stresses occur. The other material parameter, the Weibull modulus $m$, describes the narrowness of the strength distribution. The larger the value of $m$, the smaller is the variation of the fracture stress. From the stress field determined usually by the finite element method it is possible to compute the probability of failure from the expression

$$P_f = 1 - \exp\left[ -\left(\frac{1}{m}!\right)^m \left(\frac{1}{\sigma}\right)^m \left(\frac{1}{V_0}\right) \iiint_V (\sigma_1^m + \sigma_2^m + \sigma_3^m)\, dV \right] \quad (6.32)$$

where the integration is extended over the component volume. Here $\sigma_1$, $\sigma_2$ and $\sigma_3$ are the principal stresses which are included in the volume integral at each material point only if they are tensile, i.e. $\sigma_i > 0$. $\bar{\sigma}$, $V_0$ and $m$ are the material parameters explained earlier and

$$\left(\frac{1}{m}!\right) = \Gamma\left(\frac{1}{m} + 1\right) = \int_0^\infty t^{1/m} e^{-t}\, dt \quad (6.33)$$

The numerical computation of $P_f$ is integrated directly to the finite element analysis. From the volume integral in equation (6.32) it may be observed that the crucial parts of the component are those where high tensile stresses occur.

The efficiency of a diesel engine is one chief factor in the competition of the market shares in such applications as the main or auxiliary engines of ships, locomotives and land-based power plants. A standard way to improve the efficiency is the raising of the working temperature in the cylinders. Good thermal insulation and strength properties are needed for the materials in those parts which are subjected to the elevated temperatures. Ceramics are a very potential alternative for this purpose. For example, in the Wärtsilä Vasa 22 HF medium speed diesel engine a considerable benefit is expected by using a ceramic piston crown instead of the traditional cast iron construction. It is uneconomic to use a monolithic ceramic piston because at the moment the material is expensive and extreme difficulties appear in trying to achieve an accurate shape during the sintering process. Thus, a ceramic material is used only in the crown where good thermal insulation and strength properties at these high temperatures are needed. The prototype piston construction including the ceramic crown, which is the object of the shape optimization here, is shown in Fig. 6.7. From the two alternative materials, zirconia ($ZrO_2$) and silicon nitride ($Si_3N_4$), the latter was chosen mainly because the density and thermal expansion coefficient values of zirconia are too large. Silicon nitride has a considerable strength at high temperatures and in spite of its small thermal insulation capability compared with zirconia it allows the elevation of the temperature in the cylinder. In addition to the possible improvement in the efficiency, another advantage is obtained simultaneously. This is the possibility of dropping the quality of the heavy fuel used in this engine because of the better ignition properties due to the higher surface temperature.

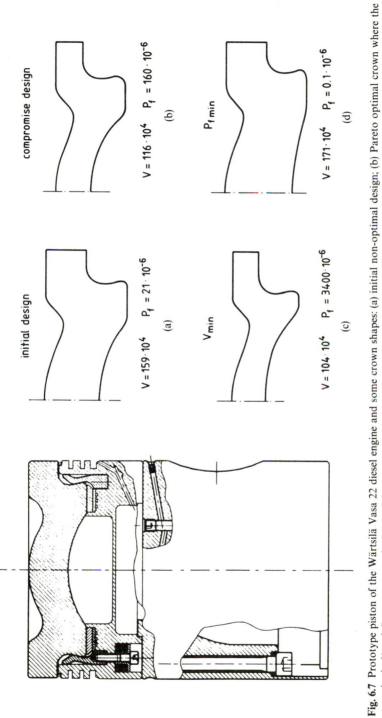

**Fig. 6.7** Prototype piston of the Wärtsilä Vasa 22 diesel engine and some crown shapes: (a) initial non-optimal design; (b) Pareto optimal crown where the two criteria $V$ and $P_f$ are compromised; (c) minimum material volume design; (d) maximum reliability design. The numerical values of the two criteria for every shape are presented for a comparison. The volume criterion is given in $mm^3$; $P_f$ has no units.

In addition, the cooling system of the piston becomes unnecessary and it can be left out.

Shape optimization has been performed for the ceramic piston crown only, leaving the other parts of the piston out. It is important to minimize the material volume of the crown for two reasons. The first is the high price of the silicon nitride itself. The second reason is associated with the inertia forces which should be at the same relatively low level as they are in the traditional cast iron pistons. The other criterion to be minimized is the probability of failure of the crown which should be as small as possible in order to avoid expensive damage during operation. The combustion conditions in the cylinder determine the shape of the upper surface of the crown whereas the shape of the lower surface and the location of the axial support, which are shown in Fig. 6.8, can be varied during optimization. The curve which defines the shape of the lower surface of the crown is represented by two $B$ splines of order 2 with the control nodes $P_0, P_1, P_2$ and $Q_0, Q_1, Q_2$. Instead of the nodal coordinates of the finite elements on the boundary it is advantageous to use the $B$ splines to describe the two parts of the lower surface and to apply the coordinates of the control nodes as the design variables. The initial shape and all the eight components of the design variable vector $\mathbf{x} = [x_1 \, x_2 \cdots x_8]^{\mathrm{T}}$ have been depicted in Fig. 6.8. It is natural to use side constraints to determine the allowable space for the component. Thus the feasible set $\Omega$ is defined by the inequality constraints imposed for the design variables, i.e. $\Omega = \{\mathbf{x} \in R^8 \,|\, \underline{x}_i \leqslant x_i \leqslant \bar{x}_i, \, i = 1, \ldots, 8\}$. From the practical viewpoint, it seems reasonable to apply the side constraints because the space limits can be reliably

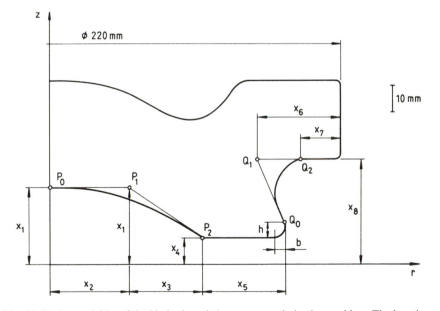

**Fig. 6.8** Design variables of the bicriterion piston crown optimization problem. The locations of the control points $P_0$, $P_1$, $P_2$ and $Q_0$, $Q_1$, $Q_2$, which define the shape of the crown, are the design variables $x_i$. The fixed curve segment near $Q_0$ determines the values $h = 5$ mm and $b = 3$ mm. The support surface must be a horizontal plane.

assessed and the designer may concentrate on the trade-off between the two conflicting physical criteria in making his final choice. The resulting bicriterion shape optimization problem for the ceramic piston crown is

$$\min [V(\mathbf{x}) \, P_f(\mathbf{x})]^T \qquad (6.34)$$

subject to

$$
\begin{array}{ll}
10 \leqslant x_1 \leqslant 35 & 10 \leqslant x_5 \leqslant 30 \\
25 \leqslant x_2 \leqslant 45 & 25 \leqslant x_6 \leqslant 35 \\
10 \leqslant x_3 \leqslant 40 & 10 \leqslant x_7 \leqslant 20 \\
0 \leqslant x_4 \leqslant 10 & 30 \leqslant x_8 \leqslant 40
\end{array}
$$

where $V(\mathbf{x})$ is the material volume of the crown and $P_f(\mathbf{x})$ is its probability of failure given in equation (6.32). Upper and lower limits for the design variables are given in millimeters. The purpose is to compute some Pareto optimal shapes for the designer who can use his own experience in the decision-making phase where the manufacturing cost, the compatibility with the rest of the piston and many other details are considered. The choice of the strictly physical criteria in the vector objective function in problem (6.34) offers a sound basis for the optimum design because they offer Pareto optimal solutions which may be utilized even if the manufacturing techniques or the surrounding construction changes during the design process.

The piston crown is subjected to several loadings during every working cycle. A large amount of heat is transferred from the combustion gas through the crown to the cylinder liner and the cooling water. This causes thermal stresses which can be added to the stresses caused by other loadings. Inertia forces are proportional to the accelerations and their value as well as the gas pressure depend on the position of the piston. A mounting ring is used to fix the crown to the intermediate steel part and it causes a static compressive force along the connection circumference. In the optimization it seems sufficient to consider only the loading case where the thermal charge and the connection force are acting simultaneously.

As a result, an axisymmetric linear static analysis problem is obtained. All the numerical analyses during the optimization, each associated with some modified shape, have been performed by the general-purpose FEM package ANSYS. Only the computation of the probability of failure must be performed outside the program because it was not included in the package. Also, the stationary heat transfer problem for the new shape at every iteration step was calculated by ANSYS in order to determine the corresponding temperature distribution. Thus for each new shape several analyses, including thermal and structural problems, must be performed. An adaptive mesh of isoparametric quadrilateral and triangular ring elements was generated for the crown. By applying a certain parametrization the mesh was adapted for the modified shapes which were obtained during the optimization.

The constraint method was used for the generation of Pareto optimal solutions to problem (6.34). The sequential quadratic programming (SQP) method, which uses the gradients of the objective and constraint functions in the iterations, was chosen and combined with the FEM program ANSYS. A few hundred finite element analyses were needed to compute one Pareto optimum by using an arbitrary starting shape. It was possible to reduce the number of iterations and analyses by choosing the former Pareto optimum as a starting point. The minimum volume and the minimum probability of failure crowns, which represent the extreme Pareto optima for problem (6.34), are shown in Fig. 6.7. A more practical Pareto optimal crown shape and the initial design have also been depicted in the same figure. These three Pareto optimal shapes only illustrate the proposed approach whereas for real decision making many more Pareto optimal crowns were generated.

### 6.7.2 Composite beam

In traditional lightweight structures, such as aircraft and spacecraft for example, composites have become very popular because of their high strength–weight ratio. Recently these advanced materials have emerged into some terrestrial applications as well mainly because light components in the design offer the possibility to make the construction cheaper elsewhere. Remarkable variations in both the mechanical and the thermal behavior of the composite can be achieved by varying layer thicknesses and fiber orientations but simultaneously making an intuitive guess of a good design becomes extremely difficult. Also, the manufacturing cost may become high and the common idea that the weight represents the material cost is not valid any more. This is due to the large variations in the prices of the different materials used in the layers. Also, in the composite design the multicriterion approach seems suitable because several conflicting criteria usually appear in practical applications.

In the present example it is demonstrated that weight and material cost are competing objectives in the composite design. A simply supported beam shown in Fig. 6.9(a) is optimized subject to failure and side constraints. Carbon and glass fiber layers are located in turn forming a symmetric eight-layer cross section according to Fig. 6.9(b). Because of the symmetry only four layer thicknesses $t_1, \ldots, t_4$ are used as the design variables whereas the fiber orientations have been fixed to $0°$ for the four inner layers (1 and 2) and $90°$ for the remaining outer layers (3 and 4). The finite element analysis model is based on the eight-node rectangular plane-stress element and the Tsai–Hill failure criterion has been applied in the stress constraints. The notations $C_c$ and $C_g$ have been used for the prices of the carbon and glass layers, respectively. In Table 6.2 these prices, which have been presented in relative rather than absolute values, as well as the material properties are given.

The total weight of the beam, denoted here by $W = m_c + m_g$, is chosen as the first criterion to be minimized. The symbols $m_c$ and $m_g$ represent the masses of all the carbon and glass layers, respectively. The total material cost, defined correspondingly as $C = C_c m_c + C_g m_g$, is used as the second criterion which should also be minimized. The notations $\mathbf{x} = [t_1 \, t_2 \, t_3 \, t_4]^\mathrm{T}$ and

$$f_i(\mathbf{x}) = \left(\frac{\sigma_1}{X}\right)^2 - \frac{\sigma_1 \sigma_2}{X^2} + \left(\frac{\sigma_2}{Y}\right)^2 + \left(\frac{\tau_{12}}{S}\right)^2 \qquad (6.35)$$

are used for the design variable vector and the stress function appearing in the Tsai–Hill failure constraint of layer $i$. In the latter expression the allowed stresses in tension ($X_t$ and $Y_t$) and in compression ($X_c$ and $Y_c$) usually have different

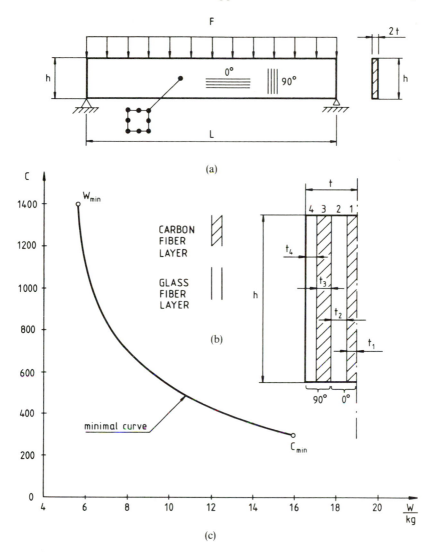

**Fig. 6.9** Composite beam problem where weight $W$ and material cost $C$ are chosen as criteria: (a) structure, loading, fiber directions ($0^0$ for layers 1 and 2, $90^\circ$ for layers 3 and 4) and plane-stress element used in the analysis; (b) half of the symmetric cross section and the four design variables $t_1, t_2, t_3, t_4$; (c) minimal curve in the criterion space. The load resultant $F = 180\,\text{kN}$, $L = 1200\,\text{mm}$ and $h = 200\,\text{mm}$.

values. In Table 6.3 the numerical values for them as well as for the limit shear stress $S$ are given. Subscripts 1 and 2 as well as strength notations $X$ and $Y$ in equation (6.35) refer to the longitudinal and transverse directions of the layer. Also, the allowable values $\underline{t}$ and $\bar{t}$, which are the same for all the design variables, are shown in the table.

The optimization problem consists of minimizing the two chosen criteria

**Table 6.2** Prices and material constants of carbon and glass layers

| Material | $C$ $(\mathrm{kg}^{-1})$ | $\rho$ $(\mathrm{kg\,m}^{-3})$ | $E_1$ (GPa) | $E_2$ (GPa) | $G_{12}$ (GPa) | $\nu_{12}$ |
|---|---|---|---|---|---|---|
| Glass layer | 18 | 1600 | 22 | 8.7 | 5.7 | 0.278 |
| Carbon layer | 250 | 1600 | 125 | 6.2 | 4.8 | 0.270 |

Subscripts 1 and 2 refer to the fiber and the transverse directions of the layer, respectively.

**Table 6.3** Allowable values used in the constraints of the composite beam problem

| Material | $X_t$ | $X_c$ | $Y_t$ | $Y_c$ | $S$ | $\underline{t}$ | $\bar{t}$ |
|---|---|---|---|---|---|---|---|
| Glass layer | 500 | 480 | 37 | 176 | 51 | $5 \times 10^{-3}$ | 50 |
| Carbon layer | 1750 | 1100 | 35 | 140 | 62 | $5 \times 10^{-3}$ | 50 |

Stresses are given in megapascals and thicknesses in millimeters. $X$ refers to the fiber and $Y$ to the transverse direction of the layer. Subscripts t and c are used for tension and compression, respectively.

subject to the Tsai–Hill failure constraint which is applied separately for every layer in each element. Also, the lower and upper limits for the plate thicknesses have been imposed. Thus the following bicriterion problem is considered:

$$\min [W(\mathbf{x})\,C(\mathbf{x})]^{\mathrm{T}} \qquad (6.36)$$

subject to

$$f_i(\mathbf{x}) \leqslant 1 \qquad i = 1, 2, \ldots, 4$$
$$\underline{t} \leqslant t_i \leqslant \bar{t} \qquad i = 1, 2, \ldots, 4$$

The results have been depicted in the criterion space in Fig. 6.9(c). This minimal curve confirms the assumed conflict between the two criteria immediately when two materials with different prices and strengths are used. Naturally, this property can be expected also in more complicated problems where fiber orientations are additional design variables. A trade-off between these two criteria $C$ and $W$ really seems useful in the decision-making phase because the variations of the criteria values are considerable along the minimal curve.

### 6.7.3 Nonconvex frame problem

In the preceding design examples the minimal curve has been smooth in the criterion space and the corresponding region of the attainable set has been convex. Thus the Pareto optimal set of every problems could be generated by using any of the basic techniques. Sometimes the attainable set may, however, possess peculiar properties. Especially, the case where that part of the set $\Lambda$, which is connected to the minimal curve, has concave parts is interesting. Then some of the basic methods, the linear weighting method for example, can fail in generating the Pareto optimal set. These cases have been reported also in structural optimization (Koski, 1985) but, generally speaking, they seem to be relatively rare. Next a simplified plane frame problem, where all the inessential complexity is reduced as much as possible, is treated to illustrate the nonconvexity phenomenon.

A three-member frame under one loading condition and the kinematic analysis model, which includes also the axial deformation of the members, are shown in Fig. 6.10. The three member areas are the design variables, i.e. $\mathbf{x} = [A_1 \, A_2 \, A_3]^\mathsf{T}$,

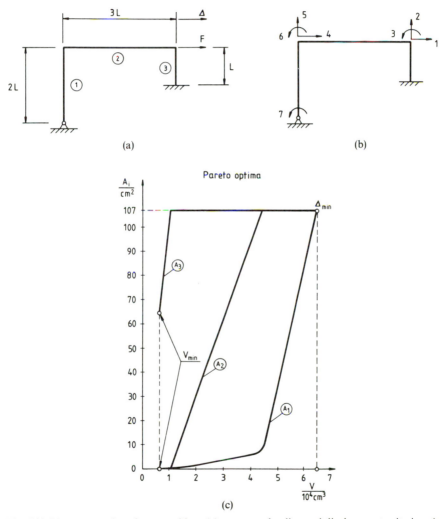

**Fig. 6.10** Nonconvex plane frame problem: (a) structure, loading and displacement criterion $\Delta$; (b) nodal degrees of freedom; (c) Pareto optimal member areas. The numerical design data for the problem are as follows: $F = 80\,\text{kN}$, $L = 100\,\text{cm}$, $E = 200\,\text{GPa}$, $\bar{\sigma} = -\underline{\sigma} = 140\,\text{MPa}$, $\bar{A} = 107\,\text{cm}^2$, $\underline{A} = 0\,\text{cm}^2$.

and the stress as well as the side constraints have been imposed for the problem. In the latter constraints the lower limits of the member areas are zero, which allows the removal of the members. In order to reduce the number of design variables to three the well-known relations

$$I_i = 2.1 A_i^2, \qquad i = 1, 2, 3$$

$$Z_i = 1.1 A_i^{3/2}, \qquad i = 1, 2, 3 \tag{6.37}$$

are used for the section moment of inertia and for the section modulus, respectively. Both the material volume of the frame and the horizontal displacement of the loaded node, denoted by $V$ and $\Delta$, should be minimized simultaneously. As a result, optimization problem

$$\min [V(\mathbf{x}) \Delta(\mathbf{x})]^{\mathrm{T}}$$

$$\underline{\sigma} \leqslant \sigma_i \leqslant \bar{\sigma}, \qquad i = 1, 2, 3$$

$$0 \leqslant A_i \leqslant \bar{A}, \qquad i = 1, 2, 3 \tag{6.38}$$

where the constraint limits are given in the figure legend, is obtained.

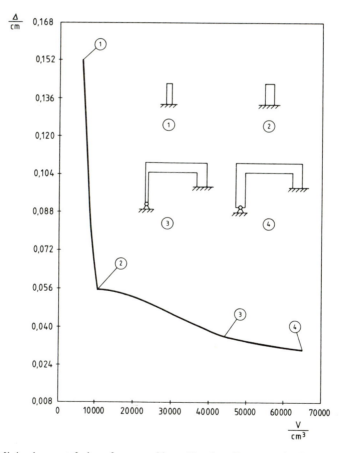

**Fig. 6.11** Minimal curve of plane frame problem. Also four Pareto optimal structures corresponding to minimal points 1...4 are shown. A different scale is used for the lengths and the widths of the members.

The Pareto optimal member areas are shown in Fig. 6.10(c) and the corresponding minimal curve is presented in Fig. 6.11. Also, some Pareto optimal structures associated with the most interesting minimal points have been depicted in the latter figure. Only the end points of the curve segment 2–3 can be achieved by the weighting method whereas the intermediate part is lost. On the other hand, the curve segments 1–2 and 3–4 may be generated by this method. Another interesting feature found here is point 2 on the minimal curve, which corresponds to the change of the topology of the frame. According to the numerical computations this seems to be a vertex where the derivative $\partial\Delta/\partial V$ has a discontinuity. This in turn may be troublesome for the decision maker who has not generated the whole curve and is not expecting a jump in the trade-off number. If for example only one minimal solution located near point 2 has been determined, the trade-off study at this Pareto optimum is misleading.

## 6.8 CONCLUSION

The history of multicriterion structural optimization is relatively short starting from the few articles published in the late 1970s. Now that an early stage of this discipline is still at hand, it seems reasonable to include all the existing publications to represent the recent advances in the field. From this viewpoint it has been natural to concentrate on those fundamental matters which seem especially important in the future and which have been used throughout the published applications. Instead of repeating different cases treated in the articles, the whole decision-making process, starting from the problem formulation and ending at the choice of the best compromise solution, has been discussed in this chapter. The computation of Pareto optima and some characteristic features of a minimal set have been illustrated by several structural design examples.

The vector optimization problem and the Pareto optimality concept offer a clear and sound basis for any further development in the multicriterion optimum design theory. The proper control of the decision-making process becomes possible in this way because then the choice can be restricted to designs which are optimal in an undisputed mathematical sense. Also, the comparison of different methods is facilitated by these concepts. For example the question of which technique should be applied in a certain problem to generate good designs, may be often answered by comparing how many Pareto optima different methods can achieve. This matter was explained in section 6.3 and in Fig. 6.5 the functioning of the most popular methods in different problems was geometrically described. Whatever method the designer may choose, he should check whether it computes Pareto optima or something else. In the latter case the method is useless, provided that the problem formulation corresponds well to reality, and in the former case it should be able to generate as many Pareto optima as possible. If the designer wants to generate good points by just varying the constraint limits in a scalar problem, he has several possibilities. Both equality and inequality constraints can be applied in many different ways. The choice among these alternative methods can be made reliably only by checking which technique gives Pareto optima. The choice of the method by intuition may lead to undesired designs. Sometimes it has been suspected that the solution of a multicriterion problem depends on the chosen decision-making procedure. This is true in the sense that usually experience is being acquired by the decision maker during the design process. If the decision-maker's preferences remain unchanged during the process and the Pareto optimality concept is used, then any available procedure should give the same result provided that the designer makes consistent decisions. The natural requirement is that the chosen method must be able to generate also the best compromise solution. Thus in the case of unchanged preferences the result can be regarded as a function of the person making decisions rather than a function of the design procedure.

The future of the multicriterion approach looks promising also in structural optimization where it has recently reached industrial applications. A multicriterion module, which generates Pareto optimal solutions, can be expected to be an essential part in any new optimization-oriented finite element package. The computation of a covering collection of Pareto optima may be expensive for large-scale problems because several structural and sensitivity analyses must be performed for each solution. This matter together with the graphic or numerical representation of the results, which should be modified into a form that is suitable for the decision making, will be one challenge in the field. Most existing applications deal with physical criteria only but also some economic measures, such as the manufacturing cost for example, might be expected in the vector objective function. New tempting designs become available along with the multicriterion optimization which, on the other hand, requires consistent and competent input information, thus emphasizing the role of the decision maker.

# REFERENCES

Adali, S. (1983a) Pareto optimal design of beams subjected to support motions. *Computers and Structures*, **16**, 297–303.
Adali, S. (1983b) Multiobjective design of an antisymmetric angle-ply laminate by nonlinear programming. *Journal of Mechanisms, Transmissions, and Automation in Design*, **105**, 214–19.
Baier, H. (1983) Structural optimization in industrial environment, in *Optimization Methods in Structural Design* (eds H. Eschenauer and N. Olhoff), Proceedings of the Euromech-Colloquium 164, University of Siegen, Bibliographisches Institut A6, Zürich, pp. 140–5.
Bendsøe, M.P., Olhoff, N. and Taylor, J.E. (1983–1984) A variational formulation for multicriteria structural optimization. *Journal of Structural Mechanics*, **11**, 523–44.
Bühlmeier, J. (1990) Design of laminated composites under time dependent loads and material behaviour, in *Engineering Optimization in Design Processes* (eds H.A. Eschenauer, C. Mattheck and N. Olhoff), Proceedings of the International Conference, Karlsruhe Nuclear Research Centre, Lecture Notes in Engineering 63, Springer, Berlin, pp. 107–25.
Carmichael, D.G. (1980) Computation of Pareto optima in structural design. *International Journal for Numerical Methods in Engineering*, **15**, 925–9.
Cohon, J.L. (1978) *Multiobjective Programming and Planning*, Academic Press, New York.
Dhingra, A.K., Rao, S.S. and Miura, H. (1990) Multiobjective decision making in a fuzzy environment with applications to helicopter design. *AIAA Journal*, **28**, 703–10.
Diaz, A. (1987a) Sensitivity information in multiobjective optimization. *Engineering Optimization*, **12**, 281–97.
Diaz, A. (1987b) Interactive solution to multiobjective optimization problems. *International Journal for Numerical Methods in Engineering*, **24**, 1865–77.
Eschenauer, H. (1983) Vector optimization in structural design and its application on antenna structures, in *Optimization Methods in Structural Design* (eds H. Eschenauer and N. Olhoff), Proceedings of the Euromech-Colloquium 164, University of Siegen, Bibliographisches Institut A6, Zürich, pp. 146–55.

Eschenauer, H., Kneppe, G. and Stenvers, K.H. (1986) Deterministic and stochastic multiobjective optimization of beam and shell structures. *Journal of Mechanisms, Transmissions, and Automation in Design*, **108**, 31–7.

Eschenauer, H.A., Schäfer, E. and Bernau, H. (1988) Application of interactive vector optimization methods with regard to problems in structural mechanics, in *Discretization Methods and Structural Optimization – Procedures and Applications* (eds H.A. Eschenauer and G. Thierauf), Proceedings of a GAMM-Seminar, Siegen, Lecture Notes in Engineering 42, Springer, Berlin, pp. 110–17.

Eschenauer, H.A. and Vietor, T. (1990) Some aspects on structural optimization of ceramic structures, in *Engineering Optimization in Design Processes* (eds H.A. Eschenauer, C. Mattheck and N. Olhoff), Proceedings of the International Conference, Karlsruhe Nuclear Research Centre, Lecture Notes in Engineering 63, Springer, Berlin, pp. 145–54.

Eschenauer, H., Koski, J. and Osyczka, A. (eds) (1990) *Multicriteria Design Optimization – Procedures and Applications*, Springer, Berlin.

Fu, G. and Frangopol, D.M. (1990) Reliability-based vector optimization of structural systems. *ASCE Journal of Structural Engineering*, **116**, 2143–61.

Fuchs, W.J., Karandikar, H.M., Mistree, F. and Eschenauer, H.A. (1988) Compromise: an effective approach for designing composite conical shell structures, in *Advances in Design Automation* (ed. S.S. Rao), ASME Design Technology Conference, **14**, 279–86.

Gerasimov, E.N. and Repko, V.N. (1978) Multicriterial optimization. *Soviet Applied Mechanics*, **14**, 1179–84.

Hajela, P. and Shih, C.J. (1990) Multiobjective optimum design in mixed integer and discrete design variable problems. *AIAA Journal*, **28**, 670–5.

Jendo, S., Marks, W. and Thierauf, G. (1985) Multicriteria optimization in optimum structural design. *Large Scale Systems*, **9**, 141–50.

Koski, J. (1979) *Truss Optimization with Vector Criterion*, Tampere University of Technology, Publication No. 6.

Koski, J. (1981) Multicriterion optimization in structural design, in Proceedings of the 11th ONR Naval Structural Mechanics Symposium on Optimum Structural Design, University of Arizona, Tucson, AZ.

Koski, J. (1984a) Multicriterion optimization in structural design, in *New Directions in Optimum Structural Design* (eds E. Atrek, R.H. Gallagher, K.M. Ragsdell and O.C. Zienkiewicz), Wiley, New York, pp. 483–503.

Koski, J. (1984b) Bicriterion optimum design method for elastic trusses. Acta Polytechnica Scandinavica, Mechanical Engineering Series No. 86, Dissertation, Helsinki.

Koski, J. (1985) Defectiveness of weighting method in multicriterion optimization of structures. *Communications of Applied Numerical Methods*, **1**, 333–6.

Koski, J. (1993) Multicriterion structural optimization – state of the art, in Proceedings of the NATO Advanced Study Institute: Optimization of Large Structural Systems, September 23–October 4, 1991, Berchtesgaden, Kluwer, Dordrecht.

Koski, J. and Silvennoinen, R. (1982) Pareto optima of isostatic trusses. *Computer Methods in Applied Mechanics and Engineering*, **31**, 265–79.

Koski, J. and Silvennoinen, R. (1987) Norm methods and partial weighting in multicriterion optimization of structures. *International Journal for Numerical Methods in Engineering*, **24**, 1101–21.

Koski, J. and Silvennoinen, R. (1990) Multicriteria design of ceramic piston crown. *Engineering Costs and Production Economics*, **20**, 175–89.

Leitmann, G. (1977) Some problems of scalar and vector-valued optimization in linear viscoelasticity. *Journal of Optimization Theory and Applications*, **23**, 93–9.

Lógó, J. and Vásárhelyi, A. (1988) Pareto optima of reinforced concrete frames, *Periodica Polytechnica, Civil Engineering*, **32**(1–2).

Nafday, A.M., Corotis, R.B. and Cohon, J.L. (1988a) Multiparametric limit analysis of frames: part I – model. *ASCE Journal of Engineering Mechanics*, **114**, 377–86.

Nafday, A.M., Corotis, R.B. and Cohon, J.L. (1988b) Multiparametric limit analysis of frames: part II – computations. *ASCE Journal of Engineering Mechanics* **114**, 387–403.

Olhoff, N. (1989) Multicriterion structural optimization via bound formulation and mathematical programming. *Structural Optimization*, **1**, 11–17.

Osyczka, A. (1981) An approach to multi-criterion optimization for structural design, in Proceedings of the 11th ONR Naval Structural Mechanics Symposium on Optimum Structural Design, Tucson, AZ.

Osyczka, A. (1984) *Multicriterion Optimization in Engineering*, Ellis Horwood, Chichester.

Pareto, V. (1896–1897) *Cours d'Economic Politique*, Volumes 1 and 2, Rouge, Lausanne.

Post, P.U. (1990) Optimization of the long-term behaviour of composite structures under hygro-thermal loads, in *Engineering Optimization in Design Processes* (eds H.A. Eschenauer, C. Mattheck and N. Olhoff), Proceedings of the International Conference, Karlsruhe Nuclear Research Centre, Lecture Notes in Engineering 63, Springer, Berlin, pp. 99–106.

Rao, S.S. (1984) Multiobjective optimization in structural design with uncertain parameters and stochastic processes. *AIAA Journal*, **22**, 1670–8.

Rao, S.S. (1987) Game theory approach for multiobjective structural optimization. *Computers and Structures*, **25**, 119–27.

Rao, S.S. and Hati, S.K. (1986) Pareto optimal solutions in two criteria beam design problems. *Engineering Optimization*, **10**, 41–50.

Rao, S.S., Venkayya, V.B. and Khot, N.S. (1988) Optimization of actively controlled structures using goal programming techniques. *International Journal for Numerical Methods in Engineering*, **26**, 183–97.

Rozvany, G.I.N. (1989) *Structural Design via Optimality Criteria*, Kluwer, Dordrecht.

Stadler, W. (1977) Natural structural shapes of shallow arches. *Journal of Applied Mechanics*, **44**, 291–8.

Stadler, W. (1978) Natural structural shapes (the static case). *Quarterly Journal of Mechanics and Applied Mathematics*, **31**, 169–217.

Stadler, W. (1987) Initiators of multicriteria optimization – recent advances and historical development of vector optimization, in Proceedings of an International Conference on Vector Optimization (eds J. Jahn and W. Krabs), Darmstadt, August 1986, Lecture Notes in Economics and Mathematical Systems 294, Springer, Berlin.

Stadler, W. (ed.) (1988) *Multicriteria Optimization in Engineering and in the Sciences*, Mathematical Concepts and Methods in Science and Engineering 37, Plenum, New York.

Stadler, W. and Krishnan, V. (1989) Natural structural shapes for shells for revolution in the membrane theory of shells. *Structural Optimization*, **1**, 19–27.

Tseng, C.H. and Lu, T.W. (1990) Minimax multiobjective optimization in structural design, *International Journal for Numerical methods in Engineering*, **30**, 1213–28.

Yoshimura, M., Hamada, T., Yura, K. and Hitomi, K. (1984) Multiobjective design optimization of machine-tool spindles. *Journal of Mechanisms, Transmissions, and Automation in Design*, **106**, 46–53.

# 7

# Multicriteria design optimization by goal programming

ERIC SANDGREN

## 7.1 INTRODUCTION

A good design is certainly optimal in some sense, but it may not always be possible to express the attributes of such a design within the context of a single objective, nonlinear programming problem. The traditional formulation of a nonlinear programming problem requires the specification of a scalar objective function with all other factors being included as constraints. Nonlinear programming methods also force the computer to operate on a mathematical abstraction of the 'real' design problem with no knowledge of the trade-offs which are being made during the optimization process. The human mind can operate easily in an abstract mode and as such can consider a wide range of design alternatives. In contrast, the use of conventional computer-aided design and analysis tools forces the user to deal with design issues at a very specific level. If one considers design as a process which proceeds from the general to the specific, then the computer cannot have the desired level of impact early in the process where the benefits are potentially the greatest. The result is generally a solution which is optimal with respect to the mathematical formulation applied but far from optimal in a practical design sense. This deficiency has been a major factor in limiting the number of practical applications of design optimization which is unfortunate as the potential of the concept is significant.

The difficulty associated with a conventional, single objective formulation may be observed by considering a hypothetical example. If a structure is optimized for minimum weight subject to constraints imposed on stress, maximum deflection, buckling and natural frequency, the result will be a design in which several of the constraints are active. This means that the final design will border on several failure modes at the solution, subject to the safety factors imposed. In addition, the solution will very likely be extremely sensitive to changes in loading or restraints. When imposed design loads and restraint specifications are only estimates of actual conditions, one must question the validity of a design produced by this process. The value added by computer-aided design lies in the ability to investigate a number of alternate designs which would not be possible by hand in the time available. This means that potential trade-offs which might be extremely valuable to the final design must not be removed from consideration by either the formulation of the problem or the solution process. For example a design which has a maximum stress $1\%$ higher than that allowed by the formulation may produce an additional $10\%$ weight reduction with virtually the same safety factors but this point would never be discovered. Most real design optimization problems are multimodal. There are generally a number of different designs that can adequately perform the intended task which are scattered throughout the design space. Any solution process which simply refines the starting design estimate in a local region will not consider other, perhaps far more attractive design alternatives.

A more natural approach to the formulation of design optimization problems is required which allows potential trade-offs to be evaluated in the widest possible context. Multiobjective optimization provides additional freedom compared with a conventional nonlinear programming approach and therefore is a logical candidate for use as a realistic design tool. A valuable computational design tool should allow the user to function in a manner which closely parallels the actual design process. When a nonlinear programming algorithm is coupled with a finite element code, a specific design is input and a design which is similar in nature to this starting design is output. A more natural process would be to define the desired aspects of performance and then to select a design with both specific and general characteristics which best meet these performance levels. In addition to a multiobjective capability, the design environment should address topological issues as well as deal effectively with the difference between hard and soft constraint specifications.

Topological issues are those design factors which extend beyond a single classification. A topological choice is more than a simple geometric change in a design variable. They generally represent discrete choices rather than continuous ones. In a structural design involving trusses, for instance, the alteration of cross-sectional areas or even the movement of nodal locations does not constitute a topological change. Topological issues involve fundamental aspects such as the number of truss elements used in the structure, the use of beam instead of truss elements or a change in material choice. These changes can have a dramatic impact on design performance and often move the optimization search to different areas of the design space. For example when a material change from steel to aluminum is made, the dimensions of the design structure will increase when subject to the same load conditions. The end result is that topological choices tend to introduce local minima throughout the design space. Of course, many design problems have local minima, but topological choice generally magnifies the problem. This makes the optimization problem more difficult to solve in the global sense, but the resulting design improvement is most often worth the effort.

The difference between hard and soft constraints is subtle but very important. A hard constraint specification involves a condition that absolutely must be satisfied. On the other hand, a soft constraint specification represents a condition which is desirable but may be violated by a small amount without drastically altering the value of the design. Traditional linear and nonlinear programming methods treat all constraints as hard constraints. This is an inherent limitation in the formulation which is required for the mathematically derived solution procedures but one which often delegates the optimum located to be different from what is desired. The issue is easily seen by considering a simple example.

A failure criterion such as a yield stress in one of the members would be correctly considered as a hard constraint. The problem arises when a safety factor is imposed on the yield stress. Say that a safety factor of 2 is prescribed so that the stress constraint allows the maximum stress found in any member to be only one-half of the yield stress of the material. The question to be asked is whether or not the constraint is still a hard constraint. The answer is that it is not. When a safety factor is included, the stress constraint becomes a soft constraint. Whether the safety factor is 2 or only 1.95 is actually of little consequence, particularly if a small sacrifice in the safety factor is very beneficial to the final design. A traditional, single objective formulation with hard constraints has difficulty making this trade-off. In reality, the majority of constraints used in design optimization problem formulations are soft rather than hard constraints. Treating them as hard constraints will generally lead to suboptimal results.

Many different approaches to multiobjective optimization are possible. The intent herein is to present the topic in the widest possible context and then to deal with specific ways of implementing the concept. Before the multiobjective problem is dealt with, however, a discussion of the various classifications of optimization problems is necessary and how these problem classes are important in the multiobjective formulation and solution process. The inclusion of noise in the formulation is also addressed in order to deal with uncertainty in the model. Several multiobjective problem solution strategies are presented, but concentration is focused on a multilevel, nonlinear goal programming approach. The first level deals with the topological aspects of the design and locates promising solution regions. The next level of the process deals with the trade-offs which are present in each of these design regions in order to achieve the best final design. The procedure is demonstrated on several simple examples. Two of the examples considered involve the design of structures with truss elements, and one deals with the design of a planar mechanism.

## 7.2 PROBLEM CLASSIFICATION

Design is basically a decision process and as such it can be represented in a tree structure. A possible generic design structure is shown in Fig. 7.1. The tree represents the design of a system which may be composed of various subsystems. The branching process is used to decompose the design so that each subsystem is reduced to its lowest functional level. The system and subsystem level nodes allow information to be exchanged which in turn allows system as well as component objectives to be optimized. Design specifications are imposed as required at any level of the tree. Any node which joins subsystem level nodes is an assembly node which contains information on the various ways the subsystems may be joined.

The level directly below the subsystems in the tree structure represents candidate designs which are the various topological options selected for meeting the subsystem design specifications. Each candidate design has a set of design parameters associated with it which may either be fixed or allowed to become design variables in the subsequent optimization. Candidate designs may have subclasses associated with them and these subclasses may themselves span several levels in the tree. Each subclass has material and manufacturing process nodes

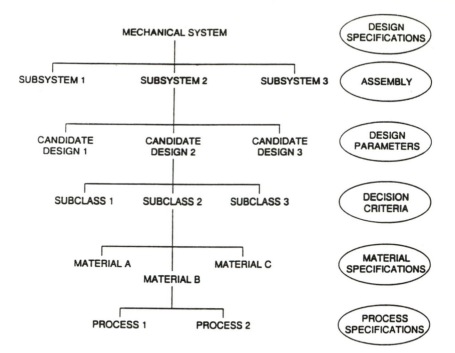

**Fig. 7.1** Generic design tree structure.

associated with it. These nodes contain the specifications, including tolerances, for each material and manufacturing process.

Finally, a set of decision criteria is imposed. The decision criteria are those factors which will be used to evaluate the candidate designs. These criteria, or goals as they become in the solution process, may also be associated with any level of the tree. Together with the design parameters for each candidate design, the decision criteria form the basis of the optimization problem formulation which in turn allows the bottom level nodes to be compared on a common basis. The relationship between the parent and child nodes may be specified as any functional expression involving design variables and parameters. The design and parameter values may be selected at any level of the design tree and lower-level nodes may or may not inherit these values depending on the relationship imposed.

When viewed in this manner it becomes possible to develop classifications for the variety of problems commonly encountered. The design, material, assembly and process specifications in conjunction with the decision criteria form a set of design objectives or goals and possibly a constraint set as well. The design parameters form the set of design variables which define how a design can be modified in order better to achieve the goals. The key here is that these variables must represent discrete or topological choices as well as continuous choices. In order to reduce the scope of the discussion, the emphasis will be concentrated on the aspects of shape. The shape optimal design problem may be defined as the determination of the form which best satisfies the design objectives subject to the imposed design constraints. Now the concept of problem classification may be reduced to that of defining the different levels of shape or form. These levels are best described within an application area, thus the initial discussion will be directed at structural design. This will be followed by a more general view of problem classification.

When the term shape optimization is used, the general inclination is immediately to think in terms of structural optimization. Certainly it is true that a large body of experience has been developed in structural design over the last 30–40 years and the definition of shape plays a dominant role in such design applications. The term structural optimization is itself a very general term and is often used in various contexts to refer to different classes of optimization problems. These classes are best defined by the types of design variables encompassed by the formulation. The most elementary level, and that most widely considered, is the modification of the geometric parameters of the elements themselves. This requires the prespecification of the number and type of each element used as well as their placement in space. This is quite a limited framework within which to operate but it is often dictated by the computational effort required to produce a solution. In order to provide maximum flexibility, however, all aspects of structural design must be considered. These aspects include topological as well as geometric factors. In order to understand the problem better in its totality, a brief description of each of the aspects of shape optimization is in order.

Geometric optimization refers to either the direct modification of specific dimensions which define a design or an indirect modification of shape by altering the boundary form. In structural design, the lowest level of geometric optimization is cross-sectional optimization. Cross-sectional optimization refers to the specific geometric dimensions for a preselected design class such as the thickness of a plate or the diameter of a circular truss element. This class of problems has been under investigation since the pioneering work of Schmidt in the early 1960s (Schmidt, 1960). Applications include the design of automotive structural components (Bennett and Botkin, 1981) and truss structures such as electrical transmission towers (Sheppard and Palmer, 1972). For problems involving truss structures, the cross-sectional areas of the elements form the entire set of design variables. For structures involving beam elements, specific cross-sectional dimensions which are sufficient to describe the principal moments of inertia are required in addition. With plate or shell structures, a thickness dimension is needed, either at the element level or for groups of elements (Abuyounes and Adeli, 1986; Adeli and Chompooming, 1989). Generally, a preselected member class is chosen such as an I-beam or channel section in order to relate design variables representing specific dimensions to cross-sectional areas and principal moments of inertia. Variable linking is often employed in order to limit the number of design variables as solution time is a strong function of the number of variables. Design sensitivities which may be used to generate gradient information are now commonly available in many commercial finite element codes.

Other forms of geometric optimization introduce design variables which allow for movement of the boundary in some fashion. The boundary may be altered either by the movement of actual node locations or by moving control points on a spline curve or surface representing the boundary. Because of the increase in complexity over cross-sectional optimization, the geometric change is generally limited to a small region of the design such as around a fillet or a hole. The shape change is also closely linked to the analysis technique employed. Both the finite element method (Pramila and Virtanen, 1986) and the boundary element method (Sandgren and Wu, 1989) have been successfully applied to the geometric shape optimization problem. A geometric change will often require continual remeshing for the analysis routine and this greatly complicates the task of generating valid gradient based search directions. While geometric optimization has been applied with some success, much work remains in the areas of geometric modeling, analysis and design optimization methods before geometric shape optimization can become an integral part of the design process.

The third class of shape optimization problems involves topological issues as well as geometric and cross-sectional ones. Relatively little work has been directed toward topological optimization despite the importance of the concept. Optimizing element cross sections or nodal geometry does little more than produce a local optimum based on the preselected topology. That is to say the optimal ten-bar truss for a defined set of loading conditions may be a good design compared with other ten-bar truss designs but it may be a relatively poor design when compared with other topologies such as a five- or six-bar truss. This issue becomes particularly significant when dealing with new materials or manufacturing processes where no design history is available to help select an appropriate topology. When consideration is being given to a material substitution, it is quite likely that the basic topology of the part must be significantly altered in order to reap potential weight and cost benefits. Traditional nonlinear programming approaches have met with limited success in dealing with topological issues. This fact seems to indicate that a nontraditional optimization approach may be required in order to deal with the structural design problem in the broadest sense (Adeli and Mak, 1988). A method which is capable of taking advantage of new computing hardware architecture such as distributed or parallel processors is also preferred over one relegated to conventional sequential processing (Adeli and Kamal, 1992). This can have an enormous impact on solution time and directly affect the size and complexity of the problem being addressed.

The geometric and topological design of other mechanical systems is possible as well. For example in the design of a mechanism, geometric aspects include items such as link lengths and pivot locations while topological aspects involve the selection of the class of mechanism (i.e. four- or six-link mechanism). For the design of a cam to drive a dynamic system, geometric aspects include the size of the cam (base circle radius), the radius of the follower and even the profile of the cam itself while topological choices include the type of follower (flat faced, roller, etc.). The same identification process may be applied to virtually any shape related problem. Similarly a set of design objectives and constraints may be defined for each application or application class. What remains to be developed is a robust solution algorithm which can deal with the wide variety of practical shape optimization problems.

## 7.3 PROBLEM FORMULATION

The conventional nonlinear programming problem formulation may be stated as follows:

$$\text{minimize } f(\mathbf{x}); \mathbf{x} = [x_1 \, x_2 \, x_3, \dots, x_N]^{\mathrm{T}} \qquad (7.1)$$

subject to

$$g_j(\mathbf{x}) \geqslant 0;\; j = 1, 2, \ldots, J \qquad (7.2)$$

$$h_k(\mathbf{x}) = 0;\; k = 1, 2, \ldots, K \qquad (7.3)$$

and

$$x_i^{(l)} \leqslant x_i \leqslant x_i^{(u)};\; i = 1, 2, \ldots, N \qquad (7.4)$$

Here, $f(\mathbf{x})$ represents a scalar objective function which defines the merit of a particular selection of design variables. The design variables contained in the vector $\mathbf{x}$ are those parameters which may be adjusted in order to improve the design. The functions $g(\mathbf{x})$ and $h(\mathbf{x})$ form a set of hard constraints which delimit the feasible design space. Upper and lower bounds are also imposed if required.

The problem formulation represented by equations (7.1)–(7.4) is depicted graphically in Fig. 7.2. In this situation, design variables are inputs to the system model which results in a value for the objective function and each of the constraints. This is a very simplistic view of the design optimization process. A more realistic view of the process is presented in Fig. 7.3. Note that multiple design objectives are now allowed. Also, in addition to the design variables, there are both inner and outer noises present. Outer noises represent uncontrollable variations in design parameters while inner noises are unavoidable but in part controllable variations. In a structural optimization context, inner noises might include factors such as tolerances or material properties while outer noises might include the loads applied to the structure. The difference is that some control over tolerances and material properties is possible but the loads applied in the

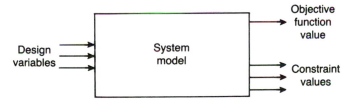

**Fig. 7.2** Standard nonlinear programming model.

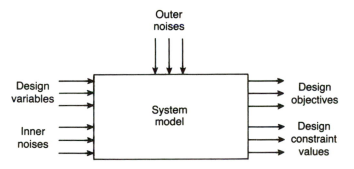

**Fig. 7.3** Desired nonlinear programming model.

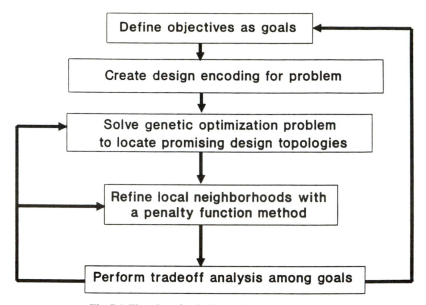

**Fig. 7.4** Flowchart for design optimization procedure.

use of the structure are not generally controllable. Both inner and outer noises can play an important role in the success or failure of a design. The generalization which can be made is that a design which is insensitive to inner and outer noises is generally a better or at least a more robust design.

The multiobjective formulation, in conjunction with the presence of inner and outer noises, more closely parallels the actual design process in that it deals with the trade-offs associated with each design evaluated. By formulating design issues as objectives rather than constraints, soft constraints may be included in a natural fashion. Since it is difficult to encode the exact character of the 'real' problem in the computer, it is appropriate that the user should remain an integral part of the optimal design process. This interaction is difficult to implement in many of the multiobjective formulations derived to date, but it is essential that the user be able to overcome the precise mathematical nature of the formulation to ensure the process converges to a practical design. The solution approach constructed for solution of the problem formulation, shown in Fig. 7.4, is a nonlinear goal programming algorithm which operates initially at a topological level through the use of a genetic optimization procedure and then switches to a penalty function approach in order to refine the best designs located in the topological search.

## 7.4 MULTIOBJECTIVE OPTIMIZATION

Multiobjective optimization is not new as it has been in existence since the early 1950s (Kuhn and Tucker, 1951) and has been under continual development ever since. Various approaches are available with which one can address a multiobjective formulation. These approaches may be divided into two fundamental categories where the primary differentiation is made as to whether the problem is treated as the optimization of a vector of design objectives or a single scalarized equation which contains contributions from all of the objectives. Reviews of multiobjective techniques are available in a number of sources (Stadler, 1979; Evans, 1984; Hwang and Masud, 1979). Most of the techniques which deal with the problem in vector form search for a Pareto optimal point for which no decrease in any objective is possible without a subsequent increase in one or more of the other objectives. In the scalarized form, a commonly implemented approach is to form a weighted sum of the individual objectives. Several of the widely used techniques for multiobjective optimization will be reviewed but the primary concentration will be placed on the method of goal programming which has its original roots in the area of linear programming. This is not to say that goal programming is superior to the outer approaches. It is, however, a straightforward and easily understood approach and makes the examples presented easier to follow.

The first class of methods to be reviewed consists of the vector optimization techniques. The nonlinear programming formulation for this problem class becomes

$$\text{minimize } f(\mathbf{x}) = \{f_1(\mathbf{x}), f_2(\mathbf{x}), \ldots, f_m(\mathbf{x})\}^{\mathrm{T}} \tag{7.5}$$

subject to the same constraint sets of the traditional nonlinear programming problem described by equations (7.2)–(7.4). Possible approaches for solving the problem include the min–max method and the $e$ constraint method. The min–max method simply minimizes the maximum deviation of the design from all design objectives subject to the additional constraint that the solution is a Pareto optimal point. The $e$ constraint method seeks to minimize one of the selected objectives with all other objectives constrained to stay within a distance $e$ from their desired goal values.

The techniques which form the problem as a scalar objective include game theory, compromise programming and goal programming. Game theory treats each design objective as a player seeking to optimize his own payoff. The players may be noncooperative where each cares only about his own standing or cooperative where players are willing to compromise their objectives if it improves the state of other players. Compromise programming is based on locating a set of solutions which lie as close as possible to an ideal point, where the ideal point is defined as having each objective at a value equal to that which it would have if it were minimized as a single objective function. Finally, goal programming seeks to minimize the under- or overachievement of a set of objectives or goals which are minimized as a weighted sum in the objective function.

## 7.5 GOAL PROGRAMMING

In a goal programming formulation, the design objectives or goals are defined and a priority is assigned to each one. The algorithm then attempts to satisfy as many of the goals as possible, starting with the highest priority goal. The inclusion of priority takes the algorithm a step beyond a simple weighted sum of objectives. This approach removes much of the difficulty of problem formulation and more closely follows the normal design pocess. The technique is also capable of operating when there is no feasible design space. It allows for an evaluation of trade-offs in order to determine a more realistic set of goals. The application of goal programming to multiobjective structural design has been demonstrated by several investigators (Shupe and Mistree, 1987; El Sayed, Ridgley and Sandgren, 1989).

The formulation of a goal programming problem is different from that of a nonlinear programming problem. In a goal programming formulation, there is no exact counterpart to the objective function in a nonlinear programming formulation. Design constraints may be included as goal constraints or as hard constraints. The goal constraints are best thought of as multiple objectives, while hard constraints are directly equivalent to the constraints in a nonlinear programming formulation. The solution process revolves around the minimization of the positive and/or negative deviations from the specified goals.

Goal constraints may take the form of inequality constraints (either $\geqslant 0$ or $\leqslant 0$) or equality constraints. The common forms of goal constraints are listed below.

$$G(x_j) + d_i^- = b_i \tag{7.6}$$

$$G(x_j) - d_i^+ = b_i \tag{7.7}$$

$$G(x_j) + d_i^- - d_i^+ = b_i \tag{7.8}$$

The $x_j$ are the design variables and $G(x_j)$ are nonlinear functions relating the design variables to the desired goal values, $b_i$. The $d_i^-$ and $d_i^+$ are the under- and overachievement deviational variables which may be thought of as a form of slack variables. With this analogy, the first two goal constraint forms are seen to represent less than or equal constraints respectively. If all deviational variables are only allowed to assume positive values, the specific form of each goal constraint is easy to verify. In the first constraint, the underachievement deviational variable represents the amount the current design falls below the right-hand side value $b_i$. Since only values below or equal to $b_i$ are possible, the constraint is equivalent to the $\leqslant$ form. The same logic may be applied to the second constraint form. Since the positive deviation is subtracted in the equation, it allows the constraint to attain only values greater than or equal to the right-hand side, $b_i$, and is therefore equivalent to a $\geqslant$ form.

The final goal constraint form represents an equality constraint. This form may be likened to a fuzzy equality constraint. It allows for both positive and negative deviations from the right hand side value as it contains both a $d_i^-$ and a $d_i^+$ variable. How much of a deviation is allowed is governed by the objective function form and the scaling present among the goal constraints. It should be noted that only one of the deviational variables will be allowed to be simultaneously nonzero in any goal constraint equation. Using these three forms of goal constraints, virtually any problem can be formulated. Traditional hard inequality and equality constraints, may be added as well to help define the feasible region when appropriate.

In a goal programming formulation, only deviational variables are allowed to appear in the objective function. The actual design variables $(x_j)$ are indirectly included by the fact that the values of the design variables determine the over- or underachievement of each goal constraint. The objective function, like the goal constraints, may take on a number of forms depending on how priority levels and weighting factors within priority levels are assigned.

The most basic form of the objective function is given as

$$\text{minimize } f(\mathbf{x}) = \sum_{i=1}^{I} (d_i^- + d_i^+) \tag{7.9}$$

Here, the subscript $i$ represents a particular goal constraint and $I$ is the total number of goal constraints. This objective function would be used when the deviational variables are not distinguished by priority or weighting. In effect, this formulation has the unitary goal of minimizing the total deviation from the specified goal levels. It should be noted that both the $d_i^-$ and $d_i^+$ deviational variables need not be included in the objective function for each constraint. If only a positive or only a negative deviation results in a suboptimal design, then only the corresponding deviational variable is included for that goal.

The next level of complexity would be to include priorities among the goal constraints. This results in an objective function of the form

$$\text{minimize } f(\mathbf{x}) = \sum_{i=1}^{I} P_k \{ d_i^- + d_i^+ \} \text{ for } k = 1, 2, \ldots, K \tag{7.10}$$

This form allows $K$ different priority levels within the $I$ goal constraints. Here, $P_k$ represents the rank ordering of the goal priorities. Allowances are now possible for a ranking or ordering of goals (i.e. some goals are more important than others). The deviational variables within each priority level have the same weighting and any number of goals may be grouped at any priority level.

The most general form of objective function allows weighting factors to be assigned within individual priority levels. This form is represented by

$$\text{minimize } f(\mathbf{x}) = \sum_{i=1}^{I} W_{kj} P_k \{ d_i^- + d_i^+ \} \quad \text{for } k = 1, 2, \ldots, K; j = 1, 2, \ldots, J \tag{7.11}$$

Here for priority level $k$, a total of $j$ individual weights ($W_{kj}$) may be assigned within that priority. This form provides the most flexibility by allowing goals at the same priority to have different weights such as positive deviational variables vs negative deviational variables. The solution algorithm still operates on a mathematical abstraction of the real problem, but the flexibility allowed in the goal formulation allows this abstraction to represent better the true design situation. The formulation of a goal programming problem is considered in each of the three example problems presented following the development of the genetic optimization algorithm.

## 7.6 GENETIC OPTIMIZATION

The solution procedure employed to solve a multiobjective or goal programming problem formulation is generally built around a conventional linear or nonlinear programming algorithm. These methods have undergone continual development over the past 30 years but the algorithms available to date are not particularly robust and no one algorithm performs uniformly well on broad classes of problems. In addition, as a general design tool, nonlinear programming methods do not possess the characteristics of intelligent methods in that they fail to make use of or to learn from information generated at previous stages in the search (Adeli and Balasubramanyam, 1988). This is very different from the iterative, manual design process performed by an engineer who is constantly gathering and updating information throughout the process. Virtually all successful algorithms make use of gradient information to guide the search and as such they tend to be trapped by the first local minimum encountered.

Procedures designed to locate global optima such as simulated annealing have been developed (Vecchi and Kirkpatrick, 1983), but an efficiency trade-off is required and the approach is quite problem specific. The methods which hold some promise as an intelligent design optimization tool are termed genetic optimization methods. These methods emulate the natural selection process of nature and operate on a principle of survival of the fittest. Genetic algorithms have been applied to a wide variety of problems over the past ten years (Goldberg, 1987; Davis and Ritter, 1987; Fourman, 1985). Their utility in solving mechanical and structural problems has recently been demonstrated by a number of investigators (Sandgren and Venkataraman, 1989; Sandgren and Jensen, 1992). The goal of a genetic algorithm is to discover the fundamental building blocks of a good design and to combine these building blocks in such a way as to create the best possible design. To accomplish this task, a design must be encoded in such a way so that the representation of the design is closely linked to its physical structure or function. This turns out to be a relatively straightforward task for structural design applications. The actual topology of the design is encoded and operated on in order to produce new designs based on features of existing designs. Not only is the process stable, but it requires no gradient information and produces multiple optima rather than a single, local optimum. It relies on the randomness present in natural selection, but quickly exploits information gathered in order to produce an efficient design procedure. Additional speed may be achieved as a result of the parallel nature of the genetic process.

A design to a genetic algorithm is an abstract representation termed a chromosome which is directly analogous to a chromosome in a living organism. The chromosome is composed of a number of genes, each of which may assume one of a number of possible values or states. While in an organism, a gene might represent sex or hair color, the gene in the design encoding represents a feature of the design itself. The genetic algorithm operates by manipulating the coding of the set of gene values. This is a fundamental distinction from traditional methods as the coding of the parameter set is altered rather than the parameter values themselves. Other differences between genetic and nonlinear programming methods are that the genetic algorithm operates on a set or population of designs rather than a single design point and the rules which govern the transition from one set of designs to the next are probabilistic rather than deterministic. These differences may not seem that great but they produce an algorithm which exhibits certain aspects of intelligence which might be demonstrated by a human designer. The randomness in the transition from generation to generation also allows for the consideration of factors such as tolerances and uncertainty.

The overall suitability of a chromosome, that is the performance of a specific design, is termed its fitness. This fitness property, which may be related to any function or functions of the design, determines the probability of the chromosome being a parent for the next generation of designs. The chromosomes possessing the greatest fitness have the highest chance of being selected as parents. Parents are combined using various genetic operators to produce new designs for the next generation. The process is continually repeated with the expectation that the fitness of the individual designs as well as the average fitness of the population will gradually improve until the optimum design or designs are produced. In order to initiate the algorithm, a coded representation of the design must be generated from which the fitness may be evaluated. Additionally, an initial population of designs must be generated and how parents are selected and combined in order to form offspring must be defined. Special operators such as mutation are also introduced in order to guard against the loss of important design information which is particularly important when using a relatively small population. The collection of all of these processes form an algorithm and one such algorithm is described in the following section.

A flowchart depicting the major operations taking place within a genetic algorithm is pictured in Fig. 7.5. A multitude of possible implementations are possible and often a small modification can significantly enhance or degrade the performance on a particular problem class. The algorithm described here will be a very basic one but one that is capable of solving a large number of very real problems. The first issue to be considered is the encoding of the design. This

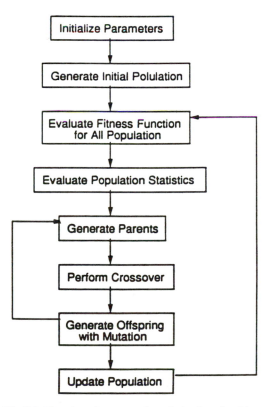

**Fig. 7.5** Flowchart for a genetic optimization algorithm.

section will be followed by a discussion of the fitness function, the generation of the initial population and the principle genetic operators.

### 7.6.1 Design encoding

The representation of a design or the structure of the chromosome which embodies the design in a genetic algorithm may assume a number of different forms. A chromosome is an ordered matrix of values which in some fashion represents the fundamental nature of the design. Traditionally, the chromosome is a one-dimensional vector consisting solely of the binary digits zero or one. This convention allows for a statistical evaluation of the probabilities of various length strings of values within a chromosome to be propagated into future generations of designs. The structure considered here will contain additional flexibility in both the dimensionality and in the number of values which may be associated with a chromosomal position. These extensions allow for a more natural design representation for most structural design problems with a reasonably small chromosome. On the other hand, the ability to perform a statistical analysis on the structures passed on to future generations is more limited. This trade-off is justified on the basis of the results achieved in experimentation on a wide variety of problem formulations.

A one-dimensional example of a chromosome structure would be for a parameter which can assume any integer value between 0 and 31 (a number of choices equal to 2 raised to a power is ideal). This parameter could be represented by a binary string of five digits. Each positional value has only two states and as such, the encoding exactly parallels the manner in which a number is represented by a computer. For instance, a parameter value of 21 would be represented by the string 10101. If another parameter is included, its binary representation could be appended to that of the first parameter. A more effective approach, however, is to couple the encoding of the design as closely as possible with the structure of the design. This may be accomplished by linking the dimensionality of the chromosome to the dimensionality of the problem.

Consider the tapered cantilever beam pictured in Fig. 7.6. The beam is fixed on the left end and has its length divided into five elements. Each of the elements must be assigned a thickness between prescribed minimum and maximum values. A possible chromosomal representation of the beam may be formed by discretizing the thickness of each element into a fixed number of uniform levels. It should be noted that genetic optimization is being used to locate the regions in the design space which hold potentially good designs and not to produce a final design with exact values of all design variables. In this respect, the discretization need not be too fine. If the number of discrete thickness values is selected to be eight, then the beam topology may be completely described by a string of integers ranging from zero to seven with zero signifying the minimum thickness and seven representing the maximum thickness. The traditional one-dimensional chromosome would consist of five blocks of three binary digits each:

$$XXX \; XXX \; XXX \; XXX \; XXX$$

A two-dimensional chromosome for the same design representation would be a

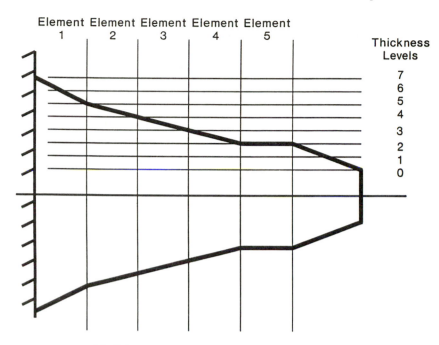

**Fig. 7.6** Tapered cantilever beam example for encoding.

matrix of the form

$$
\begin{matrix}
XXX \\
XXX \\
XXX \\
XXX \\
XXX
\end{matrix}
$$

Alternatively, the design could be encoded with multiple value chromosomal positions as

$$X X X X X$$

where each position assumes a value from the set of integers ranging from zero to seven. In all three forms, there is a direct link between the coding of the chromosome and the physical shape of the beam. The difference between the various encodings will be in the way the algorithm generates new designs or offspring from two parent designs.

### 7.6.2 The fitness function

The fitness function in a genetic algorithm closely parallels the objective function in a nonlinear programming formulation. It is used to characterize chromosomes in the population so that the best designs have the highest probability of being selected as parents for the next generation. This means that the traits of the fittest designs will be combined in order to form better designs. The fitness function may be a single scalar evaluator or it may be any weighted combination of factors. Within this framework it is a fairly straightforward process to implement a number of multiobjective techniques. Additionally, constraints may be factored in through the use of a penalty function. The most basic form of a fitness function would then be given as

$$\text{base value} - \text{evaluator value} - P[R, g(\mathbf{x}), h(\mathbf{x})] \qquad (7.12)$$

Here, the base value is used in order to keep the fitness level positive, the evaluator value is the design objective and $P[R, g(\mathbf{x}), h(\mathbf{x})]$ is a penalty term to handle inequality and equality constraints. The form presented in equation (7.12) will minimize the evaluator value as the most common form of a genetic optimization algorithm performs the maximization of a function. For structural design problems the fitness function will handle objectives or goals on weight, stress, displacement, natural frequency and buckling, when coupled with a finite element code.

As with any penalty function method, the scaling of the constraints relative to the evaluator function is important. Constraint scaling for structural problems is a fairly straightforward and well-understood issue. In a genetic optimization algorithm, however, there is the additional issue of the range of fitness function values within a population. If there is a significant range in the best to worst value, the probability of being selected as a parent for the next generation tends to be skewed to very likely candidates over very unlikely candidates. This will lead to domination by a few design encodings which is not necessarily a good strategy, particularly in the early portion of the optimization process. In order to allow that each chromosome in the current population has at least a reasonable chance of being selected as a parent, the range in the fitness function is scaled by some appropriate procedure. One possibility would be to compute the scaled fitness as

$$\text{fitness} = [0.25 + (\text{evaluator value} - \text{worst value})/(\text{best value} - \text{worst value})]^2$$
$$(7.13)$$

The best value and worst value are the lowest and highest evaluator values in the current population. This means that the range of fitness values will lie in the interval of

$$(0.25)^2 \leqslant \text{fitness} \leqslant (1.25)^2 \quad \text{or} \quad 0.0625 \leqslant \text{fitness} \leqslant 1.5625 \qquad (7.14)$$

Additionally, it is not uncommon to set a limit on the allowable worst value to prevent too wide a discrepancy among the practical designs. This is particularly true when dealing with an evaluator value which is formed from a penalty function where a design which violates a constraint may have a value which is several orders of magnitude different from that of other feasible design points.

## 7.6.3 The initial population

For well-posed problems, virtually any combination of values in a chromosome corresponds to a valid design. In structural applications there are exceptions for topological design optimization when one must guard against unconnected elements, but this may be accounted for by a preprocessing operation before any analysis is performed. In order to span the design space completely, every possible combination of chromosomes must be able to be formed by combining traits from members of the initial population. This is generally not possible, as the population size required to accomplish this task would be enormous, leading to unacceptable solution times. The best approach, then, is to generate the initial population randomly. That is, each position of each chromosome for each design is assigned a value at random from its permissible range. This provides a broad range of initial design possibilities, although the likelihood of many good designs being present in the initial population derived in this manner is low. Since locating a good design topology is the primary goal of the optimization, this is of little consequence. In any case, the larger the population, the more representative the members become of the complete design space. The difficulty, particularly with a computationally expensive analysis required for each design, is that only a small population is realistic. Fortunately, a mutation operator can be applied in order to maintain a finite probability for the creation of any arbitrary combination of variables in the design space. The mutation operator is discussed in the following section.

## 7.6.4 Genetic operators

The specific types of operators used in the implementation of a genetic optimization algorithm depend on both the encoding scheme of the chromosome and on the particular type of problem being solved. Regardless of these factors, several basic genetic operators which are always present include the selection of parents, crossover and mutation. The selection of parents has been alluded to previously and is probabilistically based on the fitness function value. The term crossover refers to how the parent chromosomes are combined in order to generate offspring (new designs). Finally, mutation is a means of allowing purely random alterations in the chromosome values with a small but finite probability in order to introduce design characteristics into the current population which either were lost or were never present in any previous population. The entire genetic optimization process is heuristic in nature which allows a wide range of choices in a particular implementation. The operators described here are fairly elementary in nature and better results may be obtained with more complicated approaches. The basic flow of the algorithm is, however, much easier to comprehend when these basic operator forms are considered.

The parent selection process may be likened to the spinning of a weighted roulette wheel. The weighting is utilized so that the fittest members of the population occupy the largest area on the wheel and the total population fitness is scaled so that the sum of all slot areas fills the wheel. This gives every member of the population a finite chance of being selected but the probability of being selected is directly proportional to the fitness of the design. The wheel is spun twice and two population members are identified as parents. This parent set may be used to generate any number of offspring, but the generally accepted procedure is to produce two offspring from two parents. This means that the parent selection operation is performed one half as many times as the number of chromosomes in the population for each stage of the algorithm. It may be beneficial to limit the number of times a single chromosome may be a parent in any one generation in order to avoid dominance by a particular design. This may also be accomplished by other selection schemes such as by tournament selection (Goldberg, 1990). The way in which the children are formed from the parent chromosome values is the crossover operation.

The crossover operation is a function of the dimensionality of the chromosome. For a one-dimensional chromosome, a position within the length of the string is selected randomly and the parent chromosomes are switched from this point on. For example, if the parent chromosomes are strings of five values as

12345 and 54321

and the crossover point is selected as three, then the offspring generated have the same first three chromosome values as the parent designs, but the remaining two positions are switched. This results in the offspring

12321 and 54345

Whether or not these designs are better than the parent designs will determine their chance of being selected as parents for future generations. Many implementations have a means of propagating the best designs from one generation unchanged into the next generation and this is generally a good idea. Additionally, a single chromosome is not allowed to be selected as both parents.

For two- and three-dimensional chromosomes, both a matrix position and a crossover direction are selected randomly in order to generate offspring. There are four crossover directions for the two-dimensional case (NW, NE, SE, SW) and eight (octant directions) for the three-dimensional case. The reason for using higher-dimensional chromosomes in a shape optimal design problem is to couple the dimensionality of the chromosome with that of the encoding. This means that, for a three-dimensional problem, the crossover operation is actually combining the parent designs in three-dimensional blocks. While no extensive testing has been conducted to date on the effect of the dimension of the chromosome encoding, the rule of thumb to follow is always to link the encoding to the physical structure of the design topology as closely as possible.

The final genetic operator is that of mutation. Mutation randomly alters a position in a chromosome with a fairly low probability. This can at times destroy a good design but it can also introduce design characteristics which could not be achieved through crossover with the current population. A high mutation probability will slow convergence but too low a probability will truncate the design space in a potentially harmful way. The selection of all algorithm parameters is problem dependent and requires some experimentation. This is no different than with a traditional nonlinear programming algorithm. Additional detail on genetic algorithms is presented by Goldberg (1989).

---

## EXAMPLE 7.1 A THREE-BAR TRUSS

The best way to demonstrate the difference between a traditional nonlinear programming problem approach and a multiobjective programming problem approach is to consider an example. This problem involves the design of a

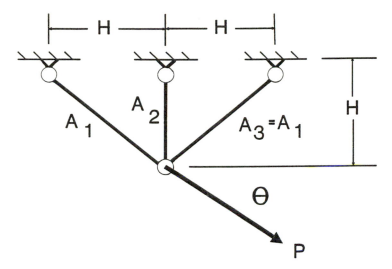

**Fig. 7.7** Three-bar truss.

three-bar truss and is used to demonstrate how a multiobjective formulation can drive the solution from one topology to another which might not be located by a single objective formulation. The issue of uncertainty in the magnitude and direction of the applied loading is also considered. The three-bar truss as shown in Fig. 7.7 is to be designed to withstand a specified load of 20 units with the minimum amount of material. The allowable stress in each bar is constrained to be less than 20. The nodal locations are fixed, so the design variables are the cross-sectional area of each truss element. In order to maintain symmetry, both bars 1 and 3 are required to have the same cross-sectional area. This reduces the problem to two design variables which is convenient from the point of view that the results can be displayed graphically. The problem in this form will be addressed as a nonlinear programming problem, and then an uncertainty in the exact load magnitude and direction will be introduced and a multiobjective formulation and solution will be considered.

The nonlinear programming problem formulation for the original problem is expressed as follows:

$$\text{minimize volume} = 2\sqrt{2}x_1 + x_2; \mathbf{x} = \begin{Bmatrix} A_1, & A_3 \\ & A_2 \end{Bmatrix}$$

subject to

$$g_1(\mathbf{x}) = 20 - |\text{stress}_1| \geqslant 0$$
$$g_2(\mathbf{x}) = 20 - |\text{stress}_2| \geqslant 0$$
$$g_2(\mathbf{x}) = 20 - |\text{stress}_3| \geqslant 0$$

Here, $A_i$ represents the cross-sectional area of the $i$th truss element and stress$_i$ represents the stress in the $i$th element. The solution to this problem formulation, generated by a standard penalty function algorithm, is $\mathbf{x}^* = \{0.788 \ 0.408\}^T$ with a resulting volume of 2.638 and stresses in the bars of 20, 14.65 and $-5.35$ units.

The stresses are easily computed by assembling the element stiffness matrices and solving for the deflection of the free node and then calculating the element forces and stresses. It would be a straighforward procedure to include deflection, frequency and buckling constraints, but these additions would serve only to complicate the example. The problem is simple but yet it is representative of a wide class of problems which are routinely solved.

The normal solution process would terminate at this point since the optimum design for this particular formulation has been obtained. The question which should be asked at this point, however, is whether or not the solution generated is indeed a good solution. One important issue is the effect on the design of a change in the magnitude or direction of the applied load. This is often a critical factor as loads applied to finite element models are rarely more than approximations or estimates of the actual loading. For this example, consider what happens as the magnitude of the load is altered from 15 to 25 units and the loading direction changes from 30° to 60° from the horizontal. If the load magnitude and direction are altered independently from the baseline of the original problem and the resulting new problem formulation is reoptimized, the new solutions generated are those plotted in Fig. 7.8. The weight of the design as well as the cross-sectional areas of the elements of the resulting design change on the order of ± 25%. Even though the change in the direction of the load had a minor impact on the weight for the optimal design, the resulting cross-sectional areas underwent considerable change.

The next question to ask is which, if any, of these designs is optimal? The answer could easily be either all of them or none of them. Each is an optimal design with respect to a specific problem formulation. On the other hand, none is a true optimal design for the real issues in designing the truss. At best, the results point out the region of potential design solutions as represented by the rectangle in Fig. 7.8. The designer can use the information derived from the optimization runs, along with his insight into the problem, and make the final design decisions.

For this two-dimensional problem which is not computationally expensive, it is easy to investigate the solution space around each potential optimum. For larger problems, this is not the case and the procedure becomes difficult if not impossible to apply. In addition, in this particular example, the uncertainty in the loading was well defined. If the uncertainty is known then this runs contrary to the concept of uncertainty. What would be far more beneficial would be a more natural approach to optimization which can deal with uncertainty as well as other important design issues.

In the absence of any specific information concerning the possible range of design uncertainty, there are several courses of action one might take. Two such courses are either to include a margin of safety in the design or to try and make the design as insensitive to changes in uncertain parameters as possible. The inclusion of a safety factor is the traditional approach, but, as was mentioned earlier, this converts the hard stress constraints to soft constraints which are not dealt with effectively by a nonlinear programming problem formulation and solution. The second approach is attractive, but may not in itself lead to a realistic design and is again not easily incorporated into a traditional problem formulation and solution procedure. Nonlinear goal programming may be applied to

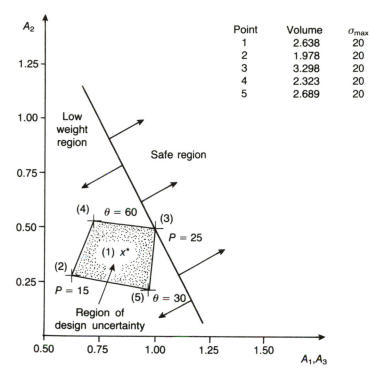

| Point | Volume | $\sigma_{max}$ |
|---|---|---|
| 1 | 2.638 | 20 |
| 2 | 1.978 | 20 |
| 3 | 3.298 | 20 |
| 4 | 2.323 | 20 |
| 5 | 2.689 | 20 |

**Fig. 7.8** Optimal design for three-bar truss (variation in load magnitude and direction).

formulate a design optimization problem which deals with either or both of these approaches to modifying the problem.

From the margin of safety point of view, one must consider what decreases the chance of a failure of the design in service. For this example, the maximum stress present in the design provides a good indicator. A design which has a very low peak stress is unlikely to fail even if the load magnitude is increased. Of course, this will produce a design which is counter to the original objective of a light weight design. The trade-off between a safe design and a lightweight structure may be evaluated easily through a multiobjective formulation as a nonlinear goal programming problem. Consider the formulation given as

$$\text{minimize } f(\mathbf{x}) = \{ W_1 d_1^+ + W_2 d_2^+ \}$$

with goal constraint of the form

$$\text{volume} + d_1^- - d_1^+ = 0$$

and

$$\max \{ \text{stress}_1, \text{stress}_2, \text{stress}_3 \} + d_2^- - d_2^+ = 0$$

This formulation will try and strike a compromise between the minimization of the volume and the minimization of the peak stress. The first goal constraint sets forth the goal of a zero weight design, the ultimate lightweight structure, and the second goal constraint establishes the goal of a zero stress design. It is obvious that no design can satisfy the goal constraints, but this is of little consequence. The objective function seeks to minimize the underachievement of the two goals. The maximum stress constraints are maintained to guarantee their satisfaction at the solution. The path the optimal design takes as a function of the relative weights placed on the design goals is shown in Fig. 7.9. When a low weight is assigned to the stress goal relative to the weight goal, the design generated will be equivalent to the nonlinear programming result. As the relative weight on the stress goal is increased, the optimum traces the path shown in Fig. 7.9. As the weight increases, the design moves farther into the safe region. This path provides a wealth of trade-off information. For problems of higher dimension, this path becomes a trade-off surface which is somewhat equivalent to a response surface in multivariate statistics.

The information produced by the first goal programming formulation provided useful design information but it did not lead to any new design region which is potentially better than that located by the straightforward nonlinear programming solution. This issue will now be addressed by considering a different goal programming solution. Instead of considering a design which has low stresses in all of its structural members, goal constraints will be included to try and make the peak design stress as insensitive as possible to changes in the applied load

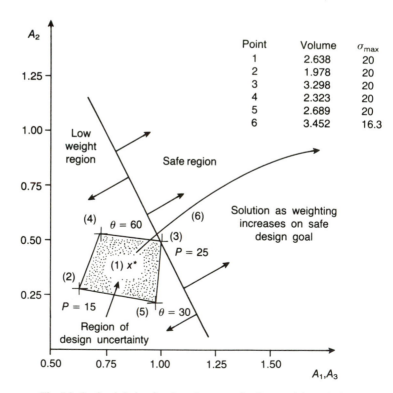

**Fig. 7.9** Optimal design for three-bar truss for first goal formulation.

(both magnitude and direction). A possible goal programming formulation for this problem is given as

$$\text{minimize } f(\mathbf{x}) = \{W_1 d_1^+ + W_2(d_2^+ + d_3^+ + d_3^-)\}$$

with goal constraints

$$\text{volume} + d_1^- - d_1^+ = 0$$

$$\max\{\text{stress}_1, \text{stress}_2, \text{stress}_3\}_{P=25} - \max\{\text{stress}_1, \text{stress}_2, \text{stress}_3\}_{P=15}$$
$$+ d_2^- - d_2^+ = 0$$

$$\max\{\text{stress}_1, \text{stress}_2, \text{stress}_3\}_{\theta=60^\circ} - \max\{\text{stress}_1, \text{stress}_2, \text{stress}_3\}_{\theta=30^\circ}$$
$$+ d_3^- - d_3^+ = 0$$

The objective function now has one weight for the volume or minimum weight goal and a common goal for the design sensitivity of the peak stress relative to a change in load magnitude or direction. Both $d_3^+$ and $d_3^-$ are included in the objective function as it is not known beforehand what sign the result will take. The first two goal constraints may only take on possitive values, so only the corresponding positive deviational variable is included in the objective function. It would be a minor matter to include different weights for the stress sensitivity

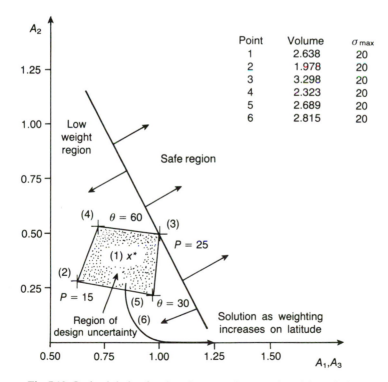

| Point | Volume | $\sigma_{max}$ |
|---|---|---|
| 1 | 2.638 | 20 |
| 2 | 1.978 | 20 |
| 3 | 3.298 | 20 |
| 4 | 2.323 | 20 |
| 5 | 2.689 | 20 |
| 6 | 2.815 | 20 |

**Fig. 7.10** Optimal design for three-bar truss for second goal formulation.

with respect to load magnitude and direction, but this would increase the number of weighting parameters.

The results from this formulation are presented in Fig. 7.10. As with the previous results, the path which the solution follows as the relative weights between the weight goal and the sensitivity goals change is plotted. It is interesting to note that a topological change in the design is made by reducing the area of the middle element to zero. This is due to the fact that by using the material from the second element to increase the cross-sectional areas of the outer elements, a design is generated which is less sensitive to load variation. This result, although perhaps intuitive, may not have been achieved without the investigation of this alternative formulation. The most interesting aspect of the result is that the two-bar truss design moves toward the safe design region with less of a weight penalty than did the three-bar truss. This is simply a reminder that topological changes may result in alternative optimal design regions.

Which of the presented formulations produces the best design is not as important as the fact that much useful design information was derived by considering the goal programming formulations. A combination of both the goal formulations would perhaps be even more productive. On the other hand, this problem is quite simple and the approach taken in the solution may not be attractive for more difficult problems. The evaluation of multidimensional trade-offs between competing design criteria is not a simple task. The flexibility of the multiobjective problem formulation, however, is apparent as is the fact that it is a mistake to concentrate on one small portion of the feasible design space. Better design solutions involving topology changes may well be available and should at least be considered as possibilities.

## EXAMPLE 7.2 MECHANISM DESIGN

The topological change considered in the first example involved an alteration in the number of structural elements present. Other classes of design problems benefit from a multiobjective formulation as well as from consideration of topological modifications. The problem considered here belongs to the class of planar mechanism design. The basic challenge is to produce a design which possesses a number of specific characteristics including position, velocity, cost and small sensitivities to inner and outer noise. The optimization of mechanisms is difficult owing to a number of factors including a high degree of nonlinearity, the possibility of divergence in the iterative solution process for determining link positions and velocities and the presence of a multitude of local minima. The issue of hard vs soft design specifications also comes into play. Hard constraints such as the requirement of assembly are always present but most other considerations are best treated as soft specifications.

The particular problem at hand involves the design of a planar mechanism which can produce motion that accurately traces a straight line path from point A with coordinates $(5, 4)$ to point B located at coordinates $(2, 2)$. In addition, the mechanism is to provide as close to a constant velocity of 100 units $s^{-1}$ during the motion from point A to B. The pivot point of the driving link for the

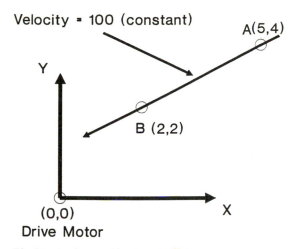

**Fig. 7.11** Design specifications for linkage design problem.

mechanism is required to be located at the origin as depicted in Fig. 7.11. The linkage is to be driven at a constant angular velocity. Other design considerations include the cost of the mechanism as well as its sensitivity to small changes in either the angular velocity of the input link or the various link geometries (i.e. tolerances on link lengths and position). The mechanism should also remain far from a position in which it will jam or not assemble. At this point in the process it is not at all obvious which design criteria will be most influential in defining the solution, and as such, it provides an excellent multiobjective optimization example.

There are an extremely large and varied assortment of mechanisms which could be investigated for their potential in meeting the design criteria. This is typical of most design endeavors in the early stage. For the purpose of this example, three mechanism classes will be considered, a four-bar mechanism as illustrated in Fig. 7.12, an inverted slider crank mechanism as shown in Fig. 7.13 and a six-bar mechanism as pictured in Fig. 7.14. There actually are many other configurations of six-bar mechanisms, but the one selected, which is commonly referred to as a Stephenson type III linkage, is assumed to represent the general performance of the class. Each of the mechanisms selected possesses a single degree of freedom in that, for a specified input link motion, the position, velocity and acceleration of each link in the mechanism are completely specified. This allows the design objectives to be evaluated in a direct fashion from the mechanism geometry.

The design variables are the link lengths, ground pivot locations and coupler point location. The coupler point is the position on one of the links which is to provide the displacement and velocity specified in the problem statement. A ground pivot is simply a point where the linkage is fixed in translation but not in rotation. The design variables are defined for each of the three selected design topologies in Figs. 7.12–7.14. This example points out the fact that the number of design variables for each design topology need not be the same as for another. In this case, the inverted slider crank has nine design variables, the four-bar has

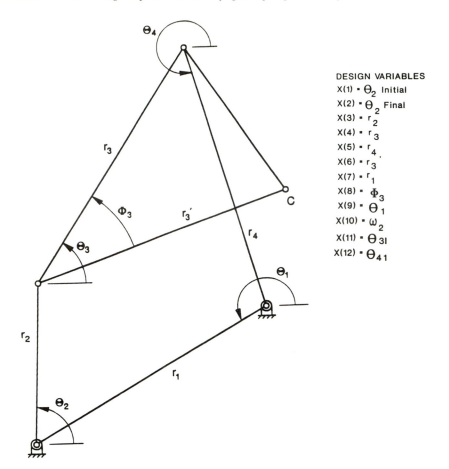

DESIGN VARIABLES
$X(1) = \Theta_2$ Initial
$X(2) = \Theta_2$ Final
$X(3) = r_2$
$X(4) = r_3$
$X(5) = r_4$,
$X(6) = r_3'$
$X(7) = r_1$
$X(8) = \Phi_3$
$X(9) = \Theta_1$
$X(10) = \omega_2$
$X(11) = \Theta_{3I}$
$X(12) = \Theta_{41}$

**Fig. 7.12** Design variables for four-bar linkage.

ten design variables and the six-bar has sixteen design variables. Cross-sectional link dimensions are not included as no force or stress analysis will be considered in this example. Additional variables were added which estimated the initial link positions for the displacement and velocity analysis which resulted in a total of eleven, thirteen and twenty design variables for the three cases.

The design objectives are formed into the goal constraints based on the criteria listed below.

1. The distance between the actual motion generated by the mechanism and the desired linear path should be as small as possible.
2. The difference between the desired velocity and the actual coupler point velocity should be as small as possible over the range of motion.
3. The coupler point should start as close to point A as possible at the beginning of the required motion range.
4. The coupler point should be located as close to point B as possible at the end of the required motion range.

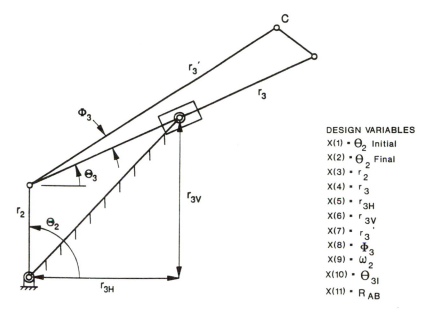

**DESIGN VARIABLES**

X(1) = $\Theta_2$ Initial
X(2) = $\Theta_2$ Final
X(3) = $r_2$
X(4) = $r_3$
X(5) = $r_{3H}$
X(6) = $r_{3V}$
X(7) = $r_3'$
X(8) = $\Phi_3$
X(9) = $\omega_2$
X(10) = $\Theta_{3I}$
X(11) = $R_{AB}$

**Fig. 7.13** Design variables for inverted slider–crank mechanism.

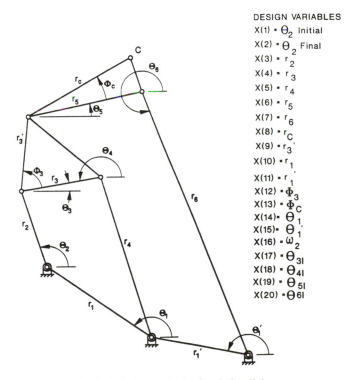

**DESIGN VARIABLES**

X(1) = $\Theta_2$ Initial
X(2) = $\Theta_2$ Final
X(3) = $r_2$
X(4) = $r_3$
X(5) = $r_4$
X(6) = $r_5$
X(7) = $r_6$
X(8) = $r_C$
X(9) = $r_3'$
X(10) = $r_1$
X(11) = $r_1'$
X(12) = $\Phi_3$
X(13) = $\Phi_C$
X(14) = $\Theta_1$
X(15) = $\Theta_1'$
X(16) = $\omega_2$
X(17) = $\Theta_{3I}$
X(18) = $\Theta_{4I}$
X(19) = $\Theta_{5I}$
X(20) = $\Theta_{6I}$

**Fig. 7.14** Design variables for six-bar linkage.

5. The mechanism should remain as far as possible from a position where the mechanism will not assemble over the required motion range.
6. The size and weight of the linkage should be as low as possible.
7. The change in velocity of the coupler point over the required range of motion should be as insensitive as possible to small fluctuations in the angular velocity of the input link.
8. The coupler point path traced should be as insensitive as possible to a small change in the length of the input link.
9. The coupler point path traced should be as insensitive as possible to a small change in the location of the coupler point position.

The goals, when defined in this manner, are easily understood by the designer and apply equally well to any of the mechanism classes selected. This is a key point as placing goals on the design defines the problem but does not limit the range of solutions to the problem. These goals are applied uniformly to each design class. Of course there may be other design goals to include or the existing goals could be defined in an alternative fashion, but the goals listed certainly capture the majority of important design criteria for this example. The goals defined for this example may appear to be rather severe, but, at this point in the process, no concern is placed over the possibility that no solution exists which can even begin to satisfy all of the design goals. Trade-off evaluations and goal redefinition should occur at a later point in the process. Keep in mind that design should proceed from the general to the specific and that any attempt to alter this philosophy will likely result in a suboptimal design.

In order to evaluate the goal constraints, a displacement and velocity analysis must be performed for each of the mechanisms considered. Many techniques exist for such an analysis, especially for a specific mechanism class such as a four-bar linkage. A general approach employing vector loop equations and a nonlinear equation solver is considered herein as it applies equally well to any selected planar, single-degree-of-freedom mechanism. Consider the four-bar linkage pictured in Fig. 7.12. The objective of the analysis is to determine the position and the velocity of the coupler point C, for a given input link position, $\theta_2$. Considering each link as a vector, the loop equation for the closed loop formed by links $r_2, r_3, r_4$ and the ground link $r_1$ is as follows:

$$r_2 + r_3 + r_4 + r_1 = 0$$

The bold print is used to denote the terms as vectors. Simply put, this equation states that if one were to follow each link in the mechanism consecutively, then one would arrive back at the starting point. The vector equation may now be broken down into two scalar equations. This produces the $x$ and $y$ component equations shown below:

$$r_2 \cos \theta_2 + r_3 \cos \theta_3 + r_4 \cos \theta_4 + r_1 \cos \theta_1 = 0$$

and

$$r_2 \sin \theta_2 + r_3 \sin \theta_3 + r_4 \sin \theta_4 + r_1 \sin \theta_1 = 0$$

where the individual angles are measured in the counterclockwise sense from the horizontal as shown in Fig. 7.12. Since the input link position is known and all link lengths are known as well as as the angle of the ground link, the only unknowns are $\theta_3$ and $\theta_4$. Solving two equations for two unknowns is a straightforward process, with the only difficulty being the fact that the equations are nonlinear. A Newton–Raphson technique is employed to solve the equations. A starting estimate of the unknown angles is required and, if these estimates are not good enough, the solution process can diverge. Program checks must be inserted in order to flag this situation.

Once angles $\theta_3$ and $\theta_4$ are determined, it becomes a simple matter to locate the position of the coupler point C. A vector loop equation can be written beginning from the input link and ending at the coupler point. This equation takes the following form:

$$C = r_2 + r'_3$$

where $C$ refers to the location of the coupler point relative to the origin. Again this vector equation may be broken down into its $x$ and $y$ component equations which produces

$$C_x = r_2 \cos \theta_2 + r'_3 \cos (\theta_3 + \phi_3)$$

and

$$C_y = r_2 \sin \theta_2 + r'_3 \sin (\theta_3 + \phi_3)$$

Since there are no unknowns on the right-hand sides of the equations, the $x$ and $y$ coordinates of the coupler point are completely determined.

The determination of the velocity of the coupler point is achieved by differentiating the $x$ and $y$ component equations resulting from the vector loop equation for the mechanism. This produces

$$- r_2 \omega_2 \sin \theta_2 - r_3 \omega_3 \sin \theta_3 - r_4 \omega_4 \sin \theta_4 = 0$$

and

$$r_2 \omega_2 \cos \theta_2 + r_3 \omega_3 \cos \theta_3 + r_4 \omega_4 \cos \theta_4 = 0$$

The only two unknowns in these equations are the angular velocities $\omega_3$ and $\omega_4$. The input velocity, $\omega_2$, is constant and is known. These equations are linear in the unknowns, so the solution is trivial. Once the angular velocity of all links is known, the velocity of the coupler point may be determined by differentiation of the component equations which define the position of the coupler point. This results in

$$V_{x\text{point C}} = - r_2 \omega_2 \sin \theta_2 - r'_3 \omega_3 \sin (\theta_3 + \phi_3)$$

and

$$V_{y\text{point C}} = r_2 \omega_2 \cos \theta_2 + r'_3 \omega_3 \cos (\theta_3 + \phi_3)$$

Again, all right-side terms are known which allows for the direct solution of the velocity components.

The solution processes for the displacement and velocity for the inverted slider crank mechanism and the six-bar mechanism follow a similar vector loop process. It is always possible to write a vector loop displacement equation which will result in only two unknowns. For the six-bar linkage, two such vector loops are required for a displacement solution, but this adds little to the difficulty. Once the displacement equations are solved, direct differentiation allows for a velocity solution. The process is repeated for a number of input link positions ranging from the initial position to the final position of interest. A total of ten intermediate positions was arbitrarily selected for this example. This provides sufficient velocity and displacement information for the evaluation of any selected geometric link dimensions for any of the mechanism classes considered.

The first goal constraint concerning the specific linear motion requirement was formed by summing the distance of the coupler point from the line segment connecting points A and B for all ten intermediate positions. The second goal constraint which specifies the constant velocity requirement is formed by summing the square of the difference in the coupler point velocity and the desired velocity at each intermediate position. The next two goals which state that the coupler point should start at point A and end at point B are formed as the distance of the coupler point from point A at the initial motion position and from point B at the end of the specified motion. The desired goal value for each of these constraints is zero which would produce a mechanism with the exact positional and velocity characteristics.

The fifth goal constraint states that at each intermediate position the linkage should be reasonably far from a position in which it would not assemble. This keeps the mechanism away from a position where it either might lock up or would require a large force to maintain the desired operation. For ease of computation, this constraint is specified as a distance from a nonassembly position. In the case of the four-bar linkage, this distance is simply the difference in the sum of the length of links 3 and 4 and the distance from the ground point of link 4 and the current position of the end of link 2. As long as this difference is positive, the mechanism will assemble. The desired goal value for this constraint was specified as a minimum of one distance unit for the worst case over the range of motion. The satisfaction of this goal directly effects the transmission angles in the mechanism as well. The assembly goal for the inverted slider crank mechanism is that link 3 will always extend beyond the slider pivot point by at least one distance unit. For the six-bar mechanism, two separate assembly conditions are imposed, one for each vector loop used in the displacement solution.

The sixth goal constraint is imposed as a complexity or, in a sense, a cost measure. For the same level of satisfaction of the other goal constraints, a smaller mechansim or one with fewer links is preferred. This type of linkage will require less material and as such should have a cost advantage over a larger and more complex mechanism. The goal constraint is computed as the sum of the lengths of all links, including coupler links for each design. A more precise cost goal could be assigned, but this formulation will serve to point out potential cost differences among the design choices. A desired goal value of zero is assigned to the goal signifying a smaller is better condition.

The final three goal constraints are intended to encourage a design which has performance which is as insensitive as possible to small changes in dimensions or the input velocity of the driving motor. Any inexpensive motor will have some speed fluctuation which is dependent on the load being driven. The dimensional tolerances of the links can be controlled accurately but there is a cost associated with the accuracy required. If a design is found which is insensitive to these changes it is a robust design and this characteristic is highly desirable. The goals are computed as the difference in the velocity accuracy as specified by goal 2 caused by a 5% increase in the angular velocity of the input link, the length of the input link and in the position of the coupler point. The desired goal level for the sensitivity constraints is set to zero which specifies the ideal design to be completely insensitive to these changes. The sensitivity with respect to other link dimensions and ground attachment points could be considered as well but the nine goal constraints specified are adequate to define a rather challenging design problem.

The fact that no design will be able to meet all of the design criteria is of no consequence to the approach. This is a fundamental difference in comparison with traditional nonlinear programming methods. These methods would have the null space defined for the search and would fail to find a feasible point. This would not allow for an easy evaluation of trade-offs involved in meeting the design goals. All that remains for the goal formulation is the selection of a set of goal priorities and weights. The goals are separated into three levels of priorities. The initial and final position requirements for the coupler point have the highest goal priority. The linear motion and constant velocity requirements have the second and third highest priorities. Finally, all remaining goals are set at the lowest priority. The selection of individual goal weights is a scaling issue and the idea is to keep the contribution of each goal constraint approximately

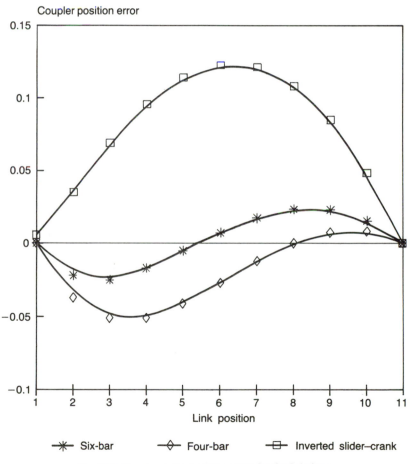

**Fig. 7.15** Coupler point position error for final designs.

equal at the starting point selected. The weights may be fine tuned at a later point to modify the final design configuration as deemed appropriate.

The assignment of priorities was not performed in an arbitrary fashion but was in fact meant to mimic the path a human designer might follow. The design search would most likely be for a coupler point on the mechanism which had good displacement characteristics, followed by a modification to achieve the desired velocity requirement. At this point, the designer would consider factors such as weight, cost and design sensitivity. In truth, it is unlikely that a human designer would get past the second or third goal levels as very few design topologies would be investigated because of the time required and the difficulty involved in the process. This is not true, however, of the multiobjective computational process where thousands of designs may be investigated.

For this particular problem both a genetic optimization algorithm and a gradient based nonlinear programming algorithm were employed. Since the link dimensions are continuous, the nonlinear programming algorithm is more efficient in locating a minimum than the genetic algorithm. The genetic method, however, is not trapped by local minima and as such it is used to perform the topological search which defines the mechanism class as well as the starting

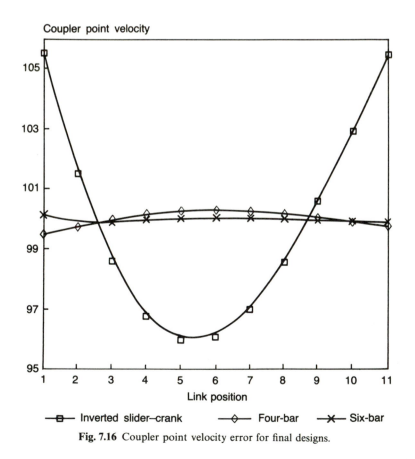

**Fig. 7.16** Coupler point velocity error for final designs.

**Table 7.1** Unweighted goal values for linkage design problem

| Goal | Inverted slider crank | Four bar | Six bar |
|------|------|------|------|
| 1 | 0.804 | 0.235 | 0.156 |
| 2 | 126.733 | 9.144 | 2.931 |
| 3 | $5.621 \times 10^{-3}$ | $3.000 \times 10^{-4}$ | $2.861 \times 10^{-4}$ |
| 4 | $5.376 \times 10^{-5}$ | $5.229 \times 10^{-4}$ | $8.692 \times 10^{-4}$ |
| 5 | 2.138 | 3.206 | 0.512 |
| 6 | 40.531 | 23.615 | 26.480 |
| 7 | 220.942 | 182.322 | 182.605 |
| 8 | 0.306 | 0.118 | 0.196 |
| 9 | 1.032 | 0.785 | 0.512 |

dimensions for the link lengths which are passed to the nonlinear programming code. Even the initial angle estimates for the iterative solution of the vector loop equations were determined by the genetic algorithm which gave the various alternate assembly positions an equal chance of being investigated. The inclusion of the initial position estimates as variables accounts for an increase in two design variables for the four-bar and inverted slider crank mechanisms and four variables for the six-bar linkage. All goal constraints were treated as soft constraints to allow the maximum amount of freedom in the solution.

**Table 7.2** Optimal design variable values for linkage design problem

| Design variables | Inverted slider crank | Four bar | Six bar |
|------|------|------|------|
| $X(1)$ | 97.430 | 78.745 | 50.910 |
| $X(2)$ | 131.762 | 113.643 | 87.602 |
| $X(3)$ | 5.229 | 4.266 | 2.833 |
| $X(4)$ | 15.000 | 4.098 | 5.298 |
| $X(5)$ | 4.919 | 5.114 | 3.065 |
| $X(6)$ | 13.640 | 4.171 | 1.274 |
| $X(7)$ | 5.803 | 5.966 | 5.440 |
| $X(8)$ | 291.664 | 273.392 | 4.022 |
| $X(9)$ | 21.248 | 224.015 | 1.967 |
| $X(10)$ | 56.506 | −16.856 | 0.662 |
| $X(11)$ | 10.138 | 84.081 | 1.919 |
| $X(12)$ | | 306.416 | 204.147 |
| $X(13)$ | | | 323.663 |
| $X(14)$ | | | 155.269 |
| $X(15)$ | | | 168.177 |
| $X(16)$ | | | 17.725 |
| $X(17)$ | | | 270.118 |
| $X(18)$ | | | 112.954 |
| $X(19)$ | | | 36.451 |
| $X(20)$ | | | 275.024 |

In short, the combined genetic–nonlinear goal solution provided several excellent solutions to what would be considered a most difficult design problem by any other approach. One solution from each of the selected mechanism classes is presented. Fig. 7.15 plots the position error produced by the final design for each class while Fig. 7.16 displays the velocity error over the range of motion. From these results it is clearly evident that there are indeed designs which satisfy the positional and velocity goals very precisely. The six-bar mechanism achieves the best results, followed closely by the four-bar mechanism and finally by the inverted slider crank mechanism.

When the remaining goals are factored into the evaluation, it is still fairly easy to distinguish among the various design alternatives. Table 7.1 presents the final values of the unweighted goal values for each of the three final designs. The six-bar mechanism performed well in all meeting all goals with the only caution arising in the assembly condition which is met but not by as much as for the other designs. The performance of the four-bar is certainly adequate unless the mechanism is required to have a very high level of precision. The design variables for the final designs are presented in Table 7.2. It should be noted that other designs were located in addition to the three documented in the final results. Two other six-bar mechanisms and one other four-bar mechanism which had excellent performance characteristics were obtained in the final population as well. Additional trade-offs could be made for each of the final designs to fine tune them as well.

---

## EXAMPLE 7.3 A TEN-BAR TRUSS

The literature concerning structural optimization contains a multitude of problem solutions, particularly in the area of truss structures. Several examples have appeared time and time again to the point that the solutions presented are fairly well accepted as being correct. The ten-bar truss presented in Fig. 7.17 is one such example. The design objective is to minimize the weight of the structure which is subjected to two equal vertical loads applied to nodes 2 and 4 of 10 000 units. The maximum allowable stress in a member (tension or compression) is 25 000. A maximum displacement constraint of 2 units is imposed for nodes 2 and 4 as well. The design variables are simply the cross-sectional area of each of the ten truss members. Lower variable bound limits are placed on all areas to be greater or equal to 0.1. The solution to this problem formulation, generated by a gradient based penalty function method, is presented in the first column of Tables 7.3 and 7.4. The question to be addressed, however, is whether this is indeed the solution or even represents a good design.

If the minimum weight structure is desired and the topology of the structure is held constant (i.e. ten members with fixed nodal location), then this is indeed the optimal point. The solution may be verified by regenerating the solution from a number of starting points. While the formulation presented fits very nicely into the standard structural optimization format, the solution raises several important design issues. First of all, why was a ten-bar truss selected? From the loading specified it seems plausible that several members could be removed and

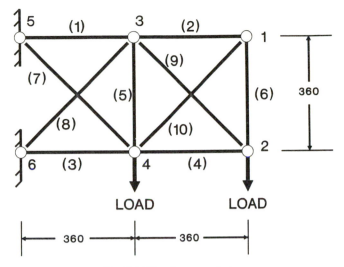

**Fig. 7.17** Ten-bar truss design.

**Table 7.3** Unweighted goal values for structural design problem

| Goal | Design 1 | Design 2 | Design 3 | Design 4 | Design 5 |
|------|----------|----------|----------|----------|----------|
| 1 | 5062 | 6689 | 2317 | 3067 | 3082 |
| 2 | 25000 | 8616 | 24940 | 12490 | 12510 |
| 3 | 2.000 | 2.000 | 2.002 | 1.502 | 1.499 |
| 4 | 3303 | 742 | 2973 | 2082 | 2080 |
| 5 | 0.255 | 0.257 | 0.407 | 0.338 | 0.336 |
| 6 | 7409 | 2319 | 5174 | 1648 | 1708 |
| 7 | 10 | 10 | 6 | 6 | 3 |

**Table 7.4** Optimal design variable values for structural design problem

| Variable | Design 1 | Design 2 | Design 3 | Design 4 | Design 5 |
|----------|----------|----------|----------|----------|----------|
| Area (1) | 30.154 | 23.720 | 13.550 | 17.700 | 17.960 |
| Area (2) | 0.100 | 12.440 | – | – | – |
| Area (3) | 23.440 | 24.620 | 10.590 | 13.730 | 13.640 |
| Area (4) | 15.397 | 16.100 | 9.890 | 12.800 | 13.640 |
| Area (5) | 0.100 | 9.430 | – | – | – |
| Area (6) | 0.493 | 8.240 | – | – | – |
| Area (7) | 7.402 | 17.450 | 2.010 | 2.920 | 2.920 |
| Area (8) | 20.878 | 17.950 | 3.840 | 3.470 | 2.920 |
| Area (9) | 21.764 | 16.340 | 13.300 | 17.700 | 17.960 |
| Area (10) | 0.100 | 12.780 | – | – | – |
| Node 5($Y$) | 360.00 | 360.00 | 510.24 | 560.00 | 560.00 |
| Node 6($Y$) | 0.00 | 0.00 | – 151.72 | – 200.00 | – 200.00 |
| Node 3($X$) | 360.00 | 360.00 | 367.38 | 393.00 | 389.85 |
| Node 3($Y$) | 360.00 | 360.00 | 317.34 | 311.00 | 314.47 |

not detract from the ability of the structure to meet all design requirements. This perception is warranted by the solution which has three of the member areas set at the lower limit. Whether or not the removal of members would produce a lower weight design, however, remains unknown. It also seems possible that allowing the node points to move in space could be beneficial to the solution. Added to this is the fact that the stress and displacement constraints are treated as hard constraints. What if an additional 10% weight reduction could be achieved by increasing the limit to 2.1 units or increasing the allowable stress to 26 000 units? Alternately, if safety is a consideration, what weight penalty would be imposed if the maximum stress were reduced? Finally the sensitivity of member stresses to small changes in load magnitude or direction is unknown. All of these design issues may be critically important, but they remain uniformly unresolved with the current solution.

Approaching the problem from a strict design point of view with no predisposition as to what the final design should be, the following goals deserve consideration.

1. The weight of the structure should be as low as possible.
2. The maximum stress of any member should be as low as possible.
3. The maximum displacement of any node should be as small as possible.
4. The maximum stress should be as insensitive as possible to a change in loading magnitude or direction.
5. The maximum displacement should be as insensitive as possible to a change in loading magnitude or direction.
6. The level of stress in each member should be as uniform as possible.
7. The number of different beam sections should be as few as possible.

These seven goals encompass the majority of objectives for structural design. The first goal is equivalent to the minimum weight objective in a conventional formulation. It is formed by summing the volume of material used in each member and multiplying the sum by the specific weight of the material. The desired goal value of this constraint was set at zero which is a lighter is better formulation. The specific weight was set at 0.100 and the elastic modulus, $E$, was set at $10^6$.

The second and third goal constraints are equivalent to the inequality constraints imposed in the traditional formulation. The advantage of formulating them as goal constraints is that they now become soft constraints and can be violated slightly if other goals can benefit significantly from the violation. These two goal constraints are normalized by dividing by the desired goal values (25 000 for stress and 2 units for displacement) and subtracting the ratio from 1. In this form, if the stress or displacement goal value is exceeded, the goal constraint becomes negative.

The fourth and fifth goal constraints are to ensure a robust design in that both the maximum stress and displacement should be as insensitive as possible to reasonable changes in the loading. This is an important requirement as exact loading conditions are seldom known precisely and the load seen in actual usage may be far from the load case imposed on the problem. These constraints are computed as the change in peak stress and displacement for a 10% increase in the load. The directional uncertainty of the load was also included by adding a 10 000 unit load in the positive $x$ direction as well. The desired goal level for both these constraints is zero which says the perfect design is completely insensitive to a load change.

The sixth goal constraint seeks to locate a design in which all members are subject to as equal a stress as possible. From a safety point of view this means that no one member is more likely to fail than any other. The goal is calculated by computing the standard deviation of the stresses in all members. The desired goal value is zero which can be achieved if all members have identical stress levels.

The final goal constraint in one which is rarely considered in structural optimization but is of great practical importance. In constructing a truss structure, common sized components are beneficial from both a cost and an assembly point of view. For example consider assembling a 200-bar electrical transmission tower with 200 different sizes of beams. This constraint seeks to produce the fewest groups of elements with common cross-sectional areas. No length consideration is given but it could be imposed if desired (assuming the nodal locations are allowed to float). The constraint is formed by adding the number of elements having cross-sectional areas differing by more than 2%. The desired goal value is one which states that the best design has all elements with the same area.

Other goals are possible including one which keeps the number of truss elements as low as possible or one which maintains the peak stress at a reasonable level even if one element fails. Actually these goals are somewhat accounted for in the current formulation. The minimum number of elements is somewhat equivalent to the minimum weight constraint and the peak stress under the failure of a member would require a larger number of analysis runs for each design although the uniform stress goal has some impact on this criteria. The advantage of this formulation is that all goals may be explained in plain English and they are all computationally inexpensive to evaluate.

Each analysis run was performed using a finite element formulation for truss elements. The solution time required for each analysis of all goal constraints for a design was approximately 10 s on a SUN 4–470 workstation. This allowed many different designs to be investigated. As in the second example, a hybrid solution method was employed with the genetic code used to generate the number of truss elements and the number of different cross-sectional areas to consider and a gradient based penalty function method to perform the local search.

The goal priorities were set with the first goal having the highest priority, the second and third goals having the second priority, the forth and fifth goals having the third priority and the last two goals having the lowest priority. A number of different solutions were generated by making design trade-offs during the solution process. For some cases, the nodal locations were fixed and for others were allowed to float. When the nodal locations were allowed to float, a limit of $\pm 200$ units was imposed as a hard constraint and the ground nodes were only allowed to move in a vertical direction. This allowed for structural optimization with consideration of cross-sectional, geometric and topological design parameters.

Five separate solutions are presented in Tables 7.3 and 7.4. Compared with the results for the traditional ten bar truss design, several of the designs generated by the multiobjective approach are better in every aspect. The majority of the designs turned out to have six truss members in the configuration pictured in Fig. 7.18. This seems reasonable, given the loading applied. The stresses and displacements are lower and the weight is significantly reduced (over 50% in

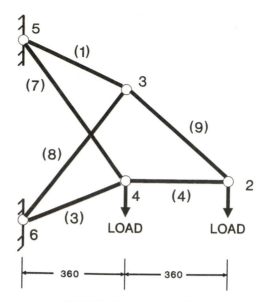

**Fig. 7.18** Six-bar truss design.

some cases). Even when only three different cross-sectional areas were used, the total weight was still quite attractive. Precise tuning of the design solution is possible, but would require a more detailed understanding of the intended application. The important point to note is that the multiobjective approach did indeed generate solutions which were far superior to the previously published results. The goal formulation allows the designer to concentrate on real design issues rather than a rigid prescribed mathematical formulation.

## 7.7 SUMMARY AND CONCLUSIONS

The need for computer-aided design tools which can have a significant impact on a design at an early stage of the design process is evident. Traditional design optimization via a single objective nonlinear programming formulation does not always service this need in the best fashion. A goal programming, multiobjective problem formulation and solution methodology have been presented and demonstrated on three design examples. The key attributes of the procedure are the ability to handle multiple, competing design objectives and that the process is applicable at an early stage of the design process where it can have the greatest impact on shaping the design solution. The hybrid approach of a genetic algorithm for handing topological choices and a traditional nonlinear programming algorithm for refining locally optimum solutions is shown to be an effective design optimization tool. The inclusion of topological design issues is important as multiobjective formulations often have many locally optimal solutions spread over the design space.

In the nonlinear goal programming formulation, the user defines a set of design objectives or goals and decides which goals have the highest priority. Goals at the same priority can be differentiated by weighting factors. Design constraints may be imposed as goals where they can be treated as hard or soft design specifications or as inequality or equality constraints where they are treated as hard constraints. Design variables involving topological decisions are separated from the rest and a genetic algorithm is applied to handle these issues. The remaining variables are manipulated by a nonlinear goal programming algorithm which is built on a gradient based penalty function method. Whether or not a feasible design space is defined has no effect on the ability to generate a solution. Starting with the highest priority goals, the algorithm satisfies them to the best extent possible. From this point, the lower priority goals are factored in and satisfied to the extent that they do not significantly effect the higher priority goals.

The user plays a key role in the optimization process, not only in defining goals, priorities and weighting factors, but in providing insight to the solution process which would be difficult or impossible to codify. Various graphical techniques may be employed to help evaluate and modify solutions. Mathematically precise multiobjective optimization methods which seek a Pareto optimal solution can make poor design trade-offs. This is due to the fact that, in most applications, design objectives or goals have priorities and increasing the goal satisfaction of a high priority goal at the expense of one or more lower priority goals may be beneficial even though the result is not a Pareto optimal point. Also in the formulation of design constraints it is important to recognize the difference between hard and soft specifications. Traditional methods deal almost exclusively with hard constraints which may again mask important design trade-offs which could be made.

The three-bar truss example was used to demonstrate that a multiple objective formulation could include criteria such as design sensitivities which in turn could lead to topological design changes with the appropriate weighting. The second example involving the design of a planar mechanism points out how topological issues can be combined with a gradient based goal programming algorithm to generate valid design solutions to an extremely difficult problem. The final example of a ten-bar truss design is used to highlight the fact that formulating and solving a conventional design optimization problem does not deal effectively with important design issues which when considered can generate a far superior design. Again the inclusion of topological issues in the goal formulation is the key to success.

The hybrid method presented does have some drawbacks. The solution time required is generally well beyond that of a traditional single or multiobjective optimization solution. Generally, many designs must be evaluated and the topological solution is nongradient based to achieve a truly global search and cannot benefit from design sensitivity information supplied by the analysis code. On the positive side, the algorithm is parallel in nature and can achieve virtually a linear time reduction with the number of processors available. Parallel machines can be configured with computing power equal to a supercomputer at a fraction of the cost, but the power of such a machine can only be accessed by software which can run efficiently in parallel. Few conventional optimization techniques can take full advantage of such a machine.

In the final analysis, it is not so much a question of how much computer time was used in generating a solution, but how good a solution was generated. The inclusion of topological issues allows a diverse set of designs to be investigated. This is critical as a poor design optimized will generally result in a poor design. The methodology presented allows for a formulation which is natural and parallels the manual design process better than conventional methods do. The approach is continually undergoing development and is still in the infancy. The potential of the method from the problems solved to date however, is very high.

# REFERENCES

Abuyounes, S. and Adeli, H. (1986) Optimization of steel plate girders via general geometric programming. *Journal of Structural Mechanics*, **14**(4), 501–24.

Adeli, H. and Balasubramanyam, K.V. (1988) A synergic man–machine approach to shape optimization of structures. *Computers and Structures*, **30**(3), 553–61.

Adeli, H. and Chompooming, K. (1989) Interactive optimization of nonprismatic girders. *Computers and Structures*, **31**(4), 505–22.

Adeli, H. and Kamal, O. (1992) Concurrent optimization of large structures: part I – algorithms. *ASCE Journal of Aerospace Engineering* (in press).

Adeli, H. and Mak, K.Y. (1989) Application of a coupled system for optimum design of plate girder bridges. *Engineering Applications of AI*, **2**, 72–6.

Bennett, J.A. and Botkin, M.E. (1981) Automated design for automotive structures. Progress on Engineering Optimization, ASME Technical Conference, Hartford, CT.

Davis, L. and Ritter, F. (1987) Schedule optimization with probabilistic search. Proceedings of the 3rd IEEE Conference on Artifical Intelligence Applications, 1987, pp. 231–6.

El Sayed, M., Ridgley, B. and Sandgren, E. (1989) Structural design by nonlinear goal programming. *Computers and Structures*, **32**, 69–73.

Evans, G.W. (1984) An overview of techniques for solving multiobjective mathematical programs. *Management Science*, **30**(11), 1268–82.

Fourman, M.P. (1985) Compaction of symbolic layout using genetic algorithms. Proceedings of International Conference on Genetic Algorithms and their Applications, 1985, pp. 141–53.

Goldberg. D.E. (1987) Computer-aided gas pipeline operation using genetic algorithms and machine learning parts I and II. *Engineering with Computers*, 35–45.

Goldberg, D.E. (1989) *Genetic Algorithms in Search, Optimization and Machine Learning*, Addison-Wesley, Reading, MA.

Goldberg, D.E. (1990) A note on Boltzman tournament selection for Genetic algorithms and population-oriented simulated annealing. *Complex Systems*, (4), 445–60.

Hwang, C.L. and Masud, A.S.M. (1979) *Multiple Objective Decision-making Methods and Applications*, Springer, Berlin.

Kuhn, H.W. and Tucker, A.W. (1951) Nonlinear programming. Proceedings of the Second Berkeley Symposium on Mathematical Statistics and Probability (ed. J. Neyman), University of California, Berkeley, CA, 1951, pp. 481–91.

Pramila, A. and Virtanen, S. (1986) Surface of minimum area by FEM. *International Journal of Numerical Methods in Engineering*, **23**, 1669–77.

Sandgren, E. and Jensen, E. (1992) Automotive structural design employing a genetic optimization algorithm. *SAE Tech. Pap. Ser. 920772*.

Sandgren, E. and Venkataraman, S. (1989) Straight line path generation for a planar robotic manipulator. Proceedings of the First National Applied Mechanisms and Robotics Conference, Cincinnati, OH, November 1989 Vol. 1, pp. 3A-2-1–3A-2-7.

Sandgren, E. and Wu, S.J. (1989) Shape optimization using the boundary element method with substructuring. *International Journal of Numerical Methods in Engineering*, **26**(9), 1913–24.

Schmidt, L.A. (1960) Structural design by symmetric synthesis. Proceedings of the Second ASCE Conference on Electronic Computation, Pittsburgh, PA, 1960, pp. 105–22.

Sheppard, K.H. and Palmer, A.C. (1972) Optimal design of a transmission tower by dynamic programming. *Computers and Structures*, **2**(4), 445–68.

Shupe, J.A. and Mistree, F. (1987) Compromise: an effective approach for the design of damage tolerant structural systems. *Computers and Structures*, **27**(3), 407–15.

Stadler, W. (1979) A survey of multicriteria optimization or the vector maximum problem. *Journal of Optimization Theory and Applications*, **29**, 1–52.

Vecchi, M.P. and Kirkpatrick, S. (1983) Global wiring by simulated annealing. *IEEE Transactions on Computer Aided Design*, **2**(4).

# 8

# Optimization of controlled structures

N.S. KHOT

## 8.1 INTRODUCTION

The topic of this chapter is to introduce the subject of designing structural and control systems simultaneously by using an integrated approach. The fields of structural optimization and optimum control have reached a stage of maturity. The methods developed under these fields have been extensively used in different engineering fields. However, the development of a design procedure to combine structural and control design has been quite recent and it is emerging as a new field in the integrated approaches of the design process. The literature shows that there are various formulations proposed by different investigators. The problem is highly nonlinear because of the nature of the objective functions and the constraints and therefore it can be solved only by using iterative numerical techniques even for a small problem with few design variables. For fairly large problems, obtaining a solution becomes computationally intensive. The optimum solutions obtained by an integrated process may be nonunique because of the nonlinearity of the problem.

The integration of the structural and control design processes is necessary for flexible structures where the effects of control forces react with the dynamic behavior of the structure and it is required to improve overall performance of the system. For rigid structures, integration of the design process might not yield any advantage. From the references cited in this chapter it can be observed that the integrated design approaches are being developed for civil engineering, mechanical engineering, and aerospace engineering applications even though the maximum work may be found related to flexible space structures. These structures must have minimum weight and need an active control system to suppress vibration during maneuvering. In an increasingly competitive world, it is not enough to design a system which can perform well but it is essential to have the best operating system, and this can be achieved only by using integrated methods, making maximum use of the available numerical design tools and the use of computers.

The conventional approach in designing a controlled structure is to optimize the structure alone and then to design a control for the optimum structural design. This process can then be reiterated until the acceptable design is obtained. If the structure is rigid, then it may not be necessary to iterate the design process because the control system would not significantly interact with the structural dynamic behavior. This can be achieved by limiting the upper frequency of the controller bandwidth to be less than the fundamental frequency of the structural vibrations. However, if the structure is flexible then the control system and structure would interact and the control system would affect the vibration characteristics of the structure and also introduce loads which would create stresses and displacements beyond the allowable limits.

The objective in structural optimization generally is to design a structure with the objective of minimizing the cost or weight with constraints on stresses, displacements, buckling load, frequency distribution, minimum and maximum size requirements etc. subjected to static and dynamic loads. The nature of the loading depends on the functionality of the structure. The design variables would be geometry, cross-sectional areas of the members and materials. In the case of the control design, the constraints would be on the closed-loop frequency distribution, active damping, total available energy, maximum output response, robustness characteristics, and other quantities affecting the performance of the control. The control design variables would be gains, number and location of actuators and sensors, and weighting parameters. A proper selection of the control approach is one of the important design considerations. In the integrated design formulation, the objective functions, constraints and design variables would be selected from both disciplines and treated at the same time.

Most of the structures at present are designed by using finite element methods leading to the large degrees of freedom. The main input to the control design from the structural response is the vibration frequencies and the modes. The control is normally designed for a small number of degrees of freedom. This makes it essential to take into consideration the effect of low order control on the higher structural vibration modes. This is known as the spillover effect.

In the case of a flexible structure, the principal source of concern is the interaction between structure and control due to the control bandwidth overlapping the structural modes. One of the solutions would be to design a stiffer structure, but this adds weight to the system. The best or optimal balance between the structural and control design is essential. To find this balance, the conventional approach of compartmental design is not efficient. A design method which simultaneously takes into consideration the objectives of the structural and control design and satisfies requirements for both disciplines is necessary.

The process of obtaining an efficient design can be considered as a mathematical optimization problem, where single or multiple objective functions are to be minimized and constraints satisfied. The number of design variables and constraints may be large depending on the complexity of the problem.

The standard constrained optimization problem can be stated as follows. Find a vector $\{a\} = (a_1\, a_2 \cdots a_{n_1})$ of $n_1$ design variables in order to minimize a cost function

$$f(\boldsymbol{a}) = f(a_1, a_2, \ldots, a_{n_1}) \tag{8.1}$$

subject to $p_1$ equality constraints

$$h_j(\boldsymbol{a}) = h_j(a_1, a_2, \ldots, a_{n_1}) = 0 \quad j = 1, \ldots, p_1 \tag{8.2}$$

and $m_1$ inequality constraints

$$g_i(\boldsymbol{a}) = g_i(a_1, a_2, \ldots, a_{n_1}) \leqslant 0 \quad i = 1, \ldots, m_1 \tag{8.3}$$

where $p_1$ is the total number of equality constraints and $m_1$ is the total number of inequality constraints. Any design problem can be transcribed into a mathematical model. The method of obtaining the solution depends on the linearity and nonlinearity of the objective function and the constraints. The sensitivities of these quantities with respect to the design variables are required. If they cannot be evaluated analytically, it may be necessary to use a finite difference scheme. Except for a few cases, the constraints can be expressed as an algebraic expression in terms of the design variables. Most of the design problems need elaborate numerical calculations before the objective function, constraints or sensitivities can be calculated. In the definition of the problem in equations (8.1)–(8.3), it was assumed that there is only one objective function to be minimized. One of the purposes of optimization is to enhance one's understanding of the real problem to be modeled.

The first step in formulating a real problem is the selection of the proper objective functions and equality and inequality constraint functions that must be used to define the design problem. In practical design problems, it may be even necessary to consider more than one objective function. In this case, methods such as Pareto optimization or goal programming might have to be used. A detailed discussion on optimization methods may be found in books by Morris (1982), Arora (1989), and Haftka, Gurdal and Kamat (1990).

The following section contains a summary of the literature published in this field. The section on analysis contains basic dynamic equations of a structure, the linear quadratic regulator (LQR) control theory and the concept of robust control. Books by D'Azzo and Houpis (1988) and Meirovitch (1989) contain in-depth discussion on the control analysis. The sensitivities of the closed-loop eigenvalues and the spectral radius are then derived. The application of the integrated approach is illustrated by solving two problems. This is followed by a short discussion on the optimization programs NEWSUMT-A (Thareja and Haftka, 1985), ADS (Vanderplaats, Sugimoto and Sprague, 1983), IDESIGN (Arora, Thanedar and Tseng, 1985), and VMCON (Crane, Hillstrom and Minkoff, 1980) available in the public domain and are frequently used for simultaneous design of the structure and control systems.

## 8.2 LITERATURE REVIEW

A literature search on the simultaneous design of the structure and control system indicates that most of the papers in this field have been published within the last fifteen years. Research work reported in these references differs in the selection of the objective functions, nature of constraints, control theory and the approximations made in developing the algorithms. Most of the research work is oriented towards understanding of the problem and solving it within a limited scope of the objective functions, number of constraints and design variables, and the type of control.

Kirsch and Moses (1977) have presented the idea of structural optimization with control. For optimization, four different types of control systems were studied: (a) passive system without control forces, (b) passive system with prestressing forces, (c) passively controlled system, and (d) actively controlled system. Each method is applied to a 'continuous' beam with intermediate springs and supports. The beam cross-sectional area is minimized subject to compatibility, equilibrium and stress constraints. The paper is intended for civil engineering beam structures.

Haftka, Martinovic and Hallauer (1984) minimized the single rate feedback viscous damping coefficient subject to limits on the real part of the closed-loop system eigenvalues and constraints on the magnitude of design variable changes. A finite element model was developed for a laboratory structure consisting of a beam and cables. Sensitivities of eigenvalues and damping rates were determined analytically with respect to changes in concentrated mass and element thicknesses. The experimental and theoretical damping and frequency values were compared qualitatively for changes in the mass.

A two-step method of determining the optimal control cost for a given set of design variables and minimizing the linear control cost by iteratively modifying the design variables was presented by Messac and Turner (1984). A modal formulation with reduced order modeling was implemented. The constant mass was the constraint requirement on the design. The method was applied to 20 extensional (only) finite elements of equal length to study the effect of reduced order modeling and to allow physical interpretation of the results. The dynamic response of the optimal design was found to be smoother than the initial design with all design variables of the same size.

Khot, Venkayya and Eastep (1984) discussed the effect of optimum structural modifications on the dynamic behavior of the ACOSS-FOUR structure with active controls. The nominal design of the ACOSS-FOUR was modified by using two optimization approaches. In the first case, the static displacements associated with the line-of-sight error were minimized by changing the distribution of the cross-sectional areas of the members for the same weight of the structure as the nominal design. In the second case, a constraint was imposed on the fundamental frequency and the weight of the structure was minimized. The dynamic responses of different optimum designs were then compared. The weighting matrix for state variables used for the LQR control is a function of the structural frequencies. The design, with a constraint on the displacements, showed improvement in the response of the structure to the initial disturbance.

Hale, Lisowski and Dahl (1985) have considered an idealized four-boom structure attached at the center to a rigid hub. The structure was modeled by finite elements to obtain an optimal design for a rest-to-rest (two-point boundary value problem) single-axis rotational maneuver (slew maneuver) due to reduced order optimal control. For the optimization formulation, a weighted nonnegative structural cost (independent of control parameters) and the state vector cost were augmented by penalties due to control derivatives. The optimum structural parameters are obtained by overall cost minimization with control forces and their derivatives treated to be independent of the parameters. The combined structural and control optimization was solved iteratively for the finite time maneuver.

A set-theoretic analysis procedure was described by Hale (1985) for the integrated structural–control synthesis. The procedure approximates box (absolute magnitude) constraints on the state vector, output, and control forces by ellipsoidal sets. Matrices in the ellipsoidal sets are governed by a matrix differential equation; the steady state solution is correspondingly solved by a Lyapunov equation. The input disturbance is assumed to be bounded with an amplitude bound. The set-theoretic analysis procedure providing steady state bounds directly replaces the computations required in a transient analysis procedure. The procedure was applied to a single finite element with structural and nonstructural end masses undergoing only axial motion. The design variable was chosen as the cross-sectional area which is analytically related to the first flexible natural frequency.

Bodden and Junkins (1985) systematically presented the pole placement type eigenvalue optimization formulation for a linear output feedback including the position, velocity and acceleration measurements. The sensitivities – first and second derivatives – of the closed-loop eigenvalues were obtained. The method was applied to the Draper–RPL configuration with four flexible beams attached to a hub at the center. A constrained optimization strategy was developed. Three cases of eigenvalue placement were considered. The objective of all the cases was to minimize the control gains subjected to constraints.

Structural optimization of an optimally controlled structure was performed by Khot, Venkayya and Eastep (1986) using an LQR control law and the VMCON optimization subroutine based on Powell's algorithm. The objective was to minimize the structural weights with equality constraints on the closed-loop damping ratio and lower bounds on cross-sectional areas. The procedure was applied to a two-bar and the ACOSS-FOUR trusses. The gradients of the weight were determined analytically, whereas those of the constraints were evaluated by using a finite difference scheme. The line-of-sight (LOS) error and performance index (PI) time histories show considerable improvements as a result of optimally controlling an optimum structure. Lower modes are more heavily damped.

Eastep, Khot and Grandhi (1987) studied the effects of different percentages of passive damping on an actively controlled structure with minimum weight objective with specific application to the ACOSS-FOUR model. Imaginary parts of the closed-loop eigenvalues and closed-loop modal damping factors for LQR modal formulation were constraints. Analytical sensitivities were incorporated in NEWSUMT-A optimization software. The results include the LOS time histories for various values of passive damping. The effect of percentage damping on the minimum performance index shows an inverse relationship.

A study was performed for weight and/or Frobenius norm minimization on a two-bar and the ACOSS-FOUR trusses by Khot, Grandhi and Venkayya (1987). Constraints were imposed on the damping factors and imaginary part of the closed-loop eigenvalues. Analytical sensitivity expressions were used in the VMCON, NEWSUMT and IDESIGN optimization routines. Comparative evaluation of these routines was tabulated. The main objective of this paper was to indicate the different performance of various optimization packages when used as black boxes. NEWSUMT-A was found to give a better optimum.

Lim and Junkins (1987) presented a design algorithm using robustness measures. The three cost functions, total mass, stability robustness and eigenvalue sensitivities were minimized with design variables which included structural as well as control parameters. The locations of actuators and sensors were also considered as design variables. The robustness measure proposed by Patel and Toda was used. The optimization algorithm based on the homotopy approach with sequential linear programming algorithms was utilized. A hypothetical structure consisting of a free–free flexible beam with rigid body attached to the center of the beam by a pinpoint and torsional spring was designed with different objective-functions and constraints.

Miller and Shim (1987) performed minimization of combined structural mass and controlled system energy. The objective functions were weighted by positive scalar weights to balance the dimensions. The major emphasis was on the gradient projection and penalty function technique. A ten-bar cantilevered truss was designed for two initial conditions and different constraints on the lowest open loop frequency.

The actual cost of the structure and control system was considered as a non-dimensional objective function by Onoda and Haftka (1987) with constants representing a normalizing mass and weightings. A beam-like spacecraft experiencing a steady state white noise disturbance force was optimized for stiffness distribution, location of controller and control gains. The NEWSUMT-1 program was employed for the unconstrained minimization. The sensitivity derivatives were calculated by finite differences. Both noninertial and inertial disturbances were chosen. The results showed improvements as a result of the simultaneous optimization.

Bendsoe, Olhoff and Taylor (1987) formulated a system design problem based on the criterion to minimize the spillover effect with respect to the stiffness and distribution and variations in feedback gains and actuator positions. Examples of simple mass–spring–damper systems were analytically solved to illustrate the basic approach of structural and control design. It was shown that in the distributed parameter structure the technique of modal control gave rise to design problems which are similar to eigenfrequency optimization. The sensitivities were calculated by using analytical expressions.

Junkins and Rew (1988) discussed the sources of uncertainties or imperfectness in the modeling process. These are (1) ignorance of actual properties, geometric parameters and boundary conditions, (2) discretization and truncation error, (3) ignoring the nonlinearities, (4) ignorance of disturbances – random or deterministic, and (5) sensors and actuator modeling. With this background, the robust eigenstructure assignment methods were discussed. A multicriterion approach was presented to study the trade-offs between robustness and competing performance measures.

Mesquita and Kamat (1988) considered the effect of structural optimization on the improvement of the control design. A laminated composite plate structure was optimized with constraints on the frequency distribution and the structural weight. Two optimization objectives were considered. In one case the fundamental frequency was maximized with separation constraints on other frequencies. In the second case the sum of the square of the differences between the structural frequencies and a certain target frequency was minimized. The design variables were the number of plies in a specified orientation and the areas of the stiffeners. The problem was solved with VMCON.

Belvin and Park (1988) showed that the LQR control with stiffeners and mass dependent weighting matrices to the state and control variable and the full rank actuator influence matrix have a closed-form solution to the Riccati equation. The (undamped) approximate cost function was explicitly stated in terms of structural parameters, two scalar gain variables and the assumed model force coefficients. This objective function was found to be inversely proportional to the frequency cubed suggesting that the low frequency modes contribute most to the cost index. Minimization of the function with constraints on mass was considered for a clamped–pinned beam and a cantilever beam model The constrained optimization was performed using the ADS program.

Khot (1988a) presented two weight minimization problems differing in the constraints and the structure used for the analysis. One, with constraints on the first two closed-loop eigenvalues and four constraints on the closed-loop damping, was applied to a tetrahedral truss, and another with constraints on closed-loop eigenvalues as well as the control gain norm, specifically the Frobenius norm, was applied to a two-bar truss.

Khot (1988b) considered minimization of weight of the structure subjected to constraints on closed-loop modal damping parameter and imaginary part of the closed-loop systems is achieved. A modal formulation with LQR control law is used. Structural optimization is performed by using VMCON routine. Sensitivity derivatives required for the procedure are evaluated analytically. Comparison of nominal and optimum integrated design is made by observing LOS transient response and performance index time history. In addition, frequencies, closed-loop eigenvalues, and the modal damping parameters are also compared.

Khot *et al.* (1988) developed a simultaneous structural–control optimization procedure in which weight as well as the control effort measure, Frobenius norm, were minimized subject to constraints on the closed-loop eigenvalues, with lower bound on cross-sectional areas and upper bound on the norm. A linear full state feedback was used with the state space formulation. Cross-sectional areas and gains were considered as design variables to be evaluated by the simultaneous minimization of the weight and Frobenius norm. The gradients of the objective functions and the constraints were evaluated analytically. The procedure was applied to a two-bar truss.

Khot and Grandhi (1988) discussed the existence of a nonunique solution to this highly nonlinear problem in this paper. The minimization of weight or Frobenius norm was conducted with the LQR control law and modal state space form. Four different optimum designs of the ACOSS-FOUR model were obtained differing in the constraints and objective function. It was indicated that different designs could be obtained satisfying all the constraints with different distributions of design variables. NEWSUMT-A was used as a black box for optimization.

A combined optimization of the structure and control system was presented by Salama *et al.* (1988) with the sum of the weight and the control system performance index as an objective function. An LQR optimal control was scaled to match the dimensions of the objective function for combined optimization. With this operation, the problem is locally converted to the minimization of a function with only the structural design variables present. The constrained optimization was converted to the unconstrained optimization by the use of Lagrange multipliers. The gradients are all evaluated analytically. The procedure was applied to a cantilever beam with a tip control force.

Cheng and Pantelides (1988) discussed structural optimization with (i) non-optimal control, (ii) optimal control, (iii) critical mode control, and (iv) optimal location of controllers applied to the civil engineering/building models. The control was achieved through the active tendons and active mass damper, and the structure was subjected to a random earthquake acceleration. For the non-optimal control, a transfer matrix approach was used in the frequency domain. The optimal control–structural optimization implementation scheme was classified into (i) instantaneously open loop in which the controller nullifies the earthquake excitation in the time domain, (ii) instantaneous closed loop in which the control forces are regulated by the full state feedback at the instant, and (iii) instantaneous open–closed loop control in which the effect of excitation is nullified and the instantaneous control-response performance index is minimized. The problem of weight minimization with the displacement and control forces was solved for an eight-story building using the three different instantaneous controls.

Rao (1988) considered the problem of combined structural and control optimization with different objective functions. The objective functions selected for these parametric studies were structural weight, quadratic performance index, Frobenius norm and the effective damping response time. Designs for a two-bar truss and ACOSS-FOUR were obtained by minimizing one objective function and constraining other functions. The sensitivities of the objective functions to the design variables were investigated. The optimization was carried out by using the VMCON subroutine.

A cooperative game theory approach was described by Rao, Venkayya and Khot (1988) for a combined structure–control optimization problem. The multiobjective functions were structural weight, the quadratic performance index for LQR, the Frobenius norm of the control effort, and the effective damping response time. The objective was to obtain a Pareto-optimal set combining these functions for specified constraints on the closed-loop damping ratios and bounds on the cross-section areas of truss members. The method was applied to a two-bar truss and a twelve-bar ACOSS-FOUR truss. The required sensitivities were determined numerically. The optimization was carried out by using VMCON.

Structural and controller parameter optimization was considered by Lim and Junkins (1989) in the presence of system uncertainty with three independent cost functions. These were expressed as (i) eigenvalue sensitivity minimization, (ii) stability robustness maximization, and (iii) minimization of mass. Direct output feedback controllers were exercised on a rigid body with an attached flexible beam. The eigenvalue sensitivity objective function was expressed as the weighted quadratic sum of the individual eigenvalue sensitivities. The upper bound of the robustness measure was defined as the minimum of the negative real part of the closed-loop state matrix eigenvalues. The design variables were the thickness of the flexible beam and the mass density of the rigid body. The constraints were imposed on the actuator location plant parameters, eigenvalues and local step size allowables for the structural parameters.

Manning and Schmit (1990) considered the composite objective function involving the structural weight and control effort wth behavior constraints and constraints on weight and control effort separately for the two cases. A second-order modal formulation including excitation forces and control forces was used and the equations were integrated by the Wilson-O explicit time integration method. The first two cases of the planar grillage involved weight minimization with control effort constraint, whereas the last two were for control effort minimization with constraint on weight. The constraints were imposed on the displacements and accelerations at some locations. The limitations on dynamic response and actuator force levels were satisfied by treating them as direct behavior constraints.

Onoda and Watanabe (1989) proposed a method to design an LQG type optimal controller with regulator and observer the reduce the spillover effect. A method was proposed to design an LQG type controller with a regulator and observer by minimizing the performance index involving controlled modes and residual modes. The approach consisted of two steps; first, the maximum value of the real parts of the eigenvalues of the closed-loop matrix is minimized, then the minimization of the performance index is carried out. The integrated approach to design the control and structural system was presented in the end where a total cost objective function, which represents the sum of the control cost and structural cost, was minimized with constraints on the response amplitude. The design variables were the observer and regulator gains, and the cross-sectional areas.

Thomas and Schmit (1989) studied the effect of noncollocated sensor–actuator pairs by applications to a cantilever beam and the Draper–RPL (four beams with central hub) structure. A direct output feedback type of controller was used which essentially modifies the effective damping and stiffness properties of the structure leading to a free vibration type of problem. The minimization of either mass or the control effort was performed along with the constraints on the control force, open-loop frequency, closed-loop eigenvalues, dynamic displacements and the damping ratio, appropriately chosen for different cases. Structural parameters and control gains were treated as the design variables.

Grandhi (1989) studied the performance of optimization programs VMCON, NEWSUMT and IDESIGN to design the ACOSS-FOUR structure with constraints on closed-loop eigenvalues and damping. The objective function was the weight of the structure. The objective of this paper was to study the performance of various optimization programs.

Rao, Pan and Venkayya (1990) discussed a multiobjective optimization for improving the robust characteristics of an actively controlled structure. This was achieved by structural modifications with a modal space formulation and LQR control law. Both the performance and stability robustness were considered and appropriate measures defined. The objective functions chosen were the stability robustness, the performance robustness, and the total structural weight which were simultaneously optimized.

Haftka (1990) presented an overview of some of the relevant issues of the integrated structure–control optimization of space structures, namely the accuracy of the sensitivity derivatives, effects due to reduced ordered models, different objective functions and integrated design formulations.

Thomas and Schmit (1990) proposed the integrated design problem for the objective function equal to the weighted sum of the masses, a sum of the amplitudes of the dynamic responses and the sum of the amplitude of the control forces. A noncollocated direct output feedback control was applied to a cantilever beam and a structure having four arms and a central rigid hub. The approximate optimization problem was solved using CONMIN. The design variables used were the thicknesses of different elements, concentrated masses and the controller gains and positions.

Padula *et al.* (1990) applied controls–structures optimization methodology to a general space structure synthesizing the analysis, optimization and control software. The problem was decomposed into a structural optimization, control optimization and system optimization in which both the optimal structural and control steps were included leading to a minimization of mass and power consumption. The constraints included limits on the frequency spacing, fundamental bending frequency, buckling loads, damping factors etc. The design variables were treated at different levels.

Zeiler and Gilbert (1990) proposed a method based on Sobieski's multilevel decomposition approach. This decomposed the problem into subsystem optimization and integrated the effects for the structure–control optimization. The formulation incorporated a structural nodal form and LQR control. The ADS optimization code was used for the top level and structural subsystem, and control law subsystem optimization was achieved using the ORACLS software. The method was applied to the ACOSS-FOUR structure to investigate the feasibility of the proposed method.

Grandhi, Haq and Khot (1990) considered the optimization of structural parameters and control gains (LQR) by maximizing the stability robustness measure in the presence of parametric uncertainty. The measure was proposed by Qiu and Davidson and is based on the closed-loop stability under structured perturbations. The inverse of the robustness was chosen to be proportional to the maximum spectral radius of a matrix related to the closed-loop nominal matrix. The NEWSUMT-A optimization software was used with the finite difference approach to calculate the sensitivities.

Sepulveda and Schmit (1990) formulated a standard form of the modal state space equations with an optimal regulator and observer, and time-dependent external forces. A composite objective function was decomposed into a control subproblem along with a standard structural optimization on the outer loop. The multiobjective functions involving target values of mass control efforts and number of actuators was considered. The optimal placement of actuators and sensors was determined by the (0, 1) design variables in the control input and measurement matrices. The concepts were applied to a cantilever beam problem and an antenna structure.

McLaren and Slater (1990) solved the problem of minimizing the mass of the structure subject to a set of prescribed stochastic disturbances with limitations on the available control energy and a set of allowable output responses. The three controller types considered were (1) direct output feedback control, (2) output filter feedback control, and (3) positive real output feedback control. The mathematical optimization method based on continuation methods to impose nonlinear constraints coupled with sequential linear programming was used to solve the illustrative problems.

Khot and Veley (1990) considered integrated structure–control optimization under the presence of structured uncertainties in the plant matrix. A full state LQR control law was implemented with the formulation given in modal form. The sensitivity of various quantities with respect to the design variables was expressed analytically. Weight minimization, with constraints on the imaginary part of closed-loop eigenvalues, damping factors, spectral radii bound, and cross-sectional areas, was performed by using the NEWSUMT-A software. The method was applied to the tetrahedral truss.

Suzaki and Matsuda (1990) applied a goal programming approach to an active flutter suppression system to perform combined structure–control optimization. A typical wing section with the control surface undergoing a plunging pitching motion was modeled in an unsteady aerodynamic field and a random gust velocity disturbance. A sensitivity analysis for all the dependent variables such as eigenvalues, gains and covariance matrix with respect to the design variables was performed analytically. The design variables were the design airspeed, elastic axis to mass center distance, hinge line to control surface mass center, elastic axis location and the hinge line location. The constraints were imposed on the open-loop characteristics, closed-loop characteristics, control surface deflection angle and structural design changes.

Grandhi (1990) performed weight minimization of the ACOSS-FOUR subject to constraints on the imaginary part of closed-loop eigenvalues and closed-loop modal damping parameters. The objective was to study the quadratic performance index, optimum weight, and active control effort as a function of the passive damping. The range of passive damping chosen was 0–10%.

A model error sensitivity suppression method was implemented for a structure–control optimization problem by Khot (1990) with weight as the primary objective function. The reduced order LQR optimal control problem was then solved for weight minimization with constraints on the closed-loop damping factors, frequencies and structural parameters. The NEWSUMT-A computer package was used to solve the constrained minimization problem. The sensitivities of the weight, damping factor, closed-loop eigenvalues, closed-loop matrix, Riccati matrix, structural frequencies and eigenvectors were evaluated analytically.

Khot and Veley (1991) studied the effect of structured uncertainties on the performance of an optimally controlled closed-loop structure. Weight minimization with constraint on the imaginary part of the closed-loop eigenvalues and on the spectral radius was achieved by using the NEWSUMT-A program. The cross-sectional areas were considered as structural design variables and parameters multiplying the weighting matrices in the quadratic performance index were treated as control design variables. Sensitivity expressions with respect to these design variables were calculated analytically.

Dracopoulos and Öz (1992) proposed an integrated approach for the aeroelastic control of laminated composite lifting surfaces. The lifting surface was modeled as a laminated composite cantilevered plate in an airstream. The geometry of the plate and the ply thickness were assumed constant with ply orientation angles as the structural design variables, and the control design speed was used as a control variable. The optimization problem was solved by using optimization routines, IDESIGN and NEWSUMT-A. The problem formulation incorporated the Rayleigh–Ritz energy method and two-dimensional incompressible unsteady aerodynamic theory.

## 8.3 ANALYSIS

The basic equations needed for the dynamic analysis of a structure, control design, and sensitivities of the response functions are derived in this section.

### 8.3.1 Dynamic equations of motion

The dynamic equations of motion for an elastic system with finite degrees of freedom with no external disturbance can be written as

$$\mathbf{M}\ddot{u} + \mathbf{E}\dot{u} + \mathbf{K}u = \mathbf{D}f \qquad (8.4)$$

where **M** is the mass matrix, **E** is the proportional damping matrix, and **K** is the total stiffness matrix. These matrices are $n = n$ where $n$ is the number of degrees of freedom of the structure. In equation (8.4), $u$ and $f$ are the displacement and control force vectors of dimensions $n$ and $p$ respectively, and are functions of time. The matrix **D** is $n \times p$ and its elements consists of direction cosines and zeros. This matrix relates the control force vector $f$ in its local coordinates to the global coordinate system in which the equilibrium equations are written.

Introducing the coordinate transformation

$$u = [\phi]\eta \tag{8.5}$$

where $\eta$ is the modal coordinates and $[\phi]$ is the $n \times n$ modal matrix, equation (8.4) can be transformed into $n$ uncoupled equations. The vector $\eta$ is a function of time. The columns of the square matrix $[\phi]$ are the normalized eigenvectors of the homogeneous set of equations obtained by setting the right-hand side equal to zero in equation (8.4) and assuming the dampling matrix **E** is also zero. The uncoupled equations can be written as

$$\bar{\mathbf{M}}\ddot{\eta} + \bar{\mathbf{E}}\dot{\eta} + \bar{\mathbf{K}}\eta = [\phi]^{\mathrm{T}}\mathbf{D}f \tag{8.6}$$

where

$$\bar{\mathbf{M}} = \mathbf{I} = [\phi]^{\mathrm{T}}\mathbf{M}[\phi] \tag{8.7}$$

$$\bar{\mathbf{E}} = [2\zeta\omega] = [\phi]^{\mathrm{T}}\mathbf{E}[\phi] \tag{8.8}$$

$$\bar{\mathbf{K}} = [\omega^2] = [\phi]^{\mathrm{T}}\mathbf{K}[\phi] \tag{8.9}$$

The matrices $\bar{\mathbf{M}}$, $\bar{\mathbf{E}}$ and $\bar{\mathbf{K}}$ are diagonal square matrices. $\omega$ is the vector of structural frequencies which is equal to the eigenvalues of the homogeneous equations of motion. $\zeta$ is the vector of modal damping. Equations (8.6) are second-order uncoupled equations.

### 8.3.2 Control analysis

A system in control theory is defined as an assemblage of components that act together to perform a certain function. The dynamics of the system is the response of the system to the specified disturbance. The system is also known as a plant and the disturbance is termed as input and the response of the system as an output. A typical uncontrolled system which generally occurs in nature is shown in Fig. 8.1, where we have no control of the output. A good example of the uncontrolled system is the behavior of a building during an earthquake, where the building represents the plant and the motion of the foundation is the input. In a controlled system, shown in Fig. 8.2, the desired output acts as an input to the controller and the output from the controller acts as an input to the plant.

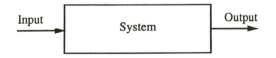

**Fig. 8.1** Uncontrolled system.

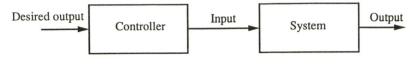

**Fig. 8.2** Controlled system.

A good example of this open-loop control system is a heating system in a building which is set to start or shut down at specific times without consideration of the temperature inside the building. A feedback or closed-loop heating system is one in which starting and shutting down of the system is controlled by temperature sensors. The closed-loop system is shown in Fig. 8.3.

In control theory, the equations describing the behavior of the system are written in terms of state variables. The state of a system is a mathematical statement containing state variables, such that the initial values of these state variables and the system inputs at a specified time are sufficient to uniquely describe the behavior of the system in the future. The state equations of the system are a set of first-order differential equations equal to the number of independent states.

The second-order uncoupled equations (8.6) can be reduced to first-order equations by using the transformation

$$x_{2n} = \begin{bmatrix} \eta \\ \dot{\eta} \end{bmatrix}_{2n} \tag{8.10}$$

where $x$ is the state variable vector of size $2n$. This gives

$$\dot{x} = \mathbf{A}x + \mathbf{B}f \tag{8.11}$$

where $\mathbf{A}$ is a $2n \times 2n$ plant matrix and $\mathbf{B}$ is a $2n \times p$ input matrix and $f$ is $p \times 1$ control vector. The plant and input matrices are given by

$$\mathbf{A} = \begin{bmatrix} \mathbf{0} & \mathbf{I} \\ \hline -\omega^2 & -2\zeta\omega \end{bmatrix} \tag{8.12}$$

$$\mathbf{B} = \begin{bmatrix} \mathbf{0} \\ \hline \boldsymbol{\phi}^T\mathbf{D} \end{bmatrix} \tag{8.13}$$

In equation (8.12) $\mathbf{0}$ is a null matrix and $\mathbf{I}$ is a diagonal matrix with elements equal to unity. The bottom left quadrant consists of diagonal elements equal to

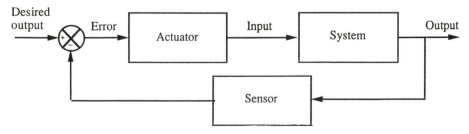

**Fig. 8.3** Closed-loop system.

the negative of squared frequencies. The right bottom quadrant also contains only diagonal elements equal to the negative of the product of the frequencies and associated modal damping parameters multiplied by 2. In equation (8.13) **0** is also a null matrix. Equation (8.11) is known as the state input equation.

In a closed-loop system, the control forces depend on the system behavior. The behavior can be interpreted as measured outputs. The relation between the output vector $y$ of dimension $q$ and input vector $x$ can be expressed as

$$y = \mathbf{C}x + \mathbf{D}f \tag{8.14}$$

where **C** and **D** are $q \times 2n$ and $q \times p$ matrices respectively. Most of the time the output does not depend on the input so that equation (8.14) reduces to

$$y = \mathbf{C}x \tag{8.15}$$

If the number of sensors measuring $y$ and actuators generating forces $f$ are equal and they are collocated, then $q = p$ and

$$\mathbf{C} = \mathbf{B}^{\mathrm{T}} \tag{8.16}$$

A system is completely state controllable if for any initial time, each initial state can be transferred to the final state in a finite time with an unconstrained input vector. Mathematically, a system is completely controllable if the rank of the controllability matrix

$$\mathbf{C} = [\mathbf{B} \vdots \mathbf{AB} \vdots \mathbf{A}^2\mathbf{B}^1 \vdots \cdots \vdots \mathbf{A}^{n-1}\mathbf{B}] \tag{8.17}$$

has $n$ independent columns implying it is of rank $n$. A system is considered to be observable if every initial state can be exactly determined from the measurements of the output over the final time. Similar to the mathematical definition of controllability, a system is completely observable if the rank of the observability matrix defined as

$$\phi = [\mathbf{C}^{\mathrm{T}} \vdots \mathbf{A}^{\mathrm{T}}\mathbf{C}^{\mathrm{T}} \vdots (\mathbf{A}^{\mathrm{T}})^2\mathbf{C}^{\mathrm{T}} \vdots \cdots \vdots (\mathbf{A}^{\mathrm{T}})^{n-1}\mathbf{C}^{\mathrm{T}}] \tag{8.18}$$

has $n$ independent columns implying $\phi$ is of rank $n$.

The stability of the system is determined by the real part of the eigenvalues of the plant matrix for the open-loop and closed-loop system. The real eigenvalues can be considered as complex eigenvalues with zero imaginary parts. In order for the system to reach stable equilibrium, all eigenvalues must have negative real parts. Even if one eigenvalue has a positive real part, the system would be unstable. A motion in the $x$ directon cannot be controlled by a force in the $y$ direction is a simple example of controllability. Similarly, a motion in the $x$ direction cannot be detected by a sensor in the $y$ direction illustrates the condition of observability.

### 8.3.3 Optimal control

The objective of optimal feedback control is to determine the control which can transfer the system from initial to final state minimizing the performance index (PI) defined as

$$J = \int_0^\infty (x^{\mathrm{T}}\mathbf{Q}x + f^{\mathrm{T}}\mathbf{R}f)\,\mathrm{d}t \tag{8.19}$$

where $\mathbf{Q}$ and $\mathbf{R}$ are the state and control weighting matrices respectively. The matrix $\mathbf{Q}$ must be positive semidefinite $(x^T\mathbf{Q}x \geqslant 0)$ and $\mathbf{R}$ must be positive definite $(f^T\mathbf{R}f > 0)$. The dimensions of $\mathbf{Q}$ and $\mathbf{R}$ depend on the size of vectors $x$ and $f$ respectively. The selection of the elements of $\mathbf{Q}$ and $\mathbf{R}$ determine the amount of closed-loop damping and the time required to control the disturbances. The performance index represents a compromise between minimum error and minimum energy criteria. The result of minimizing the performance index and satisfying equation (8.11) gives the state feedback control law

$$f = -\mathbf{G}x \qquad (8.20)$$

where $\mathbf{G}$ is the optimum gain matrix given by

$$\mathbf{G} = \mathbf{R}^{-1}\mathbf{B}^T\mathbf{P} \qquad (8.21)$$

where $\mathbf{P}$ is a positive definite matrix called the Riccati matrix and is obtained by the solution of the algebraic Riccati equation

$$0 = \mathbf{Q} + \mathbf{P}\mathbf{A} + \mathbf{A}^T\mathbf{P} - \mathbf{P}\mathbf{B}\mathbf{R}^{-1}\mathbf{B}^T\mathbf{P} \qquad (8.22)$$

The open-loop system is given by

$$\dot{x} = \mathbf{A}x \qquad (8.23)$$

and the closed-loop system is given by

$$\dot{x} = \bar{\mathbf{A}}x \qquad (8.24)$$

where

$$\bar{\mathbf{A}} = \mathbf{A} - \mathbf{B}\mathbf{G} \qquad (8.25)$$

Equation (8.24) can be obtained by substituting equation (8.20) in equation (8.11). The eigenvalues of the closed-loop matrix $\bar{\mathbf{A}}$ are a set of complex conjugate pairs written as

$$\lambda_i = \tilde{\sigma}_i \pm j\tilde{\omega}_i \qquad i = 1,\ldots,n \qquad (8.26)$$

where $j = (-1)^{1/2}$. The sign of $\tilde{\sigma}_i$ must be negative for all $i$ in order to make the system asymptotically stable. The closed-loop damping factors are given by

$$\xi_i = -\frac{\tilde{\sigma}_i}{(\tilde{\sigma}_i^2 + \tilde{\omega}_i^2)^{1/2}} \qquad (8.27)$$

The magnitudes of $\xi_i$ associated with each mode determine the extent to which those modes are damped.

For the initial condition $x(0)$, the solutions to equations (8.23) and (8.24) are given by

$$x(t) = \exp(\mathbf{A}t)x(0) \qquad (8.28)$$

and

$$x(t) = \exp(\bar{\mathbf{A}}t)x(0) \qquad (8.29)$$

where

$$\exp(\mathbf{A}t) = 1 + \frac{\mathbf{A}t}{1!} + \frac{(\mathbf{A}t)^2}{2!} + \cdots \qquad (8.30)$$

and

$$\exp(\bar{\mathbf{A}}t) = 1 + \frac{\bar{\mathbf{A}}t}{1!} + \frac{(\bar{\mathbf{A}}t)^2}{2!} + \cdots \qquad (8.31)$$

Equations (8.28) and (8.29) can be used to determine the transient response of the open-loop and closed-loop systems.

The performance index for the given initial condition $x(0)$ is given by

$$PI = x(0)^T p x(0) \qquad (8.32)$$

The behavior of the control system depends on the distribution of the closed-loop eigenvalues and the damping parameters. For a structural–control system, these values are functions of the weighting matrices $\mathbf{Q}$ and $\mathbf{R}$ of optimal control and the distribution of structural frequencies in the $\mathbf{A}$ matrix which in turn depends on the distribution of the stiffness of the structure or the cross-sectional areas of the members.

### 8.3.4 Measure of robustness

The control design is called robust if it can perform in an acceptable manner in the presence of uncertainties. The uncertainties are generally divided into two categories, structural and unstructural. The structural uncertainties are due to the variation in the elements of the plant due to the inaccuracies of the structural frequencies and assumed passive damping. These may be due to the elastic properties of the material used in the formulation of the stiffness matrix or the assumptions made in the derivation of the mass and damping matrices. The unstructured uncertainties are due to the unmodeled dynamics. This is due to the control system being designed for a reduced-order system and higher structural frequencies being ignored in the assembly of the plant matrix. In this chapter, we will discuss one of the measures of the structural uncertainty. A discussion on the effect of unmodeled dynamics on the behavior of the structural and control system is beyond the scope of this chapter.

The perturbed closed-loop state space model can be written as

$$\dot{x} = (\bar{\mathbf{A}} + \varepsilon)x \qquad (8.33)$$

where $\bar{\mathbf{A}}$ is a stable matrix and $\varepsilon$ is a perturbation matrix. The perturbed system would be stable if the elements $\varepsilon_{ij}$ of matrix $\varepsilon$ satisfy the relation

$$\varepsilon_{ij} < \frac{1}{\sup_{p \geqslant 0} \rho[|(jp\mathbf{I} - \bar{\mathbf{A}})^{-1}|\mathbf{U}_e]} U_{e_{ij}} = \frac{1}{\rho_s} U_{e_{ij}} \qquad (8.34)$$

according to the robustness measure proposed by Qiu and Davison (1986). In Equation (8.34) $|(\cdot)|$ denotes an absolute matrix and $\rho[\cdot]$ denotes the spectral radius of the matrix $[\cdot]$. $U_{e_{ij}}$ are the elements of the perturbation identification matrix $\mathbf{U}_e$. SUP represents the supremum of the matrix over a range of $p$. The elements of $\mathbf{U}_e$ have assigned values depending on the relative perturbations allowed for those elements of $\bar{\mathbf{A}}$. The spectral radius is equal to the maximum modulus of the complex eigenvalues for a specified operating frequency $p$. The maximum spectral radius value amongst all possible values of $p$ gives the critical value of $p_s$. The peaks of the spectral radius for small values of damping occur at the values of $p$ equal to the modulus of the closed-loop frequencies of $\bar{\mathbf{A}}$.

The elements of the perturbation matrix $\varepsilon$ are proportional to the elements of the perturbation identification matrix $\mathbf{U}_e$ and inversely proportional to the spectral radius $\rho_s$. The matrix $\mathbf{U}_e$ can be defined as the normalized closed-loop plant matrix $\bar{\mathbf{A}}$. This gives

$$U_{e_{ij}} = \frac{|\bar{A}_{ij}|}{|\bar{A}_{pq}|} \qquad i > n \tag{8.35}$$

where $|\bar{A}_{pq}|$ is the absolute value of a specific element of $\bar{\mathbf{A}}$. The ratio of $\varepsilon_{ij}$ and $|\bar{A}_{ij}|$ defined as $\varepsilon_r$ for all the elements is given by

$$\varepsilon_r = \frac{1}{\rho_s |\bar{A}_{pq}|} \tag{8.36}$$

The percent allowable deviation in the elements of $\bar{\mathbf{A}}$ is $100\varepsilon_r$. The percent allowable deviation is independent of the selection of the element $|\bar{A}_{pq}|$.

## 8.4 SENSITIVITY CALCULATIONS

Sensitivity analysis is the determination of changes in the response functions due to changes in the design variables. This can be achieved either by using the finite difference approach which is generally computationally expensive or by calculating partial derivatives. The sensitivities are needed using most optimization programs. If these gradients cannot be calculated by using analytical expressions, then the finite difference approach can be used. The behavior of the structural and control system depends on the eigenvalues and damping parameters of the closed-loop system defined in equations (8.26) and (8.27). In this section, the expressions for the sensitivities of these functions with respect to the structural design variables are given. The structural design variables are the cross-sectional areas of the members. Any change in the cross-sectional areas affects the mass and stiffness matrices in equation (8.4), and thus the eigenvalues $\omega_j$ and eigenvectors $\phi$ of the homogeneous set of equations. This change alters the plant matrix $\mathbf{A}$ and the input matrix $\mathbf{B}$ in the state equation (equation (8.11)).

The sensitivity of the closed-loop eigenvalues $\lambda_i$ with respect to the design variables $A_l$ is given by

$$\lambda_{i,l} = \boldsymbol{\beta}_i^T \bar{\mathbf{A}}_{,l} \boldsymbol{\alpha}_i \tag{8.37}$$

The closed-loop matrix given in equation (8.25) can be rewritten as

$$\bar{\mathbf{A}} = \mathbf{A} - \mathbf{XP} \tag{8.38}$$

where

$$\mathbf{X} = \mathbf{BR}^{-1}\mathbf{B}^T \tag{8.39}$$

In equation (8.37) $\boldsymbol{\beta}_i^T$ and $\boldsymbol{\alpha}_i$ are the left-hand and right-hand eigenvectors of $\bar{\mathbf{A}}$ defined by the solution to the eigenvalue problems

$$\boldsymbol{\beta}_i^T \bar{\mathbf{A}} = \lambda_i \boldsymbol{\beta}_i^T \tag{8.40}$$

and

$$\bar{\mathbf{A}} \boldsymbol{\alpha}_i = \lambda_i \boldsymbol{\alpha}_i \tag{8.41}$$

respectively. The eigenvectors $\boldsymbol{\beta}_i^T$ and $\boldsymbol{\alpha}_i$ are normalized such that

$$\boldsymbol{\alpha}_i^T \boldsymbol{\alpha}_i = 1 \tag{8.42}$$

and

$$\boldsymbol{\beta}_i^T \boldsymbol{\alpha}_i = 1 \qquad (8.43)$$

These eigenvectors satisfy the following relations

$$\boldsymbol{\alpha}^T \boldsymbol{\beta} = \mathbf{I} \qquad (8.44)$$

$$\boldsymbol{\beta}^T \bar{\mathbf{A}} \boldsymbol{\alpha} = \boldsymbol{\lambda} \qquad (8.45)$$

$$\boldsymbol{\beta}^T \boldsymbol{\alpha} = \mathbf{I} \qquad (8.46)$$

$$\boldsymbol{\alpha}^T = \boldsymbol{\beta}^{-1} \qquad (8.47)$$

where $\boldsymbol{\beta}^T$ and $\boldsymbol{\alpha}$ are the matrices with each column equal to the eigenvector of the corresponding eigenvalue problem. $\mathbf{I}$ is the identity matrix and $\boldsymbol{\lambda}$ is the square diagonal matrix with elements equal to the eigenvalues.

The partial derivatives of $\bar{\mathbf{A}}$ with respect to the structural design variables $A_l$ can be obtained by differentiating equation (8.38) with respect to $A_l$. This gives

$$\bar{\mathbf{A}}_{,l} = \mathbf{A}_{,l} - \mathbf{X}_{,l}\mathbf{P} - \mathbf{X}\mathbf{P}_{,l} \qquad (8.48)$$

The matrices $\mathbf{A}$ and $\mathbf{X}$ in this equation are functions of the structural frequencies $\omega_j$ and the modal matrix $\boldsymbol{\phi}$. The sensitivities of $\omega_j^2$ and $\phi_j$ are given by

$$\omega_{j,l}^2 = \frac{1}{A_l} \boldsymbol{\phi}_{j,l}^T (\mathbf{k}_l - \omega_j^2 \mathbf{m}_l) \boldsymbol{\phi}_{j,l} \qquad (8.49)$$

and

$$\boldsymbol{\phi}_{j,l} = \sum_{i=1}^{n} \alpha_{ijl} \boldsymbol{\phi}_i \qquad (8.50)$$

where

$$\alpha_{ijl} = \frac{1}{A_i(\omega_i^2 - \omega_j^2)} \boldsymbol{\phi}_{j,l}^T (\mathbf{k}_l - \omega_i^2 \mathbf{m}_l) \boldsymbol{\phi}_{i,l} \qquad i \neq j \qquad (8.51)$$

and

$$\alpha_{iil} = -\frac{1}{A_l} \boldsymbol{\phi}_{i,l}^T \mathbf{m}_l \boldsymbol{\phi}_{i,l} \qquad (8.52)$$

In equation (8.49), $\mathbf{k}_l$ and $\mathbf{m}_l$ represent the element stiffness and mass matrix respectively of the $l$th element. In deriving equations (8.49)–(8.52), it is assumed that the element matrices are linear functions of the cross-sectional areas. This is true for the rod elements which are subjected to axial force only. Using equations (8.49)–(8.52), the sensitivities of $\mathbf{A}$ and $\mathbf{X}$, required in equation (8.40), can be written. Equation (8.48) also needs the sensitivities of the Riccati matrix $\mathbf{P}$. Differentiating equation (8.22) with respect to the design variables $A_l$ and using equation (8.38) gives

$$\bar{\mathbf{A}}^T \mathbf{P}_{,l} + \mathbf{P}_{,l}\bar{\mathbf{A}} = \tilde{\mathbf{B}} \qquad (8.53)$$

where

$$\tilde{\mathbf{B}} = -\mathbf{A}_{,l}^T \mathbf{P} - \mathbf{P}\mathbf{A}_{,l} + \mathbf{P}\mathbf{X}_{,l}\mathbf{P} \qquad (8.54)$$

The sensitivity of the Riccati matrix $\mathbf{P}_{,l}$ is given by the solution to the Lyapunov equation (8.53). The solution can be obtained by using available subroutines in control software or using the following procedure.

Premultiplying equation (8.53) by $\boldsymbol{\alpha}^T$ and postmultiplying by $\boldsymbol{\alpha}$ and using the relations in equations (8.44)–(8.47) gives

$$\lambda^T Y + Y\lambda = \tilde{B}^* \tag{8.55}$$

where

$$Y = \alpha^T P_{,l} \alpha \tag{8.56}$$

and

$$\tilde{B}^* = \alpha^T \tilde{B} \alpha \tag{8.57}$$

Premultiplying equation (8.56) by $\beta$ and postmultiplying by $\beta^T$ gives

$$P_{,l} = \beta Y \beta^T \tag{8.58}$$

where the elements of matrix $Y$ are given by

$$Y_{ij} = \frac{\tilde{B}^*_{ij}}{\lambda_i + \lambda_j} \tag{8.59}$$

Thus, using equation (8.37), the sensitivities of the complex eigenvalues $\lambda_i$ can be obtained.

The sensitivities of the spectral radius can be obtained by assuming that the sensitivities are calculated for a specified critical operating frequency $p_s$ and are invariant for small changes of the design variables. This is an approximation. Without this assumption, the analytical partial derivatives cannot be derived since the first derivative of the spectral radius is a discontinuous function of the operating frequency $p$.

The largest eigenvalue of $|(jp_s I - \bar{A}^{-1})| \cdot U_e$ for the specified critical operating frequency $p_s$ can be written as

$$\lambda_p = \lambda_p^r + j\lambda_p^i \tag{8.60}$$

where $\lambda_p^r$ and $\lambda_p^i$ are the real and imaginary parts of $\lambda_p$. Then the spectral radius $\rho_s$ can be written as

$$\rho_s = [(\lambda_p^r)^2 + (\lambda_p^i)^2]^{1/2} \tag{8.61}$$

The sensitivity of $\rho_s$ with respect to the $l$th design variable can be written as

$$\rho_{s,l} = \frac{\lambda_p^r \lambda_{p,l}^r + \lambda_p^i \lambda_{p,l}^i}{\rho_s} \tag{8.62}$$

The sensitivity of the eigenvalues $\lambda_p$ can be written as

$$\lambda_{p,l} = \beta_p \tilde{\bar{A}}_{p,l} \alpha_p \tag{8.63}$$

where

$$\tilde{\bar{A}}_p = \bar{A}_p \cdot U_e \tag{8.64}$$

where

$$\bar{A}_p = |(jp_s I - \bar{A})^{-1}| \tag{8.65}$$

In equation (8.63), $\beta_p$ and $\alpha_p$ are the left and right eigenvectors of $\tilde{\bar{A}}_p$. Differentiating equation (8.64) with respect to the design variable $A_l$ gives

$$\tilde{\bar{A}}_{p,l} = \bar{A}_{p,l} \cdot U_e + \bar{A}_p \cdot U_{e,l} \tag{8.66}$$

The elements of matrix $\bar{A}_p$ can be written as

$$\bar{A}_{p_{ij}} = [(a_{p_{ij}}^r)^2 + (a_{p_{ij}}^i)^2]^{1/2} \tag{8.67}$$

where $a_{p_{ij}}^r + ja_{p_{ij}}^i$ are the elements of the complex matrix $A_p = (jp_s I - \bar{A})^{-1}$. The sensitivity of $\bar{A}_{\rho_{ij}}$ can be written as

$$\bar{A}_{\rho_{ij,l}} = \frac{a^r_{\rho_{ij}} a^r_{\rho_{ij,l}} + a^i_{\rho_{ij}} a^i_{\rho_{ij,l}}}{\bar{A}_{\rho_{ij}}} \tag{8.68}$$

The sensitivity of $\mathbf{A}_\rho$ is given by

$$\mathbf{A}_{\rho,l} = -\mathbf{A}_\rho \mathbf{A}_{\rho,l}^{-1} \mathbf{A}_\rho \tag{8.69}$$

where

$$\mathbf{A}_{\rho,l}^{-1} = [jp_s \mathbf{I} - \bar{\mathbf{A}}]_{,l} \tag{8.70}$$

For a specified value of $\rho_s$

$$\mathbf{A}_{\rho,l}^{-1} = -\bar{\mathbf{A}}_{,l} \tag{8.71}$$

The sensitivity of $\bar{\mathbf{A}}_{,l}$ is given by equations (8.46). Taking into consideration the sign of the elements of $\bar{\mathbf{A}}$ and $\bar{\mathbf{A}}_{ij,l}$, the sensitivity of the elements of $\mathbf{U}_e$ can be written:

$$U_{e_{ij,l}} = \frac{|\bar{A}_{ij}|_{,l}|\bar{A}_{pq}| - |\bar{A}_{ij}||\bar{A}_{pq}|_{,l}}{(|\bar{A}_{pq}|)^2} \qquad i > n \tag{8.72}$$

## 8.5 OPTIMIZATION STEPS

The major steps involved in the simultaneous design of a structure and control system are as follows:

1. For the specified initial cross-section areas of the members, calculate the structural frequencies $\omega_j$ and the vibration modes $\phi_j$.
2. The plant matrix $\mathbf{A}$ and the input matrix $\mathbf{B}$ are determined.
3. The control problem is solved and the closed-loop eigenvalues and damping parameters are calculated.
4. The sensitivities of the objective function and constraints are calculated.
5. The design variables are modified by using a suitable optimization program such as NEWSUMT-A.
6. With a new values of the design variables, steps 1–5 are repeated until the optimum solution satisfying the constraints are obtained.

Steps 1–5 must be repeated because of the nonlinear nature of the problem.

## 8.6 NUMERICAL EXAMPLES

The details of two optimization problems are given in this section for illustration of the optimum design of a controlled structure.

---

## EXAMPLE 8.1 TWO-BAR TRUSS

The finite element model of the truss is shown in Fig. 8.4. The elements of the structure are represented by bar elements that allow only axial deformation. The

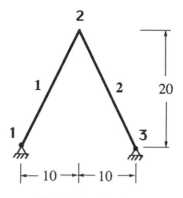

**Fig. 8.4** Two-bar truss.

dimensions of structure are defined in unspecified consistent units. The elastic modulus of the members is equal to 1.0 and the density of the structural material $\rho$ is equal to 0.001. A nonstructural mass of two units is attached at node 2. An actuator and a sensor are collocated in element 1 connecting nodes 1 and 2. The weighting matrices $\mathbf{Q}$ and $\mathbf{R}$ in equation (8.19) are assumed to be identity matrices. The design variables are the cross-sectional areas of the two members. The objective is to determine the cross-sectional areas of the members which will give a minimum weight structure and satisfy constraints on the closed-loop eigenvalue $\tilde{\omega}_1$ and damping parameter $\xi_1$. The areas of members 1 and 2 for the initial design are assumed to be 1000 and 100 units.

The optimization problem for the two-bar truss is specified as follows: minimize the weight

$$W = \sum_{i=1}^{2} \rho_i A_i l_i \tag{8.73}$$

such that

$$\tilde{\omega}_1 \geqslant 1.7393$$
$$\xi_1 \geqslant 0.029\,16 \tag{8.74}$$

where $\tilde{\omega}_1$ and $\xi_1$ are the imaginary parts of the lowest closed-loop eigenvalue and the associated damping respectively.

The details of the initial design with cross-sectional areas of elements 1 and 2 equal to 1000 and 100 respectively are as follows:

The weight of the initial design is 24.59 units. The mass, stiffness and load distribution matrices are

$$\mathbf{M} = \begin{bmatrix} 2 & 0 \\ 0 & 2 \end{bmatrix} \quad \mathbf{K} = \begin{bmatrix} 9.839 & 16.10 \\ 16.10 & 39.35 \end{bmatrix} \quad \mathbf{D} = \begin{bmatrix} -0.447 \\ -0.894 \end{bmatrix} \tag{8.75}$$

The mass matrix $\mathbf{M}$ is written by taking into consideration only the nonstructural mass. The squares of the structural frequencies for the initial design are

$$\omega_1^2 = 1.378 \qquad \omega_2^2 = 23.22 \tag{8.76}$$

The modal matrix normalized with respect to the mass matrix is given by

$$\boldsymbol{\phi} = \begin{bmatrix} 0.6472 & 0.2847 \\ -0.2847 & 0.6472 \end{bmatrix} \tag{8.77}$$

The plant matrix and the input matrix are

$$\mathbf{A} = \begin{bmatrix} 0.0 & 0.0 & 1.0 & 0.0 \\ 0.0 & 0.0 & 0.0 & 1.0 \\ 1.378 & 0.0 & 0.0 & 0.0 \\ 0.0 & -23.22 & 0.0 & 0.0 \end{bmatrix} \tag{8.78}$$

$$\mathbf{B} = \begin{bmatrix} 0.0 \\ 0.0 \\ -0.0348 \\ -0.7063 \end{bmatrix} \tag{8.79}$$

The plant matrix **A** is obtained by assuming that the structural damping parameter $\zeta$ is equal to zero in equation (8.12).

The state and control weighting matrices are

$$\mathbf{Q} = \begin{bmatrix} 1 & 0 & 0 & 0 \\ 0 & 1 & 0 & 0 \\ 0 & 0 & 1 & 0 \\ 0 & 0 & 0 & 1 \end{bmatrix} \tag{8.80}$$

$$\mathbf{R} = [1.0] \tag{8.81}$$

The Riccati matrix obtained by solving equation (8.22) is given by

$$\mathbf{P} = \begin{bmatrix} 5.2120 \times 10^1 & -5.6461 \times 10^{-2} & 3.6199 \times 10^{-1} & -8.4474 \times 10^{-2} \\ -5.6461 \times 10^{-2} & 3.3646 \times 10^1 & 1.4253 & 2.1444 \times 10^{-2} \\ 3.6199 \times 10^{-1} & 1.4253 & 3.7862 \times 10^1 & -6.0925 \times 10^{-3} \\ -8.4474 \times 10^{-2} & 2.1444 \times 10^{-2} & -6.0925 \times 10^{-3} & 1.4463 \end{bmatrix} \tag{8.82}$$

The closed-loop plant matrix is given by

$$\bar{\mathbf{A}} = \begin{bmatrix} 0.0000 & 0.0000 & 1.0000 & 0.0000 \\ 0.0000 & 0.0000 & 0.0000 & 1.0000 \\ -1.3766 & -2.2523 \times 10^{-3} & -4.5682 \times 10^{-2} & -3.5530 \times 10^{-2} \\ 3.3240 \times 10^{-2} & -2.3264 \times 10^1 & -9.2731 \times 10^{-1} & -7.2124 \times 10^{-1} \end{bmatrix} \tag{8.83}$$

The closed-loop eigenvalues are

$$\begin{aligned} -0.3606 &\pm j4.8062 \\ -0.0228 &\pm j1.1739 \end{aligned} \tag{8.84}$$

The closed-loop damping parameters associated with these eigenvalues are

0.074 82 and 0.019 44 respectively. Comparing the constraints on the optimization problem (equations (8.74)) with the closed-loop eigenvalues and damping parameters of the initial design shows that we seek an optimum design for which $\tilde{\omega}_1$ is same as the initial design but damping increased from 0.019 44 to 0.029 16. The optimum design was obtained by using the NEWSUMT-A program as a black box. It took ten iterations and the weight of the final design was 17.58 with the cross-sectional areas of the two members equal to 684.62 and 101.98.

The details of the optimum design, with cross-sectional areas equal to 684.62 and 101.98, are as follows.

The mass and load distribution matrices would be same as those of the initial design. The stiffness matrix is given by

$$\mathbf{K} = \begin{bmatrix} 7.036 & 10.42 \\ 10.42 & 28.14 \end{bmatrix} \tag{8.85}$$

The squares of the structural frequencies for the final design are

$$\omega_1^2 = 1.378 \qquad \omega_2^2 = 16.21 \tag{8.86}$$

$\omega_1^2$ for the optimum design is same as the initial design owing to the constraint on $\tilde{\omega}_1$. The modal matrix normalized with respect to the mass matrix is given by

$$\boldsymbol{\phi} = \begin{bmatrix} 0.6541 & 0.2686 \\ -0.2686 & 0.6541 \end{bmatrix} \tag{8.87}$$

The plant matrix and the input matrix are

$$\mathbf{A} = \begin{bmatrix} 0.0 & 0.0 & 1.0 & 0.0 \\ 0.0 & 0.0 & 0.0 & 1.0 \\ -1.378 & 0.0 & 0.0 & 0.0 \\ 0.0 & -16.21 & 0.0 & 0.0 \end{bmatrix} \tag{8.88}$$

$$\mathbf{B} = \begin{bmatrix} 0.0 \\ 0.0 \\ -0.0523 \\ -0.7052 \end{bmatrix} \tag{8.89}$$

The weighting matrices $\mathbf{Q}$ and $\mathbf{R}$ should be same as the initial design since they were not functions of the design variables. The Riccati matrix is given by

$$\mathbf{P} = \begin{bmatrix} 3.4725 \times 10^1 & -9.0134 \times 10^{-2} & 3.6102 \times 10^{-1} & -1.2515 \times 10^{-1} \\ -9.0134 \times 10^{-2} & 2.3799 \times 10^1 & 1.4769 & 3.0542 \times 10^{-2} \\ 3.6102 \times 10^{-1} & 1.4769 & 2.5261 \times 10^1 & -1.3560 \times 10^{-2} \\ -1.2515 \times 10^{-1} & 3.0542 \times 10^{-2} & -1.3560 \times 10^{-2} & 1.4618 \end{bmatrix}$$

$$\tag{8.90}$$

The closed-loop plant matrix is given by

$$\bar{\mathbf{A}} = \begin{bmatrix} 0.0000 & 0.0000 & 1.0000 & 0.0000 \\ 0.0000 & 0.0000 & 0.0000 & 1.0000 \\ -1.3747 & -5.1710 \times 10^{-3} & -6.8668 \times 10^{-2} & -5.3902 \times 10^{-2} \\ 4.8909 \times 10^{-2} & -1.6281 \times 10^1 & -9.2537 \times 10^{-1} & -7.2639 \times 10^{-1} \end{bmatrix}$$

$$\tag{8.91}$$

The closed-loop eigenvalues are

$$-0.3633 \pm j4.011$$
$$-0.0343 \pm j1.1739$$

(8.92)

The closed-loop damping parameter associated with the lowest eigenvalue is equal to 0.029 16. The optimum design satisfied both constraints as nearly equality constraints. For the two-bar truss, the optimum design given above is unique in the sense that with any other initial design and the constraints specified in equation (8.74), the optimization procedure would give the same optimum design.

## EXAMPLE 8.2 TETRAHEDRAL TRUSS

The twelve-bar truss shown in Fig. 8.5 is optimized with constraints on frequencies and spectral radius. This structure has been used very frequently to illustrate the application of the optimum design of a controlled structure. The coordinates of the node points are given in Table 8.1. A nonstructural mass of 2 units is located at node points 1–4. Young's modulus and material density are the same as those for the two-bar truss. The structure has twelve degrees of freedom. The six

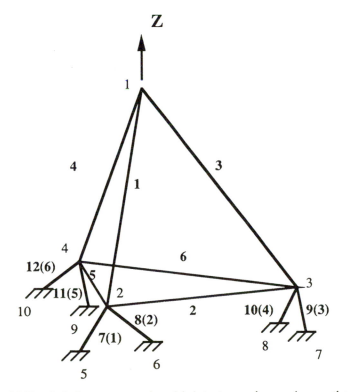

**Fig. 8.5** Tetrahedral truss structural model. Actuator numbers are in parentheses.

**Table 8.1** Node point coordinate or tetrahedral truss

| Node | X | Y | Z |
|------|------|---------|--------|
| 1 | 0.0 | 0.0 | 10.165 |
| 2 | −5.0 | −2.887 | 2.00 |
| 3 | 5.0 | −2.887 | 2.00 |
| 4 | 0.0 | 5.7735 | 2.00 |
| 5 | −6.0 | −1.1547 | 0.0 |
| 6 | −4.0 | −4.6188 | 0.0 |
| 7 | 4.0 | −4.6188 | 0.0 |
| 8 | 6.0 | −1.1547 | 0.0 |
| 9 | −2.0 | 5.7735 | 0.0 |
| 10 | 2.0 | 5.7735 | 0.0 |

actuators and sensors are colocated in the elements 7–12. In order to avoid the effect of unmodeled dynamics on the optimum design, all vibration modes are used in the plant matrix. Thus the matrices **A** and **B** would be 24 × 24 and 24 × 6 respectively. Similarly the weighting matrices **Q** and **R** would be 24 × 24 and 6 × 6 respectively. The weighting matrices are set equal to $\delta$**Q** and $\gamma$**R**. $\delta$ and $\gamma$ are treated as the control design variables. The matrices **Q** and **R** are assumed to be identity matrices. For the initial design in optimization, $\delta$ and $\gamma$ were set equal to unity.

The cross-sectional areas of the members for the initial design are given in Table 8.2. For this optimization problem, the constraints are imposed on the imaginary parts of the first two lowest frequencies of the closed-loop system and the spectral radius $\rho_s$. The constraints are

$$\tilde{\omega}_1 - 1.34 \geqslant 0 \tag{8.93}$$

$$\tilde{\omega}_2 - 1.6 \geqslant 0 \tag{8.94}$$

$$5.56 - \rho_s \geqslant 0 \tag{8.95}$$

**Table 8.2** Cross-sectional areas of members

| Initial design | | | | | | |
|---|---|---|---|---|---|---|
| Member | 1 | 2 | 3 | 4 | 5 | 6 |
| Area | 1000.0 | 1000.0 | 100.0 | 100.0 | 1000.0 | 1000.0 |
| Member | 7 | 8 | 9 | 10 | 11 | 12 |
| Area | 100.0 | 100.0 | 100.0 | 100.0 | 100.0 | 100.0 |
| Weight = 43.69 | | | | | | |
| Optimum design | | | | | | |
| Member | 1 | 2 | 3 | 4 | 5 | 6 |
| Area | 216.4 | 138.5 | 165.5 | 168.0 | 154.5 | 468.7 |
| Member | 7 | 8 | 9 | 10 | 11 | 12 |
| Area | 174.6 | 144.8 | 155.6 | 235.4 | 143.2 | 170.2 |
| Weight = 16.01 | | | | | | |

The constraint on the spectral radius is based on achieving 10% permissible deviation in all the elements of the closed-loop matrix before the system becomes unstable. The perturbation identification matrix is obtained by normalizing the closed-loop matrix $\bar{\mathbf{A}}$ for $\bar{A}_{p,q} = \bar{A}_{13,1}$ (equation (8.35)). The constraint value of $\rho_s$ in equation (8.95) is calculated by using equation (8.36) for $\varepsilon_r = 0.1$ and $|\bar{A}_{p,q}| = |\bar{A}_{13,1}| = 1.34^2$ where 1.34 would be the constraint value of $\tilde{\omega}_1$.

The initial design weighed 43.69 units. The spectral radius $\rho_s$ for this design was 44.35 giving $\varepsilon_r = 0.0126$. Thus the control design would allow 1.26% variation in the closed-loop matrix before becoming unstable according to equation (8.34).

**Table 8.3** Iteration history

| Iteration number | 1 | 2 | 3 | 4 | 5 | 6 |
|---|---|---|---|---|---|---|
| Weight | 43.69 | 32.01 | 32.01 | 25.91 | 22.12 | 19.34 |
| Iteration number | 7 | 8 | 9 | 10 | 11 | |
| Weight | 16.58 | 16.11 | 16.02 | 16.01 | 16.01 | |

**Table 8.4** Structural frequencies

Initial design

| Mode | 1 | 2 | 3 | 4 | 5 | 6 |
|---|---|---|---|---|---|---|
| $\omega_j$ | 1.342 | 1.664 | 2.890 | 2.950 | 3.398 | 4.204 |
| Mode | 7 | 8 | 9 | 10 | 11 | 12 |
| $\omega_j$ | 4.662 | 4.755 | 8.539 | 9.250 | 10.284 | 12.905 |

Optimum design

| Mode | 1 | 2 | 3 | 4 | 5 | 6 |
|---|---|---|---|---|---|---|
| $\omega_j$ | 1.355 | 1.607 | 2.865 | 3.003 | 3.654 | 5.175 |
| Mode | 7 | 8 | 9 | 10 | 11 | 12 |
| $\omega_j$ | 5.390 | 5.718 | 6.147 | 6.383 | 6.922 | 7.884 |

**Table 8.5** Closed-loop frequencies

Initial design

| | | | |
|---|---|---|---|
| 1 | $-0.0734 \pm j1.341$ | 7 | $-0.3547 \pm j\,4.649$ |
| 2 | $-0.1089 \pm j1.663$ | 8 | $-0.3440 \pm j\,4.743$ |
| 3 | $-0.2133 \pm j2.884$ | 9 | $-0.2917 \pm j\,8.534$ |
| 4 | $-0.2372 \pm j2.949$ | 10 | $-0.2758 \pm j\,9.246$ |
| 5 | $-0.2852 \pm j3.388$ | 11 | $-0.2136 \pm j10.28$ |
| 6 | $-0.3634 \pm j4.190$ | 12 | $-0.0828 \pm j12.90$ |

Optimum design

| | | | |
|---|---|---|---|
| 1 | $-0.1927 \pm j1.352$ | 7 | $-1.060 \pm j5.168$ |
| 2 | $-0.2206 \pm j1.601$ | 8 | $-1.370 \pm j5.558$ |
| 3 | $-0.3633 \pm j2.854$ | 9 | $-1.332 \pm j6.006$ |
| 4 | $-0.4593 \pm j2.992$ | 10 | $-1.259 \pm j6.252$ |
| 5 | $-1.5936 \pm j3.620$ | 11 | $-1.187 \pm j6.827$ |
| 6 | $-1.353 \pm j5.145$ | 12 | $-0.8830 \pm j7.756$ |

**Table 8.6** Closed-loop damping

| Initial design | | | | | | |
|---|---|---|---|---|---|---|
| Mode | 1 | 2 | 3 | 4 | 5 | 6 |
| $\xi_i$ | 0.0546 | 0.0653 | 0.0737 | 0.0801 | 0.0839 | 0.0864 |
| Mode | 7 | 8 | 9 | 10 | 11 | 12 |
| $\xi_i$ | 0.0760 | 0.0723 | 0.0342 | 0.0298 | 0.0207 | 0.0064 |
| Optimum design | | | | | | |
| Mode | 1 | 2 | 3 | 4 | 5 | 6 |
| $\xi_i$ | 0.1411 | 0.1364 | 0.1262 | 0.1517 | 0.1618 | 0.2544 |
| Mode | 7 | 8 | 9 | 10 | 11 | 12 |
| $\xi_i$ | 0.2009 | 0.2365 | 0.2165 | 0.1974 | 0.1713 | 0.1131 |

The optimum design weighing 16.01 units was obtained after eleven iterations. The iteration history is given in Table 8.3. The control design parameters $\delta$ and $\gamma$ for the optimum design were 3.89 and 0.257 respectively. The spectral radius $\rho_s$ of the optimum design was 5.560 and $\varepsilon_r$ was 0.9999. The cross-sectional areas of the members for the optimum design are given in Table 8.2. The structural frequencies for the two designs are given in Table 8.4. The optimum design has a narrower frequency bandwidth than the initial design. It is well known that in structural optimization with constraints on frequencies, the frequencies come closer. The closed-loop frequencies are listed in Table 8.5. For the optimum design $\tilde{\omega}_1 = 1.352$ and $\tilde{\omega}_2 = 1.601$ indicating that the two constraints on the imaginary parts of the closed-loop frequencies are satisfied. Table 8.6 contains closed-loop damping for the two designs. For the optimum design, the damping associated with all modes is larger than for the initial design.

## 8.7 OPTIMIZATION PROGRAMS

In addition to the optimization software available in the math libraries of the computer software, there are some public domain programs available for use. The capabilities of some of these programs are discussed below.

### 8.7.1 NEWSUMT-A (Thareja and Haftka, 1985)

The minimization algorithm used in this program is a sequence of unconstrained minimizations (SUMT) technique. This program can be used for inequality or equality constraints. It uses Newton's method for unconstrained minimization. An extended interior function is used for inequality constraints or an exterior penalty function for equality constraints. The sensitivities can be either calculated by using finite differences or a user-supplied analysis program.

### 8.7.2 ADS (Vanderplaats, Suginoto and Sprague, 1983)

This is a general-purpose optimization program for solution of nonlinear constrained optimization problems. The program is segmented into three levels, these being strategy, optimizer and one-dimensional search. The available strategies are sequential unconstrained minimization, the augmented Lagrange multiplier method and sequential linear programming. The optimizers include a variable metric method and the method of feasible directions as examples. The feasible direction method amongst all approaches in this software is known to be robust and frequently used. The original feasible direction program is called CONMIN.

### 8.7.3 IDESIGN (Arora, Thanedar and Tseng, 1985)

This general-purpose optimization program has been under development since the late 1970s. It contains several state-of-the-art nonlinear programming algorithms. It has a wide range of applications with many friendly features. The program can also be used in an interactive mode. Some of the algorithms available in the software are cost function bounding, Pschenichny's linearization method, sequential quadratic programming and conjugate gradient. The gradient information can be user supplied or the finite difference method based on a forward, backward or central scheme can be used.

### 8.7.4 VMCON (Crane, Hillstrom and Minkoff, 1980)

This program solves a sequence of positive definite quadratic programming subproblems. Each solution determines a direction in which a one-dimensional minimization is performed. It is an iterative algorithm designed to converge to a point that satisfies the first-order necessary condition. These conditions are the Kuhn–Tucker conditions and related conditions on the Lagrange multipliers and the constraints.

The programs discussed above can be used as black boxes. The primary information they need is the objective function, constraints and the sensitivities. Experience has shown that the performance of these programs depend on the problem and the familiarity of the user with them. The control software ORACLE (Armstrong, 1978) is available in the public domain in addition to a number of proprietary programs distributed by various companies.

## 8.8 CONCLUDING REMARKS

This chapter has reviewed the research work performed by different investigators in developing the algorithms for the simultaneous design of structures and control systems. It was observed that the primary objective of the investigations has been to understand the problem and study the feasibility of integrated techniques. A simple two-bar truss problem was solved to illustrate the application of combined structure and control system optimization. Many of the critical issues in structural and control design such as nonlinearities, dynamic characteristics of sensors and actuators, order of the structure and control modes, use of modern control theories, experimental verification of the simultaneous structure and control design, development of software and efficiencies of the optimization algorithm have still to be investigated.

## REFERENCES

Armstrong, E.S. (1978) ORACLS – A system for linear–quadratic–gaussian control law design. *NASA Technical Paper 1106.*

Arora, J.S. (1989) *Introduction to Optimum Design,* McGraw-Hill, New York.

Arora, J.S., Thanedar, P.B. and Tseng, C.H. (1985) User's manual for program IDESIGN. *Technical Report ODL 85.10,* University of Iowa, Iowa City, IA.

Belvin, W.K. and Park, K.C. (1988) Structural tailoring and feedback control synthesis: an interdisciplinary approach. In Proceedings of 29th AIAA/ASME/ASCE/AHS Structures, Structural Dynamics and Materials Conference, Part 1, AIAA-88-2206-CP, Williamsburg, VA, April 18–20, 1988, pp. 1–8.

Bendsoe, M.P., Olhoff, N. and Taylor, J.E. (1987) On the design of structure and controls for optimal performance of actively controlled flexible structures. *Mechanics of Structures and Machines,* **15**(3), 265–95.

Bodden, D.S. and Junkins, J.L. (1985) Eigenvalue optimization algorithms for structure/controller design iterations. *Journal of Guidance, Control, and Dynamics,* **8**(6), 697–706.

Cheng, F.Y. and Pantelides, C.P. (1988) Combining structural optimization and structural control. *Technical Report NCEER-88-0037,* National Center for Earthquake Engineering Research.

Crane, R.L., Hillstrom, K.E. and Minkoff, M. (1980) Solution of the general nonlinear programming problem with subroutine VMCON. *ANL-80-64,* Argonne National Laboratory, Chicago, IL.

D'Azzo, J.J. and Houpis, C.H. (1989) *Linear Control System Analysis and Design,* McGraw-Hill, New York.

Dracopoulos, T.N. and Öz, H. (1992) Integrated aeroelastic control optimization of laminated composite lifting surfaces. *Journal of Aircraft,* **29**(2), 282–8.

Eastep, F., Khot, N.S. and Grandhi, R. (1987) Improving the active vibrational control of large space structures through structural modifications. *Acta Astronautica,* **15**(6), 383–9.

Grandhi, R.V. (1989) Structural and control optimization of space structures. *Computers and Structures,* **31**(2), 139–50.

Grandhi, R.V. (1990) Optimum design of space structures and active and passive damping, *Engineering with Computers,* **15**(6), 177–83.

Grandhi, R.V., Haq, I. and Khot, N.S. (1990) Enhanced robustness in integrated structural/control systems design. In Proceedings of 31st AIAA/ASME/ASCE/AHS/ASC Structures, Structural Dynamics and Materials Conference, Long Beach, CA, April 2–4, pp. 247–57.

Haftka, R.T. (1990) Integrated structure-control optimization of space structures. In Proceedings of AIAA Dynamics Specialist Conference, Long Beach, CA, April 5–6, pp. 1–9.

Haftka, R.T., Gurdal, Z. and Kamat, M.P. (1990) *Structural Optimization,* Kluwer, Dordrecht.

Haftka, R.T., Martinovic, Z.N. and Hallaur, W.L. Jr (1984) Enhanced vibration controllability by minor structural modifications. In Proceedings of 25th AIAA/ASCE/AHS Structures, Structural Dynamics and Materials Conference, Palm Springs, CA, May 14–16, pp. 401–10.

Hale, A.L. (1985) Integrated structural/control synthesis via set-theoretic methods. In Proceedings of 26th AIAA/ASME/ASCE/AHS Structures, Structural Dynamics and Materials Conference, Orlando, FL, April 15–17, pp. 636–41.

Hale, A.L., Lisowski, R.J. and Dahl, W.E. (1985) Optimal simultaneous structural and control design of maneuvering flexible spacecraft. *Journal of Guidance, Control, and Dynamics,* **8**(1), 86–93.

Junkins, J.L. and Rew, D.W. (1988) Unified optimization of structures and controls, in *Large*

*Space Structures: Dynamics and Control* (eds S.N. Atluri and A.K. Amos), Springer, Berlin, pp. 323–53.

Khot, N.S. (1988a) An integrated approach to the minimum weight and optimum control design of space structures, in *Large Space Structures: Dynamics and Control* (eds S.N. Atluri and A.K. Amos), Springer, Berlin, pp. 355–63.

Khot, N.S. (1988b) Structural/control optimization to improve the dynamic response of space structures. *Computational Mechanics*, (3), 179–86.

Khot, N.S. (1990) On optimization of structural and control systems using a reduced order model. *Structural Optimization*, **2**(3), 185–91.

Khot, N.S. and Grandhi, R.V. (1988) Structural and control optimization with weight and Frobenius norm as performance functions, in *Structural Optimization* (eds G.I.N. Rozvany and B.L. Karihaloo), Kluwer, Dordrecht, pp. 151–8.

Khot, N.S. and Veley, D.E. (1990) Robustness characteristics of optimum structural/control design. In Proceedings of AIAA Guidance, Navigation and Control Conference, Portland, OR, August 20–22, pp. 394–403.

Khot, N.S. and Veley, D.E. (1991) Use of robustness constraints in the optimum design of space structures. *Journal of Intelligent Materials Substructures*, **2**(2), 161–76.

Khot, N.S., Venkayya, V.B. and Eastep, F.E. (1984) Structural modifications of large flexible structures to improve controllability. In Proceedings of AIAA Guidance, Navigation, and Control Conference, Seattle, WA, August 20–22, pp. 420–30.

Khot, N.S., Venkayya, V.B. and Eastep, F.E. (1986) Optimal structural modifications to enhance the active vibration control of flexible structures. *AIAA Journal*, **24**(8), 1368–74.

Khot, N.S., Grandhi, R.V. and Venkayya, V.B. (1987) Structural and control optimization of space structures. In Proceedings of 28th AIAA/ASME/ASCE/AHS Structures, Structural Dynamics and Materials Conference, Monterey, CA, April 6–8, pp. 850–60.

Khot, N.S., Öz, H., Grandhi, R.V., *et al.* (1988) Optimal structural design with control gain norm constraint. *AIAA Journal*, **26**(5), 604–11.

Kirsch, U. and Moses, F. (1977) Optimization of structures with control forces and displacements. *Engineering Optimization*, **1**(3), 37–44.

Lim, K.B. and Junkins, J.L. (1987) Robustness optimization of structural and controller parameter. In Proceedings of 28th AIAA/ASME/ASCE/AHS Structures, Structural Dynamics and Materials Conference, Monterey, CA, April 6–8, pp. 351–61.

Lim, K.B. and Junkins, J.L. (1989) Robustness optimization of structural and controller parameters. *Journal of Guidance, Control and Dynamics*, **12**(1), 89–96.

Manning, R.A. and Schmit, L.A. (1990) Control augmented structural synthesis with transient respone constraints. *AIAA Journal*, **28**(5), 883–91.

McLaren, M.D. and Slater, G.L. (1990) A covariance approach to integrated control/structure optimization. In Proceedings of AIAA Dynamics Specialist Conference, Long Beach, CA, April 5–6, pp. 189–205.

Meirovitch, L. (1989) *Dynamics and Control of Structures*, Wiley, New York.

Mesquita, L. and Kamat, M.P. (1988) Structural optimization for control of stiffened laminated composite structures. *Journal of Sound and Vibration*, **116**(1), 33–48.

Messac, A. and Turner, J. (1984) Dual structural-control optimization of large space structures. In Proceedings of Recent Experience in Multidisciplinary Analysis and Optimization Symposium, NASA Langley Research Center, April 24–26, pp. 775–802.

Miller, D.F. and Shim, J. (1987) Gradient-based combined structural and control optimization. *AIAA Journal*, **10**(3), 291–8.

Morris, A.J. (ed.) (1982) *Foundations of Structural Optimization: A Unified Approach*, Wiley, London.

Onoda, J. and Haftka, R.T. (1987) An approach to structure/control simultaneous optimization for large flexible spacecraft. *AIAA Journal*, **25**(8), 1133–8.

Onoda, J. and Watanabe, N. (1989) Integrated direct optimization of structure/regulator/observer for large flexible spacecraft. In Proceedings of 30th AIAA/ASME/ASCE/AHS/ASC

Structures, Structural Dynamics and Materials Conference, Mobile, AL, April 3–5, pp. 1336–44.

Padula, S.L., Sandridge, C.A., Walsh, J.L. and Haftka, R.T. (1990) Integrated controls-structures optimization of a large space structure. In Proceedings of 31st AIAA/ASME/ASCE/AHS/ASC Structures, Structural Dynamics and Materials Conference, Long Beach, CA, April 2–4, pp. 258–67.

Qiu, L. and Davison, E.J. (1986) New perturbation bounds for the robust stability of linear state space models. In Proceedings of the 25th Conference on Decision and Control, Athens, pp. 751–5.

Rao, S.S. (1988) Combined structural and control optimization of flexible structures. *Engineering Optimization*, **13**(1), 1–16.

Rao, S.S., Venkayya, V.B. and Khot, N.S. (1988) Game theory approach for the integrated design of structures and controls. *AIAA Journal*, **26**(4), 463–9.

Rao, S.S., Pan, T.S. and Venkayya, V.B. (1990) Robustness improvement of actively controlled structures through structural modifications. *AIAA Journal*, **28**(2), 353–61.

Salama, M., Garba, J., Demsetz, L. and Udwadia, F. (1988) Simultaneous optimization of controlled structures. *Computational Mechanics*, **3**(4), 275–82.

Sepulveda, A.E. and Schmit, L.A., Jr (1990) Optimal placement of actuators and sensors in control augmented structural optimization. In Proceedings of 31st AIAA/ASME/ASCE/AHS/ASC Structures, Structural Dynamics and Materials Conference, Long Beach, CA, April 2–4, pp. 217–40.

Suzuki, S. and Matsuda, S. (1990) Structure/control design synthesis of active flutter suppression system by goal programming. In Proceedings of AIAA Guidance, Navigation and Control Conference, Portland, OR, August 20–22.

Thareja, R. and Haftka, R.T. (1985) NEWSUMT-A. A modified version of NEWSUMT for inequality and equality constraints. VPI Report 148, Aerospace Engineering Department, Virginia Tech., Blacksburg, VA.

Thomas, H.L. and Schmit, L.A., Jr (1989) Control augmented structural synthesis with dynamic stability constraints. In Proceedings of 30th AIAA/ASME/ASCE/AHS/ASC Structures, Structural Dynamics and Materials Conference, Mobile, AL, April 3–5, pp. 521–31.

Thomas, H.L. and Schmit, L.A. (1990) Improved approximations for control augmented structural synthesis. In Proceedings of 31st AIAA/ASME/ASCE/AHS/ASC Structures, Structural Dynamics and Materials Conference, Long Beach, CA, April 2–4, pp. 277–94.

Vanderplaats, G.N., Sugimoto, H. and Sprague, C.M. (1983) ADS-1: a new general-purpose optimization program. In AIAA/ASME/ASCE/AHS 24th Structures, Structural Dynamics and Materials Conference, Lake Tahoe, NV, pp. 117–23.

Zeiler, T.A. and Gilbert, M.G. (1990) Integrated control/structure optimization by multilevel decomposition. In Proceedings of 31st AIAA/ASME/ASCE/AHS/ASC Structures, Structural Dynamics and Materials Conference, Long Beach, CA, April 2–4, pp. 247–57.

# 9

# Database design and management in engineering optimization

JASBIR S. ARORA

## 9.1 INTRODUCTION

Design of modern structural and mechanical systems requires considerable computer analysis. Also, as more computing power becomes available, the desire to design larger and more complex systems grows. Many of the systems operate in an environment that requires integration of many disciplines in their design process. These systems must be designed to be cost effective and efficient in their operations, requiring use of organized optimization techniques. These techniques are iterative, generating and using massive amounts of data. Multiple programs may be used during the optimization process that require sharing of data. Also, complexity of the design software grows as more disciplines are integrated and more complicated problems need to be solved. These requirements points to the fact that the data and the software must be organized and managed properly. This will allow the design process to be completed in an efficient manner as well as give flexibility in updating and enhancing the design software itself.

This chapter presents basic concepts related to the database design and management in optimization of engineering systems. Most of the concepts are new to the engineering community, so they are explained in some detail. Terminologies used in the database design and management field are defined and explained. Need for the database design and management is discussed. Basic components of a database management system are identified and discussed. The topic of designing a proper database is discussed and a methodology to design a database is given. Examples from the finite element analysis procedure and design of structural systems are used to illustrate various ideas and concepts.

The material presented in this chapter is derived from several references, such as Arora and Mukhopadhyay (1988), Blackburn, Storaasli and Fulton (1982), Bell (1982), Date (1977), Felippa (1979, 1980), Fulton (1987), Martin (1977), Mukhopadhyay and Arora (1987a, b), Pahl (1981), Rajan (1982), Rajan and Bhatti (1983, 1986), SreekantaMurthy and Arora (1985, 1986a, b), SreekantaMurthy et al. (1986) and Vetter and Maddison (1981). These references can be consulted for more details on some of the topics.

### 9.1.1 Computer-aided design optimization process

To understand the need for use of a database in computer-aided design optimization, one needs to understand the optimization process. Therefore this process is briefly explained here. A first step in optimum design of any system is the precise mathematical statement of the problem (Arora, 1989, 1990). This requires identification of design variables that describe the system, definition of a cost function that needs to be minimized, and identification of constraints that must be satisfied. In design of aircraft components such as stiffened panels and cylinders, the design variables are spacing of the stiffeners, size and shape of stiffeners, and thickness of the skin. In optimization of structural systems such as frames and trusses of fixed configuration, thickness of members, cross-sectional areas of bars, and moment of inertia are the design variables. If shape optimization is the objective, the design variables include parameters related to geometry of the system. In practical applications design variables are usually grouped together to reduce the size of design variable vector.

The constraints for the system are classified into the performance and size constraints. The performance constraints are on stresses, displacements, and local and overall stability requirements in the static case; frequencies and displacements in the dynamic case; flutter velocity and divergence in aeroelastic case, or a combination of these. The size constraints are the minimum and maximum value of the design variables. To evaluate the performance constraints, the system must be analyzed. Finite element and other numerical methods must be used for analysis of large and complex systems. These methods use data such as element number, nodal connectivity, element stiffness matrix, element mass matrix, element load matrix, assembled stiffness, mass and load matrices, displacement vectors, eigenvalues, eigenvectors, buckling modes, decomposed stiffness matrix, and the stress matrix. In general, data used is quite large even when symmetry of the matrices is taken into account. Hypermatrix or other special schemes can be used in dealing with large matrix equations.

In nonlinear programming, the search for the optimum design variables involves iterations. The design variable and other data at the $k$th iteration are used to compute a search direction. A step size must be calculated in the search direction to move to a new point. This calculation can involve evaluation of problem functions which requires reanalysis of the system. To calculate the search direction, gradients of cost and constraint functions with respect to the design variables are needed. This computation is called design sensitivity analysis which can be quite tedious and time consuming because most functions of the problem are implicitly dependent on the design variables. Therefore, special methods must be developed and used for each class of applications. For design of large structures, efficient design sensitivity analysis is particularly critical. For such structures, substructuring concepts can be effectively integrated into structural analysis, design sensitivity analysis, and optimal design procedures (Haug and Arora, 1979). In this concept, one deals with small-order matrices as the data can be organized substructure-wise (Kamal and Adeli, 1990). It can be seen that the number of matrices and their sizes depend on the number of substructures and their sizes.

In most practical applications, interactive computations and graphics can be profitably employed in design optimization (Arora, 1989; Arora and Tseng, 1988; Al-Saadoun and Arora, 1989; Park and Arora, 1987). At a particular iteration, the designer can study the data of design variables, active constraints, cost function, search direction, sensitivity coefficient etc. Judgements can be made regarding suitability of a particular algorithm, change of system parameters, and perhaps about the problem formulation. The problem conditions can be redefined to achieve convergence to optimal design.


### 9.1.2 Need for database and its management

It can be seen from the foregoing discussion that sophisticated engineering design optimization methods use large a amount of data and require substantial computer analysis. Numerical simulation techniques, such as the finite element method, must be used. The data generated during the simulation phase must be saved in a database for later use in defining constraints and performing design sensitivity analysis. Once design sensitivity analysis has been completed, a search direction determination subproblem is defined and solved. Note that the size of this subproblem at each iteration depends on the number of active constraints. Therefore, sizes of data sets change from iteration to iteration, making the nature of data to be dynamic. We should be able to create large data sets dynamically, to manipulate them during the iteration, and to delete some of them at the end of the iteration. Useful trend information from each iteration must be saved for processing in later iterations. Note that a row of the history matrix (such as design variable values) is generated at each iteration. However, to use the trend information for a quantity (e.g. a design variable), we need to look at its value at the previous iterations. This implies that we should look at a column of the history matrix, i.e. we should be able to create data in one form and view it in another form. The entire process is frequently interactive where the designer needs to interrupt it, to analyze the data, and to make design decisions. Therefore the designer needs control over the data and the program to guide the iterative process properly towards acceptable designs.

It is concluded that significant improvement in design capability can be achieved with effective management of the design software system and the database. A properly designed database and a database management system when used with interactive computer graphics can be an invaluable tool for the engineer involved in the design process. These capabilities are also essential to allow for proper growth and refinement of the design software system to accommodate new developments in the field.


## 9.2 TERMINOLOGY AND BASIC DEFINITIONS

The terminology used in database design and management for describing various ideas differs considerably from one group to another and even from one period to another within the same organization. It is therefore necessary to explain various terminologies that are taken from most widely accepted sources (Date, 1977; Martin, 1977; Vetter and Maddison, 1981; Felippa, 1979, 1980; Rajan, 1982; SreekantaMurthy and Arora, 1986a, b). This will also facilitate in the description of various concepts in subsequent sections. They are grouped into three categories – hardware terminology, logical data terminology and physical storage terminology.

### 9.2.1 Hardware terminology

*Auxiliary storage*
Storage facilities of large capacity and lower cost but slower access than main memory are called auxiliary storage. They are also referred to as peripheral or secondary storage devices. They are usually assessed via data channels, in which case data is stored and retrieved by physical record blocks. They include magnetic tape and disk units, drums and other devices used to store data.

*Cell*
This is used as a generic word to mean either track, cylinder, module or other zone delimited by a natural hardware boundary such that the time required to access data increases by a step function when data extends beyond a cell boundary.

*Cylinder*
An access mechanism may have many reading heads. Each head can read one track. A cylinder refers to a group of tracks that can be read without moving the access mechanism.

*Direct access storage device*
In this device, access to a position for storage or retrieval of data is not dependent on the position at which data was previously stored or retrieved. It is also called random access device.

*Input/output (I/O) device*
This is an auxiliary storage device connected to the central processing unit (CPU) by a data channel.

*Main memory*
This is fast, direct access, electronic memory hardwired to the central processing unit. It holds machine instructions and data that can be accessed in a time of the order of nanoseconds. It is also referred to as core, main storage, or internal memory.

*Module*
A module of the peripheral storage device is a section of hardware which holds one volume, such as one spindle of disk.

*Storage device – logical (logical file, memory device, named space, or logical address space)*
This is a subset of the storage space that is treated as a named entity by the operating system for purpose of allocating and releasing storage resources during the execution of a run unit (task). The term is most often applied to auxiliary storage facilities.

*Storage facility*
This is hardware available to store data at a computer installation.

*Track*
A track on a direct access device contains data that can be read in a single reading without the head changing its position.

*Volume*
A volume is normally a single physical unit of any peripheral storage medium such as tapes, disk packs, or cartridges.

## 9.2.2 Logical data terminology

*Arithmetic data*
An arithmetic data item has a numeric value with characteristics of base, scale mode and precision, e.g. fixed point data (integer), and floating point data (real and double precision).

*Attribute*
Properties of an entity are called its attributes. Columns of a two-dimensional table (a relation) are referred to as its attributes. Attributes associate a value from a domain of values for that attribute with each entity in the entity set. For the entity set 'finite elements', NODES for the element, MATERIAL, cross-sectional SHAPE etc. are its attributes. Nodes for an element may be anywhere from 1 to 100, so this defines the domain for the attribute NODES. The attribute MATERIAL may have its domain as several steel and aluminum grades such as (STEEL.1, STEEL.2, ..., ALUM.1, ALUM.2, ...).

*Creation*
This involves adding new files to the database, initializing the files (i.e. file table definition), data validation, deciding file types etc.

*Data aggregate*
A data aggregate is a collection of data items within a record. Data aggregates may be vectors or repeating groups.

- A vector is a one-dimensional ordered collection of data items, e.g. node numbers of a structure.
- A repeating group is a collection of data that occurs repeatedly within a data aggregate. For example, degrees of freedom for an element appear in multiples of node numbers:

$$D_{1i} \quad D_{2i} \quad D_{3i}, \quad i = 1, \ldots, n$$

where $i$ is the node number, $D_{ki}$ the $k$th degree of freedom at node $i$, and $n$ the number of nodes for the element.

*Data definition language (DDL)*
This is a set of commands that enables users of a database management system (DBMS) to define data structures to store the data. All data that is to be managed by the DBMS must follow the rules laid down in data definition language. A DBMS must provide a DDL to specify the conceptual scheme and some of the

details regarding the implementation of the conceptual scheme by a physical scheme. It describes relationships among types of entities for a particular data model.

### Data independence
This refers to the independence between physical and logical data structures. Physical data structure can change without affecting the user's view of the data when we have data independence. Similarly, logical data structure can change without affecting the physical data structure.

### Data item
A data item is the smallest unit of named data. It is also referred to as the data element or field. Each data item has a unique representation. The data item can be any of the following types: arithmetic (integer, real or double precision real), or character string (character, bits).

### Data library
This is a named collection of data sets residing on a permanent storage device. It is the most complex data structure upon which a database management system operates.

### Data manipulation language (DML)
This is a set of permissible commands that are issued by users or application programs to the DBMS to carry out storage, retrieval or manipulation of data. The DML represents the interface between the application program and the database management system. Thus the data managed by the DBMS can be accessed and processed through the use of DML. It can be an extension of the host language, such as FORTRAN.

### Data model
Data model is a representation of the conceptual scheme for the database. Generally a data definition language (a higher-level language) is used to describe the data model. Examples of data model are hierarchical, network and relational.

### Data set
This is an ordered collection of logically related data items arranged in a prescribed manner. Each data set has some control information that can be accessed by a programming system.

### Data structure
This is a logical arrangement of data as viewed by the users or applications programmers.

### Database
A database is a collection of interrelated data stored together without unnecessary redundancy to serve multiple applications. It can also be viewed as a collection of the occurrences of multiple record types, containing relationships between records, data aggregate and data elements. Thus, it is a collection of data files stored on a storage device. The data are stored so that they are independent of

programs which use them. A common and controlled approach is used in storing new data and in modifying and retrieving existing data within the database. A database allows a user or application programmer to retrieve or write data without actually making calls to the input/output device; that chore is left to the DBMS.

*Database administrator (DBA)*
This is the brain of the database management system that provides interfaces between the various parts of the system, does error recovery, and enforces security measures.

*Database management system (DBMS)*
This is the software that allows one or many persons to use and/or modify the database is called a DBMS. DBMS also deals with security, integrity, synchronization and protection of the database.

*Database system*
The set of all databases maintained on a computer installation (or computer network) which are administered by a common database manager is called the database system.

*Domain*
A domain is the set of eligible values for a quantity. For example consider the entity set 'finite elements'; element name, element material type, and length are its attributes. Domains for these attributes can be defined as

$$\text{element name} = (\text{BEAM, TRUSS}, \ldots)$$
$$\text{element material type} = (\text{STEEL, ALUMINUM}, \ldots)$$
$$\text{length} = x : x \geqslant 0 \text{ and } x \leqslant 100$$

*Entity*
An entity may be 'anything having reality and distinctness of being in fact or in thought', e.g. a finite element, a relation, a pre-processor, or a post-processor for a structural analysis program.

*Entity identifier*
This uniquely identifies an entity. The identifier is needed by the programmer to record information about a given entity. Also it is needed by by the computer to identify and have means of finding the entity in a storage unit. Entity identifier must be unique, e.g. element number.

*Entity key*
Entity key is an attribute having different values for each occurring entity and provides unique identification of a tuple (an ordered list or a record). An entity represents a compound key if it corresponds to a group of attributes. It is also called the candidate key.

*Entity set*
Collection (group) of all similar entities is referred to as an entity set, e.g. a set of finite elements (ELEMENTS) and a set of nodes (NODES) are called entity sets.

*Garbage collection*
The process of locating all pages of the memory that are no longer in use and adding them to the list of available space is called garbage collection.

*Group*
This is a data set containing a special 'owner' or 'master' record (the group directory) and a set of member records.

*Instances*
The current contents of a database is called an instance of the database.

*Interrogation*
This deals with identification, selection and extraction of data from the database for further processing. It can be divided into two phases:

- the process of selection and identification of needed data and extracting it;
- the processing part which involves computation, display or any other manipulation required including updating parts of the database.

*Logical data structure*
Data in a particular problem consist of a set of elementary items of data. An item usually consists of single element such as an integer, a bit, a character and a real, or a set of such items. The possible ways in which the data items are structured define different logical data structures. Therefore, it is the data structure as seen by the user of the DBMS without any regard to details of actual storage schemes.

*Memory management system (MMS)*
This is a system that allocates the available memory to the different entity sets in a program and makes it appear as if more memory is available than what the computer has. It partitions the memory allocated to the DBMS into pages and manages the information contained in them. If the data required is not in the memory it retrieves pages from the secondary storage.

*Module*
This is a program that performs an identifiable task.

*Physical database*
This is actual data or information residing in a file in the form of bits, bytes or words. It is the database that actually exists either in the computer memory or on some secondary storage device.

*Programming language*
This is a language that an application programmer may use, e.g. FORTRAN.

*Program library*
This is a collection of subroutines that perform primitive functions.

*Property*
Property is a named characteristic of an entity, e.g. finite element name, and element material type. Properties allow one to identify, characterize, classify and relate entities.

*Property value*
This is an occurrence of a property of an entity, e.g. 'element name' has property value BEAM.

*Primary key*
The entity identifier is referred to as the key of the record group or strictly it is a primary key, e.g. element number.

*Query language*
Query is the process of question and answer that can be accomplished using the query language, i.e. query the database. The commands are generally quite simple and can be used by nonprogramming as well as programming users. These can be interactive commands as well as utilities that can be called from an application program.

*Record*
A record is a named collection of data elements or data aggregates. When an application program reads data from a database, it may read one complete record at a time.

*Relation*
A relation is defined as a logical association of related entities or entity sets. It is represented as $R(A1, A2, \ldots, Ai)$, where $A1, A2$, etc. are the attributes of the relation R. Data in a relation can be represented in the form of a table. Each column of the table is called its attribute. Attributes may be integers, reals, characters, fixed length vectors, variable length vectors or matrices.

*Schemes*
When a database is to be designed, we develop plans for it. Plans consist of an enumeration of the types of entities that the database deals with, the relationships among the entities, and the ways in which the entities and relationships at one level of abstraction are expressed at the next lower level. The term scheme (schema) is used to refer to plans, so we talk about conceptual schemes and physical schemes. The plan for a 'view' is referred to as a subscheme (subschema).

*Secondary key*
The database may also use a key which does not identify a unique record but identifies all those which have certain properties. This is called the secondary key.

*Storage address (address)*
This is a label name or number that identifies the place where data is stored in a storage device, or the part of a machine instruction that specifies the allocation of an operand or the destination of a result.

*String data*
String data are either of the type character or bit. The length of the string data item is equivalent to the number of characters (for a character string) or the number of binary digits (for a bit string) in the item.

*Subscheme (view)*
The map of a programmer's view of the data is called a subscheme. It is derived from the global logical view of the data – the schema, and external schema. It is an abstract view of a portion of the conceptual database or conceptual scheme. A scheme may have several subschemes. These are defined using the data definition language (DDL).

*Systems programmer*
This is a person responsible for installation and maintenance of computer programs.

*Vectors*
These are one-dimensional ordered collections of data items, all of which have identical characteristics. The dimension of a vector is the number of data items contained in it.

*Word*
This is the standard main storage allocation unit for numeric data. A word consists of a predetermined number of byte characters or bytes, which is addressed and transferred by the computer circuitry as an entity.

### 9.2.3 Physical data storage terminology

*Address*
This is a means of assigning data storage locations and subsequently retrieving them on the basis of key for the data.

*Bit*
This is an abbreviation of binary digit. The term is extended to the actual representation of a binary digit in a storage medium through an encoded two state device.

*Byte*
This is a generic term to indicate a measurable portion of consecutive binary digits and is the smallest main storage unit addressable by hardware. In machines with character addressing, byte and character are synonymous.

*Character*
This is a member of a set of elementary symbols that constitute an alphabet interpretable by computer software. It is also a group of consecutive bits that is used to encode one of the above symbols.

*File*
A file is a named collection of all occurrences of a given type of logical records. It is also a collection of data sets.

*Page*
This is a basic unit of primary storage; also basic transaction unit between primary and secondary storage.

*Paging*
In virtual storage systems, the computer memory is made to appear larger than it is by transferring blocks (pages) of data or programs into memory from external storage when they are needed. This is called paging.

*Pointer*
This is the address of a record (or other data grouping) contained in another record so that a program may access the former record when it has retrieved the latter record. The address can be absolute, relative or symbolic.

*Physical data structure*
It is important to distinguish explicitly between logical data structures and the ways in which these structures are represented in the memory of a particular computer. This may be dictated by specific hardware and software systems. The way in which a particular logical data structure is represented in the memory or secondary storage of a computer system is known as storage or physical data structure.

*Sequential access*
A serial access storage device can be characterized as one that relies strictly on physical sequential positioning and accessing of information.

*Storage*
This is the process of assigning specific areas of storage to specific type of data.

*Transaction*
An operation performed on a file or physical database is called a transaction, such as create/insert a record, delete a record, update a record, and find a record.

*Virtual memory*
This is the simulation of large capacity main storage by a multilevel relocation and paging mechanism implemented in the hardware.

### 9.2.4 Data structures

A structure whose elements are items of data, and whose organization is determined both by the relationship between the data items and by the access functions that are used to store and retrieve them, is called a data structure (Baron and Shapiro, 1980). Examples of data structures are a list, a tree, a vector of fixed length, a vector of variable length, different types of matrices, a relation, etc. Some of the data structures that can be useful for design optimization are discussed below.

*Linear structure*
A linear structure is one that is stored in the order in which the data are processed. These data can then be used to manipulate information such as insertion and deletion of elements of information. Two such structures are linear list and linked list.

- A **linear list** is a finite ordered list of elements. For example, records in a file containing the design variable value, the upper bound and the lower bound, constitute a linear list. Another example is a vector that is often used to store information whose size (number of elements) is not known *a priori* or one that cannot be created all at once. In this case, the vector is used as a buffer and the buffer is filled with one element at a time. The buffer is emptied once it is full. A disadvantage with a linear list is that if elements in the list must be deleted or inserted then either the entire list or a portion of the list must be modified.
- In a **linked list**, each element contains a pointer to its successor. Only a linked list where each element contains a pointer to the next element (singly linked list) is discussed here. The process of insertion is carried out by modification of one pointer – the pointer preceding the new entry (the pointer value of the new entry now is the same as the old pointer value of the preceding entry). Deletion follows the same process – the pointer value of the preceding entry must be updated.

*Trees*

A tree is a data structure that is used to represent hierarchical relationships among data items. A tree is described by nodes; the top level node is the root node. Each node may be connected to a node at the lower level that is the child of the parent node. Figure 9.1 shows a tree structure. As an example of data retrieval from node 17, one must go to nodes 4 and 9 first. Trees form a basis for hierarchical data models.

*Matrices*

These are the most widely used data structures in engineering programs. They are usually stored in memory in contiguous blocks and are accessed by a simple accessing function. A one-dimensional array is called a vector. Arrays can be of two, three or higher dimensions and are stored either in the column order or row order. Matrices can be of different forms, e.g. rectangular, symmetric, banded, upper triangular, lower triangular, sparse, diagonal, etc. Usually a matrix contains

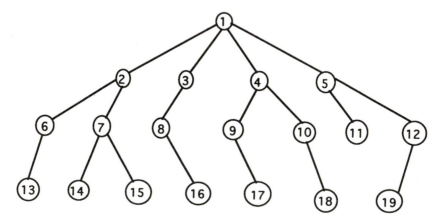

**Fig. 9.1** Hierarchical database model.

only one type of data, such as reals or integers. Another similar data structure that is now commonly used is a table where mixed data types can be stored. This forms the basis of the relational data model that is discussed later.

## 9.3 DATABASE MANAGEMENT CONCEPTS FOR DESIGN OPTIMIZATION

Data management systems in business applications such as accounting, inventory control and task scheduling, are quite sophisticated. However, due to the complex nature of engineering applications, growth of their data management systems has been very slow, although advancements in computer-aided design (graphics) have led to the development of some database systems (DBMS) and concepts (Ulfsby, Steiner and Oian, 1981; Fischer, 1979; Comfort and Erickson, 1978; Fulton and Voigt, 1976; Browne, 1976). Recently the problem has been investigated relative to computer-aided structural analysis and design applications (Rajan, 1982; Rajan and Bhatti, 1983, 1986; SreekantaMurthy and Arora, 1986a,b; Mukhopadhyay and Arora, 1987a, b). This has led to the development of some concepts that are particularly suitable for engineering applications.

### 9.3.1 Elements of a database

A database can be viewed at many levels of abstraction, such as user's view, conceptual view and physical view. Only the physical database exists on some secondary storage device. The implementation of a physical database is based on a conceptual scheme that is defined in an application program using a data definition language (DDL). The external view represents the data as seen by the interactive terminal users and application programmers. The conceptual view deals with inherent nature of data occurring in the real world and represents a global view of the data. The data organization describing the physical layout is dealt at the internal level. These views – external, conceptual and internal – have been suggested by ANSI/SPARC (American National Standards Institute/Standards Planning and Requirements Committee). A major advantage of using the database capability is that the logical relationships of data are uncoupled from the physical storage scheme; the user need not make calls to files on the computer storage device.

### (a) Conceptual data model

Conceptual model (or conceptual scheme) describes the layout of all types of data that need to be stored in a database. It represents real world data independent of any computer constraints and therefore provides a theoretical basis for organizing data of finite element analysis and design optimization problems. A logical approach for conceptualization of data is through information collection and its analysis (Date, 1977). Conceptual model can be derived either by first forming global views and then deriving local views, or by aggregating local views to form a global view. The first approach involves entity identification, relation formation, and name assignment. The other approach is based on combining segregated views of usage and information contents of individual perspectives.

Information collection is through identification of entities based on their properties. To illustrate how this may be done in design optimization, consider the entity PLATE. The entity has properties of two side dimensions, thickness, and material constant $E$. The associated property values are, say, 100, 50, 0.5, and $10^7$, respectively. These property values belong to a certain domain, for example, domain of property 'dimension' is $x$, $0 \leqslant x \leqslant 200$. Entity sets are formed by considering entities so identified. Unique names are assigned to the entities and the attributes so formulated. Analysis of information about conceptual objects – entities, attributes and relations – leads to the formation of a conceptual model which replaces the real world information.

### (b) Internal model

Logical organization of data to be stored on physical storage media is described by an internal model. It is basically an organization of elementary relations or parts of them, and storing them as a unit to reduce the number of accesses. One approach to define an internal model is by means of relations and use of normalization criteria to obtain relations consistent with the conceptual model (Date, 1977; Martin, 1977). Such a collection of relations reduces redundancy, eliminates undesired anomalies in storage operations, and ensures database integrity. Well-organized procedures need to be used to design the internal data model.

### (c) External model

Data structure as seen by an application program or interactive user is called an external data model. Data retrieved from actual physical storage in the database undergoes transformation until it reaches the user. Transformation involves rearrangement of data from internal level to external level into a form acceptable to the application program.

One of the important requirements of a database is to provide a facility for data retrieval by different application programs and users depending on their needs. Different application programs can have different views of a database. To illustrate this consider the needs of two finite element analysis programs. Let one finite element analysis program use a skyline approach to assemble and solve the governing equations. Another program using a hypermatrix approach to perform a similar task needs to use the same database. Basically, the two application programs using some common data such as geometry, material and other finite element idealization data should be able to derive them through an existing database. This aspect of catering to the needs of different applications is possible through an external model.

### 9.3.2 Elements of a database management system

A database management system (DBMS) is a software system consisting of several programs or subprograms that manages a database created by it. It has capabilities to creat new data sets, delete data sets, modify existing data sets, add or delete data items from data sets and respond to queries about the data in the database. In this section, basic organization and components of a DBMS are presented and explained.

Three types of DBMS can be identified.

- A context-free DBMS is designed to work as a stand-alone package. An application program executes under the control of DBMS, i.e. it acts as the main program. The data manager must be modified or extended when a new application is introduced.
- An application-dependent DBMS is designed for a particular application. Their data definition, data manipulation and query languages use syntax of the application. It is not possible to use the DBMS for other applications unless extensive modifications are done.
- An application-independent DBMS, designed based on the concept of a library of subroutines, is not tied to any particular application. Any application program can call DBMS facilities to define, manipulate and query its database. This type of DBMS is more suitable for design optimization applications.

For more details on different types of DBMS and capabilities of some existing systems, the article by SreekantaMurthy and Arora (1985) should be consulted.

At a higher level of abstraction, a DBMS has two major components as shown in Fig. 9.2: data language interface (DLI), and data storage interface (DSI). These components can have many subcomponents to accomplish their tasks.

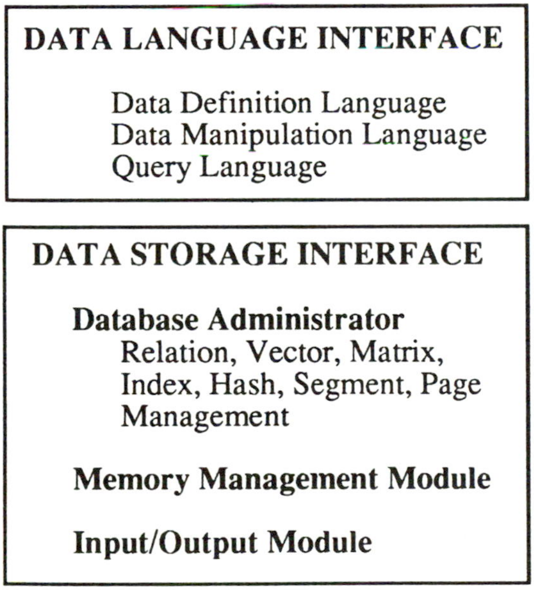

---

**DATA LANGUAGE INTERFACE**

Data Definition Language
Data Manipulation Language
Query Language

---

**DATA STORAGE INTERFACE**

**Database Administrator**
    Relation, Vector, Matrix,
    Index, Hash, Segment, Page
    Management

**Memory Management Module**

**Input/Output Module**

---

**Fig. 9.2** Components of a database management system.

### (a) Data language interface (DLI)

The data language interface is the external interface which can be called directly from a programming language. The high-level data language is embedded within the DLI and is used as the basis for all data definition and manipulation. The language can be in the form of interactive commands, subroutine calls, or some higher-level language constructs. The DLI usually has three subcomponents.

- Data definition language (DDL) is a means to describe data types and logical relations among them. It allows one to assign unique names to data types, to specify the sequence of occurrence, to specify the keys, to assign length of data items, to specify the dimension of a matrix and to specify the password for database security. Since data definition is continuously redefined in an optimization program, DDL must have features to define data dynamically. Also provisions for query of schema must be provided.
- Data manipulation language (DML) contains commands for storing, retrieving and modifying data in the database. These commands should be simple and callable from a higher-level language (such as FORTRAN) as they are frequently used in an application program. They also include utility and schema information commands. Utility commands are used for opening and closing a database, and printing error messages. Schema commands are useful in verifying data definition in situations where they are continuously changing.
- Query language allows a user to interrogate and update the database. Even though some of the design optimization algorithms are automated, it is necessary to provide flexibility and control to the user for modifications during the design process. This control is useful to execute decisions that either cannot be automated or are based on designer's intuition and judgement. For example, a designer may want to change lower and upper limits on the design variables during the iterative process. The query commands should be general and simple enough to be understood by a nonprogramming user. Some typical queries are FIND, LIST, SELECT, PLOT, CHANGE, ADD, DELETE, RENAME, OPEN and CLOSE. Query of large matrices requires special conditional clauses so that data may be displayed in parts. Formatted display of data is essential while dealing with floating point numbers.

    In design optimization, a convenient query language is essential. Using the query commands, the applications programmer can formulate his design optimization problem. Cost and constraint functions can be defined by querying the database. Active set of constraints and their values can be obtained using appropriate commands. This use of query language has been demonstrated in structural design optimization (Rajan, 1982).

### (b) Data storage interface (DSI)

Data storage interface is an internal interface that handles all management chores and accesses to data items. It manages space allocation, storage buffers, transaction consistency, system recovery etc. It also maintains indexes on selected fields of relations and pointer chains across relations.

The DSI should be designed in such a way that new data objects or new indexes can be created at any time or existing ones destroyed without exiting or modifications of the system, and without dumping and reloading the data.

One should be able to redefine objects, i.e. change dimensions of matrices or add new fields to relations. Existing programs which execute DSI operations on data aggregates should remain unaffected by the addition of new fields. The DSI has three major subcomponents as shown in Fig. 9.2.

- The database administrator is the brain of the system containing all the administrative information. It has facilities to manage relations, vectors, matrices, pages, indexes etc. Proper algorithms need to be investigated for these tasks for efficient DBMS operations.
- Each DBMS has an internal memory block (often called buffer) to store management information as well as data. This memory needs to be managed judiciously for efficiency of the database operations. The memory management module dynamically controls the allocation of available memory space. The memory is organized into a number of pages, each having the same size. The size of a page is set to a multiple of the physical record. The performance is better with larger page size; however, the space may be wasted if there are too many partially full pages. Small page size leads to increased page replacement activity and maintenance of a larger page table. Variable length pages require more programming effort.
- The input/output module is responsible for all disk operations, i.e. writing and reading of data records on some secondary storage device. For efficiency of operations, system software facilities may be used, making the DBMS system dependent.

### 9.3.3 Data models

Data in a database can be defined in terms of data sets. Other common approaches are through data models, e.g. hierarchical, network and relational. These data models that represent user's view of the data are described in the sequel with reference to finite element analysis and design optimization data. In all the models, the data definition language is used to define the data structures for the DBMS.

### (a) Data set approach

In this approach, the data is organized using uniquely named data sets. Data sets are grouped to form a data library. How the contents of the data sets are managed is completely up to the application programs. Since many of the engineering data are unstructured, the data set offers a simple solution to describe the user's view of the data. Further improvement in this type of organization can be done by defining ordered data sets. For example, row, column or submatrix order may be used to deal with matrix data. The data libraries formed by this approach may be classified according to project or their usage. For example, data of substructures in finite element analysis may be grouped substructure-wise, each in a separate library. This type of data modeling, however, has high redundancy. Also, it is not suitable for interactive use.

### (b) Hierarchical model

Hierarchical model organizes data at various levels using a simple tree structure (Date, 1977). This structure appears to fit data of many design problems modeled

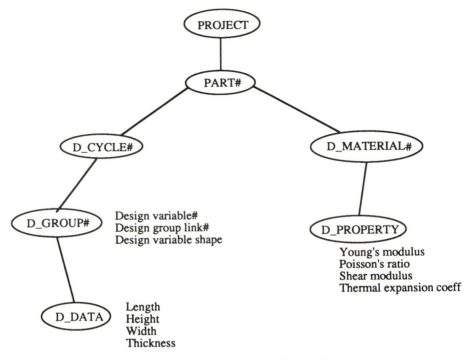

**Fig. 9.3** A hierarchical model for design variable data.

using finite elements (Pahl, 1981; Elliott, Kunii and Browne, 1978; Lopez et al., 1978). To illustrate an application of this model, consider the design variable data as shown in Fig. 9.3. The root level in the model represents project name, followed by part number in the immediately next lower level. Part number has two dependent nodes having design cycle (D_CYCLE) and material number (D_MAT) data. At the fourth level design variable group data (D_GROUP) consisting of design variable number, design group link, and design variable shape data are indicated. Material properties data (D_PROPERTY) is also set up at the same level. Lowermost level consists of detailed design data (D_DATA). To retrieve design variable data for a component, one must go through the hierarchy of nodes PART#, D_CYCLE#, D_GROUP#, and D_DATA.

## (c) Network model

A collection of arbitrarily connected logical relations is called a network. A data model defined by such a network is called the network model. It is more general than the hierarchical model as it allows many-to-many relations. In the hierarchical model, a parent node can have many child nodes, but a child node must have only one parent. In the network model, a child can also have multiple parent nodes. Disadvantages of the model are in its complexity and the associated data definition language.

**Table 9.1** Relation CORD for node coordinate
data

| NODE# | X | Y | Z |
|-------|------|------|-----|
| 1 | 720. | 0. | 0. |
| 2 | 720. | 360. | 0. |
| 3 | 360. | 0. | 0. |
| 4 | 360. | 360. | 0. |
| 5 | 0. | 0. | 0. |
| 6 | 0. | 360. | 0. |

## (d) Relational model

Both the hierarchical and network data models have proven to be functionally
inefficient. Therefore, the relational model has been developed (Date, 1977;
Martin, 1977; Vetter and Maddison, 1981). This model is constructed using a
tabular representation of the data which provides a convenient way of
representing many engineering data. Each row of the table represents an entity
and each column represents an attribute of the relation. Table 9.1 shows an
example of a relation CORD for the node coordinate data. This relation has
four attributes: NODE#, and X, Y and Z coordinates of the node. There is a
distinct relationship between a row entry and the corresponding column entry,
and redundant data are not allowed. The mathematical concept of a relation
requires the definition of a domain, Cartesian product, and tuples. A set of values
in one column is the domain of that attribute, e.g. attribute X for the relation
CORD has domain (0., 360., 720.). If there are $m$ domains, then their Cartesian
product defines a set of $m$-tuples (there are as many tuples as there are columns
in the table), e.g. each entity (row) of the relation CORD has 4 tuples. In this
model, the relational operations, such as JOIN, INTERSECT and PROJECT,
can be used on some elementary relations to form new relations.

The relational model is quite appropriate for design optimization applications,
since retrieval of data requires smaller preconceived paths. Applications generally
require a complete set of related items simultaneously. Retrieving parts of
information is not useful. In such a case, the relational model which is set oriented
provides a suitable way to organize the design data.

## (e) Numerical model

Most of the computations in design optimization involve operations on matrices
such as matrix addition, multiplications, solution of simultaneous equations, and
eigenvalue calculations. The data models presented earlier are not tailored to
handle matrix data effectively. It is necessary to provide a user-friendly facility
for defining such numerical data and manipulating a numerical database (Daini,
1982; Rajan and Bhatti, 1983). It is possible to provide such a facility by defining
a new data model called numerical model. This numerical model is basically a
variation of hierarchical data model having two levels of data representation. At
the first level, information pertaining to type, size, and other attributes of the
matrix is placed. The second level contains the actual numerical data.

(f) Generalized relational model

Whereas most physical models in business applications can be represented conveniently in the database as relations, engineering applications require both numerical and relational data types. Most large matrices form temporary or semipermanent data private to a program. Relations are either permanent data in public domain used by different users or final results of a program to the end user. Therefore a scheme is needed to represent both relations and matrices in a unified way for integrated engineering applications.

A relation (table) can be imagined as simply a two-dimensional array in which each column has a unique definition. This way, the concept of a matrix is generalized for relations and the resulting model is called the generalized relational model (Mukhopadhyay and Arora, 1987a, b). This scheme is supported by primitive and structured data types. Using these data types users can define their own data models.

The novelty of this model is that the relation is derived from a matrix. In previous attempts, e.g. RIM database management system (Comfort and Erickson, 1978), one tried to extend the relation for matrix data type. That led to clumsy and inefficient handling of numerical data. With the generalized relational model, matrix is the basic data type. The matrix can have elements with composite data structures. Relations are derived from the matrices as vectors of records.

### 9.3.4 Choice of data model

It is seen that the data models described in previous paragraphs can be used for design optimization applications. The hierarchical data model is suitable where the data to be organized occurs in a truly hierarchical fashion. It has been tried by Lopez (1974) and Pahl (1981) for finite element analysis applications. The model, however, requires complex maneuvers through a chain of pointers to access particular data. Also, it has a fixed structure and offers little flexibility to change to alternate structures. Another drawback of the model is the complexity of database design requiring a tedious process of establishing links between data. If new kinds of data need to be added or new information must be generated from the database, it is necessary to add new links. Generally, this process requires a redesign of the database. Network models have similar problems, although addition of new item is much easier compared with the hierarchical model.

Relational data model provides maximum flexibility of all the three basic data models. Moreover, the model is easier to understand as users find it natural to organize data in the form of tables. A major advantage of the model is the ease with which database can be changed. As the design evolves new attributes and relations can be added, and existing ones deleted easily. The model is more appropriate for design applications, since data storage and retrieval uses a less preconceived path. It is possible to support a simple query structure using this model. Fishwick and Blackburn (1982) have tried this model for finite element analysis and optimization applications with success. Their use of the model was, however, limited to interfacing various programs using an available relational DBMS. A major drawback of the relational model is that it does not provide means to handle large matrix data very efficiently which occurs frequently in engineering applications. A numerical model which is basically a variation of the hierarchical model appears to be quite effective in representing matrix data structures. One can also use the generalized relational model that can treat relations and numerical data in an integrated manner. A prototype DBMS based on this model has been developed and implemented (Mukhopadhyay and Arora, 1987a,b). It has been successfully used in structural optimization applications (Al-Saadoun and Arora, 1989; Park and Arora, 1987; Spires and Arora, 1990; SreekantaMurthy and Arora, 1987). Therefore this model is recommended for engineering applications.

### 9.3.5 Global and local databases

Computer-aided design of complex structural systems uses several application programs during the design process. Many of these programs require common information such as geometry of the structure, finite element idealization details, material properties, loading conditions, structural stiffness, mass and load distributions, and responses resulting from the analysis runs. Also, it is common that data generated by one program is required for processing in subsequent programs in a certain predetermined pattern. These data do not include transitory information such as intermediate results generated during an analysis run. The transitory information is usually unstructured and its usage pattern is known only to applications that use them. Generally, the transitory information is deleted at the end of a run. Therefore, there is a need for systematic grouping of the data.

A network of databases offers a systematic approach to support data of multiple applications (Jumarie, 1982; Blackburn, Storaasli and Fulton, 1982). A network of databases consists of a global database connected to a number of local databases through a program data interface. Application programs which use them may be thought of as links connecting the databases. A global database contains common information required for all applications whereas a local database contains only application dependent transitory data. Data in a global database is structured and integrity of the database is maintained carefully. Data in a local database, however, is extremely flexible and integrity is not of importance.

The network of databases offers considerable aid in the structural design process. Any changes made to the data in the global databases are immediately available for use in other applications of the system. Any new application program can be added to share the common data. The data views in global databases are clear to all applications and any modified views can be easily incorporated to suit a new application. Local databases are dependent on application programs and are highly efficient in data transfer operations since no overhead is involved in maintaining complicated data structures.

## 9.4 DATABASE DESIGN TECHNIQUES

The problem is how to organize data in a database, what kind of information is to be stored, what kind of database management system is suitable, and how data is manipulated and used. In this regard, sophisticated techniques are available in business data management area to deal with complex data organization problems. The paper by Koriba (1983) describes several of these approaches and their suitability to computer-aided design (CAD) applications. Buchmann and Dale (1979) and Grabowski and Eigner (1982) have also studied these approaches relative to CAD applications. Most commonly known approaches are ANSI/SPARC, CODASYL (Committee on Data System Languages, Database Task Group of ACM), relational, hierarchical, and network. Among them, the ANSI/SPARC approach which recognizes three levels of data views (conceptual, internal and external), provides a generalized framework and basis for a good database design.

Designing a good database is important for successful implementation of finite element analysis and structural design optimization methods (SreekantaMurthy and Arora, 1986a,b). The design procedure should follow well-defined steps. The basic problem is that once all the data items have been identified, how should they be combined to form useful relations. The first step is the extraction of all the characteristics of the information that is to be represented in the database. Analysis of the information to form associations and their integration into one conceptual model is the second step. The conceptual data model obtained by this process is abstract. It is independent of any computer restraint or database management software support. In order for the conceptual model to be useful, it must be expressed in terms that are compatible with a particular DBMS by considering efficiency of storage space and access time. An internal model is developed for this purpose which is compatible with the conceptual data model. Finally, the database design requires accommodation of different users of the database by providing an external data model. The systematic process by which one traverses the different steps of database design and performs the mapping from one level of abstraction to the next is called a database design methodology.

In this section, a methodology based on the ANSI/SPARC approach to design databases for finite element analysis and structural design optimization applications is presented. The methodology considers the following aspects: (i) three views of data – conceptual, internal, and external; (ii) entity set, relationship set, and attributes to form syntactic basic elements of the conceptual model; (iii) relational data model; (iv) matrix data; (v) processing requirements; (vi) normalization of data for relational model. The material for the section is derived from an article by SreekantaMurthy and Arora (1986b).

### 9.4.1 General concepts

Before presenting the database design methodology a few general concepts that are useful in the design process are described. The idea of an entity–relationship model is explained. Various forms of dependencies between the attributes to form relations are ·explained. The idea of normalization of data is presented and explained.

### (a) Entity–relationship

The idea of entities and relationships between them is important in analyzing data and their associations. To explain this concept, consider the entities PRE-

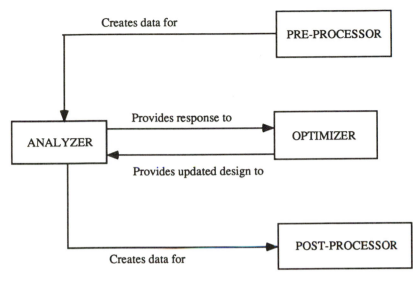

**Fig. 9.4** Entity–relationship concept.

PROCESSOR, ANALYZER, POST-PROCESSOR, and OPTIMIZER. Figure 9.4 shows relationships between these entities. PRE-PROCESSOR 'creates data for' the ANALYZER; the ANALYZER determines the structural response and 'provides response to' the OPTIMIZER, and the OPTIMIZER 'provides updated design to' the ANALYZER. The ANALYZER also 'creates data for' the POST-PROCESSOR which provides visual displays for the response quantities. The entity–relationship model for an application is useful in identifying data dependencies and sequence of data usage. This idea of entity–relationship can be also used to identify 'objects' in designing object-oriented design systems.

### (b) Functional dependence

An attribute $A$ is functionally dependent on the attribute $B$ of a relation $R$ if at every occurrence of a $B$ value is associated with no more than one $A$ value. This is denoted as $R.B. \rightarrow R.A.$ As an example, consider the relation ELEMENT (EL#, EL-NAME, AREA), where EL# is the element number, EL-NAME is its name and AREA is its cross-sectional area. EL-NAME is functionally dependent on EL#(EL# $\rightarrow$ EL-NAME). AREA is functionally dependent on EL#(EL# $\rightarrow$ AREA). EL# is not functionally dependent on EL-NAME(EL-NAME $\nrightarrow$ EL#), because more than one element could have the same name. Similarly, EL# is not functionally dependent on AREA(AREA $\nrightarrow$ EL#).

An attribute can be functionally dependent on a group of attributes rather than just one attribute. For example, consider the relation CONNECTION for nodal connectivity of triangular finite elements:

$$\text{CONNECTION(EL\#, NODE1\#, NODE2\#, NODE3\#)}$$

Here EL# is functionally dependent on three nodes NODE1#, NODE2#, and

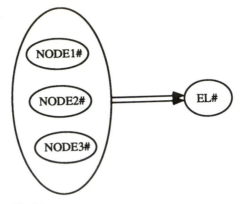

**Fig. 9.5** Example of full functional dependency.

NODE3#. Given any one of NODE1#, NODE2#, or NODE3# it is not possible to identify EL#.

### (c) Full functional dependency

An attribute or a collection of attributes $A$ of a relation $R$ is said to be fully functionally dependent on another collection of attributes $B$ of $R$ if $A$ is functionally dependent on the whole of $B$ but not on any subset of $B$. This is written as $R.B. \Rightarrow R.A$. In Fig. 9.5, for example, EL# in the relation CONNECTION of a triangular finite element is fully functionally dependent on concentrated attributes NODE1#, NODE2#, and NODE3# because three nodes combined together define an element. NODE1#, NODE2#, or NODE3# alone does not identify the EL#.

### (d) Transitive dependence

Suppose $A, B$ and $C$ are three distinct attributes or attribute collections of a relation $R$. Suppose the following dependencies always hold: $C$ is functionally dependent on $B$ and $B$ is functionally dependent on $A$. Then $C$ is functionally dependent on $A$. If the inverse mapping is nonsimple (i.e. if $A$ is not functionally dependent on $B$ or $B$ is not functionally dependent on $C$), the $C$ is said to be transitively dependent on $A$ (Fig. 9.6). This is written as

$$R.A \rightarrow R.B, \quad R.B \nrightarrow R.A, \quad R.B \rightarrow R.C$$

Then, we can deduce that

$$R.A \rightarrow R.C, \quad R.C \nrightarrow R.A$$

As an example, consider the relation EL_DISP between element number, element type and degrees of freedom per node:

$$EL\_DISP(EL\#, EL\text{-}TYPE, DOF/NODE)$$

Here

$$EL\# \rightarrow EL\text{-}TYPE, \quad EL\text{-}TYPE \nrightarrow EL\#$$
$$EL\text{-}TYPE \rightarrow DOF/NODE$$

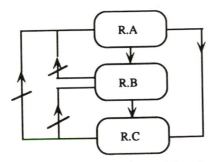

**Fig. 9.6** Transitive dependence of *C* on *A*.

Therefore

EL# → DOF/NODE (transitively dependent)
DOF/NODE ↛ EL#

### 9.4.2 Normalization of data

It is seen that data items must be grouped together to form associations. The question is, how to decide what data items should be grouped together? In particular, using a relational model, what relations are needed and what should their attributes be? As a database is changed, older views of data must be preserved so as to avoid having to rewrite the programs using the data. However, certain changes in data associations could force modification of programs which could be extremely disruptive. If grouping of data items and keys is well thought out originally, such disruptions are less likely to occur.

Normalization theory (Date, 1977; Martin, 1977; Vetter and Maddison, 1981) provides certain guidelines to organize data items together to form relations. The theory is built around the concept of normal forms. A relation is said to be in a particular normal form if it satisfies a certain specified set of constraints. Three normal forms – first, second and third – are described below.

### (a) First normal form (1NF)

A relation is said to be in the first normal form if and only if it satisfies the constraint of having atomic values. As an example, Fig. 9.7 shows the relation CONNECTION between four attributes: element number (EL#), element name (EL-NAME), node numbers (NODES#) and degrees of freedom per node (DOF/NODE) with domains $D_1, D_2, D_3$ and $D_4$, respectively. The relation is first shown not in the 1NF and then in the 1NF.

### (b) Second normal form (2NF)

A relation is in second normal form if and only if it is in 1NF and every nonkey attribute is fully functionally dependent on each candidate key. Let us see whether the relation CONNECTION of Fig. 9.7 that is in the 1NF is also in the 2NF.

| Domain $D_1$ | Domain $D_2$ | Domain $D_3$ | Domain $D_4$ |
|:---:|:---:|:---:|:---:|
| EL# | EL-NAME | NODES# | DOF/NODE |
| 1<br>2<br>3 | BEAM3<br>TRUSS3<br>PLATE | 1<br>2<br>3<br>4<br>5 | 6<br>3<br>2 |

**CONNECTION**

| Key | | Key | |
|:---:|:---:|:---:|:---:|
| EL# | EL-NAME | NODES # | DOF/NODE |
| 1 | BEAM3 | 1<br>2 | 6 |
| 2 | TRUSS3 | 3<br>5 | 3 |
| 3 | PLATE | 2<br>3<br>4<br>5 | 2 |

| Key | | Key | |
|:---:|:---:|:---:|:---:|
| EL# | EL-NAME | NODES# | DOF/NODE |
| 1 | BEAM3 | 1 | 6 |
| 1 | BEAM3 | 2 | 6 |
| 2 | TRUSS3 | 3 | 3 |
| 2 | TRUSS3 | 5 | 3 |
| 3 | PLATE | 2 | 2 |
| 3 | PLATE | 3 | 2 |
| 3 | PLATE | 4 | 2 |
| 3 | PLATE | 5 | 2 |

Not in 1NF                                          In 1NF

**Fig. 9.7** First normal form for the relation CONNECTION.

Consider a nonkey attribute EL-NAME:

$$(EL\#, NODES\#) \rightarrow EL\text{-}NAME$$
$$EL\# \rightarrow EL\text{-}NAME$$
$$NODE\# \nrightarrow EL\text{-}NAME$$

Therefore $(EL\#, NODES\#) \nrightarrow EL\text{-}NAME$, i.e. EL-NAME is not fully functionally dependent on $(EL\#, NODES\#)$. Similarly for the nonkey attribute DOF/NODE:

$$(EL\#, NODES\#) \rightarrow DOF/NODE$$
$$EL\# \rightarrow DOF/NODE$$
$$NODE\# \nrightarrow DOF/NODE$$

Therefore $(EL\#, NODES\#) \nrightarrow DOF/NODE$. Since neither EL-NAME nor DOF/NODE is fully functionally dependent on candidate key $(EL\#, NODES\#)$, the relation CONNECTION is not in 2NF.

Conversion of the relation CONNECTION to 2NF consists of replacing it by two of its projections as shown in Fig. 9.8 (note: ← implies PROJECT operation):

NAM_DOF ← CONNECTION (EL#, EL-NAME, NODES#, DOF/NODE)
EL_NODE ← CONNECTION(EL#, EL-NAME, NODES#, DOF/NODE)

Relation EL-NODE does not violate 2NF because its attributes are all keys.

(c) Third Normal Form (3NF)

A relation is in the third normal form if it is in second normal form and its every nonprime attribute is nontransitively dependent on each candidate key of the relation. For example, consider the relation NAM-DOF (Fig. 9.8) to see whether

NAM_DOF

| EL# | EL-NAME | DOF/NODE |
|-----|---------|----------|
| 1 | BEAM3 | 6 |
| 2 | TRUSS3 | 3 |
| 3 | PLATE | 2 |

ELMT_NODE

| EL# | NODE# |
|-----|-------|
| 1 | 1 |
| 1 | 2 |
| 2 | 3 |
| 2 | 5 |
| 3 | 2 |
| 3 | 3 |
| 3 | 4 |
| 3 | 5 |

**Fig. 9.8** Second normal form for the relation CONNECTION.

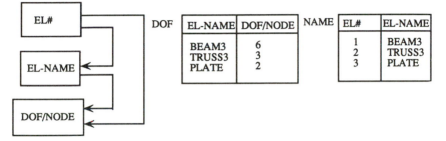

DOF

| EL-NAME | DOF/NODE |
|---------|----------|
| BEAM3 | 6 |
| TRUSS3 | 3 |
| PLATE | 2 |

NAME

| EL# | EL-NAME |
|-----|---------|
| 1 | BEAM3 |
| 2 | TRUSS3 |
| 3 | PLATE |

**Fig. 9.9** Third normal form for the relation NAM_DOF.

it is in third normal form. It still suffers from a lack of mutual independence among its nonkey attributes. The dependency of DOF/NODE on EL#, though it is functional, is transitive (via EL-NAME). Each EL# value determines an EL-NAME value and in turn determines the DOF/NODE value. This relation is reduced further into relations NAME and DOF in Fig. 9.9. These relations are in the third normal form.

### 9.4.3 Methodology to develop a conceptual model

Analysis of the data used in finite element analysis and structural design optimization is necessary to develop a conceptual data model. In the analysis, the information in use or needed later is identified, classified, and documented. This forms the basis for a conceptual data model to represent structural design data and the design process as a whole.

The following steps can be used to develop a conceptual data model.

1. Identify all the conceptual data objects of structural analysis and design optimization.
2. Data identified is stored in a number of relations. The data is reduced to elementary relations representing inherent association of data.
3. More elementary relations are derived from the ones formed in step 2. This step uncovers more relationships between the basic data collected in step 1.

4. Redundant and meaningless relations obtained in step 3 are removed to obtain a conceptual data model.

The conceptual model obtained by this process is abstract, representing the inherent nature of structural design data and is independent of any computer restraint or database management software support. These steps are now discussed in detail.

### (a) Identification of conceptual data objects

The following steps can be used to identify the conceptual data objects used in structural design. Entity sets and attributes are considered to be the syntactic basic elements of the model. Domain definition can be extended to include vectors and matrices.

- *Step 1.* Identify each type of entity and assign a unique name to it.
- *Step 2.* Determine the domains and assign unique names to them. This step identifies the information that will appear in the model, such as attributes.
- *Step 3.* Identify the primary key for each type of entity depending on the meaning and use.
- *Step 4.* Replace each entity set by its primary key domains. Determine and name relations corresponding to the association between the primary key domain and other domains. This step gives a collection of relations forming a rough conceptual data model.

---

### EXAMPLE 9.1

We consider a sample structural design problem to describe these steps.

- *Step 1.* The following entity sets can be identified for the structure:

  STRUCTURE, BEAM, TRUSS(TR), MEMBRANE-TRI(TRM),
  MEMBRANE-QD(QDM), NODE, ELEMENT(EL)

- *Step 2.* We can identify the following domains:

| | |
|---|---|
| STRUCTURE# | Structure identification number (integer) |
| BEAM# | Beam element identification number (integer) |
| TR# | Truss element identification number (integer) |
| TRM# | Triangular membrane element identification number (integer) |
| QDM# | Quadrilateral membrane element identification number (integer) |
| NODES# | Node number (integer) |
| EL# | Element number (integer) |
| EL-TYPE | Element type {BEAM2, BEAM3, TR2, TR3, TRM2, TRM3, QDM2, QDM3} |
| MATID | Material identification code, for example, {STEEL.1, STEEL.2, ALUM.5, COMP.1}. It also refers to a relation |

or table of material properties; for example, STEEL.1 refers to relation STEEL and material subtype 1

MATPRO    Material property $\{E, \mu, G, \ldots\}$

CSID    Cross-section-type identification code; for example, $\{$THICK.1, THICK.2, RECT.1, CIRC.5, ISEC.6, LSEC.15$\}$. It also refers to a relation of cross-sectional details. For example, RECT.1, refers to a relation RECT and a cross-section subtype 1

CSPRO    Cross-sectional property $\{H, W, T, R, \ldots\}$; $\{$height, width, thichness, radius, $\ldots\}$

DOF#    Degrees of freedom numbers

LOAD-TYP    Load type $\{$CONCENTRATED, DISTRIBUTED, TEMPERATURE, ACCELERATION$\}$

X    $X$ coordinate (real)

Y    $Y$ coordinate (real)

Z    $Z$ coordinate (real)

DESCRIPTION    Description (characters)

VEC    Vectors {integer, real, and double precision vectors}

MATX    Matrices {integer, real, and double precision matrices}

VECID    Vector identification code $\equiv \{x \cdot y | x =$ vector description, $y =$ number$\}$; for example, FORCE.5, LOAD.10

MAXID    Matrix identification code $\equiv \{x \cdot y | x =$ matrix description, $y =$ number$\}$; for example, EL-STIFF.10, EL-MASS.5

- *Step 3.* The following entity keys are identified:

STRUCTURE#    for entity set structure
BEAM#    for entity set beam
TR#    for entity set truss
TRM#    for entity set TRM
QDM#    for entity set QD
EL#    for entity set element

- *Step 4.* In the association between entity sets and domain the entity sets from step 1 are replaced by their primary keys. Attribute names are derived from domain names to provide role identification. The following relation TRM for the triangular membrane element is identified for the entity set TRM:

TRM(TRM#, EL#, EL-TYPE, MATID, E, NODE1#, NODE2#, NODE3#, CSID, T, LOAD-TYP, LOAD#, VECID, VEC, MAXID, MATX)

A triangular membrane element is identified by TRM#. Element number EL# uniquely identifies the finite elements of a structure. Attributes NODE1#, NODE2#, and NODE3# are derived from domain NODES#. Similarly, E is the role name for domain MATPRO. CSID identifies the cross-section property T, the thickness. Vectors and matrices associated with the element are identified through VECID and MAXID, respectively. Similarly, the relations TRUSS, BEAM, QDM are obtained.

(b) Reduction to elementary relations

In the previous section we described a method to identify entities, domains, and relations to produce a rough conceptual model of the structure. Our idea is to develop a conceptual model that contains all the facts and each fact occurring only once. In order to produce a conceptual data model, we transform the rough model into a better model by using a set of elementary relations (Vetter and Maddison, 1981). Using the concept of functional dependencies, full functional dependencies, and transitive dependencies, we can establish rules for reducing a relation to an elementary relation. The following steps are identified to form elementary relations.

- *Step 1.* Replace the original relations by other new relations to eliminate any (nonfull) functional dependencies on candidate keys.
- *Step 2.* Replace the relations obtained in step 1 by other relations to eliminate any transitive dependencies on candidate keys.
- *Step 3.* Go to step 5 if (a) the relation obtained is all keys, or (b) the relation contains a single attribute that is fully functionally dependent on a single candidate key.
- *Step 4.* Determine the primary key for each relation that may be a single attribute or a composite attribute. Take projections of these relations such that each projection contains one primary key and one nonprimary key.
- *Step 5.* Stop when all elementary relations are obtained.

---

EXAMPLE 9.2

To see how these steps are used, we consider the relation TRM that was given earlier and reduce it to elementary relations:

- *Step 1.*

| | |
|---|---|
| ER1(TRM#, EL#) | ER2(TRM#, EL-TYPE) |
| ER3(TRM#, NODE1#) | ER4(TRM#, NODE2#) |
| ER5(TRM#, NODE3#) | ER6(EL#, TRM#) |

- *Step 2.*

| | |
|---|---|
| ER7(TRM#, MATID) | ER8(MATID, E) |
| ER9(TRM#, CSID) | ER10(CSID, T) |
| ER11(TRM#, VECID) | ER12(VECID, VEC) |
| ER13(TRM#, MAXID) | ER14(MAXID, MATX) |

- *Step 3.* The preceding relations contain a single attribute, so go to step 5.
- *Step 4.* Skip.
- *Step 5.* ER1–ER14 are elementary relations.

The steps can be applied to the rest of the relations identified earlier to obtain a set of elementary relations for the sample structural problem.

**Table 9.2** Transitive closure for elementary relations

| Derived relations | Dependencies | Composition | Semantically meaningful |
|---|---|---|---|
| ER15 | EL# → EL-TYPE | EL# → TRM# → EL-TYPE | Yes |
| ER16 | EL# → NODE1# | EL# → TRM# → NODE1# | Yes |
| ER17 | EL# → NODE2# | EL# → TRM# → NODE2# | Yes |
| ER18 | EL# → NODE3# | EL# → TRM# → NODE3# | Yes |
| ER19 | EL# → MATID | EL# → TRM# → MATID | Yes |
| ER20 | EL# → CSID | EL# → TRM# → CSID | Yes |
| ER21 | EL# → LOAD-TYP | EL# → TRM# → LOAD-TYP | Yes |
| ER22 | EL# → MAXID | EL# → TRM# → MAXID | Yes |
| ER23 | TRM# → E | EL# → MAXID → E | No |
| ER24 | TRM# → T | EL# → CS-TYP → T | No |
| ER25 | TRM# → VEC | EL# → VECID → VEC | No |
| ER26 | TRM# → MATX | EL# → MAXID → MATX | No |

## (c) Determination of transitive closure

While deriving a large number of relations for obtaining a conceptual data model it is prossible that some relations might have been missed. In general, one can derive further elementary relations from any incomplete collection of such relations. To explain in a simple way how such additional relations can be derived, consider two relations ER1(A, B) and ER2(B,C), which imply functional dependencies: A → B and B → C. Taking the product of these functional dependencies, we obtain A → C. Therefore, from suitable pairs of elementary relations representing functional dependencies, further elementary relations can be derived. Deriving all such relations from the initial collection of elementary relations yields a transitively closed collection of elementary relations called the transitive closure (Vetter and Maddison, 1981).

There are problems associated with interpreting relations in transitive closure. For example, consider relations ER1(TRM#, MATID) where TRM# → MATID, and ER7(MATID, E) where MATID → E. Transitive closure for this set yields the relation ER(TRM#, E) which implies TRM# identifies E. This relation, however, does not represent true information since material property E is dependent only on the material number and not on the element number. The relation could be wrongly interpreted. Therefore such semantically meaningless dependencies must be eliminated. It is possible to determine transitive closure by using directed graphs and the connectivity matrix (Vetter and Maddison, 1981).

The transitive closure for the example produces additional dependencies as given in Table 9.2. We have eliminated meaningless dependencies from the list.

## (d) Determination of minimal covers

We need to remove redundant elementary relations to provide a minimal set of elementary relations. A minimal cover is the smallest set of elementary relations from which transitive closure can be derived (Vetter and Maddison, 1981). The following points are noted: (i) minimal cover is not unique, (ii) deriving several

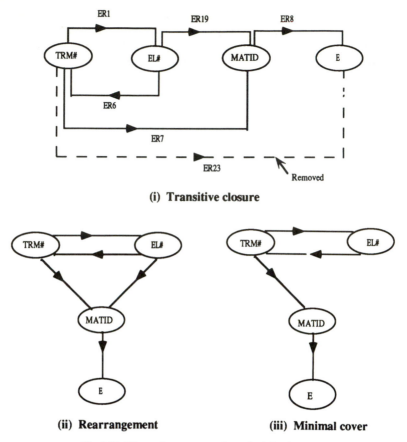

**(i) Transitive closure**

**(ii) Rearrangement**          **(iii) Minimal cover**

**Fig. 9.10** Diagraph representation of minimal cover.

alternative minimal covers from a transitive closure guarantees that every possible minimal cover is found, and (iii) we can select a minimal cover that best fits the structural design process needs.

An example of finding a set of minimal covers from the transitive closure derived in previous sections is given in Fig. 9.10. A set of minimal cover for this transitive closure is {ER1, ER6, ER7, ER8} and {ER1, ER6, ER19, ER8}, out of which one set may be chosen to suit requirements.

The preceding procedure can be applied to other transitive closures derived in the previous section. Thus we can obtain further sets of minimal covers. Each minimal cover is a nonredundant list of elementary relations and is an appropriate conceptual model of the structural design data.

### 9.4.4 Database design to support an internal model

An internal model deals with the logical organization of data to be stored on physical storage devices. Once a conceptual model has been developed, the internal model needs to be developed so that the data can be stored in files. The methodology to design such a model considers two important aspects: (i) normalization of data, and (ii) processing requirements of data. Normalization of data avoids anomalies in storage and update operations (insert, modify, delete) of data (Date, 1977). The processing requirements are important because they specify how various attributes should be derived from their underlying domains and combined together to from a relation.

An example of element stiffness matrix generation is considered to describe the methodology to design the internal model. Consider the conceptual model given by the following elementary relations:

$$ER1(TRM\#, EL\text{-}TYPE) \quad ER2(TRM\#, NODE1\#)$$
$$ER3(TRM\#, NODE2\#) \quad ER4(TRM\#, NODE3\#)$$
$$ER5(TRM\#, MATID) \quad ER6(MATID, E)$$
$$ER7(TRM\#, CSID) \quad ER8(CSID, T)$$
$$ER9(NODE\#, X) \quad ER10(NODE\#, Y)$$
$$ER11(NODE\#, Z) \quad ER12(TRM\#, MAXID)$$
$$ER13(MAXID, MATX) \quad ER14(EL\#, NODES\#)$$
$$ER15(TRM\#, EL\#)$$

Data needed for the generation of the element stiffness matrix is derived from various domains and represented in a single relation TRM_D as shown in Table 9.3. The domains used are TRM#, MATID, CSID, NODES, X, Y, Z, MAXID, MATX, EL#, EL-TYPE, CSPRO and MATPRO. Our main intention is to obtain all the data required for generation of stiffness matrices for triangular membrane elements in one access or a minimum number of accesses. It is observed that the relation in Table 9.3 is not in the first normal form. Therefore this unnormalized relation should be replaced by a semantically equivalent relation in 1NF as shown in Table 9.4. The advantage of 1NF over the unnormalized relation is that operations required for application programs are less complicated and easy to understand.

To check consistency of this model, first we identify the key attributes. Candidate keys are compound consisting of (EL#, NODES#) and (TRM#, NODES#). The primary key is selected as (TRM#, NODES#). Secondary keys are TRM#, EL#, MATID, CSID, NODES#, and MAXID. These key attributes of the relation are consistent with those in the elementary relation. Second, we need to identify whether all the attributes in the internal model and dependencies between them are consistent with the conceptual model. It can be observed that attributes NODE1#, NODE2#, and NODE3# do not appear in the relation. Therefore these there attributes should be included in the relation. The relation TRM_D is now written as:

$$TRM\_D(TRM\#, EL\#, EL\text{-}TYPE, E, CSID, T, NODE1\#,$$
$$NODE2\#, NODE3\#, X, Y, Z, MAXID, MATX)$$

The functional dependencies reflected by elementary relations ER1 to ER15 are satisfied in the internal model with the values shown in Table 9.4. Therefore at this instant the internal model is consistent with the conceptual model. However, it would be no longer consistent if arbitrary changes in the values of the table are made. Also, note that many values in the relation TRM_D are redundant.

**Table 9.3** A tentative internal model TRM_D

| TRM# | EL# | EL-TYPE | MATID | E | CSID | T | NODE# | X | Y | Z | MAXID | MATX |
|---|---|---|---|---|---|---|---|---|---|---|---|---|
| 1 | 12 | TRM3 | STEEL.2 | 1.0E8 | THICK.2 | .1 | 5 | 1. | 6. | 8. | STF.1 | [•] |
|  |  |  |  |  |  |  | 6 | 2. | 5. | 9. |  |  |
|  |  |  |  |  |  |  | 9 | 3. | 6. | 7. |  |  |
| 2 | 15 | TRM3 | ALUM.2 | 0.9E7 | THICK.3 | .2 | 15 | 6. | 8. | 10. | STF.2 | [•] |
|  |  |  |  |  |  |  | 16 | 4. | 7. | 9. |  |  |
|  |  |  |  |  |  |  | 17 | 5. | 9. | 2. |  |  |

**Table 9.4** Relation TRM_D in 1NF

| TRM# | EL# | EL-TYPE | MATID | E | CSID | T | NODE# | X | Y | Z | MAXID | MATX |
|---|---|---|---|---|---|---|---|---|---|---|---|---|
| 1 | 12 | TRM3 | STEEL.2 | 1.0E8 | THICK.2 | .1 | 5 | 1. | 6. | 8. | STF.1 | [•] |
| 1 | 12 | TRM3 | STEEL.2 | 1.0E8 | THICK.2 | .1 | 6 | 2. | 5. | 9. | STF.1 | [•] |
| 1 | 12 | TRM3 | STEEL.2 | 1.0E8 | THICK.2 | .1 | 9 | 3. | 6. | 7. | STF.1 | [•] |
| 2 | 15 | TRM3 | ALUM.2 | 0.9E7 | THICK.3 | .2 | 15 | 6. | 8. | 10. | STF.2 | [•] |
| 2 | 15 | TRM3 | ALUM.2 | 0.9E7 | THICK.3 | .2 | 16 | 4. | 7. | 9. | STF.2 | [•] |
| 2 | 15 | TRM3 | ALUM.2 | 0.9E7 | THICK.3 | .2 | 17 | 5. | 9. | 2. | STF.2 | [•] |

These inconsistencies and redundancies occur because of the anomalies in the 1NF. Thus it is not desirable to use the relation in Table 9.4 to represent the internal model. Modification to 2NF is necessary to avoid the anomalies in the storage operations (Date, 1977). The relation TRM_D should be converted into a set of semantically equivalent relations as follows:

> TRM_D1(TRM#, EM#, EL-TYPE, NODE1#, NODE2#, NODE3#,
>     MATID, E, CSID, T, MAXID, MATX)
> TRM_D2(NODE#, X, Y, Z)
> TRM_D3(EL#, NODE#)

The preceding three relations TRM_D1, TRM_D2, and TRM_D3 are all in 2NF because the first two relations do not possess any compound candidate key and the third relation has all keys. Note that by splitting the relation TRM_D no information is lost and the relations are still consistent with the conceptual model. However, the TRM_D1 relation is still not satisfactory since it can lead to anomalies in storage operations. Modification of the relation is necessary to 3NF to avoid anomalies in storage operation. Nonkey attributes must be nontransitively dependent on candidate keys to avoid these anomalies. It can be observed from the relation TRM_D1 of Table 9.5 that attributes E, T, and MATX are transitively dependent on TRM# through MATID, CSID, and MAXID, respectively. Removing these transitive dependencies, we obtain the following relations:

> TRM_D4(TRM#, EL#, EL-TYPE, NODE1#, NODE2#, NODE3#,
>     MATID, CSID, MAXID)
> TRM_D5(MATID, E)
> TRM_D6(CSID, T)
> TRM_D7(MAXID, MATX)

The preceding three relations together with TRM_D2 and TRM_D3 constitute the internal model for element stiffness matrix generation purpose. This internal model is consistent with the conceptual model identified earlier. Also, note that the number of relations in the internal model is only 6 as compared with 15 elementary relations in the conceptual model.

In summary, the following steps are necessary to derive an internal model that is consistent with the conceptual model. Normalization procedures have to be adopted at each step to reduce redundancy and to eliminate undesired anomalies in storage operation. This ensures integrity of the stored values in the database. At each step unsatisfactory relations are replaced by others.

- *Step 1*. Form relations with attributes derived from a set of domains.
- *Step 2*. Eliminate multiple values at the row–column intersection of the relation table. Vectors and matrices are considered to be single data items for this step.
- *Step 3*. The relations are in the 1NF as a result of step 2. Take projections of 1NF relations to eliminate any nonfull functional dependencies and get relations in the 2NF.
- *Step 4*. Take projection of relations obtained in step 3 to eliminate transitive dependencies to form relations in the 3NF. Thus a set of relations in the 3NF is the internal model.

**Table 9.5** Relations in 2 NF

TRM_D1

| TRM# | EL# | EL-TYPE | NODE1# | NODE2# | NODE3# | MATID | E | CSID | T | MAXID | MATX |
|---|---|---|---|---|---|---|---|---|---|---|---|
| 1 | 12 | TRM3 | 5 | 6 | 9 | STEEL.2 | 1.0E8 | THICK.2 | .1 | STF.1 | [-] |
| 2 | 15 | TRM3 | 15 | 16 | 17 | ALUM.2 | 0.9E7 | THICK.3 | .2 | STF.2 | [-] |

TRM_D2

| NODE# | X | Y | Z |
|---|---|---|---|
| 4 | 1. | 6. | 8. |
| 6 | 2. | 5. | 9. |
| 9 | 3. | 6. | 7. |
| 15 | 6. | 8. | 10. |
| 16 | 4. | 7. | 9. |
| 17 | 5. | 9. | 2. |

TRM_D3

| EL# | NODE# |
|---|---|
| 12 | 5 |
| 12 | 6 |
| 12 | 9 |
| 15 | 15 |
| 15 | 16 |
| 15 | 17 |

### 9.4.5 Some aspects to accommodate an external model

One of the important requirements of a database is to provide facility for data retrieval by different application programs depending on their needs. Different application programmers can have different views of a database. Transformations are required involving rearrangement of data from the internal level to the external level into a form acceptable to the application program. Some constraints have to be observed while designing an external model. Constraints arise while rearranging data from internal data structure to an external data structure. Any retrieval and storage operations specified on the external model must be correctly transformed into corresponding operations on the internal model, and at the same time, the internal model must be consistent with the conceptual data model. An example of how an external model is derived from an internal model is given subsequently.

Suppose a particular user would like to know the coordinates of nodes of each triangular finite element for generation of element stiffness matrices. This means that the external model

EL_CORD(TRM#, EL#, EL-TYPE, X1, Y1, Z1, X2, Y2, Z2, X3, Y3, Z3)

has to be provided for that particular user. Note that the external view EL-CORD contains data item from two different relations – TRM_D4, TRM_D2. Therefore a procedure is required to transform the internal data model (relations TRM_D4 and TRM_D2) to the external data model (relation EL_CORD). This can be done using JOIN and PROJECT operations (Date, 1977) as follows (note: ← indicates PROJECT; * indicates JOIN):

TRM_A(TRM#, EL#, EL-TYPE, NODE1#) ← TRM_D4
TRM_B(TRM#, EL#, EL-TYPE, NODE2#) ← TRM_D4
TRM_C(TRM#, EM#, EL-TYPE, NODE3#) ← TRM_D4
TRM_D(TRM#, EL#, EL-TYPE, X1, Y1, Z1) = TRM_A * TRM_D2
TRM_E(TRM#, EL#, EL-TYPE, X2, Y2, Z2) = TRM_B * TRM_D2
TRM_F(TRM#, EL#, EL-TYPE, X3, Y3, Z3) = TRM_C * TRM_D2
EL_CORD(TRM#, EL#, EL-TYPE, X1, Y1, Z1, X2, Y2, Z2, X3, Y3, Z3)
    = TRM_D * TRM_E * TRM_F

It can be seen from the algorithm that we did not modify the original relations TRM_D4 and TRM_D2 to retrieve the data required for a particular inquiry. The relations TRM_D4 and TRM_D2 are still consistent with the conceptual model. Therefore pure retrieval operations for rearrangement of data do not cause any inconsistency in data values.

Now, consider the reverse process of transforming external data structure to internal data structure. Suppose a particular user wants to insert the nodal coordinates of a finite element using the external view EL_CORD. Here relation EL_CORD has the only key TRM# and has no reference to the node number to which the element is connected. Insertion is not consistent with the conceptual model because it requires the coordinates of nodes which are dependent on keys NODE#. This restriction is also reflected in the internal model – TRM_D2 that requires NODE3 as key values for insertion. Therefore the transformation of relation EL_CORD into the internal model is not possible. From this example it follows that there are restrictions for rearranging data from the external model to the internal model.

### 9.4.6 Methodology to incorporate matrix data into a database

In finite element analysis and structural design optimization, we encounter the problem of storage of large-order matrices. This data is unique to the application, so no attempts have been made to design such databases in the business database management area. However, some studies have been reported in the recent engineering literature to deal with large-order matrices. The methodology for representing large matrices can be based on the conceptual, internal and external views.

Conceptually, a matrix is a two-dimensional array of numbers. These numbers appear in a certain pattern; for example, square, sparse, symmetric, diagonal, banded, lower triangular form, upper triangular form, tridiagonal form, hypermatrix form, and skyline form. A matrix is uniquely identified by a name. Rows and columns of the two dimensional array are used for identification of data elements in the matrix. A conceptual view of a matrix can be represented by the following elementary relations:

ER1(NAME, MATRIX TYPE)
ER2(NAME, NUM-OF-ROWS)
ER3(NAME, NUM-OF-COLUMNS)
ER4(NAME, ROW, COLUMN, DATA-ELEMENT-VALUE)
ER5(NAME, NUM-OF-HYPER ROWS)
ER6(NAME, NUM-OF-HYPER COLUMNS)
ER7(NAME, HYPER-ROW, HYPER-COLUMN, ROW, COLUMN,
    DATA-ELEM-VALUE)
ER8(NAME, BAND-WIDTH)
ER10(NAME, SUB-MAT-ROW-SIZE)
ER11(NAME, SUB-MAT-COLUMN-SIZE)
ER12(NAME, VECTOR OF SKYLINE-HEIGHT)
ER13(NAME, HYPER-ROW, HYPER-COLUMN, NULL-OR-NOT)

The attributes of these elementary relations are self-descriptive. These elementary relations completely define a matrix and provide the conceptual structure of the matrix.

An internal (storage) structure for large-order matrices has to be developed that is consistent with the conceptual structure. Storage schemes have to be developed based on efficiency and processing considerations. The special nature of the matrix, that is, sparse, dense, or symmetric, should be used to provide storage efficiency. We can classify various matrix types into two basic types – sparse and dense. Many possible storage schemes are available to store dense and sparse matrices. Conventional storage schemes – row-wise, column-wise, submatrix-wise – are useful for storing dense matrices. Choice among these storage schemes should be based on consideration of several aspects – storage space, processing sequence, matrix operation, page size, flexibility for data modification, ease of transformation to other schemes or user's views, number of addresses required to locate rows or submatrices, and availability of database management system support.

*Storage space*
The row storage scheme can be used for square, banded, and skyline matrix types. However, this scheme is not appropriate for the hypermatrix. Symmetric,

triangular, and diagonal properties of a square matrix can be used in saving storage space if the variable length of rows is used. Similar schemes can be used for banded and skyline matrices to store data elements that appear in a band or skyline column. Submatrix storage can be used for all matrix types. The submatrix is most appropriate for hypermatrix data. Both schemes have disadvantages when zero elements within a row or submatrix have to be stored.

*Processing sequence*
Row storage requires that the assembly of matrices, storage, and retrieval be made only row-wise. This becomes inefficient if row-wise processing cannot be done. The submatrix approach is suitable for all types of processing sequences – row-wise, column-wise, or in any arbitrary order.

*Matrix operations*
Operations such as transposition, addition, multiplications, and solutions of simultaneous equations are frequently carried out at various stages of the structural design process. The row storage scheme is inefficient for matrix transpose when column-wise storage is required. During multiplication of two matrices A and B, a column of B can be obtained only by retrieving all of the rows of B. Therefore, the row storage scheme becomes inappropriate for such an operation. However, the submatrix storage scheme does not impose any such constraints in matrix operation, thus providing a suitable internal storage scheme.

*Page size*
A page is a unit or block of data stored or retrieved from memory to disk. For a fixed page size, only a number of full rows or a number of full submatrices together with fractional parts of them can be stored or retrieved at a time. It is clear that fragmentation of rows or submatrices takes place depending on the size of rows or submatrices. Large row size will overlap more than one page in memory and cause wastage of space. The submatrix scheme has the advantage of providing flexibility in choosing the submatrix size to minimize fragmentation of pages.

*Flexibility for data modification*
For modification of rows of a matrix, both row and submatrix storage schemes are suitable. However, the row scheme would be more efficient than the submatrix storage scheme. For modification of a few columns of a matrix, the row storage scheme may require a large number of I/O.

*Transformation to other schemes*
The submatrix storage scheme requires a minimum number of data accesses to transform to the column-wise or row-wise storage scheme.

*Address required*
The submatrix storage requires a lower number of addresses to locate data than the row or column storage scheme, provided that submatrices are reasonably large.

For internal storage of large-order matrices in a database, the preceding aspects should be carefully considered. It appears that both submatrix and row storage schemes can be appropriate for various applications.

In order that the internal storage scheme be consistent with the conceptual model, we need to store additional information about the properties of the matrix. That additional information is given by the elementary relations ER1–ER3, ER5, ER6, ER9–ER13. These can be combined and stored in a relation.

So far we have considered schemes for internal organization of large matrices. Since different users may view the same matrix in different forms – banded, skyline, hypermatrix, triangular, or diagonal – it is necessary to provide external views to suit individual needs. The unit of transactions on various views of a matrix may be row-wise, column-wise, submatrix-wise, or data element wise. If the internal scheme is submatrix-wise, the external view need not be submatrix wise. Therefore transformations are necessary to convert the internal matrix data into the form required by a particular user.

Next, we consider the sparse matrix storage scheme. Several storage schemes have been suggested by Pooch and Nieder (1973) and Daini (1982). They are the bit-map scheme, address map scheme, row–column scheme, and threaded list scheme. Out of these, the row–column scheme is simple and easy to use. Also, the row–column scheme can be easily incorporated into the relational model. Therefore, this scheme can be considered for storing sparse matrices encountered in finite element analysis and design applications.

The row-column storage scheme consists of identification of row and column numbers of nonzero elements of a sparse matrix and storing them in a table. This scheme provides flexibility in the modification of data. Any nonzero value generated during the course of a matrix operation can be stored or deleted by simply adding or deleting a row in the stored table. The external view of the row–column storage scheme can be provided through suitable transformation procedures.

Many procedures in structural design, such as element stiffness matrix routines, generate huge amounts of data. Generally, it is not preferable to store such data in a database at the expense of disk space and data transportation time. This inefficiency can be avoided by storing only a minimum amount of data needed to generate the required information (element stiffness matrix). In general, a data model can be replaced by (i) an algorithm that generates the user-requested information, and (ii) a set of (minimum) data that will be used by an algorithm to generate user-requested information.

## 9.5 CONCLUDING REMARKS

Various concepts of database design and management applicable to engineering design optimization are discussed and illustrated with examples. Conceptualization of design data, and its internal and external views, are discussed. The user's view of data is described through hierarchical, network, numerical and generalized relational models. Even though needs of database management in engineering and business applications are different, there is some commonality between them, and we can take advantage of these developments. One major difference is the need to manage dynamically large amounts of numerical data in engineering and scientific applications. Numerical data models can be used for this purpose. Alternatively, a recently developed generalized relational model can be used to treat various types of data in a unified way (Arora and Mukhopadhyay, 1988).

A methodology to design databases for finite element analysis and structural design is described. The methodology considers several good features of the available database design techniques, such as three views of data organization – conceptual, internal, and external. Tabular and matrix forms of data are included. The relational data model is shown to be quite useful in providing a simple and clear picture of data in the database design. Entities, relations, and attributes are considered to form a conceptual view of data. First, second, and third normal forms of data are suggested to design an internal model. More recently developed normal forms can also be used. Several aspects such as processing, iterative needs, multiple views of data, efficiency of storage and access time, and transitive data are considered in the methodology. Examples used for discussing the methodology are relevant and can be extended to the actual database design for computed-aided analysis and optimization of structural systems.

The need to use organized databases in engineering and scientific applications has increased in recent years. Large and complex systems in multidisciplinary environment need to be designed. Software systems for such applications consisting of perhaps several independent programs can be quite complex. Properly designed databases and the associated database management system are essential in the design process as well as in extending and modifying the software modules. The database capabilities can be used at two levels:

- At a global level, the database can store permanent information about the system being designed, such as the finite element data, element connectivity, node coordinates, material properties, loading conditions, and other such data. These are permanent types of data that may be accessed by different application programs.
- Each application program may generate huge amount of data and save it for later use. These data are usually not needed at the end of the run, so they may be kept in a temporary database and deleted at the end of the run.

Therefore, we see that it is essential for the database management system to have facilities to manage static as well as dynamic data during execution of the program. Concepts of global and local databases provide a means to organize data at various levels in the design optimization environment.

Finally, it is important to note that use of a well-designed database and the associated database management system can speed up the development time for a software system. This has been demonstrated in the literature through the development of a few prototype engineering design systems (Al-Saadoun and Arora, 1989; Park and Arora, 1987; Rajan and Bhatti, 1986; Spires and Arora, 1990; SreekantaMurthy and Arora, 1987).

# REFERENCES

Al-Saadoun, S.S. and Arora, J.S. (1989) Interactive design optimization of framed structures. *Journal of Computing in Civil Engineering*, **3** (1), 60–74.

Arora, J.S. (1989) *Introduction to Optimum Design*, McGraw-Hill, New York.

Arora, J.S. (1990) Computational design optimization: a review and future directions. *Structural Safety*, **7**, 131–48.

Arora, J.S. and Mukhopadhyay, S. (1988) An integrated database management system for engineering applications based on an extended relational model. *Engineering with Computers*, **4**, 65–73.

Arora, J.S. and Tseng, C.H. (1988) Interactive design optimization. *Engineering Optimization*, **13** (3), 173–88.

Baron, R.J. and Shapiro, L.G. (1980) *Data Structures and Their Implementation*, Van Nostrand, New York.

Bell, J. (1982) Data modelling of scientific simulation programs. International Conference on Management of Data, 2–4 June, Association for Computing Machinery, Special Interest Group on Management of Data (ACM-SIGMOD), pp. 79–86.

Blackburn, C.L., Storaasli, O.O. and Fulton, R.E. (1982) The role and application of database management in integrated computer-aided design. American Institute of Aeronautics and Astronautics 23rd Structures, Structural Dynamics and Materials Conference, New Orleans, LA, May 10–12, pp. 603–13.

Browne, J.C. (1976) Data definition, structures, and management in scientific computing. Proceedings of Institute for Computer Applications in Science and Engineering (ICASE) Conference on Scientific Computing, pp. 25–56.

Buchmann, A.P. and Dale, A.G. (1979) Evaluation criteria for logical database design methodologies, *Computer-Aided Design*, **11** (3), 121–6.

Comfort, D.L. and Erickson, W.J. (1978) RIM – A prototype for a relational information management system. *NASA Conference Publication 2055*.

Daini, O.A. (1982) Numerical database management system: a model. International Conference on Data Management, Association for Computing Machinery, Special Interest Group on Management of Data (ACM-SIGMOD), pp. 192–9.

Date, C.J. (1977) *An Introduction to Database Systems*, Addison-Wesley, Reading, MA.

Elliott, L., Kunii, H.S. and Browne, J.C. (1978) A data management system for engineering and scientific computing. *NASA Conference Publication 2055*.

Felippa, C.A. (1979) Database management in scientific computing – I. General description. *Computers and Structures*, **10**, 53–61.

Felippa, C.A. (1980) Database management in scientific computing – II. Data structures and program architecture. *Computers and Structures*, **12**, 131–45.

Fischer, W.E. (1979) PIDAS – A database management system for CAD/CAM software. *Computer-Aided Design*, **11** (3), 146–50.

Fishwick, P.A. and Blackburn, C.L. (1982) The integration engineering programs using a relational database scheme, in *Computers in Engineering*. International Computers in Engineering Conference, Vol. 3, ASME, pp. 173–81.

Fulton, R.E. (ed.) (1987) Managing Engineering Data: The Competitive Edge. Proceedings of the Symposium on Engineering Database Management: Critical Issues, American Society of Mechanical Engineers, Computers in Engineering Conference and Exhibition, August 9–13, New York.

Fulton, R.E. and Voigt, S.J. (1976) Computer-aided design and computer science technology. Proceedings of the 3rd ICASE (Institute for Computer Applications in Science and Engineering) Conference on Scientific Computing, pp. 57–82.

Grabowski, H. and Eigner, M. (1982) A data model for a design database. File structures and databases for CAD. Proceedings of the International Federation of Information Processing, pp. 117–44.

Haug, E.J. and Arora, J.S. (1979) *Applied Optimal Design*, Wiley, New York.

Jumarie, G.A. (1982) A decentralized database via micro-computers a preliminary study, in *Computers in Engineering*. International Computers in Engineering Conference, Vol. 4, American Society of Mechanical Engineers, p. 183.

Kamal, O. and Adeli, H. (1990) Automatic partitioning of frame structures for concurrent processing. *Microcomputers in Civil Engineering,* **5** (4), 269–83.

Koriba, M. (1983) Database systems: their applications to CAD software design. *Computer-Aided Design,* **15** (5), 277–88.

Lopez, L.A. (1974) FILES: automated engineering data management system in *Computers in Civil Engineering.* Electronic Computation Conference, American Society of Civil Engineers, pp. 47–71.

Lopez, L.A., Dodds, R.H., Rehak, D.R. and Urzua, J.L. (1978) Application of data management to structures, in *Computing in Civil Engineering,* American Society of Civil Engineers, pp. 477–496.

Martin, J. (1977) *Computer Data-base Organization,* Prentice-Hall, Englewood Cliffs, NJ.

Mukhopadhyay, S. and Arora, J.S. (1987a) Design and implementation issues in an integrated database management system for engineering design environment. *Advances in Engineering Software,* **9** (4), 186–93.

Mukhopadhyay, S. and Arora, J.S. (1987b) Implementation of an efficient run-time support system for engineering design environment. *Advances in Engineering Software,* **9** (4), 178–85.

Pahl, P.J. (1981) Data management in finite element analysis, in *Nonlinear Finite Element Analysis in Structural Mechanics* (eds W. Wunderlich, E. Stein and K.J. Bathe), Springer, Berlin, pp. 714–16.

Park, G.J. and Arora, J.S. (1987) Role of database management in design optimization systems. *Journal of Aircraft,* **24** (11), 745–50.

Pooch, U.W. and Nieder, A. (1973) A survey of indexing techniques for sparse matrices. *Computing Surveys,* **5** (2), 109–33.

Rajan, S.D. (1982) SADDLE: a computer-aided structural analysis and dynamic design language. Ph.D. Dissertation, Civil Engineering, The University of Iowa, Iowa City, IA 52242.

Rajan, S.D. and Bhatti, M.A. (1983) Data management in FEM-based optimization software. *Computers and Structures,* **16** (1–4), 317–25.

Rajan, S.D. and Bhatti, M.A. (1986) SADDLE: a computer-aided structural analysis and dynamic design language. *Computers and Structures,* **22** (2), 185–212.

Spires, D.B. and Arora, J.S. (1990) Optimal design of tall RC-framed tube buildings. *Journal of Structural Engineering,* **116** (4), 877–97.

SreekantaMurthy, T. and Arora, J.S. (1985) A survey of database management in engineering. *Advances in Engineering Software,* **7** (3), 126–33.

SreekantaMurthy, T. and Arora, J.S. (1986a) Database management concepts in design optimization. *Advances in Engineering Software,* **8** (2), 88–97.

SreekantaMurthy, T. and Arora, J.S. (1986b) Database design methodology and database management system for computed-aided structural design optimization. *Engineering with Computers,* **1,** 149–160.

SreekantaMurthy, T. and Arora, J.S. (1987) A structural optimization program using a database management system, in *Managing Engineering Data: The Competitive Edge* (ed. R.E. Fulton). Proceedings of the Symposium on Engineering Database Management: Critical Issues, American Society of Mechanical Engineers, Computers in Engineering Conference and Exhibition, August 9–13, 1987, New York, pp. 59–66.

SreekantaMurthy, T., Shyy, Y.-K. and Arora, J.S. (1986) MIDAS: management of information for design and analysis of systems. *Advances in Engineering Software,* **8** (3), 149–58.

Ulfsby, S., Steiner, S. and Oian, J. (1981) TORNADO: a DBMS for CAD/CAM systems. *Computer-Aided Design,* **13** (4), 193–7.

Vetter, M. and Maddison, R.N. (1981) *Database Design Methodology,* Prentice-Hall, Englewood Cliffs, NJ.

# 10

# Optimization of Topology*

## G.I.N. ROZVANY, M. ZHOU and O. SIGMUND

## 10.1 INTRODUCTION: AIMS AND SIGNIFICANCE OF TOPOLOGY OPTIMIZATION

**Topology of a structural system** means the spatial sequence or configuration of members and joints or internal boundaries. The two main fields of application of topology optimization are layout optimization of grid-like structures and generalized shape optimization of continua or composites.

A **grid-type structure** has the basic feature that it consists of a system of intersecting members, the cross-sectional dimensions of which are small in comparison with their length, and hence the members can be idealized as one-dimensional continua. Consequences of this feature are that

- the influence of member intersections on strength, stiffness and structural weight can be neglected, and
- the specific cost (e.g. structural weight per unit area or volume) can be expressed as the sum of the costs (weights) of the members running in various directions.

Examples of grid-type structures are trusses, grillages (beam systems), shell-grids and cable nets.

**Layout optimization** of grid-type structures consists of three simultaneous operations, namely

- topological optimization involving the spatial sequence of members and joints,
- geometrical optimization involving the coordinates of joints, and
- sizing, i.e. optimization of cross-sectional dimensions.

The above concepts are explained on an example in Fig. 10.1, in which all three trusses have the same topology, whilst the trusses in Figs 10.1(b) and 10.1(c) have the same geometry but different cross-sectional dimensions.

Prager and Rozvany (1977b) regarded layout optimization as the most challenging class of problems in structural design because there exists an infinite number of possible topologies which are difficult to classify and quantify; moreover, at each point of the available space potential members may run in an infinite number of directions. At the same time, layout optimization is of

---

*In order to understand this chapter more easily, the reader should study sections 2.6 and 2.7 of Chapter 2.

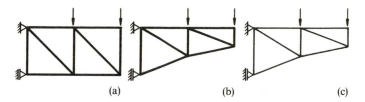

**Fig. 10.1** Example illustrating topological, geometrical and cross-sectional properties of a grid-type structure.

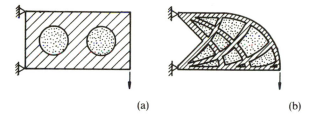

**Fig. 10.2** Example illustrating generalized shape optimization.

considerable practical importance, because it results in much greater material savings than pure cross section (sizing) optimization.

The other important field of application of topology optimization is **generalized shape optimization**, in which a simultaneous optimization of both topology and shape of boundaries is required for continua and of interfaces between materials for composites. Figure 10.2(a), for example, shows the initial boundary shape and topology and Fig. 10.2(b) a hypothetical optimal shape and topology for a composite plate in plane stress, where the dotted regions denote the less stiff, weaker and lighter material. For a cellular structure (here, perforated plate), the dotted regions denote cavities (or 'holes').

## 10.2 OPTIMAL LAYOUT THEORY FOR GRID-TYPE STRUCTURES – BASIC ASPECTS

The theory of optimal structural layouts has been covered extensively in the past, in particular in principal lectures at NATO meetings (Iowa, 1980 (Rozvany, 1981); Troia, 1986 (Rozvany and Ong, 1987); Edinburgh, 1989 (Rozvany, Gollub and Zhou, 1992); Berchtesgaden, 1991 (Rozvany, Zhou and Gollub, 1993); Sesimbra, 1992 (Rozvany, 1993c)), and also in greater detail at a CISM course in Udine in 1990 (Rozvany, 1992b) as well as in books (Rozvany, 1976, 1989) and in book chapters (e.g. Rozvany, 1984). Although the above exposure in the technical literature may appear excessive, the development in this field has also been both rapid and extensive, even during the last few years. For this reason, apart from a brief review of the basic principles and earlier work, mostly new material can be reported in this chapter.

Optimal layout theory, developed in the seventies by Prager and the first author (e.g. Prager and Rozvany, 1977a,b) as a generalization of Michell's (1904) theory for trusses, was originally formulated for grid-like structures on the basis of the simplifying assumptions listed in section 10.1. These simplifications were removed in a more advanced version of layout theory (e.g. Rozvany *et al.*, 1982, 1987), which was applied to structures in which a high proportion of the available space was occupied by material. One could also say that the original, 'classical' layout theory is concerned with structures having a low 'volume fraction' (i.e. material volume/available volume ratio), whilst advanced layout theory deals with structures having a high volume fraction. In the latter, first the microstructure is optimized for given ratio of the stiffnesses or forces in the principal directions and, in a second operation, the optimal macroscopic distribution of micro-structures is determined using methods of the layout theory.

The clasical layout theory is based on two underlying concepts, namely

• the structural universe (in numerical methods: ground structure), which is the union of all potential members or elements, and
• continuum-type optimality criteria (COC), expressed in terms of a fictitious system termed the adjoint structure, which were discussed in Chapter 2 (sections 2.6 and 2.7).

Since the above optimality criteria also provide adjoint strains for vanishing members, their fulfillment for the entire structural universe represents a necessary and sufficient condition of layout optimality if the problem is convex and certain additional requirements (e.g. existence) are satisfied. The above condition of convexity is fulfilled for certain so-called 'self-adjoint' problems, the analytical treatment of which will be discussed in sections 10.3–10.5.

On the basis of optimal layout theory, two basic approaches have been developed, using

• analytical methods for deriving closed form continuum-type solutions representing the exact optimal layout, and
• numerical, discretized iterative methods for deriving approximate (but usually highly accurate) optimal layouts.

Fundamental differences between these two methods are listed in Table 10.1. Layout optimization methods used by the numerical school (e.g. Kirsch, 1989;

**Table 10.1** A comparison of analytical and numerical methods based on layout theory

| Computational Method | Analytical | Numerical |
|---|---|---|
| Structural model | Continuum | Discretized (finite elements) |
| Procedure | Simultaneous solution of all equations | Iterative solution |
| Structural universe | Infinite number of members | Finite but large number of members (several thousands) |
| Prescribed minimum cross-sectional area | Zero | Nonzero but small ($10^{-8}$ to $10^{-12}$) |

Kirsch and Rozvany, 1993) are usually based on the following two-stage procedure:

- first the topology is optimized for a given geometry (i.e. given coordinates of the joints), and then
- for this selected topology the geometry is optimized.

A drawback of this procedure is, of course, that for the new, optimized geometry the old topology may not be optimal any more. Until the introduction of COC–DCOC methods, however, the two-stage procedure was necessary because of the limited optimization capability of other methods, particularly for realistic problems with active stress constraints for a very large number of members in the structural universe.

The new optimality criteria methods (COC–DCOC), which are discussed in Chapter 2, enable us to carry out a simultaneous optimization of topology and geometry, because the number of elements in the structural universe is either infinite (analytical methods) or very large (numerical COC–DCOC methods) and hence topological optimization achieves, in effect, also geometrical optimization.

We shall close this introductory section on layout optimization with some personal notes by the first author.

First, it is often claimed that the layout theory discussed here always results in a 'continuum-type' solution, consisting of a dense grid of members of infinitesimal spacing, whereas numerical layout studies yield 'practical' solutions consisting of a few members only. It will be seen from subsequent sections that this notion is wrong, because for one or several point loads the exact layout consists in general of a few members only (Fig. 10.8, for example), with some notable exceptions (Figs 10.18 and 10.42(a)).

Second, it is generally believed that the interest of the first author's research team is restricted to analytical solutions, mostly in the grillage field. In actual fact, during the last four years the first two authors developed together some of the most efficient numerical methods for both sizing and layout optimization.

Finally, the first author explored recently with relative ease exact optimal topologies of several rather complicated new classes of layout problems, involving grillages and trusses (new solutions under sections 10.4 and 10.5). In this sort of analytical work, one requires a certain intuitive insight as well as an intimate knowledge of optimal layout theory, in order first to guess correctly these new topologies and then to prove their optimality. During the above activity, it occurred to him that, since the tragic deaths of William Prager (Brown University) and Ernest Masur (University of Illinois) and the retirement of William Hemp (Oxford) and Marcel Save (Mons), he is rapidly becoming a sort of a 'last Mohican' of analytical layout optimization. This situation has arisen as a result of the growing popularity of discretized methods in structural optimization, which is fully justified in view of (i) the rapidly increasing computational capability and (ii) the necessity for discretization in most practical problems. Explicit analytical solutions are, however, indispensable in reliably checking the validity and convergence of numerical methods. Too much reliance on automated discretized computer procedures may also affect mentally forthcoming generations of practitioners – one cannot help drawing an analogy with H.G. Wells' 'Time Machine', in which man in some future century loses his ability to think.

### 10.3 OPTIMAL LAYOUT THEORY: ANALYTICAL SOLUTIONS FOR SOME SIMPLE SELF-ADJOINT PROBLEMS

#### 10.3.1 General formulation

The analytical procedure based on the above layout theory usually consists of the following steps.

- Set up a structural universe consisting of all potential members.
- Determine the specific cost function and continuum-type optimality criteria (COC) for the members.
- Construct an adjoint displacement field satisfying kinematic boundary and continuity conditions.
- Determine on the basis of the optimality criteria and the adjoint strain field the location and direction of nonvanishing (optimal) members.
- Check whether the external loads and the stress resultants along optimal members constitute a statically admissible set.

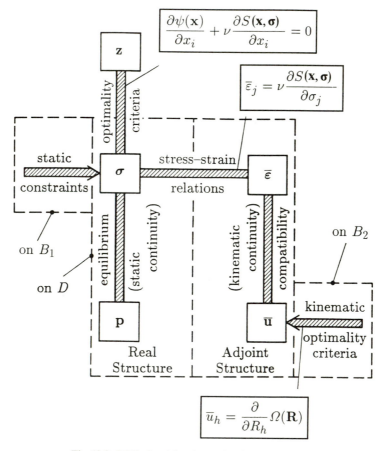

$$\frac{\partial \psi(\mathbf{x})}{\partial x_i} + \nu \frac{\partial S(\mathbf{x}, \boldsymbol{\sigma})}{\partial x_i} = 0$$

$$\bar{\varepsilon}_j = \nu \frac{\partial S(\mathbf{x}, \boldsymbol{\sigma})}{\partial \sigma_j}$$

$$\bar{u}_h = \frac{\partial}{\partial R_h} \Omega(\mathbf{R})$$

**Fig. 10.3** COC algorithm for optimal plastic design.

• Check whether the adjoint strains along vanishing (non-optimal) members satisfy the optimality criteria (usually inequalities) for such members.

A general schematic representation of continuum-based optimality criteria (COC) for linearly elastic structures with stress and displacement constraints was given in Fig. 2.21. We shall discuss two classes of problems in this section: optimal plastic design and optimal elastic design for a compliance constraint.

In the case of optimal plastic design, the real structure is only required to fulfill static admissibility. Since there are only stress constraints but no displacement constraints, the adjoint structure is subject only to prestrains due to active stress constraints, without adjoint loads or generalized stresses. It follows that we only have the kinematic part of the adjoint structure, and hence Fig. 2.21 reduces to the scheme shown in Fig. 10.3. This can be simplified further for one load condition if we express the specific cost directly in terms of the generalized stresses $\psi = \psi(\sigma)$, as was done by Prager and Shield (1967) in their pioneering contribution. The corresponding simplified scheme is shown in Fig. 10.4, in which the relations between the real stresses and adjoint strains are termed 'static–kinematic optimality criteria' (Rozvany, 1989).

In the case of optimal elastic design with a so-called compliance constraint,

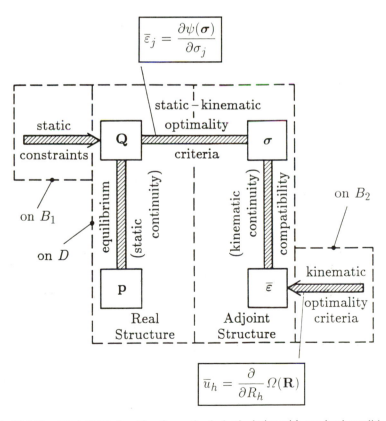

Fig. 10.4 Simplified COC algorithm for optimal plastic design with one load condition.

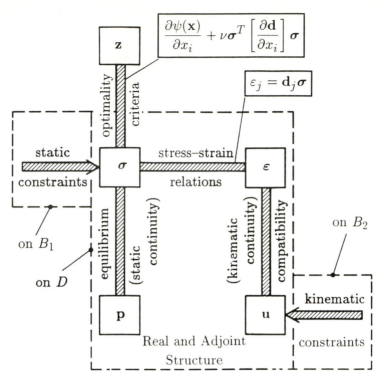

**Fig. 10.5** COC algorithm for elastic design with a compliance constraint.

the latter can be represented as

$$\int_D \mathbf{p}\mathbf{u}\,\mathrm{d}\xi \leqslant C \tag{10.1}$$

where $C$ is a given constant and all other symbols were defined in section 2.6. In this case, the 'weighting factors' for the elastic deflections $\mathbf{u}$ are given by the real load $\mathbf{p}$, which implies

$$\mathbf{p} \equiv \bar{\mathbf{p}}, \quad \boldsymbol{\sigma} \equiv \bar{\boldsymbol{\sigma}}, \quad \boldsymbol{\varepsilon} \equiv \bar{\boldsymbol{\varepsilon}}, \quad \mathbf{u} \equiv \bar{\mathbf{u}} \quad \text{(for all } \xi) \tag{10.2}$$

This means that it is only necessary to consider one half of the scheme in Fig. 2.21, as can be seen from Fig. 10.5. It follows that for a certain simple class of problems (sections 10.3.2 and 10.3.3) optimal plastic design and optimal elastic design with a compliance constraint yield the same solution within a constant multiplier.

### 10.3.2  First application: optimal plastic design

In this subsection, we consider cost functions of the form

$$\psi = k|\sigma|, \quad \Phi = \int_D k|\sigma|\,\mathrm{d}\xi \tag{10.3}$$

where $\psi$ is the member weight per unit length, $k$ is a given constant, $\sigma$ is a generalized stress (stress resultant), $\Phi$ is the total weight, $\xi$ is a spatial coordinate and $D$ is the structural domain (with $\xi \in D$).

For trusses we have (i) $\sigma = N$ where $N$ is the axial member force and (ii) $k = \rho/\sigma_0$ where $\rho$ is the specific weight of the truss material and $\pm \sigma_0$ the yield stress in tension and compression.

For beams of variable width but given depth $h$, we have (i) $\sigma = M$ where $M$ is the bending moment and (ii) $k = 4\rho/h^2\sigma_0$ where $\pm \sigma_0$ is again the yield stress.

For piecewise differentiable specific cost functions (as in equations (10.3)), the partial derivatives in Fig. 10.4 are replaced by subgradients (e.g. Rozvany, 1989), which means that at slope discontinuities any convex combination of the adjacent slopes can be taken. This implies that the adjoint strains for the specific cost function in relations (10.3) become

$$\bar{\varepsilon} = k \operatorname{sgn} \sigma \quad (\text{for } \sigma \neq 0), \quad |\bar{\varepsilon}| \leqslant k \quad (\text{for } \sigma = 0) \tag{10.4}$$

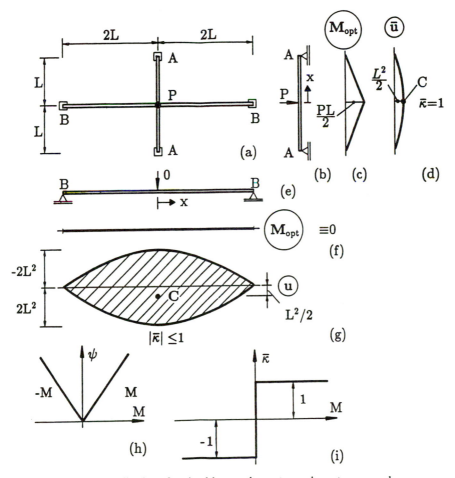

**Fig. 10.6** Application of optimal layout theory to an elementary example.

where $\bar{\varepsilon}$ is the adjoint axial strain in the members. Relations (10.3) and (10.4) are represented graphically for $k = 1$ in Figs 10.6(h) and 10.6(i).

### 10.3.3 Second application: optimal elastic design for given compliance

We consider the class of problems characterized by

$$\psi = cx, \quad [\mathbf{d}] = 1/rx \tag{10.5}$$

where $c$ and $r$ are given constants, $[\mathbf{d}]$ is the generalized flexibility matrix (section 2.6) and $x$ is a cross-sectional variable.

For trusses, for example, $\psi$ is the member weight per unit length, $x$ is the cross-sectional area and $c = \rho$ is the specific weight. Moreover, $r = E$ is Young's modulus. For beams of constant depth $h$ but variable width $x$ we have $c = h\varrho$ and $r = h^3 E/12$.

By the optimality condition at the top of Fig. 10.5 we have for this class of problems

$$c - v\sigma^2/rx^2 = 0, \quad x = (v/rc)^{1/2}|\sigma|, \quad \varepsilon = \sigma/rx = (c/rv)^{1/2} \operatorname{sgn} \sigma \quad \text{(for } \sigma \neq 0) \tag{10.6}$$

Moreover, it was shown previously (Rozvany, Zhou and Gollub, 1993) that for vanishing members (with $x = \sigma = 0$) we have the optimality criterion

$$|\varepsilon| \leqslant (c/rv)^{1/2} \quad \text{(for } \sigma = 0) \tag{10.7}$$

The reader is reminded that in equations (10.6) and (10.7) the symbol $\sigma$ denotes a stress resultant and not a conventional stress. It can be seen from equations (10.4), (10.6) and (10.7) that the adjoint strain fields for both classes of problems considered in sections 10.3.2 and 10.3.3 are identical within a constant multiplier.

It was also shown in the above-mentioned paper that the optimal weights in plastic design ($\Phi_P$) and elastic compliance design ($\Phi_C$) have the following relation:

$$\Phi_C = \frac{c}{k^2 rC} \Phi_P^2 \tag{10.8}$$

where $C$ is the prescribed maximum compliance value in equation (10.1).

### 10.3.4 An elementary example: a beam system consisting of two beams

Although this example has been used before, it will be presented in an extended form here again because of its extreme simplicity.

We consider a structural universe consisting of two simply supported beams, with a point load $P$ at the intersection of the two beams (Fig. 10.6(a)). The rather obvious optimal solution to this problem is given in Figs 10.6(b) and 10.6(e), in which the entire load is carried by the shorter beam AA, and the beam BB, with zero cross-sectional area, is unloaded. The corresponding moment diagrams are shown in Figs 10.6(c) and 10.6(f). The specific cost function and optimality criteria (with $k = 1$) are represented graphically in Figs 10.6(h) and 10.6(i). Since the adjoint displacement must be zero at the simple supports, for the short beam AA we have $\bar{u}(L) = \bar{u}(-L) = 0$ and by Fig. 10.6(i) or equation (10.4) with $\sigma = M$,

$\bar{\varepsilon} = \bar{\kappa}$ and $k = 1$ the adjoint curvature $\bar{\kappa} = -\,\mathrm{d}^2\bar{u}/\mathrm{d}x^2 = 1$ for the short beam. The above boundary and curvature conditions imply

$$\bar{u} = -\iint \mathrm{d}x\,\mathrm{d}x + Ax + B, \quad A = 0, \quad B = L^2/2, \quad \bar{u} = L^2/2 - x^2/2 \quad (10.9)$$

as shown in Fig. 10.6(d). For the long beam BB, the bending moment is zero through-out, and hence by equation (10.4) with $\sigma = M$ we have $|\bar{\kappa}| = |-\,\mathrm{d}^2\bar{u}/\mathrm{d}x^2| \leqslant 1$, giving the nonunique adjoint displacement field $-2L^2 + x^2/2 \leqslant \bar{u} \leqslant 2L^2 - x^2/2$ (shaded area in Fig. 10.6(g)). As the latter does include a central deflection of $L^2/2$ (point C in Fig. 10.6(g)), kinematic admissibility is also satisfied. Since the specific cost function is convex in this problem (Fig. 10.6(h)) and the constraints are linear, necessary and sufficient conditions for optimality have been fulfilled and thus the solution in Figs 10.6(b)–10.6(f) is indeed optimal.

### 10.3.5 Optimal regions in exact solutions for truss and grillage layouts

In deriving exact optimal layouts for trusses and grillages, the structural universe consists of members in all possible directions at all points of the available space. Since by equation (10.4) the maximum absolute value of the adjoint generalized strain is $k$, it follows that the direction of any nonvanishing member must coincide with a principal direction of the adjoint strain field, having the principal strain value of $\bar{\varepsilon}_1 = k$. Moreover, the absolute value of the second principal strain value $\bar{\varepsilon}_2$ must not exceed $k$, that is, $\bar{\varepsilon}_2 \leqslant k$. Adopting $k = 1$, these requirements allow the following optimal regions for trusses and grillages at all points of the available space where loads or non-vanishing members are present:

$$
\begin{aligned}
R^+: &\quad \sigma_1 > 0, \quad \sigma_2 = 0, \quad \bar{\varepsilon}_1 = 1, \quad &&|\bar{\varepsilon}_2| \leqslant 1 \\
R^-: &\quad \sigma_1 < 0, \quad \sigma_2 = 0, \quad \bar{\varepsilon}_1 = -1, \quad &&|\bar{\varepsilon}_2| \leqslant 1 \\
S^+: &\quad \sigma_1 > 0, \quad \sigma_2 > 0, \quad \bar{\varepsilon}_1 = \bar{\varepsilon}_2 = 1 \quad &&(10.10) \\
S^-: &\quad \sigma_1 < 0, \quad \sigma_2 < 0, \quad \bar{\varepsilon}_1 = \bar{\varepsilon}_2 = -1 \\
T: &\quad \sigma_1 > 0, \quad \sigma_2 < 0, \quad \bar{\varepsilon}_1 = -\bar{\varepsilon}_2 = 1
\end{aligned}
$$

where $\sigma_i = M_i$, $\bar{\varepsilon}_i = \bar{\kappa}_i$ for grillages and $\sigma_i = N_i$, $\bar{\varepsilon}_i = \bar{\varepsilon}_i$ for trusses. Symbols indicating the various types of optimal regions in relations (10.10) are shown in Fig. 10.7.

$$R^+ \qquad\qquad R^- \qquad\qquad S^+ \qquad\qquad S^- \qquad\qquad T$$

**Fig. 10.7** Symbols used for optimal regions for trusses and grillages.

### 10.3.6 An example of an exact optimal truss layout

We consider the optimal transmission of a vertical point load by a truss to supports formed by a horizontal and a vertical line (Fig. 10.8(b)). In this problem, the structural universe consists of an infinite number of members (at any given point of the available space ($x \geqslant 0$, $y \leqslant 0$) members may run in any directions).

First we must cover the available space with the optimal regions in relations (10.10) and Fig. 10.7, so that continuity conditions and kinematic boundary conditions are satisfied. The latter consist of:

$$\text{(for } x = 0 \text{ or } y = 0) \quad \bar{u} = \bar{v} = 0 \tag{10.11}$$

where $\bar{u}$ and $\bar{v}$, respectively, are the adjoint displacements in the $x$ and $y$ directions.

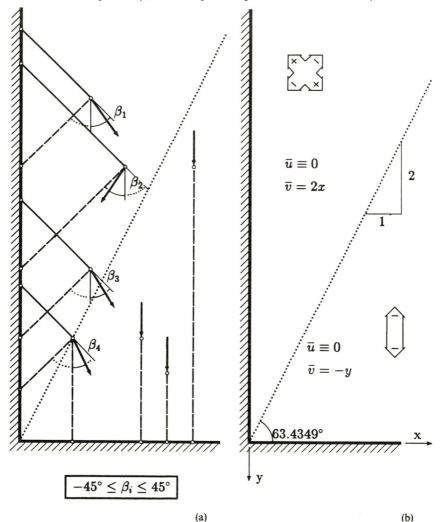

$$-45° \leq \beta_i \leq 45°$$

(a)                                                             (b)

**Fig. 10.8** Example of an exact optimal truss layout.

We shall try a topology having an $S$ region at the top and an $R^-$ region at the bottom, with a region boundary having a slope of 2:1 (Fig. 10.8(b)).

In the top region we have

$$\bar{u} \equiv 0, \quad \bar{v} = 2x, \quad \bar{\varepsilon}_x = d\bar{u}/dx = 0, \quad \bar{\varepsilon}_y = d\bar{v}/dy = 0, \quad \bar{\gamma}_{xy} = d\bar{u}/dy + d\bar{v}/dx = 2$$

$$\bar{\varepsilon}_{1,2} = \frac{\bar{\varepsilon}_x + \bar{\varepsilon}_y}{2} \pm \left[ \left( \frac{\bar{\varepsilon}_x - \bar{\varepsilon}_y}{2} \right)^2 + \frac{\bar{\gamma}_{xy}^2}{4} \right]^{1/2}, \quad \bar{\varepsilon}_1 = 1, \quad \bar{\varepsilon}_2 = -1$$

$$\alpha = \frac{1}{2} \arctan \frac{\bar{\gamma}_{xy}}{\bar{\varepsilon}_x - \bar{\varepsilon}_y} = \frac{1}{2} \arctan \infty = 45° \tag{10.12}$$

and in the bottom region

$$\bar{u} \equiv 0, \quad \bar{v} = -y, \quad \bar{\varepsilon}_x = d\bar{u}/dx = 0, \quad \bar{\varepsilon}_y = d\bar{v}/dy = -1$$

$$\bar{\gamma}_{xy} = d\bar{u}/dy + d\bar{v}/dx = 0, \quad \bar{\varepsilon}_1 = -1, \quad \bar{\varepsilon}_2 = 0, \quad \alpha = 0 \tag{10.13}$$

where $\alpha$ is the direction of the first principal strain with respect to the vertical.

It can be seen from the above results and relations (10.10) that, in the top region, the optimal bars must run at $\pm 45°$ to the vertical and must have tension and compression, respectively, in the two principal directions. In the bottom region they must be vertical (with a negative, i.e. compressive, force). It can also be checked easily that the displacement fields in Fig. 10.8(b) satisfy (i) the kinematic boundary conditions $\bar{u} = \bar{v} = 0$ along the supports and (ii) continuity conditions along the region boundary. The optimal bar directions and signs of the corresponding forces are shown by arrows in Fig. 10.8(b). Examples of admissible loads for this solution and corresponding optimal truss members are shown in Fig. 10.8(a). Whereas in the top region, the loads may enclose any angle within $\pm 45°$ to the vertical, the forces must be vertical in the bottom region of the adjoint field in Fig. 10.8(b). For loads along the region boundary, the optimal truss may consist of three bars ($\beta_4$ in Fig. 10.8(a)). The above solution was obtained in the late 1980s (Rozvany and Gollub, 1990).

The range of admissible load directions in the top region is due to the fact that one of the two bars must be in compression and the other one in tension. Since only bars in the vertical directions are admitted by the adjoint field in the bottom region, the corresponding loads must also be vertical for reasons of equilibrium. All the loads shown in Fig. 10.8(a) may act simultaneously or separately for the given optimal layouts.

*Note*

In all solutions involving trusses and grillages, continuous thick lines indicate members in tension or in positive moment and broken lines members in compression or negative moment.

## 10.4 LEAST-WEIGHT GRILLAGES

### 10.4.1 General aspects

One of the most successful applications of the exact layout theory was the optimization of grillage layouts, as can be seen from the following remark by

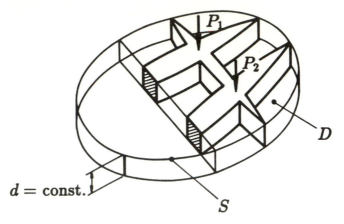

**Fig. 10.9** The grillage layout problem.

Prager: "Although the literature on Michell trusses is quite extensive, the mathematically similar theory of grillages of least-weight was only developed during the last decade. Despite its late start, this theory advanced farther than that of optimal trusses. In fact, grillages of least-weight constitute the first class of plane structural systems for which the problem of optimal layout can be solved for almost all loadings and boundary conditions" (Prager and Rozvany, 1977a). The problem of grillage optimization can be described as follows (Fig. 10.9): a structural domain $D$, bounded by two horizontal planes and some vertical surfaces, is subject to a system of vertical loads which are to be transmitted to given supports by beams of rectangular cross section having a variable width. The beams are to be contained in the structural domain and are to take on a minimum weight (or volume). The beam system is to be designed plastically for a given yield stress or elastically for a given compliance (sections 10.3.2 and 10.3.3).

As can be seen from the quotation above, Prager regarded the grillage optimization problem as particularly important because of the following unique features.

- Grillages constitute the first class of truly two-dimensional structural optimization problems for which closed-form analytical solutions are available for most boundary and loading conditions.
- Optimal grillages are more practical than Michell structures (least-weight trusses), because the latter are subject to instability which is ignored in the formulation.
- The optimal topology of ribs in solid plates and holes in perforated plates has been found to be similar to that of minimum weight grillages (e.g. Cheng and Olhoff, 1981).
- A computer algorithm is available for generating analytically and plotting optimal beam layouts for a wide range of boundary conditions (Rozvany and Hill, 1978; Hill and Rozvany, 1985).
- It has been shown that the same grillage layout is optimal for plastic design and for elastic design with a stress or a compliance or a natural frequency constraint (Rozvany, 1976; Olhoff and Rozvany, 1982).

- The optimal grillage layout is independent of the (nonnegative) load distribution if no internal simple supports are present.
- The adjoint displacement field can be readily generated and it provides an influence surface for any (nonnegative) loading (the total structural weight equals the integral of the product of loads and adjoint deflections).
- A number of additional refinements have been added to the optimal grillage theory, which are reviewed in the next subsection.

## 10.4.2 Review of earlier developments

Analytical solutions are now available for

- clamped boundaries,
- simply supported boundaries,
- internal simple supports,
- free edges,
- beam supported edges, and
- corners in the boundary.

Earlier extensions of the theory include closed form analytical solutions for plastically designed grillages with

- solutions for up to four loading conditions,
- nonuniform depth,
- partial discretization,
- allowance for cost of supports,
- bending and shear dependent cost,
- upper constraint on the beam density, and
- allowance for selfweight

as well as for elastically designed grillages with

- deflection constraints,
- natural frequency constraints, and
- a combination of stress and deflection constraints.

General theories for clamped and simply supported boundaries were developed already in the early 1970s (e.g. Rozvany, 1973; Rozvany, Hill and Gangadharaiah, 1973). Several comprehensive reviews of earlier work are available (Prager, 1974; Rozvany, 1976; Rozvany and Hill, 1976; Rozvany, 1981, 1984, 1989, Chapter 8, 1992b, 1993a). Quite recent developments are discussed briefly in the next subsection.

## 10.4.3 Recent developments

(a) Unified constructions for simply supported and clamped boundaries with allowance for cost of supports

The surprisingly simple constructions given in Figs 10.10 and 10.11 were derived quite recently (Rozvany, 1994). As indicated, clamped supports are assumed

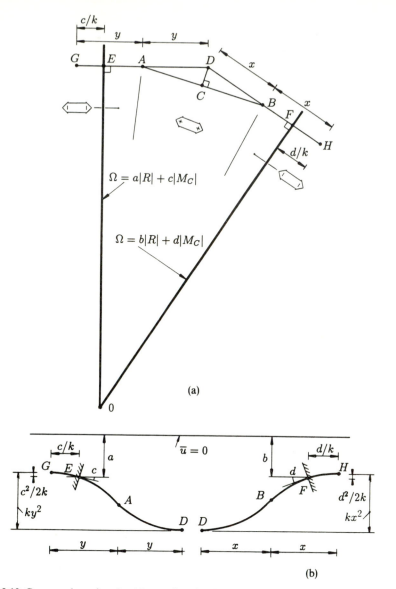

**Fig. 10.10** Construction of optimal beam direction for two clamped boundaries with support costs.

to have a cost $\Omega = a|R| + c|M_C|$ or $\Omega = b|R| + d|M_C|$ and simple supports $\Omega = b|R|$ where $a$, $b$, $c$ and $d$ are given constants, $R$ is the vertical reaction per unit length and $M_C$ the clamping moment per unit length. The distances $x$ and $y$ must also satisfy the following relations:

$$\text{(Fig. 10.10(a))} \quad a + ky^2 - c^2/2k = b + kx^2 - d^2/2k \qquad (10.14)$$

$$\text{(Fig. 10.11(a))} \quad a + ky^2 - c^2/2k = b + kx/2 \qquad (10.15)$$

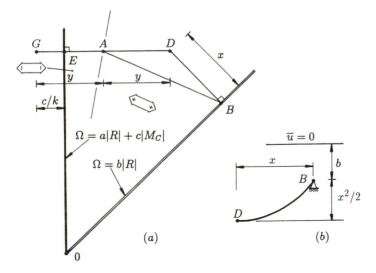

**Fig. 10.11** Construction of optimal beam direction for a clamped and a simply supported boundary with support costs.

Naturally, the above unified constructions are also valid for grillages without support cost, in which case we have $a = b = c = d = 0$.

Proof of the constructions in Figs 10.10(a) and 10.11(a) is based on the type of optimal regions in equations (10.10) and satisfaction of the kinematic boundary conditions, as well as continuity and slope continuity conditions along region boundaries.

By Fig. 10.4, the adjoint displacements along supports are given by the partial derivatives of the support cost function $\Omega$ with respect to the reaction components. For a reaction cost function $\Omega = a|R| + c|M_C|$, for example, we have the adjoint deflection $\bar{u}$ and slope $\bar{s}$ of

$$\bar{u} = a, \quad \bar{s} = c \quad \text{(for } R > 0, \quad M_C > 0) \tag{10.16}$$

The adjoint deflections satisfying all optimality conditions are shown in Figs 10.10(b) and 10.11(b) (for details, see the paper by Rozvany (1994)).

On the basis of the constructions in Figs 10.10 and 10.11, a computer program was developed at Essen University by D. Gerdes, for generating analytically and plotting optimal layouts for any combination of analytically defined (straight or circular) simply supported or clamped boundaries. Some examples of these computer-generated analytical solutions are given in Fig. 10.12, in which lines in one direction indicate beams in an $R$ region, lines in two directions at right angles denote beams in a $T$ region and black areas signify $S$ regions (in which all beam directions are equally optimal). The solutions at the top and left bottom have, respectively, two and four internal point supports.

The above-mentioned automated computer algorithm finds points (termed 'centers') for which the type of relations in equations (10.14) and (10.15) are satisfied by more than two boundary points. For example, in Fig. 10.13, the

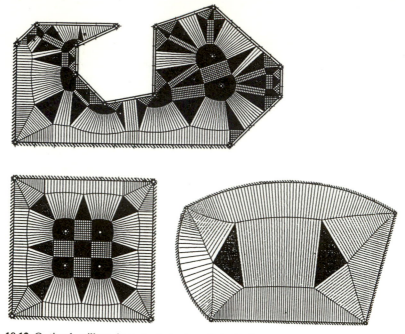

**Fig. 10.12** Optimal grillage layouts derived analytically by a computer on the basis of the constructions in Fig. 10.9 and 10.10 (without allowance for support costs).

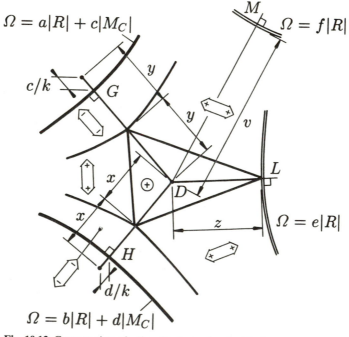

**Fig. 10.13** Construction of a 'junction' associated with three boundaries.

center D satisfies the following relation with respect to the boundary points G, H and L:

$$a + ky^2 - c^2/2k = b + kx^2 - d^2/2k = e + kz^2/2 \qquad (10.17)$$

where $\overline{DG}$, $\overline{DH}$ and $\overline{DL}$ are normal to the boundary. With respect to each boundary point (G, H, or L) in the above construction, the adjoint deflection values $\bar{u}_D$ are the same, that is $\bar{u}_{DG} = a + ky^2 - c^2/2k$, $\bar{u}_{DM} = b + kx^2 - d^2/2k$ and $\bar{u}_{DL} = e + kz^2$, with $\bar{u}_{DG} = \bar{u}_{DM} = \bar{u}_{DL}$. However, it is an important additional condition of the optimality of the layout in Fig. 10.13 that any other boundary point for which a normal can be drawn from the center D gives a greater $\bar{u}$ value than the one in equation (10.17). For example, for the boundary point M in Fig. 10.13, $\bar{u}_{DM} = f + kv^2/2$ must be greater than $\bar{u}_{DG} = \bar{u}_{DM} = \bar{u}_{DL}$. For each potential center, a large number of such tests must be carried out by the computer if the number of sides is relatively high (top of Fig. 10.12).

For each center, the construction in Fig. 10.13 defines an $S^+$ region termed a 'junction' (Rozvany, 1976, p. 198). Two such junctions (black triangular areas) are shown in the solution at the bottom right corner of Fig. 10.12.

Junctions are connected with each other and with certain boundary points (usually corners; Fig. 10.12) by so-called 'branches', which are constructed on the basis of Figs. 10.10 and 10.11. Using the above procedure, the computer selects systematically the optimal layout in an analytical representation and plots it. For a layout of such complexity as the one at the top of Fig. 10.12, this requires (e.g. on a HP 9000 work station) only a few minutes of computer time. For a reasonably accurate discretized numerical solution for the same layout, one would require an enormously large number of beam elements in the structural universe, which would make the corresponding computer time and storage requirement prohibitively high.

## (b) Grillages with combinations of free, simply supported and clamped edges

For some isolated cases of grillage geometries with free and simply supported edges, solutions were reported some time ago (e.g. Prager and Rozvany, 1977a; Rozvany, 1981). It was found at the time that, in general, the optimal topology along free edges contains a so-called beam-weave, consisting of short beams in negative bending and long beams in positive bending (Fig. 10.14(a)). The general equation for the relation between the distance $t$ along a straight free edge and the angle between the long beams and the edge is (Rozvany and Gerdes, 1994)

$$\frac{t}{a} = \exp\left[ \int_{\alpha}^{\alpha_0} \frac{\sin\gamma \, d\alpha}{\sin(\alpha + \gamma)\cos(2\alpha + 2\gamma)\sin\alpha} \right] \qquad (10.18)$$

(Fig. 10.14(b)) and the adjoint deflection at a point A (Fig. 10.14(c)) is given by

$$\bar{u} = \left( \frac{\partial \bar{u}}{\partial t} \right)_D (a - t_A) - \int_{t_A}^{\alpha} \cos(2\alpha)(t - t_A) \, dt \qquad (10.19)$$

where

$$\left( \frac{\partial \bar{u}}{\partial t} \right)_D = \frac{a \sin\gamma \sin(2\alpha_0 + \gamma)}{2 \sin^2(\alpha + \gamma)} \qquad (10.20)$$

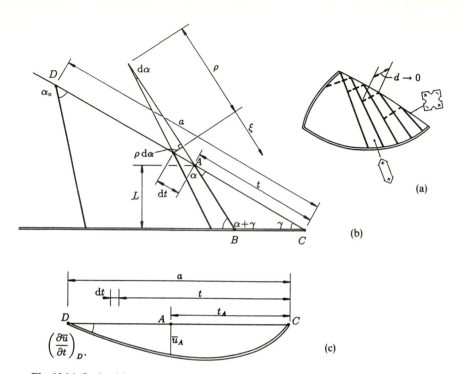

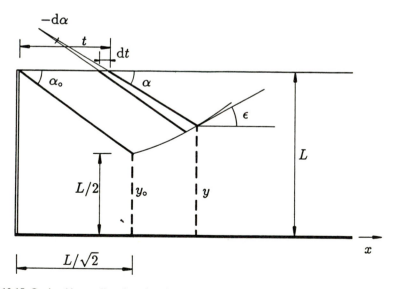

**Fig. 10.14** Optimal beam directions in grillages with free and simply supported edges.

**Fig. 10.15** Optimal beam directions in grillages with free, clamped and simply supported straight edges.

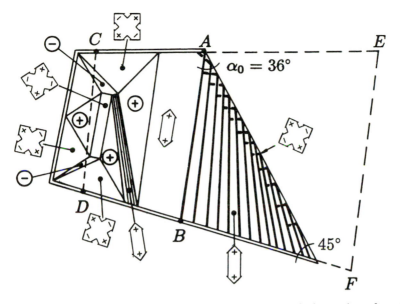

**Fig. 10.16** Optimal layout of a grillage with three simply supported edges and one free edge.

Although these general equations can at present be solved only by numerical integration, they do reduce to the known analytical solutions (Rozvany, 1981) for $\gamma = 0°$, $45°$ and $90°$. Moreover, a surprisingly simple geometrical property of the above solutions is that at any point A of the free edge, the adjoint slope in the direction parallel to the simple support is given by the explicit expression

$$\text{slope}_{A,\|} = t \sin \gamma \cot(\alpha + \gamma) \qquad (10.21)$$

For a straight free edge that is parallel to a straight clamped support, the governing equations have been shown to be (Rozvany and Gerdes, 1994)

$$dt = -\frac{L-y}{\sin^2\alpha}\ \frac{1+\sin^2\varepsilon}{2\sin^2(\alpha+\varepsilon)-1-\sin^2\varepsilon}\, d\alpha, \qquad \frac{d^2\bar{u}}{dt^2} = -\cos(2\alpha)$$

$$y = L - \sin\alpha\,[(L^2 - 2\bar{u})/(1 + \sin^2\alpha)]^{1/2}, \quad \tan\varepsilon = dy/dx \qquad (10.22)$$

The meaning of the symbols used in equation (10.22) is shown in Fig. 10.15. Equations (10.22) were solved by numerical integration for some examples and compared with discretized solutions (section 10.6.5, Fig. 10.25).

To demonstrate the complexity of grillage layouts for even relatively simple boundary conditions, a least-weight solution for a grillage with three simply supported edges and one free edge is shown in Fig. 10.16. Beams are indicated by thicker lines only in the $R^+$ region associated with the free edge.

## (c) Grillages with partially upward and partially downward loading

It was mentioned earlier that for clamped and simply supported edges the optimal grillage layout is independent of the load distribution if all loads have the same

sign. If this condition is violated, then the optimal grillage layout will become
(i) load dependent and (ii) much more complicated.

Figure 10.17(a), for example, shows one upward (negative) and three downward
point loads. The distances of the loads may appear somewhat artificial, as they
were chosen with a view to obtaining simple values for the geometry of the
optimal solution.

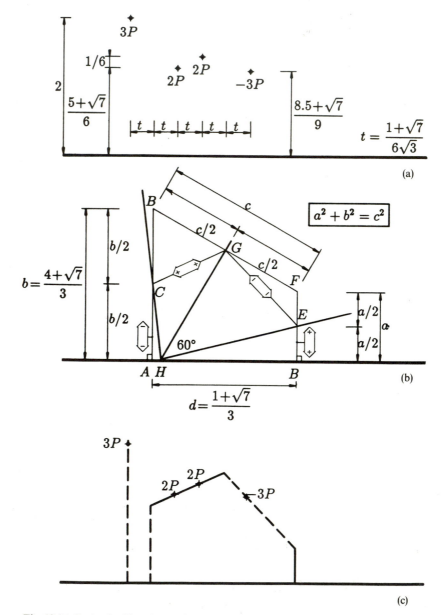

(a)

$$a^2 + b^2 = c^2$$

(b)

(c)

**Fig. 10.17** Optimal grillage layout for partially upward and partially downward loads.

As can be seen from Fig. 10.17(b), the solution for the considered class of problems can be obtained by means of introducing a fictitious simple support (HG in Fig. 10.17(b)) having a linearly varying adjoint displacement. Then the construction in Figs 10.10 and 10.11, together with slope continuity across the fictitious support, furnishes the optimal layout (Fig. 10.17(c)). A detailed treatment of the above procedure will be given elsewhere (Rozvany, 1993b). The solution in Fig. 10.17(c) was fully confirmed through numerical solutions by Sigmund.

## 10.5 LEAST-WEIGHT TRUSSES (MICHELL STRUCTURES)

As mentioned in section 10.2, the first truss solutions of least weight and a general theory for deriving them was published almost 90 years ago (Michell, 1904). The most important publications on this topic during the last 20 years were a book by Hemp (1973) and a paper by Lagache (1981).

The optimal regions for least-weight trusses are given again by relations (10.10), in which we replace $\sigma_i$ with $N_i$, i.e. the member force.

Most earlier solutions for least-weight trusses (Fig. 10.18; for a detailed treatment refer to Hemp's book of 1973) consisted of an infinite number of densely spaced members, although the truss was subject to a single point load. This has apparently created the incorrect impression that such 'truss-like continua' (Prager, 1974) constitute the rule rather than the exception.

About three years ago, the first author and his research group started a systematic exploration of optimal layouts for least-weight trusses. It was found (e.g. Rozvany and Gollub, 1990) that Michell structures usually take on such a complicated form only if the supports are statically highly restrictive (e.g. point supports or short line supports), but become relatively simple if the supports consist of longer line segments (for example, Fig. 10.8). This is demonstrated further in Figs 10.19–10.21.

Figure 10.19 shows a quadrilateral support with a point load in two different locations, together with the corresponding optimal truss layouts. Both are very simple, consisting of two bars only. For the optimal bar directions shown, explicit

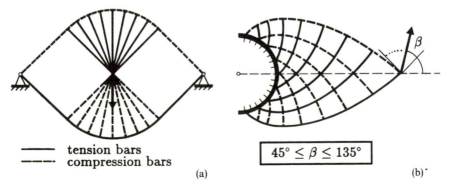

—— tension bars
---- compression bars

$$45° \leq \beta \leq 135°$$

(a)

(b)·

**Fig. 10.18** Michell structures consisting of an infinite number of members.

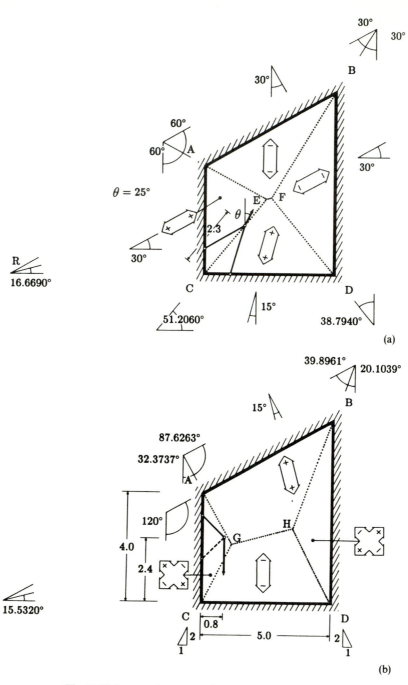

**Fig. 10.19** Least-weight trusses for a quadrilateral support.

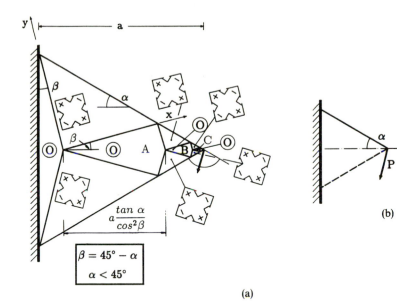

$$a\frac{\tan \alpha}{\cos^2 \beta}$$

$$\boxed{\begin{array}{l} \beta = 45° - \alpha \\ \alpha < 45° \end{array}}$$

(a)

(b)

Example:

$$\boxed{\begin{array}{l} R = \sqrt{2} \\ L = 6 \\ \omega = 22,5° \\ H = 3(1 + \sqrt{2}) \end{array}}$$

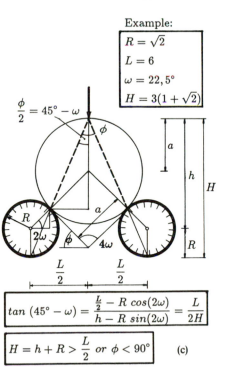

$$\frac{\phi}{2} = 45° - \omega$$

$$\boxed{\tan (45° - \omega) = \frac{\frac{L}{2} - R\,\cos(2\omega)}{h - R\,\sin(2\omega)} = \frac{L}{2H}}$$

$$\boxed{H = h + R > \frac{L}{2} \text{ or } \phi < 90°}$$

(c)

**Fig. 10.20** Optimal truss for a triangular domain with one line support and two free edges.

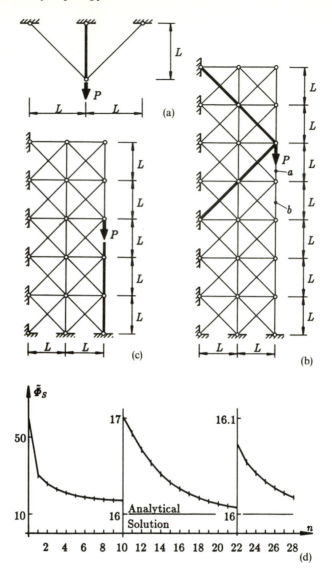

**Fig. 10.21** (a)–(c) Iterative discretized COC solutions for least-weight trusses, with (d) iteration history.

formulae are available (Rozvany and Gollub, 1990). In Fig. 10.20, the structural domain is bounded by a line support and two free edges. Although the adjoint strain field is rather complicated (Fig. 10.20(a)), the optimal layout consists of only two bars along the free edges (Fig. 10.20(b)). Finally, for two circular supports and a vertical point load, which is sufficiently far from the supports in the vertical direction, the optimal layout consists of two bars (Fig. 10.20(c)), whose centerline connects the tip of the force with the lowest point of the circles. This solution is valid only if the slope of the bars is greater than 1:1. Further details of the above two solutions are given elsewhere (Rozvany, Gollub and Zhou, 1994).

It should also be mentioned that the theory of least-weight trusses has been extended to least-weight shell grids and cable nets with only axial forces in the members (Rozvany and Prager, 1979), which were later also termed 'Prager structures'. A review of several other papers on this topic was given in a book chapter (Rozvany, 1984).

## 10.6 DISCRETIZED LAYOUT SOLUTIONS OBTAINED BY ITERATIVE COC–DCOC METHODS

The differences between analytical and numerical COC methods for layout optimization were explained in section 10.2 (Table 10.1). In this section, a number of test examples are presented.

### 10.6.1 Elementary test examples involving trusses

The first test examples solved by the iterative, discretized COC method (Fig. 10.21) concerned such elementary examples as a three-bar truss or a point load with horizontal and vertical supporting lines (Rozvany *et al.*, 1989). Figure 10.21 shows members of the structural universes in thin line and the optimal members in thick line. It can be seen that the results in Figs 10.21(b) and 10.21(c) fully confirm the analytical solutions in Fig. 10.8.

Using normalized values with $L = P = \rho = C = t = E = 1$ for the problem in Fig. 10.21(b), the analytical solution gives a weight of $\Phi = 16$. With a convergence tolerance value of $T = 10^{-14}$ (Chapter 2, equation (2.82)) and a lower limit on the cross-sectional area of $x\downarrow = 10^{-12}$, the iterative COC–DCOC method (for trusses, the COC and DCOC methods are almost identical) gave a weight value of $\Phi = 16.000\,000\,000\,048$ after 126 iterations, which represents an agreement of twelve significant digits. In the COC solutions, all non-optimal members took on a cross-sectional area of $10^{-12}$, except members a and b (Fig. 10.21(b)), which were both under $3 \times 10^{-12}$.

### 10.6.2 Truss-like continuum containing Hencky nets

Another optimal truss layout consisting of an infinite number of members is shown in Fig. 10.22(a). The truss is restricted to the rectangle ABCD and the point load $P = 1$ is to be transmitted to the supporting line $\overline{AC}$. In the solution, the triangle ACE contains no members on its interior, whilst the regions AEF and CEG contain straight radial members. The region EFHG consists of a Hencky net with curved members in both principal directions. For obvious reasons, only a finite number of members are indicated. The members along the edges of the truss (AFH and CGH) are 'concentrated' members which have a much bigger cross-sectional area than the others. The above analytical solution was discussed by Hemp (1973, pp. 97–9). The angle $\mu$ in Fig. 10.22(a) can be

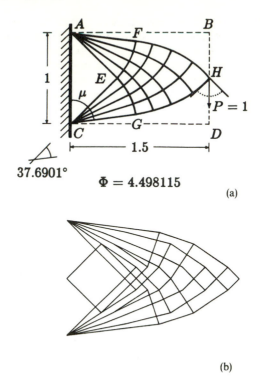

Fig. 10.22 Optimal Michell layout for a rectangular domain with one line support and three free edges: (a) analytical solution (after Hemp, 1973); (b) discretized COC solution.

calculated from Hemp's equation (4.120):

$$1.5 = \frac{1}{2} \int_0^{2\mu} [I_0(t) + I_1(t)] \, dt$$

$$= \frac{1}{2} \left[ I_0(2\mu) - 1 + 2 \sum_{n=0}^{\infty} (-1)^n I_{2n+1}(2\mu) \right] \Rightarrow \mu = 82.690\,133° \quad (10.23)$$

where $I_i$ are modified Bessel functions. Using plastic design (equation (10.3) with $k = 1$, or in Hemp's notation $\sqrt{2FR/\sigma} = 1$), the total truss weight becomes

$$\Phi = (1 + 2\mu) I_0(2\mu) + 2\mu I_1(2\mu) \quad (10.24)$$

giving the optimal truss weight

$$\Phi_{\text{opt,plastic}} = 4.498\,115 \quad (10.25)$$

For elastic design with a compliance constraint, we have by equation (10.8) with $c = k = r = C = 1$:

$$\Phi_{\text{opt,compliance}} = 4.498\,115^2 = 20.223\,042 \quad (10.26)$$

Using an iterative COC–DCOC method with structural universes containing 5055 and 12 992 members, respectively, truss weights of $\Phi = 20.540\,807$ and

$\Phi = 20.419\,699$ were obtained. By equation (10.8) these correspond to truss weights of $\Phi = 4.532\,197$ and $\Phi = 4.518\,816$, in plastic design, representing errors of 0.76% and 0.46%, respectively, compared with the analytical solution. In the above calculations a minimum cross-sectional area of $x{\downarrow} = 10^{-12}$ and a tolerance value of $T = 10^{-8}$ were used. The discretized layout, showing members having a cross-sectional area over $z^e = 0.1$ in compliance design, is shown in Fig. 10.22(b), which exhibits clear similarities with the analytical solution in Fig. 10.22(a).

### 10.6.3 Optimal grillage layout for a clamped square domain

The optimal layout for this problem was derived originally, in the context of reinforced concrete plates, by Lowe and Melchers (1972–1973) and is shown in Fig. 10.23(a). Sigmund extended the DCOC method for linearly varying elements (Sigmund, Zhou and Rozvany, 1993) and obtained for nine point loads the solution in Figs 10.23(c) and 10.23(d), using a structural universe with 624 beam

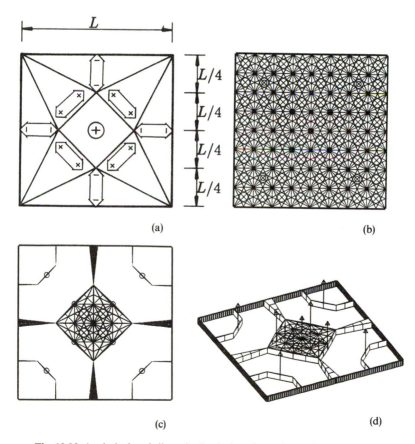

(a)

(b)

(c)

(d)

**Fig. 10.23** Analytical and discretized solutions for a clamped square grillage.

elements (Fig. 10.23(b)). The analytical and discretized solution gave almost the same nondimensional weight (0.234 619 vs 0.234 620).

### 10.6.4 Rhombic grillages with two simply supported and two free edges: beam weaves in the solution

For the above problem, the analytical solution (with $d \to 0$) is shown in Fig. 10.24(a). For a finite number of long beams, the structural weight of this layout is given by (Prager and Rozvany, 1977a)

$$\Phi = Pa^2[1 + 3c^2 + (1 - c^2)/n]/2 \tag{10.27}$$

where $n$ is the number of long beams, whilst $2a$ and $2ca$ with $c > 1$ are the dimensions of the grillage along the two axes of symmetry. Using a structural universe consisting of 620 beam elements, Sigmund obtained the solution in Fig. 10.24(c). Some of the short beams cross more than one long beam because of the nonuniqueness of the optimal solution. On the basis of a side length of 1.0 and an angle at the acute corners of 45°, we have for the above example

$$a = [1/(4 - 2\sqrt{2})]^{1/2} = 0.923\,879\,53, \quad c = \sqrt{2} - 1 = 0.414\,213\,56 \tag{10.28}$$

For $n = 10$ and $P = 1$, the relation (10.27) then yields the weight

$$\Phi_{\text{plastic}} = 0.681\,801\,95, \quad \Phi_{\text{compliance}} = \Phi^2_{\text{plastic}} = 0.464\,853\,90 \tag{10.28a}$$

The numerical solution in Fig. 10.24(c) fully confirmed the latter.

### 10.6.5 Grillage with simply supported, clamped and free edges

Two discretized solutions, with 466 and 1892 beam elements, respectively, in the structural universe, are shown in Figs 10.25(a) and 10.25(b). The corresponding 'analytical' solution, based on the numerical integrations of the differential equations in relations (10.22), is shown in Fig. 10.25(c), with obvious similarities to the discretized solutions. The above discretized solutions were actually obtained before the derivation of the analytical solution, which was based on the topology suggested by the above discretized layouts.

### 10.6.6 Square grillages with two clamped and two free edges

This combination of support conditions is one of the very few for which as yet no analytical solution is known. For this reason, a particularly detailed study of discretized solutions is being carried out with a view to determining the likely topology of the exact analytical solution.

Considering a point load at the unsupported corner, a simple beam layout consisting of two cantilevers along the free edges gives a normalized grillage weight of $\Phi = 0.250$ for elastic compliance.

Using a structural universe with 9312 elements (Fig. 10.26(a)), Sigmund obtained the solution in Fig. 10.26(b), with a structural weight of $\Phi = 0.1819$. It

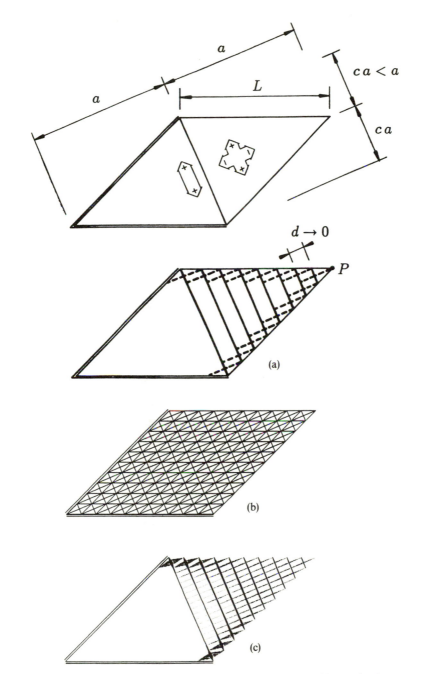

**Fig. 10.24** Analytical and discretized solutions for rhombic grillage with two simply supported and two free edges.

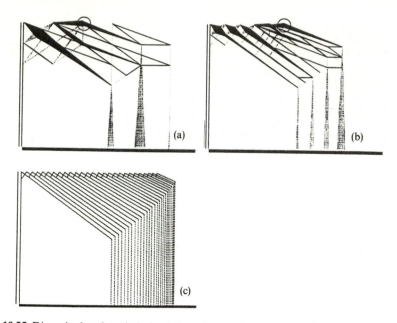

**Fig. 10.25** Discretized and analytical solutions for a grillage with simply supported, clamped and free edges.

can be seen that this solution consists of beam weaves over about one half of the free edges, supported by two heavy cantilever beams and balanced by longer beams under negative moment.

Whereas in Figs 10.26(a) and 10.26(b) beam directions are restricted to slopes of 0, 1:1 and 1:2 to the sides, the structural universe in Fig. 10.26(c) includes additional slopes of 1:3 and 2:3 but only 3752 beam elements. The weight of the corresponding solution (Fig. 10.26(d)) is somewhat higher ($\Phi = 0.1906$), but shows better the variation of the direction of the long beams which form part of the beam weave.

### 10.6.7 A 'practical' solution with stress and deflection constraints: triangular grillage with one free edge and two simply supported edges

The main advantage of discretized layout solutions by COC–DCOC is that they can readily include a combination of a variety of design conditions, such as stress, displacement, natural frequency and system buckling constraints.

The length of the simply supported edge in Fig. 10.27 is 1.0 and its slope to the axis of symmetry is 2:1. The structural universe wth 299 elements is shown in Fig. 10.27(a), and the permissible shear stresses and structural weight in the next three diagrams are as follows:

$$
\begin{aligned}
&\text{Fig. 10.27(b):} \quad \tau_a = 17.79, \quad \Phi = 0.026653 \\
&\text{Fig. 10.27(c):} \quad \tau_a = 10.00, \quad \Phi = 0.027801 \qquad (10.29) \\
&\text{Fig. 10.27(d):} \quad \tau_a = 6.00, \quad \Phi = 0.035282
\end{aligned}
$$

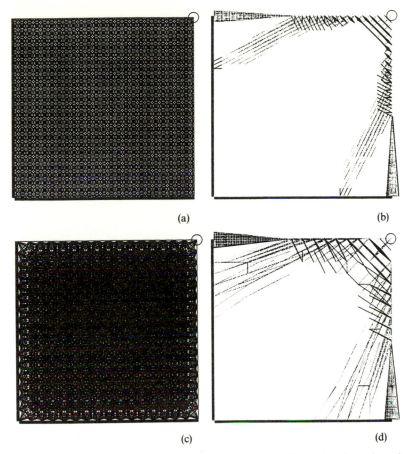

(a)                                         (b)

(c)                                         (d)

**Fig. 10.26** Discretized solutions for a square grillage with two clamped and two free edges.

The above results were obtained by Sigmund using a normalized formulation wth $P = \rho = t = 1$. If we restrict the beam elements to the free edge, then we obtain the solution in Fig. 10.27(e), with a weight of $\Phi = 0.161\,626$ ($\tau_{max} = 4.82$ inactive). This latter solution, which was proposed by Lowe and Melchers (1972–1973) as a 'practical optimum', represents over 400% of the optimal weight in this case, although it is still being claimed (Lowe, 1988) that the topology in Fig. 10.27(e) is relatively economical. Figure 10.28(a) shows the analytical solution for the same problem, together with the optimal regions. The short beam across the corner has a theoretically infinitesimal length. For a comparison, the region topology of the Lowe–Melchers solution, which clearly violates the inequality (10.4) in the direction of the axis of symmetry, is given in Fig. 10.28(b). Finally, the exact optimal solution with allowance for support costs of $\Omega = r|R|$ is shown in Fig. 10.28(c), where $r$ is a given constant and $R$ is the reaction. It can be seen that once the cost of reactions is taken into consideration, the solution becomes a transition between the first author's solution (e.g. Rozvany, Hill and Gangadharaiah, 1973)

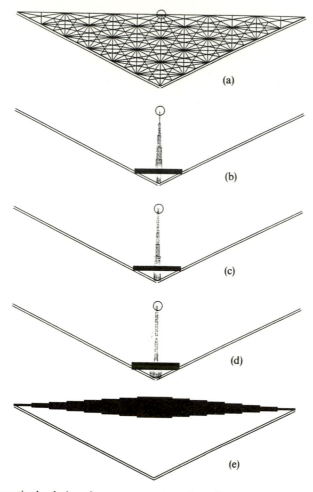

**Fig. 10.27** Discretized solutions for a corner region of a grillage with a deflection constraint and shear stress constraints.

and the Lowe and Melchers (1972–1973) solution – a very satisfactory resolution of an old controversy!

## 10.7 EXTENSION OF THE DCOC ALGORITHM TO COMBINATIONS OF STRESS, DISPLACEMENT AND NATURAL FREQUENCY CONSTRAINTS WITH ALLOWANCE FOR SELFWEIGHT AND STRUCTURAL MASS

### 10.7.1 Derivation of optimality criteria

We extend in this subsection the derivation of DCOC for stress and displacement constraints and several loading conditions (section 2.3.3) to selfweight and natural

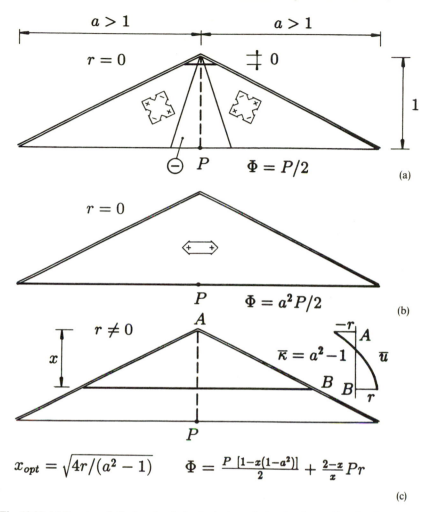

$$x_{opt} = \sqrt{4r/(a^2 - 1)} \qquad \Phi = \frac{P\,[1 - x(1 - a^2)]}{2} + \frac{2 - x}{x} Pr$$

(c)

**Fig. 10.28** (a) Exact analytical optimal plastic design solution for the problem in Fig. 10.27; (b) region topology of the Lowe and Melchers (1972–1973) solution and (c) solution with allowance for the cost of supports.

frequency constraints. It is assumed that the selfweight of an element equals the element objective function $w^e(\mathbf{x}^e)$. The relative initial displacements $\overset{\circ}{\delta}{}^{*e}$ caused by the selfweight are linear functions of $w^e$, say

$$\{\overset{\circ}{\delta}{}^{*e}\} = \{\mathbf{g}^e\}\, w^e \qquad (10.30)$$

where the vector $\{\mathbf{g}^e\}$ depends on the design variables $\mathbf{x}^e$. The natural frequency constraints can be represented by the corresponding Rayleigh quotients

$$\lambda_g \leqslant \frac{\{\check{\mathbf{U}}_g\}^T [\mathbf{K}] \{\check{\mathbf{U}}_g\}}{\{\check{\mathbf{U}}_g\}^T [\mathbf{M} + \vec{\mathbf{M}}] \{\check{\mathbf{U}}_g\}} \qquad (g = 1, \ldots, G) \qquad (10.31)$$

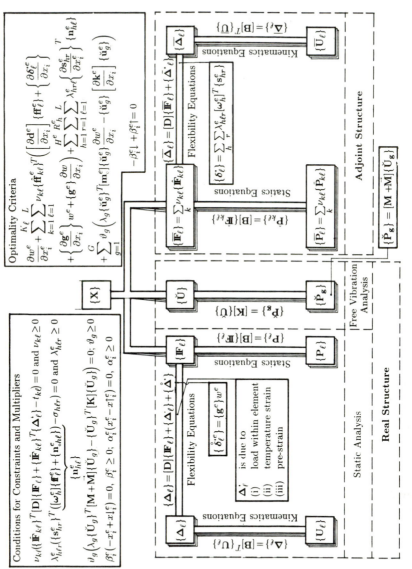

**Fig. 10.29** DCOC algorithm for linearly elastic systems with stress, displacement and natural frequency constraints and allowance for selfweight and selfmass – conceptual scheme.

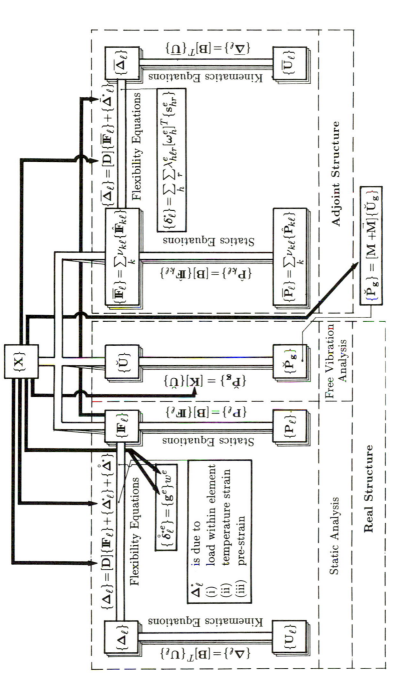

**Fig. 10.30** Conceptional diagram showing the dependency of certain relations used in the DCOC algorithm on the design variables $\{\mathbf{X}\}$ and on nodal forces $\{\mathbf{F}_{lj}\}$.

where $\lambda_g$ is the lower limit on an eigenvalue (square of natural frequency), $\{\check{U}_g\}$ is the eigenvector, $[M]$ is the structural mass matrix and $[\vec{M}]$ is the nonstructural mass matrix. The constraint (10.31) can be rewritten in the more convenient form

$$\lambda_g\{\check{U}_g\}^T[M + \vec{M}]\{\check{U}_g\} - \{\check{U}_g\}^T[K]\{\check{U}_g\} \leqslant 0 \quad (g = 1,...,G) \quad (10.32)$$

The structural mass matrix $[m^e]$ for each element can be expressed as some linear function of the element weights $w^e(x^e)$,

$$[m^e] = [m_0^e]w^e \quad (10.33)$$

We use the Lagrange multipliers $\vartheta_g (g = 1,...,G)$ for the natural frequency constraints (10.32), otherwise the notation remains the same as in section 2.3.3.

Necessary conditions for optimality (Kuhn–Tucker conditions) with respect to variation of the design variables $x_i^e (i = 1,...,I; e = 1,...,E)$ then become

$$\frac{\partial w^e}{\partial x_i^e} + \sum_{k=1}^{K_l}\sum_{l=1}^{L} v_{kl}\{\hat{\mathbf{f}}_{kl}^e\}^T\left(\left[\frac{\partial \mathbf{d}^e}{\partial x_i}\right]\{\mathbf{f}_i^e\} + \left\{\frac{\partial \boldsymbol{\delta}_i^{*e}}{\partial x_i}\right\} + \left\{\frac{\partial \mathbf{g}^e}{\partial x_i}\right\}w^e + \{\mathbf{g}^e\}\frac{\partial w}{\partial x_i}\right)$$

$$+ \sum_{h=1}^{H^e}\sum_{r=1}^{R_h^e}\sum_{l=1}^{L} \lambda_{hrl}^e \frac{\partial \theta_{hr}^e(\mathbf{n}_{hl}^e)}{\partial x_i^e} + \sum_{g=1}^{G} \vartheta_g\left(\lambda_g\{\check{\mathbf{u}}_g^e\}^T[\mathbf{m}_0^e]\{\check{\mathbf{u}}_g^e\}\frac{\partial w^e}{\partial x_i} - \{\check{\mathbf{u}}_g^e\}^T\left[\frac{\partial \mathbf{k}^e}{\partial x_i}\right]\{\mathbf{u}_g^e\}\right)$$

$$- \beta_i^e\downarrow + \beta_i^e\uparrow = 0 \quad (10.34)$$

Kuhn–Tucker conditions with respect to the real and adjoint forces give the same relations as in section 2.3.3.

A conceptual scheme for the extended DCOC algorithm considered above is given in Fig. 10.29, in which linear stress conditions $\{\mathbf{s}_{hr}^e\}^T\{\mathbf{n}_{hl}^e\}$ are instead of the nonlinear ones $\theta_{hr}^e(\mathbf{n}_{hl}^e)$ in equation (10.34). It can be seen from Fig. 10.29 that any change in the value of the design variables has an effect on the following relations in Fig. 10.29:

- the flexibility matrix $[D]$ (or stiffness matrix $[K]$) used for the real structure;
- the initial displacements $\{\boldsymbol{\Delta}_i^*\}$ caused by loads within the elements, due to changes in member flexibilities;
- the initial displacements $\{\mathring{\boldsymbol{\Delta}}^*\}$ caused by the selfweight, due to (i) changes in the member flexibilities and (ii) changes in the load caused by selfweight;
- the flexibility matrix $[D]$ (or stiffness matrix $[K]$) used for the adjoint structure;
- the stiffness matrix $[K]$ used for the free vibration analysis;
- the structural mass matrix $[M]$ used in the free vibration analysis.

The above feedback effects are represented graphically in Fig. 10.30, which also shows that the adjoint initial displacements $\{\bar{\boldsymbol{\Delta}}^*\}$ are also dependent on the real nodal forces $\{\mathbb{F}_e\}$ (through the Lagrange multipliers $\lambda_{hrl}^e$).

For problems with only natural frequency constraints, the optimality conditions (10.34) reduce to those of Berke and Khot (e.g. Berke and Khot, 1987). The detailed computational algorithm for the problems discussed in this section is given elsewhere (Sigmund and Rozvany, 1994).

## 10.7.2 Test example involving selfweight: square clamped grillage with a central point load

Considering a laminated timber grillage having a depth of 1.0 m, Young's modulus $E = 10^7 \text{kN m}^{-2}$, specific weight $\rho = 6 \text{kN m}^{-3}$, minimum width $10^{-5} \text{m}$,

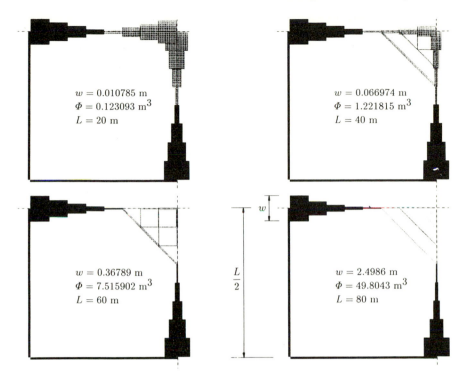

**Fig. 10.31** Optimal grillage layouts with allowance for selfweight considering various span lengths (by Sigmund).

prescribed deflection $t = L/300$ where $L$ is the span length, and a point load of 10 kN, Sigmund obtained by the extended DCOC method given in section 10.7.1 the solutions in Fig. 10.31, which also indicates the span length ($L$), the maximum beam width ($w$) in m and the total volume $\Phi$ in m³. All graphic outputs were scaled such that the maximum width indicated is the same, but the latter for the largest span is in fact over 230 times greater than that for the shortest span. It can be seen that with increasing span length more and more material is moving away from the middle of the grillage to the supports. Naturally, the prescribed minimum width would be completely unrealistic in practice, but the results indicate interesting trends.

### 10.7.3 Test example: perforated deep cantilever beams optimized for first or second natural frequency

Using the method discussed in section 10.7.1, Sigmund optimized perforated deep beams supported along the left vertical edge. The height of the beam is 1.0 and its length 4.0. At the right end a nonstructural mass is created by fixing the thickness at its maximum value over a length of 0.2. The number of bilinear square elements used is 840, with a minimum thickness of 0.01 and a maximum thickness of 1.0. In the solution in Fig. 10.32(a), the natural frequency was fixed

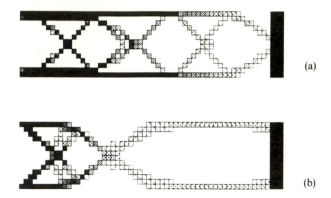

(a)

(b)

**Fig. 10.32** Least-weight, deep, perforated cantilever beams designed, respectively, for given first and second natural frequency with nonstructural mass and selfmass (by Sigmund).

at $\sqrt{\lambda_1} = 0.0707$ ($\lambda_1 = 0.005$) and the second natural frequency turned out to be $\sqrt{\lambda_2} = 0.2154$. In Fig. 10.32(b), the second natural frequency was fixed at $\sqrt{\lambda_2} = 0.346$ ($\lambda_2 = 0.12$) and the first natural frequency turned out to be $\sqrt{\lambda_1} = 0.0253$. In Fig. 10.32, black areas indicate the maximum thickness, white areas the minimum thickness, with other areas varying in darkness according to the thickness.

## 10.8 GENERALIZED SHAPE OPTIMIZATION

### 10.8.1 Historical background and problem classification

As mentioned in section 10.1, the aim of generalized shape optimization is the simultaneous optimization of both topology and shape of the boundaries of

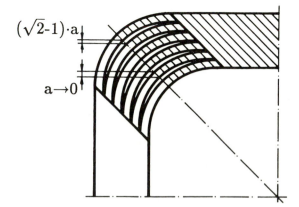

**Fig. 10.33** One of the earliest solutions in generalized shape optimization (after Kohn and Strang, 1983).

two- or three-dimensional continua or of the interfaces between different materials in composites.

It was established by Kohn and Strang (1983) in the context of plastic design for torsion of a cross section within a square area (Fig. 10.33) that generalized shape optimization may yield three types of regions in the solution, namely

- solid regions (filled with material),
- empty regions (without material), and
- porous regions (some material, with cavities of infinitesimal size).

Considering elastic perforated plates in bending or in plane stress, it was found by various mathematicians (e.g. Lurie and Cherkaev, 1984; Murat and Tartar, 1985; Kohn and Strang, 1986; Avellaneda, 1987) that one optimal microstructure for a compliance constraint consists of rank-2 laminates, i.e. ribs of first- and second-order infinitesimal width in the two principal directions.

### 10.8.2 Solid–empty–porous (SEP) solutions

Analytical solutions based on rank-2 laminates for axisymmetric perforated plates, derived by Rozvany *et al.* (1987) as well as Ong, Rozvany and Szeto (1988), have shown that

- a high proportion of the available space in exact optimal solutions consists of porous regions, and
- for low volume fractions (ratio of material volume to available volume) the solution tends to that for least-weight trusses (Michell, 1904) for plane stress and least-weight grillages (e.g. Prager and Rozvany, 1977a) for bending (this conclusion was also confirmed by Allaire and Kohn (1993)).

Following the above developments, Bendsøe (e.g. Bendsøe, 1989) opened up new avenues of research in topology optimization by investigating discretized optimal solutions derived by mathematical programming methods, using either the correct microstructure (rank-2 laminates) or a 'suboptimal' microstructure (square or rectangular holes). The above investigations have also made use of some important analytical relations regarding geometrical properties of optimal topologies (e.g. Pedersen, 1989, 1990). Extensions of this work were reported by several investigators (e.g. Bendsøe and Kikuchi, 1988; Suzuki and Kikuchi, 1991; Diaz and Bendsøe, 1992; Olhoff, Bendsøe and Rasmussen, 1991; Allaire and Kohn, 1993).

Numerical solutions using rank-2 laminates have a considerable theoretical importance because they represent (within a discretization error) an absolute limit on the structural weight for a class of compliance problems. However, other lines of research on topology optimization are justified, because solutions with rank-2 microstructures are somewhat unpractical for the following reasons:

- even an approximate version of rank-2 microstructures with finite rib widths would require very high manufacturing costs;
- rank-2 laminates in perforated structures (as distinct from composite structures) have a zero shear stiffness in one direction, which makes these designs completely unstable if the load direction is changed;

- solutions are only available for a single compliance or natural frequency constraint which are not realistic design problems. This was also demonstrated recently by Sankaranarayanan, Haftka and Kapania (1992).

Moreover, because of nonconvexity, the above solution could represent a local optimum.

Solutions with suboptimal microstructures often turn out to be more practical, because they penalize and thereby suppress porous regions.

The solutions discussed in this subsection often appear in the literature under the term homogenization, which means that an inhomogeneous structural element, containing an infinite number of discontinuities in material or geometrical properties, is replaced by a homogeneous but anisotropic element, whose stiffness is direction but not location dependent within an element. The same idea was used in a very simple form much earlier by others (e.g. Michell, 1904; Prager and Rozvany, 1977b), introducing terms such as 'truss-like continua' or 'grillage-like continua' (Prager, 1974). To engineers this is a rather obvious idealization, but its rigorous mathematical treatment is much more involved (e.g. Bensousson, Lions and Papanicolaou, 1978). It would be incorrect, however, to make the term 'generalized shape optimization' synonymous with 'shape optimization by homogenization'.

### 10.8.3 Solid–empty (SE) solutions and solid, isotropic microstructure with penalty (SIMP) for intermediate densities

The method described in this subsection has been used extensively by the authors, but it was also tried out earlier by Bendsøe (e.g. Bendsøe, 1989), as an extension of a technique employed by Rossow and Taylor (1973).

From an engineering point of view, it is more useful to aim at solutions in which porous regions are largely suppressed and then a second stage design procedure can produce a 'practical' solution consisting of solid and empty regions only, with smooth boundaries to avoid stress concentrations. This procedure was lucidly demonstrated by Olhoff, Bendsøe and Rasmussen (1991) on an example involving a simply supported beam with a central point load (Figs 10.34(a) and 10.34(b)).

It was suggested by the first author at a meeting in Karlsruhe in 1990 (Rozvany and Zhou, 1991; Zhou and Rozvany, 1991) that porous regions could be suppressed by adding to the material costs, the cost of manufacturing of holes, thereby penalizing porous regions. Once we decide that we want only solid and empty regions in the solution, any range of microstructures that includes the above two as limiting cases can be assumed in the solution process. We can therefore postulate that the specific material cost (e.g. weight) $\psi$ is proportional to the specific stiffness $s$ of perforated regions (Fig. 10.35(a)). This would also be the case if we used a plate of variable thickness (Rossow and Taylor, 1973), in which case we would have to penalize intermediate thicknesses. If the hypothetical microstructure contains holes, however, then the extra manufacturing cost would increase with the number or size of such holes, if we consider a casting or drilling process. However, for an empty region ($s = 0$), the manufacturing costs also become zero. The corresponding specific fabrication cost is shown in

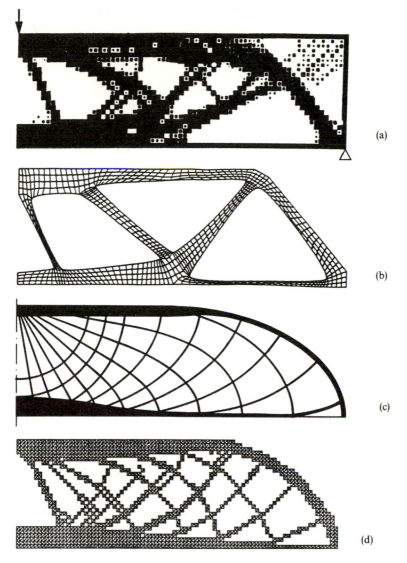

**Fig. 10.34** (a) Topology of a perforated simply supported deep beam (one half shown) using a suboptimal microstructure; (b) design after a second stage optimization (after Olhoff, Bendsøe and Rasmussen, 1991); (c) the exact topology suggested by discretized truss solutions (Zhou and Rozvany, 1991); (d) topology derived by using a solid isotropic microstructure with penalty (SIMP) for intermediate densities (Rozvany and Zhou, 1991).

Fig. 10.35(b) and the specific total cost in Fig. 10.35(c), together with an approximation of the type

$$\psi = s^{1/n} \tag{10.35}$$

where $n$ is a given constant. The above relation is identical with that suggested

Specific Material Cost

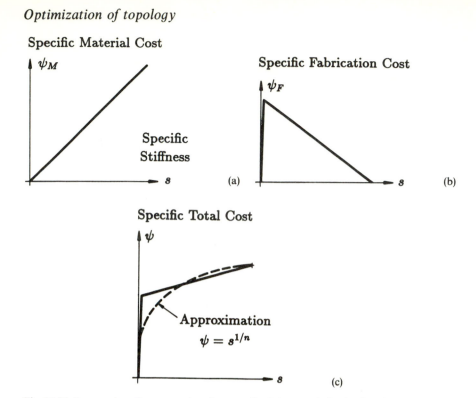

Fig. 10.35 Suppression of porous regions in generalized shape optimization by taking fabrication costs into consideration.

by Bendsøe (1989, equation (7)), although the first author was unaware of this when he proposed equation (10.35). The authors would like to emphasize, however, that this entire line of research owes its existence to the pioneering efforts of Bendsøe, and their own work should be regarded only as a small extension of Bendsøe's milestone contributions.

In selecting a hypothetical microstructure for obtaining a solid–empty (SE) solution, the following objectives should be considered:

- simplicity of analysis and optimization;
- selective suppression of porous regions by adjustable penalty;
- capability of handling a variety of design conditions (e.g. combinations of stress, displacement, natural frequency and system stability constraints for several loading conditions).

It has been found that a solid isotropic microstructure with penalty (SIMP) for intermediate densities, as explained above (Fig. 10.35), largely fulfills these objectives (Rozvany, Zhou and Birker, 1992). For example, the topology obtained by the SIMP model is shown in Fig. 10.34(d). The latter is an excellent approximation of the 'exact' topology for the same problem in Fig. 10.34(c), which was constructed on the basis of the following arguments.

As mentioned earlier, for low volume fractions the optimal topology for perforated plates in plane stress tends to that for least-weight trusses or Michell

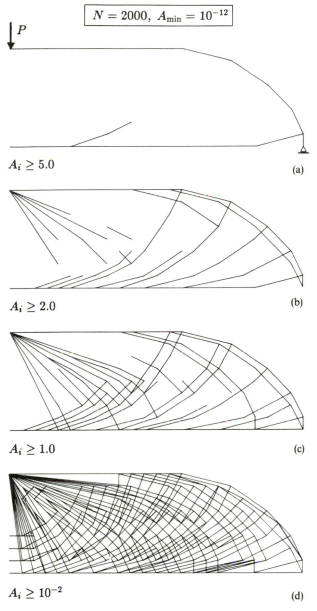

$$N = 2000, \ A_{\min} = 10^{-12}$$

$P$

$A_i \geq 5.0$      (a)

$A_i \geq 2.0$      (b)

$A_i \geq 1.0$      (c)

$A_i \geq 10^{-2}$      (d)

**Fig. 10.36** Truss layout obtained by a discretized COC method for the load and support conditions in Fig. 10.34.

structures (Rozvany *et al.*, 1987; Allaire and Kohn, 1993). The layout of a truss, with the same support and loading conditions as the beam in Fig. 10.34, was optimized by the discretized COC procedure and the results are shown in Fig. 10.36, which shows one half of the truss with various ranges of cross-sectional areas in the optimal solution. On the basis of this layout, taking some (analytical) geometrical properties of Hencky nets into consideration, the solution in Fig. 10.34(c) was constructed. In the latter, solid regions appear along the top and bottom chords of the corresponding truss, with the width proportional to the cross-sectional areas of the chords, an empty region in the top right corner and porous regions, with an infinite number of intersecting members in between the chords. The layout in Fig. 10.34(c) was also confirmed analytically (Lewinski, Zhou and Rozvany, 1993).

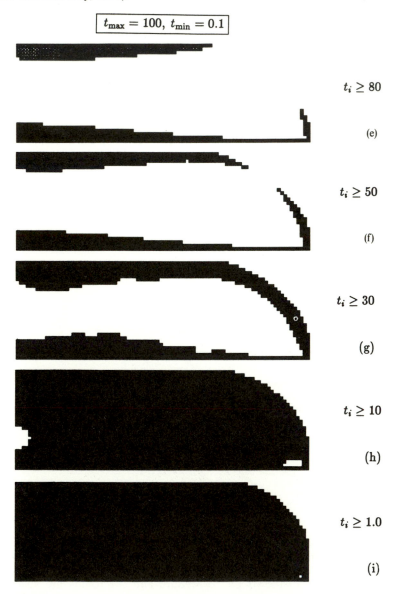

$$t_{max} = 100, \; t_{min} = 0.1$$

$t_i \geq 80$

(e)

$t_i \geq 50$

(f)

$t_i \geq 30$

(g)

$t_i \geq 10$

(h)

$t_i \geq 1.0$

(i)

**Fig. 10.37** Solution for a plate of variable thickness obtained by a discretized COC method for the load and support conditions in Fig. 10.34.

The optimal topology in Figs 10.34(c) and 10.34(d) was also made plausible by optimizing a plate of variable thickness, but without penalty for intermediate thicknesses, for the same support and load conditions. The results are shown in Fig. 10.37, which shows that solid isotropic microstructures without penalty tend to give the same average density for larger areas as the SIMP formulation but, instead of ribs, an intermediate density appears in porous regions.

The solution in Fig. 10.34(d) is difficult to compare with a solution containing rank-2 laminates, but it is easier to understand in terms of an alternate optimal microstructure for plates with a compliance constraint, which was derived recently by Vigdergauz (1992). In porous regions, the latter may start at very low volume fractions with a Michell structure or least-weight grillage (Fig. 10.34(c)), then develop roundings at the corners as we increase the volume fraction (Fig. 10.38(a)), finishing up with elliptic holes of decreasing size as the volume fraction approaches unity (Fig. 10.38(b)).

In Fig. 10.39, three types of specific cost functions are compared for a plate in plane stress or bending. The straight line represents the normalized weight per unit area of a plate of variable thickness in plane stress or bending. The next curve shows the weight of a constant thickness perforated plate with rank-2

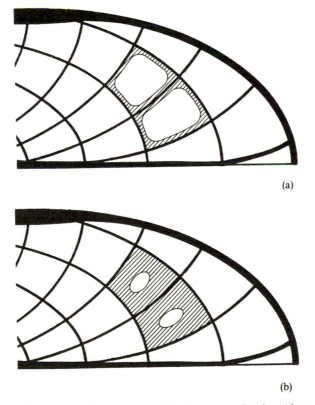

(a)

(b)

**Fig. 10.38** Alternative optimal microstructures at higher volume fractions (the spacing should be theoretically infinitesimal).

laminates having an equal stiffness in two directions (Rozvany *et al.*, 1987). Finally, the top curve represents the power type cost function of the SIMP formulation.

When Bendsøe (1989) originally considered this power-type cost function, he used the term 'direct approach' or '0–1 discrete optimization method with a suitable differentiable approximation' using an 'artificial material'. At the same time, Bendsøe expressed preference for optimal or suboptimal anisotropic micro-structures, reasoning that the solutions by the direct approach (now SIMP) are (i) highly mesh dependent and (ii) impossible to interpret physically. The authors are more optimistic in these respects, as can be seen from the following.

The SIMP model can be interpreted even for three-dimensional continua with stress and displacement constraints, assuming that both Young's modulus and permissible stress of a fictitious material (or a range of materials) are proportional to its (their) density, which varies between a given maximum value and a very small minimum value. It is then perfectly legitimate for the designer to penalize and thereby to suppress intermediate densities, if he chooses to do so. It is not necessary for such a material to exist, because we are only interested in the end result; however, it is still interesting to remark that the above relations are known to constitute a good estimate for properties of various species of timber, ranging from the very light balsa-wood to the extremely heavy, stiff and strong iron bark (from Australia). Assuming that on some Pacific island only these two tree species grew naturally and all other timbers had to be imported, then a cost optimization

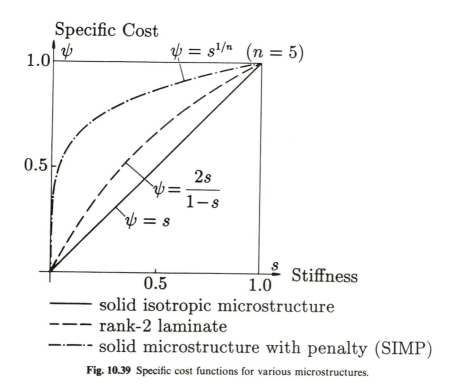

solid isotropic microstructure
rank-2 laminate
solid microstructure with penalty (SIMP)

**Fig. 10.39** Specific cost functions for various microstructures.

of various timber structures would give exactly the same results as the SIMP model.

It is shown subsequently on test examples that the SIMP formulation yields the correct topology irrespective of the FE mesh used.

### 10.8.4 Test examples based on the SIMP model

#### (a) Cantilever beam

The analytical and discretized truss solutions for this problem were shown in Fig. 10.22. A topology obtained by using a suboptimal microstructure (square holes) by Suzuki and Kikuchi (1991) is shown in Fig. 10.40(a). It was shown by the same authors that this topology is largely independent of the number of elements used (for meshes ranging from $32 \times 20$ to $80 \times 50$). It can be seen by comparing Figs 10.22(a) and 10.40(a) that in the latter the topology and shape deviate significantly from the optimal ones. It should be taken into consideration that in Fig. 10.40(a) the aspect ratio is higher (1.6 instead of 1.5) and hence the chords should have an almost horizontal tangent at the support. This condition together with other close similarities with Fig. 10.22 were fulfilled better by the SIMP solutions shown in Figs 10.40(b) and 10.40(c), which were obtained, respectively, by M. Zhou (with 10 800 constant strain triangular elements) and T. Birker (1440 isoparametric square elements). For Birker's solutions, two different strategies were used, the iteration histories of which are shown in Fig. 10.41. Whilst in one of them (Fig. 10.41(a)) the initial design consisted of a

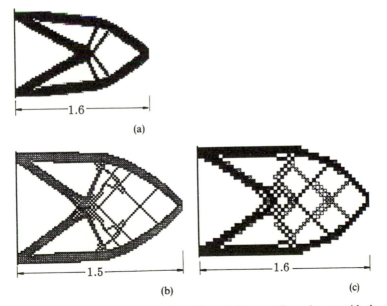

(a)

(b)                    (c)

**Fig. 10.40** Various discretized solutions for perforated deep cantilever beams with the same loading and support conditions as in Fig. 10.22.

plate of uniform thickness, in the other one (Fig. 10.41(b)), it was a plate of variable thickness optimized without penalty for intermediate densities (as the solutions by Rossow and Taylor (1973)). This latter method for the initial design, suggested by R.V. Kohn, gave a better topology and a slightly lower weight.

### (b) Clamped beam

The analytical least-weight truss solution for the above problem is shown in Fig. 10.42(a) and a discretized DCOC solution, obtained by D. Gerdes (Essen University) in Fig. 10.42(c). Finally, a solution by O. Sigmund, who used the

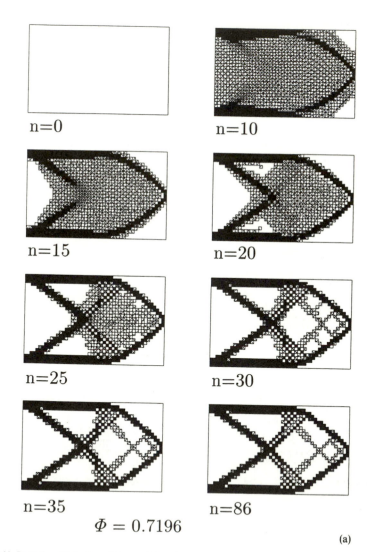

$$\Phi = 0.7196$$

(a)

**Fig. 10.41** Iteration histories of the problem in Fig. 10.40 using two different initial solutions.

SIMP model, is given in Fig. 10.42(b). The analytical solution (Fig. 10.42(a)) consists of some circular 'fans' (e.g. ACD, BCE, IGJ, IHK), 'concentrated' straight members (e.g. AD, DG, BE, FH), concentrated curved members (DF, EF, GJ, HK) and Hencky nets consisting of curved members (CDFE). In the discretized solution (Figs 10.42(b) and 10.42(c)), the narrow light fans (e.g. ACD) and heavy members (e.g. AD) are lumped together into a single member, which could only be avoided by using a much finer discretization. However, the discretized solutions in Figs 10.42(b) and 10.42(c) are much more practical, because they consist of a small number of members.

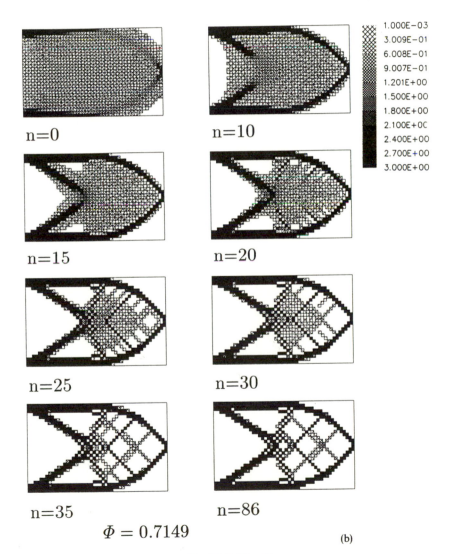

$$\Phi = 0.7149$$

(b)

**Fig. 10.41** (*Cont'd*)

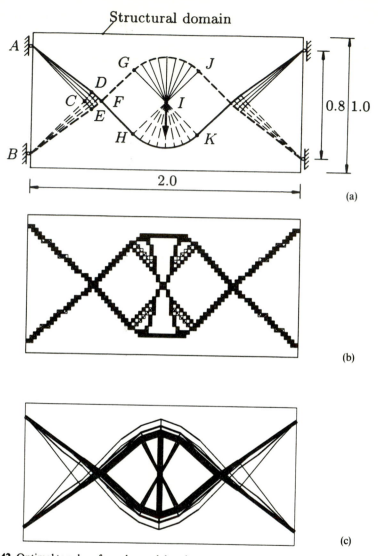

**Fig. 10.42** Optimal topology for a clamped deep beam: (a) analytical truss solution; (b) perforated plate solution using a SIMP model; (c) discretized truss solution by DCOC.

## 10.9 SOLUTIONS FOR SEVERAL LOAD CONDITIONS AND NON-SELF-ADJOINT PROBLEMS

### 10.9.1 Elastic design for compliance constraints and plastic design with several load conditions

In earlier work (e.g. Suzuki and Kikuchi, 1991; Diaz and Bendsøe, 1992), approximate solutions of the above problems were obtained, in which

- either for each element the maximum compliance value out of all loading conditions was taken and then the sum of such values was calculated for the entire system, or
- a weighted combination of the compliances (or natural frequencies) was taken as objective function.

Both methods simplify the treatment considerably, and the latter is in fact equivalent to a single compliance constraint. The solutions obtained by DCOC are based on the actual multi-load problem.

An analytical layout theory for trusses with several load conditions was developed recently (Rozvany, 1992a). In the notation of Chapter 2 herein, we have the following conditions for layout optimality:

$$\varepsilon_l^e = F_l^e/E^e A^e, \quad \bar{\varepsilon}_l^e = \left(\sum_k v_{lk} F_{lk}^e\right)/E^e A^e$$

$$\frac{E^e}{\varrho^e}\sum_l \varepsilon_l^e \bar{\varepsilon}_l^e \leqslant 1 \quad \text{(for } A^e = 0\text{)}, \quad \frac{E^e}{\varrho^e}\sum_l \varepsilon_l^e \bar{\varepsilon}_l^e = 1 \quad \text{(for } A^e \neq 0\text{)} \qquad (10.36)$$

where $l$ denotes the load condition and $k$ a displacement constraint.

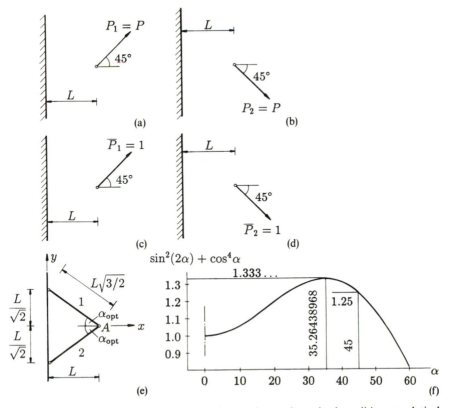

**Fig. 10.43** Least weight truss for compliance constraints and two load conditions: analytical solution.

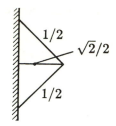

**Fig. 10.44** Optimal plastic design for the problem in Fig. 10.43.

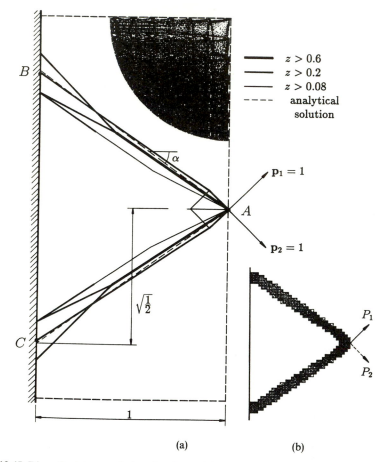

**Fig. 10.45** Discretized truss solution for the problem in Fig. 10.43 using a discretized COC procedure.

Considering two load conditions with two compliance constraints (Figs 10.43(a)–10.43(d)), the optimal solution in Fig. 10.43(e) was shown (Rozvany, 1992a) to satisfy the conditions (10.36). The actual variation of $\varepsilon_I^e \bar{\varepsilon}_I^e$ is given in Fig. 10.43(f). Since here we are dealing with a convex problem, the above proof establishes global optimality. The corresponding normalized optimal weight is $\Phi = 27/8 = 3.375$.

For a comparison, the optimal plastic design for the two loads in Fig. 10.43 is given in Fig. 10.44 (Zhou and Rozvany, 1991). It can be seen that the latter consists of three members.

Discretized COC solutions for the problem in Fig. 10.43 were obtained (Zhou and Rozvany, 1991) using 7170 and 12 202 members in the structural universe. For the former the discretized solution is shown in Fig. 10.45(a), with part of the structural universe at the top right corner. This structural universe contained only a limited number of member directions and hence it could only approximate the analytical solution (broken line in Fig. 10.45(a)) by using a fairly large number of members, with a weight value of $\Phi = 3.49295726$ (3.495% error). In the structural universe with 12 202 members, all nodes of an $11 \times 21$ grid were connected with all other nodes, and the corresponding discretized solutions consisted basically of only two members, as in the analytical solution. The corresponding weight was 3.375 668 (only 0.0198% error).

Using the SIMP procedure, T. Birker obtained the solution in Fig. 10.45(b) for a perforated plate with two load conditions and two compliance constraints. The agreement between both solutions in Fig. 10.45 and that in Fig. 10.43 is rather obvious.

## 10.9.2 Structural layouts for non-self-adjoint problems

To demonstrate the difficulties arising in the case of non-self-adjoint problems, we consider the layout problem in Fig. 10.46(a), in which we have a point load

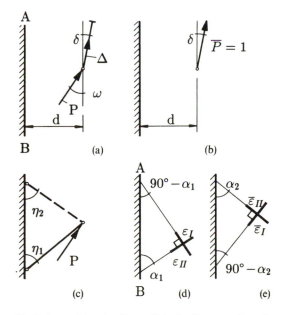

**Fig. 10.46** A non-self-adjoint problem: loading, adjoint load, assumed topology, real and adjoint strain fields.

at an angle $\omega$ and a prescribed displacement at an angle $\delta$, with $\omega \neq \delta$. The adjoint load for this problem is shown in Fig. 10.46(b). The treatment of this problem was published recently (Rozvany *et al.*, 1993). Figure 10.46(c) shows a possible topology, whilst Figs 10.46(d) and 10.46(e) the corresponding real and adjoint strain fields based on the optimality conditions (10.36). Using the latter, it was shown in the above-mentioned paper that the 'optimal' member directions are symmetrical ($\eta_1 = \eta_2$ in Fig. 10.46(c)). For example, for $\delta = 0°$, $\omega = 20°$ (Fig. 10.47), we obtain $\eta_1 = \eta_2 = 43.845\,305\,2$. Figure 10.47 (top) also shows the variation of the weight for the above displacement constraint and various values of the angle $\eta$, with cross-sectional areas optimized for a given angle. It can be seen that the solution obtained by using relation (10.36) is a local optimum (point A), the global optimum being nonstationary (ED in Fig. 10.47) and tending to zero. However, the stationary solution discussed above (point A) can be valid if additional stress conditions are introduced.

Figure 10.47 (bottom) also shows that the optimality conditions (10.36) yield two values for the strain field $\varepsilon$ and $\bar{\varepsilon}$, and the corresponding products $\varepsilon\bar{\varepsilon}$ have a maximum at the points where the weight curve has minima or maxima.

From the above example, we may draw the following conclusions.

- Layout optimization of elastic structures with deflection constraints constitutes a non-self-adjoint problem.
- Because of nonconvexity, optimality criteria yield local minima as well as local maxima with respect to some parameters.

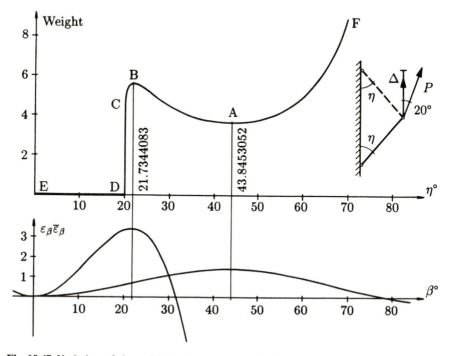

**Fig. 10.47** Variation of the weight in dependence of the bar directions and variation of the strain product $\varepsilon\bar{\varepsilon}$.

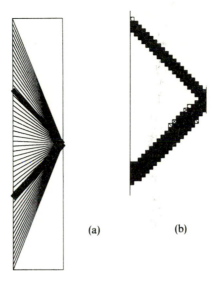

(a)                    (b)

**Fig. 10.48** Discretized truss solution and perforated plate solution for the problem in Fig. 10.46.

- The globally optimal solution for one deflection constraint is, in general, a nonstationary solution whose weight tends to zero.
- For a well-posed problem, stress constraints must also be included in the formulation.

The result in Fig. 10.47 was confirmed by using both DCOC for an optimized truss layout (Fig. 10.48(a), after Gerdes) and a SIMP formulation for a perforated plate (Fig. 10.48(b), after Birker).

## 10.10 CONCLUDING REMARKS

We can draw the following conclusions from this chapter.

- Topology optimization in structural design has two main applications at present: layout optimization of grid-type structures and generalized shape optimization of cellular continua or composite systems.
- New optimality criteria methods (COC, DCOC) are eminently suitable for topology optimization because of their high optimization capability in terms of the number of variables and the number of active constraints.
- Analytical layout solutions are now available for relatively simple design constraints (a single stress or compliance or natural frequency constraint) and structures with a single variable per cross section (e.g. trusses, grillages of given depth, shell-grids in membrane action), but for these classes of problems the exact optimum can be derived systematically for wide ranges of complicated load and support conditions.
- Discretized layout solutions can be obtained by COC–DCOC methods for many thousand potential members in the structural universe (ground structure) and for combinations of stress, displacement and natural frequency constraints.

- For non-self-adjoint, nonconvex problems, however, the above methods may yield a local optimum.
- Owing to the very fine grid in the structural universe, the proposed methods achieve in effect a simultaneous optimization of topology and geometry.
- In generalized shape optimization (simultaneous optimization of boundary shape and boundary topology), the suggested SIMP formulation (solid, isotropic microstructure with penalty for intermediate densities) seems to be highly efficient in locating SE (solid–empty) topologies which show a good agreement with analytical solutions.
- The test examples used in this chapter are somewhat academic for easy comparison with analytical solutions, but the discretized methods employed are now suitable for large systems with a variety of realistic design constraints.
- Finally, it must be emphasized that topology optimization is a new but rapidly developing field with many unsolved problems. An excellent review of basic features of optimal topologies, some involving severe computational difficulties, was given by Kirsch (1990); for a review see also Kirsch and Rozvany (1993).

## ACKNOWLEDGEMENTS

The authors are indebted to the Deutsche Forschungsgemeinschaft for financial support (Project Nos Ro 744/4-1 and Ro 744/6-1), to Sabine Liebermann (text processing), Peter Moche (drafting) and Susann Rozvany (editing and some text processing).

## REFERENCES

Allaire, G. and Kohn, R.V. (1993) Topology optimization and optimal shape design using homogenization, in *Topology Design of Structures*. Proceedings of NATO ARW, Sesimbra, Kluwer, Dordrecht, pp. 207–18.

Avellaneda, M. (1987) Optimal bounds and microgeometries for elastic two-phase composites. *SIAM Journal of Applied Mathematics*, **47**, 1216–28.

Bendsøe, M.P. (1989) Optimal shape design as a material distribution problem. *Structural Optimization*, **1**, 193–202.

Bendsøe, M.P. and Kikuchi, N. (1988) Generating optimal topologies in structural design using a homogenization method. *Computer Methods in Applied Mechanical Engineering*, **71**, 197–224.

Bensousson, A., Lions, J.-L. and Papanicolaou, G. (1978) *Asymptotic Analysis for Periodic Structures*, North-Holland, Amsterdam.

Berke, L. and Khot, N. (1987) Structural optimization using optimality criteria, in *Computer Aided Optimal Design: Structural and Mechanical Systems* (ed. C.A. Mota Soares), Springer, Berlin, pp. 271–312.

Cheng, K.T. and Olhoff, N. (1981) An investigation concerning optimal design of solid elastic plates. *International Journal of Solids and Structures*, **17**, 305–23.

Diaz, A.R. and Bendsøe, M.P. (1992) Shape optimization of multipurpose structures by a homogenization method. *Structural Optimization*, **4**, 17–22.

Hemp, W.S. (1973) *Optimum Structures*, Clarendon, Oxford.

Hill, R.H. and Rozvany, G.I.N. (1985) Prager's layout theory: a non-numeric computer alogrithm for generating optimal structural configurations and weight influence surfaces. *Computer Methods in Applied Mechanical Engineering*, **49**, 131–48.

Kirsch, U. (1989) Optimal topologies of structures. *Applied Mechanics Review*, **42**, 223–38.

Kirsch, U. (1990) On singular topologies in structural design. *Structural Optimization*, **2**, 133–42.

Kirsch, U. and Rozvany, G.I.N. (1993) Design considerations in the optimization of structural topologies, in *Optimization of Large Structural Systems* (ed. G.I.N. Rozvany), Proceedings of NATO ASI, Berchtesgaden, 1991, Kluwer, Dordrecht, pp. 121–41.

Kohn, R.V. and Strang, G. (1983) Optimal design for torsional rigidity, in *Hybrid and Mixed Finite Element Methods* (eds S.N. Atluri, R.H. Gallagher, and O.C. Zienkiewicz), Wiley, Chichester, pp. 281–8.

Kohn, R.V. and Strang, G. (1986) Optimal design and relaxation of variational problems, I, II, and III. *Communications in Pure and Applied Mathematics*, **39**, 113–37, 139–82, 353–77.

Lagache, J.-M. (1981) Developments in Michell theory, in Proceedings of International Symposium on Optimal Structure Design, Tucson, 1981 (eds E. Atrek and R.H. Gallagher), University of Arizona, Tucson, AZ, pp. 4.9–4.16.

Lewiński, T., Zhou, M. and Rozvany, G.I.N. (1993) Exact least-weight truss layouts for rectangular domains with various support conditions. *Structural Optimization*, **6**, 65–67.

Lowe, P.G. (1988) Optimization of systems in bending conjecture, bounds and estimates relating to moment volume and shape, in Proceedings of IUTAM Symposium on Structural Optimization, Melbourne, 1988 (eds G.I.N. Rozvany and B.L. Karihaloo), Kluwer, Dordrecht, pp. 169–76.

Lowe, P.G. and Melchers, R.E. (1972–1973) On the theory of optimal constant thickness fibre-reinforced plates, I, II, III. *International Journal of Mechanical Science*, **14**, 311–24; **15**, 157–70, 711–26.

Lurie, K.A. and Cherkaev, A.V. (1984) G-closure for some particular set of admissible material characteristics for the problem of bending of thin elastic plates. *Journal of Optimization Theory and Applications*, **42**, 305–16.

Michell, A.G.M. (1904) The limits of economy of material in frame-structures. *Philosophical Magazine*, **8**, 589–97.

Murat, F. and Tartar, L. (1985) Calcul des variations et homogénéisation, in *Les Méthodes de l'Homogénéisation: Théorie et Applications en Physique*, Eyrolles, Paris, pp. 319–70.

Olhoff, N., Bendsøe, M.P. and Rasmussen, J. (1991) On CAD-integrated structural topology and design optimization. *Computer Methods in Applied Mechanical Engineering*, **89**, 259–79.

Olhoff, N. and Rozvany, G.I.N. (1982) Optimal grillage layouts for given natural frequency. *Journal of Engineering Mechanics of ASCE*, **108**, 971–5.

Ong, T.G., Rozvany, G.I.N. and Szeto, W.T. (1988) Least-weight design of perforated elastic plates for given compliance: non-zero Poisson's ratio. *Computer Methods in Applied Mechanical Engineering*, **66**, 301–22.

Pedersen, P. (1989) On optimal orientation of orthotropic materials. *Structural Optimization*, **1**, 101–6.

Pedersen, P. (1990) Bounds on elastic energy in solids of orthotropic materials. *Structural Optimization*, **2**, 55–63.

Prager, W. (1974) *Introduction to Structural Optimization*, Course held in International Centre for Mechanical Science, Udine, CISM 212, Springer, Vienna.

Prager, W. and Rozvany, G.I.N. (1977a) Optimal layout of grillages. *Journal of Structural Mechanics*, **5**(1) 1–18.

Prager, W. and Rozvany, G.I.N. (1977b) Optimization of structural geometry, in *Dynamical Systems* (eds A.R. Bednarek and L. Cesari), Academic Press, New York, pp. 265–93.

Prager, W. and Shield, R.T. (1967) A general theory of optimal plastic design. *Journal of Applied Mechanics*, **34**(1), 184–6.

Rossow, M.P. and Taylor, J.E. (1973) A finite element method for the optimal design of variable thickness sheets. *AIAA Journal*, **11**, 1566–9.

Rozvany, G.I.N. (1973) Optimal force transmission of flexure-clamped boundaries. *Journal of Structural Mechanics*, **2**, 57–82.

Rozvany, G.I.N. (1976) *Optimal Design of Flexural Systems*, Pergamon, Oxford.

Rozvany, G.I.N. (1981) Optimality criteria for grids, shells and arches, in *Optimization of Distributed Parameter Structures* (eds E.J. Haug and J. Cea), Proceedings of NATO ASI, Iowa City, 1980, Sijthoff and Noordhoff, Alphen aan den Rijn, pp. 112–51.

Rozvany, G.I.N. (1984) Structural layout theory: the present state of knowledge, in *New Directions in Optimum Structural Design* (eds E. Atrek, R.H. Gallagher, K.M. Ragsdell and O.C. Zienkiewicz), Wiley, Chichester, pp. 167–95.

Rozvany, G.I.N. (1989) *Structural Design via Optimality Criteria*, Kluwer, Dordrecht.

Rozvany, G.I.N. (1992a) Optimal layout theory: analytical solutions for elastic structures with several deflection constraints and load conditions. *Structural Optimization*, **4**, 247–9.

Rozvany, G.I.N. (1992b) Optimal layout theory, in *Shape and Layout Optimization of Structural Systems and Optimality Criteria Methods* (ed. G.I.N. Rozvany), CISM Course, Udine, 1990, Springer, Vienna, pp. 75–163.

Rozvany, G.I.N. (1993a) Topological optimization of grillages: past controversies and new directions. *International Journal of Mechanical Science* (in press).

Rozvany, G.I.N. (1993b) Optimal grillage layouts for partially upward and partially downward loading. *Structural Optimization* (to be submitted).

Rozvany, G.I.N. (1993c) Layout theory for grid-type structures, in *Topology Design of Structures* (eds M.P. Bendsøe and C.A. Mota Soares), Proceedings of NATO ARW, Sesimbra, 1992, Kluwer, Dordrecht, pp. 251–72.

Rozvany, G.I.N. and Gerdes, D. (1994) Optimal layout of grillages with free, simply supported and clamped edges. *Structural Optimization* (to be submitted).

Rozvany, G.I.N. and Gollub, W. (1990) Michell layouts for various combinations of line supports, part I. *International Journal of Mechanical Science*, **32**(12) 1021–43.

Rozvany, G.I.N. (1994) Optimal layout theory – allowance for the cost of supports and optimization of support locations. *Mechanics and Structures of Machines*, **22** (in press).

Rozvany, G.I.N. and Hill, R.H. (1976) General theory of optimal load transmission by flexure. *Advances in Applied Mechanics*, **16**, 184–308.

Rozvany, G.I.N. and Hill, R.H. (1978) A computer algorithm for deriving analytically and plotting optimal structural layout. *Computers and Structures*, **10**, 295–300.

Rozvany, G.I.N. and Ong, T.G. (1987) Minimum-weight plate design via Prager's layout theory (Prager memorial lecture), in *Computer Aided Optimal Design: Structural and Mechanical Systems* (ed. C.A. Mota Soares), Proceedings of NATO ASI, Troia, 1986, Springer, Berlin, pp. 165–79.

Rozvany, G.I.N. and Prager, W. (1979) A new class of structural optimization problems: optimal archgrids. *Computer Methods in Applied Mechanical Engineering*, **19**, 127–50.

Rozvany, G.I.N. and Zhou, M. (1991) Applications of the COC method in layout optimization. in *Proceedings of International Conference on Engineering Optimization in Design Processes* (eds H. Eschenauer, C. Matteck and N. Olhoff), Karlsruhe, 1990, Springer, Berlin, pp. 59–70.

Rozvany, G.I.N., Gollub, W. and Zhou, M. (1992) Layout optimization in structural design, in *Proceedings of NATO ASI, Optimization and Decision Support Systems in Civil Engineering* (ed. B.H.V. Topping), Edinburgh, June 25–July 7, 1989, Kluwer, Dordrecht.

Rozvany, G.I.N., Gollub, W. and Zhou, M. (1994) Michell layouts for various combinations of line supports, part II. *International Journal of Mechanical Science* (to be submitted).

Rozvany, G.I.N., Hill, R.H. and Gangadharaiah, C. (1973) Grillages of least weight – simply supported boundaries. *International Journal of Mechanical Science*, **15**, 665–77.

Rozvany, G.I.N., Olhoff, N., Cheng, K.-T. and Taylor, J.E. (1982) On the solid plate paradox in structural optimization. *Journal of Structural Mechanics*, **10**, 1–32.

Rozvany, G.I.N., Olhoff, N., Bendsøe, M.P. *et al.* (1987) Least-weight design of perforated elastic plates I, II. *International Journal of Solids and Structures*, **23**, 521–36, 537–50.

Rozvany, G.I.N., Sigmund, O., Lewinski, T., Gerdes, D. and Birker, T. (1993) Exact optimal structural layouts for non-self adjoint problems. *Structural Optimization*, **5**, 204–6.

Rozvany, G.I.N., Zhou, M. and Birker, T. (1992) Generalized shape optimization without homogenization. *Structural Optimization*, **4**, 250–2.

Rozvany, G.I.N., Zhou, M. and Gollub, W. (1993) Layout optimization by COC methods: analytical solutions, in *Optimization of Large Structural Systems* (ed. G.I.N. Rozvany), Proceedings of NATO ASI, Berchtesgaden, 1991, Kluwer, Dordrecht, pp. 77–102.

Rozvany, G.I.N., Zhou, M., Rotthaus, M., Gollub, W. and Spengemann, F. (1989) Continuum-type optimality criteria methods for large structural systems with a displacement constraint – part I. *Structural Optimization*, **1**, 47–72.

Sankaranarayanan, S., Haftka, R.T. and Kapania, R.K. (1992) Truss topology optimization with simultaneous analysis and design. Proceedings of 33rd AIAA/ASME/ASCE/AHS/ASC Structural Dynamics Materials Conference, Dallas, TX, AIAA, Washington, DC, pp. 2576–85.

Sigmund, O. and Rozvany, G.I.N. (1994) Extension of discretized OC methods to combined stress, displacement and natural frequency constraints including selfweight – part I: beams and grillages. *Structural Optimization* (to be submitted).

Sigmund, O., Zhou, M. and Rozvany, G.I.N. (1993) Layout optimization of large FE-systems by new optimality criteria methods: applications to beam systems. in *Proceedings of NATO ASI Concurrent Engineering Tools and Technologies for Mechanical System Design* (ed. E.H. Haug), Iowa, 1992, Springer, Berlin.

Suzuki, K. and Kikuchi, N. (1991) A homogenization method for shape and topology optimization. *Computer Methods in Applied Mechanical Engineering*, **93**, 291–318.

Vigdergauz, S. (1992) Two-dimensional grained composites of extreme rigidity. 18th International Congress on Theoretical Applied Mechanics, Haifa.

Zhou, M. and Rozvany, G.I.N. (1991) The COC algorithm, part II: topological, geometrical and generalized shape optimization. *Computer Methods in Applied Engineering*, **89**, 309–36.

# 11

# Practical shape optimization for mechanical structures

NOBORU KIKUCHI and KAZUTO HORIMATSU

## 11.1 INTRODUCTION

Reduction of the duration of design and manufacturing processes becomes much more important than reducing the cost of raw material of a product while the required functionality is fulfilled, to reduce the overall cost and to lead to success of a new product. In order to reflect this situation to research and development in design methodology, we must reconstruct the notion of structural and mechanical design optimization. In the past most structural and mechanical design optimization techniques tried to minimize the cost of raw material under certain constraints which are implied from mechanics and manufacturing requirement. However, now, the most important matter is how easily certain design can be improved with minimal effort by design engineers rather than just considering minimization of the cost (i.e. weight) of raw material. In other words it becomes very important to examine how the optimal design can be achieved for solving the minimum cost (weight) problem. In this chapter we shall describe a method to find the optimum with less effort without constructing an excessively large system of design optimization that integrates geometric modeling, finite element analysis, sensitivity analysis, and optimization algorithms.

The standard procedure of mechanical–structural design is that, after defining the shape and topology of a structure based on the conceptual or existing design, a discrete model is developed for stress–thermal–flow analysis. Then, defining appropriate design variables such as representative physical dimensions of a structure, sensitivity of the objective and constraint functions for design optimization are computed to provide necessary information for optimization algorithms based on mathematical programming methods such as sequential linear programming (SLP), sequential quadratic programming (SQP), and other methods. In short, the procedure of design optimization consists of four modules: solid modeling for the initial design, finite element analysis, sensitivity analysis, and optimization methods.

Nowadays, it is easy to find appropriate general-purpose programs for solid modeling, finite element analysis, and optimization methods. Commercial and public domain software is available to most of structural design engineers. For example, I-DEAS for SDRC does have all of these capabilities including some capability of sensitivity analysis as an advanced engineering option, PATRAN from PDA Engineering can provide decent capability of finite element modeling, MSC/ NASTRAN from MacNeal-Schwendler Corporation is the most widely used finite element analysis program in which sensitivity analysis for limited design variables and optimization algorithms are also available as an advanced option, while ANSYS from Swanson Analysis Systems, NISA II from EMRC, and MECANICA from RASNA claim handy design optimization capability. There are also many other software packages for finite element analysis which are applicable to design optimization study. For optimization algorithms, we can find, for example, ADS and DOT from VMA, IDESIGN from University of Iowa, VMCON from Argonne National Laboratory, and others.

Most of the commercially available software is developed as an extended variation from finite element analysis programs, in particular in the United States. On the other hand, structural optimization software is developed differently in Europe. Most European systems are constructed based on sensitivity analysis or optimization algorithms, and finite element analysis is regarded as a black box type of capability in the design system. They simply call an existing finite element program such as SAP IV, ASKA, SAMCEF, MSC/NASTRAN, and others. For example, ELFINI from Dassault, OPTI/SAMCEF from University of Liege, OPTSYS from SAAB-SCANIA, OASIS from the Royal Institute of Technology, Stockholm, LAGRANGE from MBB, STARS from RAE and others have been developed for structural optimization in Europe (Hörnlein, 1987). Although all of these systems use finite element methods, the essential portion is optimization algorithms as well as sensitivity analysis. This difference between the United States and Europe seems to be created by the nature of the developing process of this software. In Europe, structural optimization systems are generated by researchers in structural optimization, while finite element software houses in the United States added design optimization capability to their existing commercially available finite element programs. Therefore, a large difference can be expected among these structural optimization software packages, in particular, in basic philosophy.

A limitation found both in European and United States' systems is that they are closed. More precisely, the state equations we can deal with in these systems are limited to the ones which are originally designed in the finite element analysis codes. Furthermore, modeling for finite element analysis is not fully integrated in most of design systems. The majority of existing design optimization is for stress analysis, and other analysis areas such as heat conduction, fluid flow, metal forming, and others are at this moment excluded from design optimization. However, as noted earlier, these wide ranging state equations must be solved in mechanical design. Thus it is required in a design system that we can deal with any kind of state equation using different analysis software i.e. a design system must be open. If other discrete methods such as boundary element and finite difference methods are applied to solve state equations, we should still be able to incorporate these different types of analysis capabilities in a design optimization system.

It is noted that structural optimization was studied mainly by structural engineers in the aeronautical and aerospace industry in the 1970s and 1980s. Most European structural optimization software packages are thus developed in the large-scale aerospace industry, and they target primarily application to aircraft or space structures. Design variables are sizes of the cross section of members of a frame structure, thickness of a plate or shell, and the orientation of fibers and the thickness of laminae of a fiber-reinforced laminate in these systems. Stress and thermal analyses are thus within linear range, i.e. they are based on the theory of linear elasticity and linear thermoelasticity. On the other hand, mechanical design requires optimal design of not only the sizes of a structure but also its shape, and it mostly deals with design of components or parts of a structure rather than a large-scale whole body of complex structure such as ships, aircraft, and other large-scale space structures. Thus the number of design variables is much smaller than that of aerospace type structures, but the shape design becomes far more important in mechanical design. Thus, analysis model development involving shape geometry is the most important in mechanical design. No matter how capable the sensitivity analysis and optimization algorithms that are developed in a design system, if an appropriate finite element model cannot be generated automatically in every design step during execution of the system, the optimum cannot be easily obtained. This implies that a design system must involve a flexible modeling capability that can reflect any geometrical design change by an optimization algorithm. In other words, a design system may be constructed based on modeling software rather than finite element or sensitivity-optimization software by adding other necessary modules for design optimization.

Although design optimization capability was added into finite element analysis programs in the United States in 1980s, shape optimization capability is still very limited even in MSC/NASTRAN, ANSYS, MECANICA, and NISA II. It is thus required to study and develop a simple methodology that can deal with both sizing and shape optimization as well as both linear and nonlinear state equations involving possibly large plastic deformation, heat conduction, and fluid flow. Nonlinearity and time dependency yield a significant change in the concept how design sensitivity is computed. Since most design problems in aerospace engineering are limited to elastic design, sensitivity analysis is rather straightforward even by using the analytical and semianalytical methods.

In this chapter we shall describe a method to construct a very small scale but sufficiently flexible optimization system that is mostly written in UNIX C-shell scripts by combining existing modeling, finite element analysis, and optimization algorithm capability using the concept of open-ended software modules in PATRAN from PDA Engineering. More precisely, we shall develop an optimization system using PATRAN's PCL (Patran Control Language) that allows communication between PATRAN and the UNIX OS so that other flexible operations can be implemented outside of PATRAN without terminating execution of PATRAN itself. It is noteworthy that the concept that we shall introduce to develop an optimization system should be general in the sense that the present approach is applicable to any software that has control flow commands to execute a variety of modules and communication commands to the operating system. These are used as a high-level computer language to write a program of design optimization. Since PATRAN is primarily for developing a finite element model by using automatic mesh generation, the present system allows us to deal with sizing and shape optimization in the same manner, while the nature of static/dynamic, or linear/nonlinear problems for analysis does not affect the system itself since the analysis is assumed to be independent of the system. In this sense, this can provide a much more flexible and powerful capability than that existing in commercially available codes for design optimization, and yet the system is far smaller than any existing programs.

## 11.2 STRUCTURAL OPTIMIZATION PROCEDURE

The standard procedure of structural optimization consists of the following four major modules:

1. developing a model for analysis;
2. finite element analysis;
3. design sensitivity analysis;
4. application of optimization algorithms.

Development of a finite element analysis model is the same with the usual preprocessing for finite element analysis in sizing optimization problems, and then it can be completely independent of the design optimization system for such problems. However, it must be a module inside the design system for shape optimization, since geometric modeling itself is subject to design change. In general, finite element methods are applied to solve the state equation, but also other analysis methods such as boundary element and finite difference methods should be applicable. In other words, the module for analysis need not be restricted to finite element methods. It is noteworthy that design sensitivity calculation is closely related to analysis. There are several methods available such as finite difference, semianalytical, and analytical methods together with application of direct and adjoint methods. An appropriate method of sensitivity analysis may be determined by the characteristics of the state equation and a choice of method in the analysis module. The last part of the design optimization is the set of optimization algorithms to solve the optimal design problem that is formulated as a constrained nonlinear programming problem.

If the expected applicability range of a design system is sufficiently small, it is not difficult to develop a FORTRAN or C program that involves all of these four modules. For example, if the cross-sectional size of members of frame structures is designed for both stress and eigenvalue analyses, it is rather easy to develop a complete design optimization FORTRAN program since it is free from the difficulty of mesh generation for a finite element model, and since its state equations are linear and definite so that sensitivity can be easily computed even by an analytical method. Development of such a program can be stright-forward for graduate students who have necessary background of finite element methods, sensitivity analysis, and some optimization algorithms. In the 1970s when the availability and capability of commercial or public domain finite element codes and optimization methods were limited, this approach of complete develop-ment of a design optimization system was very popular for a restricted class of structural optimization problem. However, there are many well-developed sophis-ticated software packages for finite element analysis and optimization at present, and they are used daily in design practice. Since many of these software packages are already available in most of industry and academia, it is better to develop a design system that can utilize these as intact existing modules so that it does not require additional effort to be familiar with analysis and optimization modules. Furthermore, since each organization has its own preference and rather extensive experience in use of a particular finite element code, the design optimization system should be able to integrate this specific one as well as other choices for other users. This means that the optimization system should be able to choose any of them according to the nature of the state equation and to the preference of a user. Since most commercially available design optimization codes are developed as an enhancement of their original finite element analysis programs, they do not possess this flexibility and openness. As mentioned earlier, state equations in structural design are not just an equilibrium of linearly elastic structures. They are sometimes heat conduction equations with convection and radiation, motion of non-Newtonian fluids, equilibrium of large deformation elastoplastic bodies, and others. Thus a design system should not be restricted by a particular finite element analysis capability. It should be designed with new concept based on the nature of mechanical design that has much larger scope than analysis of the state equation.

Since there are many available commercial and public domain packages of optimization algorithms based on the theory of mathematical programming methods for both linear and nonlinear programming problems, it is much simpler to adopt an appropriate one from existing software. In structural optimization, ADS based on the sequential linear programming and IDESIGN with sequential quadratic programming are widely used in the United States. ADS is now enhanced as DOT and is marketed by VMA.

## 11.3 SENSITIVITY ANALYSIS BY THE FINITE DIFFERENCE METHOD

In order to utilize an optimization algorithm, it is at least required to compute the values and the first derivatives of performance functions (that is the objective and constraint functions) $g$ with respect to a design variable $d$, i.e. $g(u, d)$ and $Dg/Dd(u, d)$, where $u$ is the state variable used to define the state equation (that is the equilibrium equation for stress analysis, the equation of motion is dynamics, the heat conduction equation for thermal analysis, ...) of a physical system to be optimized.

If the state equation is linear and time independent, and if it is expressed by

$$\mathbf{Ku = f} \tag{11.1}$$

then differentiation of this in a design variable $d$ yields

$$\frac{\partial \mathbf{K}}{\partial d}\mathbf{u} + \mathbf{K}\frac{\partial \mathbf{u}}{\partial d} = \frac{\partial \mathbf{f}}{\partial d} \Rightarrow \frac{\partial \mathbf{u}}{\partial d} = \mathbf{K}^{-1}\left(\frac{\partial \mathbf{f}}{\partial d} - \frac{\partial \mathbf{K}}{\partial d}\mathbf{u}\right) \tag{11.2}$$

where $\mathbf{K}^{-1}$ is the inverse of a linear operator (e.g. the global stiffness matrix in stress analysis by the finite element method). Thus the sensitivity of the performance function $g$, i.e. the first derivative of $g$ with respect to a design variable $d$, is given by

$$\frac{Dg}{Dd} = \frac{\partial g}{\partial d} + \frac{\partial g}{\partial u}\frac{\partial u}{\partial d} = \frac{\partial g}{\partial d} + \frac{\partial g}{\partial u}\left[\mathbf{K}^{-1}\left(\frac{\partial \mathbf{f}}{\partial d} - \frac{\partial \mathbf{K}}{\partial d}\mathbf{u}\right)\right] \tag{11.3}$$

This form of sensitivity requires explicit representation of a performance function in terms of the state variable $u$ and a design variable $d$.

If stress analysis is considered for a three-dimensional solid structure, and if a performance function $g$ is identified with the maximum principal stress, it is certainly a function of the displacement $u$, but its function form is not described in an explicit form, since the principal stresses are identified with the eigenvalues of the symmetric $3 \times 3$ stress tensor. Thus application of the above analytically derived sensitivity might not be useful for performance functions which are implicit functions of the state variable $u$. Even in the case that performance functions are explicitly expressed in the state and design variables, utilization of the above form of sensitivity might not be practical since we have to compute the first derivative of the linear operator (stiffness matrix in stress analysis) $\mathbf{K}$ of the state equation. For example, in a three-dimensional shell structure, the analytical partial derivative of the stiffness matrix $\mathbf{K}$ with respect to a nodal coordinate of a node may not be obtained in a simple form. For a shell formulation in the finite element method we use a local coordinate system attached to each finite element that is defined by the nodal coordinates of the four corner nodes of the element. The element stiffness matrix $\mathbf{K}^e_{\text{local}}$ is transformed to the one $\mathbf{K}^e_{\text{global}}$ in the global coordinate system and it is assembled to form the stiffness matrix, i.e.

$$\mathbf{K}^e_{\text{global}} = \mathbf{R}^{\text{T}} \mathbf{K}^e_{\text{local}} \mathbf{R} \tag{11.4}$$

where $\mathbf{R}$ is the transformation matrix from the local to the global coordinate system. Thus the first derivative of the stiffness matrix is computed by

$$\frac{\partial \mathbf{K}^e_{\text{global}}}{\partial d} = \left(\frac{\partial \mathbf{R}}{\partial d}\right)^{\text{T}} \mathbf{K}^e_{\text{local}} \mathbf{R} + \mathbf{R}^{\text{T}} \frac{\partial \mathbf{K}^e_{\text{local}}}{\partial d} \mathbf{R} + \mathbf{R}^{\text{T}} \mathbf{K}^e_{\text{local}} \frac{\partial \mathbf{R}}{\partial d} \tag{11.5}$$

this is, the first derivatives of both coordinate transformation and element stiffness matrices must be analytically obtained. It is thus not practical to compute the analytical first derivative of the stiffness matrix. This leads to a finite difference scheme to compute the first derivative of the operator $\mathbf{K}$ of the state equation:

$$\frac{\partial K}{\partial d} \approx \frac{K|_{d+\Delta d} - K|_{d-\Delta d}}{2\,\Delta d} \tag{11.6}$$

by evaluating $\mathbf{K}$ at the two perturbed designs $d + \Delta d$ and $d - \Delta d$, where $\Delta d$ is a sufficiently small design change at the current design $d$. This approximation yields the semianalytical method to compute the design sensitivity

$$\frac{Dg}{Dd} = \frac{\partial g}{\partial d} + \frac{\partial g}{\partial u}\frac{\partial u}{\partial d} \approx \frac{\partial g}{\partial d} + \frac{\partial g}{\partial u}\left[\mathbf{K}^{-1}\left(\frac{\mathbf{f}|_{d+\Delta d} - \mathbf{f}|_{d+\Delta d}}{2\,\Delta d} - \frac{\mathbf{K}|_{d+\Delta d} - \mathbf{K}|_{d-\Delta d}}{2\,\Delta d}u\right)\right] \tag{11.7}$$

The semianalytical method does not require us to know the exact explicit form of function in the state variable $u$. If the operator can be evaluated at two different designs in the analysis procedure (that is, if we can output the stiffness matrix and the load vector at two different design stages), we can compute the design sensitivity by using the semianalytical method. In other words, we need direct access to the source code of analysis to develop an efficient sensitivity analysis program using the semianalytical method.

Therefore, for general problems, it might be much simpler to apply the finite difference approximation to compute the sensitivity of a performance function $g$:

$$\frac{Dg}{Dd} \approx \frac{g|_{d+\Delta d} - g|_{d-\Delta d}}{2\,\Delta d} \tag{11.8}$$

In this way, the explicit function form of $g$ in $d$ and $u$ is not required, and even the first derivative of the linear operator of the state equation need not be calculated. This means that the finite difference approximation is applicable, both for linear and for nonlinear state equations, while the time dependency of the state equation does not affect at all the computation of the design sensitivity. The disadvantage of this method is the requirement of two analyses per design variable to compute the sensitivity. If the state equation is solved at the current design, three analyses are required. Thus, if $m$ performance functions and $n$ design variables are involved in a design problem, $(2n + 1)m$ analyses are required to compute the sensitivity and the value of performance functions.

If the number of design variables is large as in sizing problems for aerospace structures, the finite difference method is not practical at all. However, if the number of design variables is rather small, it becomes powerful. As mentioned earlier, in most mechanical design problems the number of design variables is very small, because of the requirement from cost effective manufacturing. It is nearly impossible to have varying thickness everywhere in a structure in mechanical design, while several different thicknesses may be introduced in parts. Similarly, the introduction of very arbitrary shapes of holes to reduce the total weight is unrealistic. The usual practice is to make circular holes. Such a design problem is then described to find the best locations and radii of a given number of holes. Furthermore, mechanical design requires consideration of dynamics and impact as well as static strength of a structure. This means that the state equation is far more complicated than that for most of aeronautical–aerospace structures. This nature of the mechanical design yields the consequence that analytical, or even semianalytical, methods are not practical to compute the sensitivity.

It is also noted that no special development is required for sensitivity analysis if the finite difference method is applied, while analytical and semianalytical methods need considerable code-developing effort to be implemented. If shape design of a structure is considered in a multidisciplinary setting, e.g. if the shape of a structure is optimized by minimizing the drag generated by fluid flow outside the structure as well as the maximum stress of the structure, we must calculate sensitivity in two different disciplines, i.e. two different state equations. Since analysis programs for stress and fluids are in general not integrated in a single software system by a single software house, it is not realistic for most analysis programs to request the provision of a sensitivity analysis capability based on the analytical or semianalytical methods, at least at present. The finite difference method is, however, applicable together with already existing analysis programs for different kinds of state equations to compute sensitivity without any modification or enhancement of these at all. The flexibility of the finite difference method for sensitivity is truly enormous.

Furthermore, in a distributed computing environment, it is possible to use several workstations at once to compute design sensitivity. This implies the possibility that two or three workstations are used at the same time to make the analysis in a parallel manner. More precisely, a workstation analyzes the state equation at the design $d + \Delta d$, while another workstation analyzes it at the design $d - \Delta d$. In this way, we can compute sensitivity without waiting for the long time required by sequential calculation using a single workstation. If the analyzer is using an iterative method to solve the state equation, it is also possible to take advantage of this by using the fact that the state variable at the designs $d + \Delta d$ and $d - \Delta d$ should be quite close to the one at the current design $d$. Then solutions at the designs $d + \Delta d$ and $d - \Delta d$ can be easily computed by any iterative methods by starting from the solution at the current design $d$. This indicates that research into the reduction of the computing time for the finite difference method may have more merit than developing a general-purpose program of design sensitivity analysis based on the analytical or semianalytical method.

In summary, it is concluded that the finite difference method is the most appropriate method to compute sensitivity for most practical optimization problems in structural design, since

- calculation time may not be excessive because the number of design variables is not so large, say 5–20 variables at most,
- a design optimization system can be very flexible because sensitivity is calculated using any kind of analyzer (not only an FEM analyzer) for any kind of state equation,
- the size of the software to be developed for the design system is very small and easy to maintain, and
- a workstation that forms and modifies the design model need not be the same one used for sensitivity calculation in a distributed computing environment.

Sensitivity can be calculated using the finite difference method using the following procedure:

**LOOP 1: For each design variable**
      Analysis of $g|_{d+\Delta d}$
      Analysis of $g|_{d-\Delta d}$
      **LOOP 2: For performance function**
          Calculation of Sensitivity $\dfrac{g|_{d+\Delta d} - g|_{d-\Delta d}}{2\,\Delta d}$

      **NEXT 2**
  **NEXT 1**

Here the central finite difference approximation is applied instead of forward or backward finite difference schemes, although it requires more computation time. It is noted that the central difference method is more accurate than the others. Indeed, if a performance function $g$ is sufficiently smooth, the approximation errors are given as follows: central difference,

$$\frac{g|_{d+\Delta d} - g|_{d-\Delta d}}{2\,\Delta d} - \left.\frac{Dg}{Dd}\right|_d \approx \frac{1}{3}(\Delta d)^2 \left.\frac{D^3 g}{Dd^3}\right|_d \tag{11.9}$$

forward difference,

$$\frac{g|_{d+\Delta d} - g|_d}{\Delta d} - \left.\frac{Dg}{Dd}\right|_d \approx \frac{1}{2}(\Delta d) \left.\frac{D^2 g}{Dd^2}\right|_d \tag{11.10}$$

backward difference,

$$\frac{g|_d - g|_{d-\Delta d}}{\Delta d} - \left.\frac{Dg}{Dd}\right|_d \approx -\frac{1}{2}(\Delta d) \left.\frac{D^2 g}{Dd^2}\right|_d \tag{11.11}$$

Thus the central difference scheme is a second-order accurate approximation. This means that it is not necessary to assume too small a design perturbation $\Delta d$ to calculate the sensitivity. If the central difference scheme is applied, errors in the forward and backward difference schemes can be estimated by

$$\frac{1}{2}(\Delta d)\frac{D^2 g}{Dd^2}\bigg|_d \approx \frac{1}{2}\frac{g|_{d+\Delta d} - 2g|_d + g|_{d-\Delta d}}{\Delta d} \tag{11.12}$$

This may be applicable to modify the size of design perturbation $\Delta d$ by specifying the allowable tolerance $\varepsilon$ for the approximation error in the forward or backward difference scheme. Indeed, we first estimate the second derivative of the performance function using a trial design perturbation $\Delta d_{\text{trial}}$, and then we check whether the estimated error

$$E_{\text{estimated}} = \left|\frac{1}{2}(\Delta d_{\text{trial}})\frac{g|_{d+\Delta d} - 2g|_d + g|_{d-\Delta d}}{\Delta d_{\text{trial}}^2}\right| \tag{11.13}$$

is smaller or larger than a given tolerance $\varepsilon$. If this estimated error is larger than $\varepsilon$, then we define a new perturbation $\Delta d_{\text{desired}}$ by

$$\Delta d_{\text{desired}} \leqslant \frac{2\varepsilon \Delta d_{\text{trial}}^2}{|g|_{d+\Delta d} - 2g|_d + g|_{d-\Delta d}|} \tag{11.14}$$

For the sensitivity, however, we shall apply the central difference scheme to ensure at least one order higher accuracy than this estimation based on the forward or backward difference approximation. Since it is possible to estimate the upper bound of the amount of the approximation error, the finite difference method is now applicable with more confidence to compute design sensitivity.

We shall provide an example of error estimation of the finite difference method for sensitivity analysis related to the shape of the body. To do this, the bending

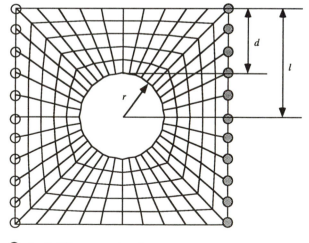

● Applied Transverse Load
○ Fixed Nodes

**Fig. 11.1** Finite element mesh and definition of design variable.

problem of a plate with a circular hole is considered shown in Fig. 11.1. The nodes on the left edge are fixed, and a transverse force is applied uniformly on the right edge. If the performance function is the maximum principal stress $\bar{\sigma}$ of the structure, and the design variable is the radius of the circular hole $r$, the finite difference approximation error is estimated by

$$E_{estimated} = \left| \frac{1}{2} \frac{\bar{\sigma}|_{r+\Delta r} - 2\bar{\sigma}|_r + \bar{\sigma}|_{r-\Delta r}}{\Delta r_{trial}} \right| \tag{11.15}$$

where the size of perturbation $\Delta r$ is determined by $\Delta r = 10^{\alpha} r (\alpha = -1, -2, -3)$. Figure 11.2 shows the estimated error of the sensitivity of the maximum principal stress with respect to the radius of the circular hole. The estimated error ratio is plotted with the design variable $r$ changed from initial value 2.0 to the upper limit 4.0. It follows from the plotting of the estimated error in Fig. 11.2 that the lowest error is obtained at $\alpha = -2$ while much worse results are obtained for $\alpha = -1$ and $-3$. In the case that $\alpha = -1$, we may say that too large a $\Delta r$ is assumed. However, for $\alpha = -3$, the roundoff error becomes much bigger than approximation error. In the finite difference approximation, the roundoff error occurs in the calculation of $g|_{d+\Delta d} - g|_{d-\Delta d}$, and thus the error is quantified by

$$E_{roundoff} = \frac{g|_{d+\Delta d} - g|_{d-\Delta d}}{g|_d} \tag{11.16}$$

In the present example, the roundoff error is $O(\Delta d) \approx 10^{-3}$ at $\alpha = -3$. This implies we would lose 3 digits in the calculation. This is the reason why the error declines when the value of performance function and the perturbation size $\Delta r$ become large.

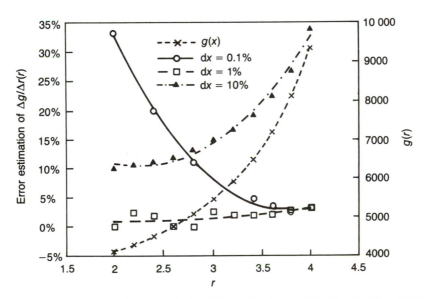

**Fig. 11.2** Estimated error in the original definition of design variable ($dx = (\Delta r/r) \times 100\%$).

In this problem, we can improve the error by the finite difference approximation, and also the distribution can be much more constant, by modifying the definition of the design variable by

$$d = l - r$$

$$\Delta d = 10^{-\alpha}(l - r) \tag{11.17}$$

Using this new definition of the design variable, the approximation error is estimated by

$$E_{\text{estimated}} = \left| \frac{1}{2} \frac{\bar{\sigma}|_{d + \Delta d} - 2\bar{\sigma}|_d + \bar{\sigma}|_{d - \Delta d}}{\Delta d_{\text{trial}}} \right| \tag{11.18}$$

and is shown in Fig. 11.3. Since the perturbation size $\Delta d$ becomes smaller as the value of performance function is larger, both approximation and roundoff error decline.

From this error analysis, the definition of the design variables is rather important to control the amount of approximation error, but the finite difference approximation is practically accurate enough to perform the optimization. In other words, the error is not too large (less than 2% at $\alpha = -2$) even in the forward or backward finite difference approximation that is a one order accurate approximation than the central difference scheme applied in our optimization study. It is also noted that the perturbation size does not need to be too small. The estimated error is less than 15% even for the case that the perturbation size

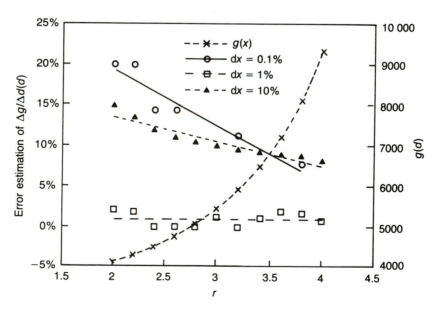

**Fig. 11.3** Estimated error in the modified design variable ($dx = (\Delta d/d) \times 100\%$).

is 10% of design variable. However, it is important that the perturbation size should not be too small because roundoff error becomes large.

## 11.4 SHAPE OPTIMIZATION IN STRUCTURAL DESIGN

Since shape optimization problems are more important in mechanical structure design, an optimization system should easily be able to deal with the shape of a structure. The characteristic of shape design is that the geometric finite element model must be modified during the optimization procedure, while sizing problems can be solved with a fixed geometrical finite element model. This nature of the adaptation of the finite element mesh has been the major difficulty in shape optimization. Many researchers have been challenged to solve this problem and it is not completely solved yet, especially for three-dimensional structures. In the early days, the direction of mesh movement was specified according to the shape change of a structure, while the topology of the finite element connectivity was fixed as the initial one. This approach failed to yield the optimum in many problems by crashing elements or destroying convexity of finite elements unless we can specify an appropriate direction of mesh movement by predicting the final shape in advance. Thus, we have reached to the notion that finite element meshes must be regenerated completely only by using boundary information that defines the shape of a structure for fully automated shape optimization methods (Bennett and Botkin, 1985; Botkin and Bennett, 1985; Botkin, Yang and Bennett, 1986; Belegundu and Zhang, 1992; Manicka and Belegundu, 1991). This means that a design optimization system must contain the capability of finite element mesh generation which can regenerate a model completely without terminating an optimization program. This suggests that the model-developing (i.e. pre-processing including automatic mesh generation) capability is the key for success in shape optimization in mechanical design, and it may be the most important module for development of a design system. Design boundary segments–patches may be expressed by appropriate splines, and the coordinate of control points of spline curves and surfaces may be the design variable in shape optimization, while this information should provide necessary information for generating finite element meshes at every design stage. This is a major difference of the mechanical design from structural optimization studied for example in aeronautical–aerospace engineering. The importance of the modeling module in a mechanical design system is thus very high. We shall now describe details of the development of shape optimization of an elastic structure. An extensive literature on this subject can be found, for example, in Haftka and Gandhi (1985, 1986), Haftka, Gurdal and Kamat (1990) and also Bennett and Botkin (1986).

The notion of shape optimization was introduced by Zienkiewicz and Campbell (1973) for design of the outside shape of a dam structure by using sensitivity analysis with respect to the coordinates of the nodes of a finite element model on the design boundary and the sequential linear programming method. Sensitivity is calculated by the semianalytical approach with the direct method. After this work, most shape optimization of a structure follows the notion that nodal coordinates of a finite element model are discrete design variables. Since they consider a rather minor shape modification of the dam profile, adaptation of the finite element model according to a design change is made only in the horizontal node movement without any sophistication. If a more general case of shape optimization is concerned, finite element model adaptation methods should be clearly described.

In the existing literature of shape optimization, we can find two major methods. The first way, shown in Fig. 11.4, is to specify the direction of node (either nodes of a finite element model or control points if the shape is defined by a spline function) movement that yields noncrashing of the mesh as much as possible, e.g. Fleury (1987). Setting of the direction of nodal adaptation requires a fair amount of advance information on the final shape of the structure. Otherwise the specification sometimes yields excessive finite element distortion or crashing of elements. The second widely applied method is the one Schnack (1979) introduced. Nodes on the design boundary are moved toward the normal direction as shown in Fig. 11.5. In this case, finite element distortion by change of boundary shape is not so large, while the possibility of contacting nodes on the design boundary may not be completely avoided. The second approach is thus far more robust than the first one for a large modification of the boundary shape.

After adapting the nodes on the design boundary, the location of the rest of nodes must be appropriately modified to maintain good quality of the finite element model. A method for the adaptation of interior nodes is the so-called mesh smoothing scheme. To relocate a node, all the finite elements connected to it are first searched, and then the nodes connected to it are identified. Using these nodal locations, we shall relocate the node to the center of 'mass' using the equation

$$(\mathbf{x}_n)_{\text{new}} = (1 - \alpha)(\mathbf{x}_n)_{\text{old}} + \alpha \frac{1}{n_{\text{max}}} \sum_{i=1}^{n_{\text{max}}} \mathbf{x}_{ni} \qquad (11.19)$$

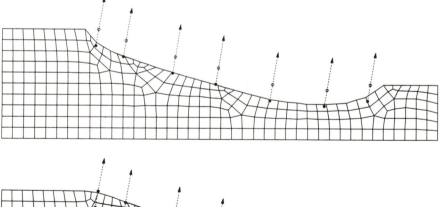

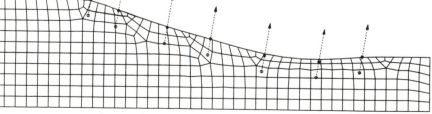

**Fig. 11.4** Adaptation to the specified direction.

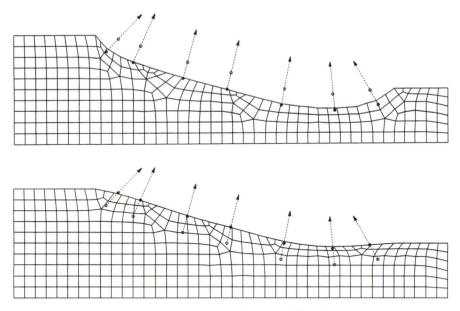

**Fig. 11.5** Adaptation to the normal direction.

where node $n$ is relocated with a specified parameter $\alpha$ such that $0 < \alpha \leqslant 1$, $n_{max}$ is the number of nodes connected to node $n$, and $\mathbf{x}_{ni}$ are the coordinates of node $ni$ (Fig. 11.6). Another relocation method, introduced by Choi (1986), may be obtained by solving

$$\mathbf{K}\,\Delta\mathbf{x} = \mathbf{f}_x \qquad (11.20)$$

where $\mathbf{K}$ is the global stiffness matrix for stress analysis, $\mathbf{f}_x$ is a specified generalized load that controls mesh density, and $\Delta\mathbf{x}$ is the amount of relocation of the nodal coordinates $\mathbf{x}$. Here $\Delta\mathbf{x}$ is specified for the nodes on the design boundary at each

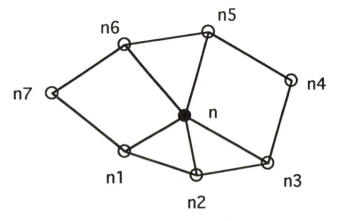

**Fig. 11.6** Node relocation scheme.

design change. Specifying $\mathbf{f}_x$ appropriately, we can control the amount of relocation in the domain. That is some portion of the domain should have a large relocation, while the other portion is subject to a small amount of relocation. It is noted that this relocation scheme is not expensive, since the LU decomposition of the stiffness matrix $\mathbf{K}$ is already obtained in the stress analysis. However, this approach may yield excessive finite element distortion as well as crashing elements. To avoid element distortion as much as possible, the center of mass approach is recommended, although it tends to assign excessively small elements in the vicinity of the concave outer boundary.

There are also two approaches to describe the shape of the design boundary. The majority of early work on shape optimization defined the design boundary by piecewise linear segments connecting nodes of a finite element model. Then the design variables are the coordinates of such nodes. In this case the number of design variables becomes significantly large if a refined finite element model is applied to ensure accuracy of the approximation. This implies the requirement of a large amount of computing effort to evaluate the design sensitivity of performance functions if mathematical programming methods are applied for optimization.

To avoid a large computing effort, the optimality criterion method is, in general, applied for such a case. If the optimality condition for the shape optimization problem involves the requirement of the satisfaction of

$$g(\mathbf{x}) = 1 \tag{11.21}$$

on the design boundary, the new boundary location (defined by the nodes) is then resized by

$$\mathbf{x}_{new} = \mathbf{x}_{old}[g(\mathbf{x}_{old})]^\alpha \tag{11.22}$$

for an appropriately given iteration factor $\alpha$ close to 1. If the optimality condition is described by the Mises equivalent stress, the function $g$ may be defined by the ratio of the Mises stress at an arbitrary point and its average value (or the specified maximum allowable Mises stress) on the design variable. If the optimality condition is given by the mutual strain energy density, $g$ is accordingly defined by using the mutual energy. The optimality criterion method does not require formal calculation of the design sensitivity of a set of performance functions. Thus if the optimization problem is written with several constraints, construction of an effective resizing algorithm is not obvious, and furthermore it may not yield the true optimum.

It is also noted that, if the optimality condition involves stress and strain, these must be accurately evaluated on the design boundary. In most finite element methods, very accurate stress and strain can be obtained at the one-order-lower Gaussian quadrature points than those applied to compute element stiffness matrices. However, these points are located inside finite elements. Thus we must introduce an extrapolation method to evaluate stress and strain on the nodes or along the element boundaries on the design boundary. A commonly applied extrapolation method is based on the least-squares method (Zienkiewicz and Taylor, 1989). Interpolating a component $q$ of stress–strain by the same shape functions $\mathbf{N}$ for displacement components

$$q = \mathbf{N}^T\mathbf{Q} \tag{11.23}$$

where $\mathbf{Q}$ is the vector of the nodal values of $q$, we find $\mathbf{Q}$ as the solution of the least-squares problem

$$\min_{Q} \frac{1}{2} \sum_{e=1}^{N_e} \int_{\Omega_e} (q - q_h)^2 \, d\Omega_e \qquad (11.24)$$

where $q_h$ is the finite element approximation of a component of stress–strain defined by the finite element approximation $\mathbf{u}_h$ of the displacement in equilibrium. Here $N_e$ represents the total number of finite elements, and $\Omega_e$ is the domain of an arbitrary finite element. If the size and orientation of finite elements moderately vary in a finite element model, this least-squares extrapolation provides quite reliable stress–strain values on the boundary. However, when a finite element model contains large and small finite elements alternately, this simple least-squares method may yield very inaccurate stress–strain values. Accurate finite element approximations of stress and strain on the boundary can also be obtained by applying the so-called mixed finite element approximation whose potential is examined by Haber (1987) in shape optimization of an elastic structure. In this case, stress and strain are independently approximated from displacement. Despite a considerable increase in the total number of degrees of freedom in a finite element model, an effective method to solve a mixed problem is studied in Zienkiewicz and Taylor (1989) using an iteration method.

Another disadvantage of the approach by defining the design boundary by a set of nodal points is that mere application of a resizing algorithm of the nodal location may yield a physically unrealistic oscillatory optimum shape, because of 'instability' of stress and strain on the boundary, as shown in Braibant and Fleury (1984). The total number of degrees of freedom of displacement should be larger than those for stress and strain in the displacement finite element approximation. However, if the least-squares method is applied to compute stress and strain on the boundary, those degrees of freedom exceed those for the displacement. This leads to instability in the approximation. Stresses and strains obtained by least squares in general possess a slightly oscillatory distribution, and then this slight oscillation is magnified during the iterations for shape optimization. Thus some smoothing algorithms for resizing or adaptive schemes to have an accurate finite element approximation of stress and strain must be introduced to provide a smooth optimum shape, (Kikuchi *et al.*, 1986).

To overcome the defect due to the definition of the boundary by nodal points, Braibant and Fleury (1984) introduced the method of *B* splines to describe the shape of design boundaries. After decomposing the design boundary into a set of design segments consisting of lines, arcs, and curves, each boundary segment is represented by an appropriate spline function. Then a finite element model is constructed independently of the number of control points of the splines. In most shape optimization problems, design boundary segments are defined by a few control points. If a cubic curve is expected, four control points are required. If an arc is desired, the location of the center and the radius are regarded as the control points. No matter how refined the finite element model introduced is the number of control points stays the same, that is, the total number of the discrete design variables can be fixed and few. Furthermore, as Braibant and Fleury (1984) showed, a smooth optimum shape can be obtained without introducing special techniques of smoothing and adaptive finite element methods. Therefore, the majority of researchers and structural engineers are now using this second approach to define the design boundary and the discrete design variables in shape optimization.

Sensitivity analysis for shape optimization has been studied by Choi and Haug (1983), Haug *et al.* (1983), and others. Since extensive treatment of sensitivity analysis can be found in Haug, Choi and Komkov (1985), we shall only describe a brief explanation of shape sensitivity. If the design boundary segment is described by the spline expression

$$\mathbf{x} = \sum_{i=1}^{N_c} \mathbf{x}_{ci}\phi_i(s) \quad 0 \leqslant s \leqslant 1 \tag{11.25}$$

where $N_c$ is the number of control points, $x_{ci}, i = 1,\dots,N_c$ are the coordinates of the control points, $s$ is the parametric coordinate, $\phi_i$ are the basis functions for spline expression. If either the analytical or the semianalytical method is applied to compute the design sensitivity, we must calculate the sensitivity

$$\frac{D g}{D\mathbf{x}_{cj}}(\mathbf{u}) = \frac{\partial g}{\partial \mathbf{u}}\frac{\partial \mathbf{u}}{\partial \mathbf{x}_{cj}} = \frac{\partial g}{\partial \mathbf{u}}\frac{\partial \mathbf{u}}{\partial \mathbf{x}(s)}\frac{\partial \mathbf{x}(s)}{\partial \mathbf{x}_{cj}} = \frac{\partial g}{\partial \mathbf{u}}\frac{\partial \mathbf{u}}{\partial \mathbf{x}(s)}\frac{\partial}{\partial \mathbf{x}_{cj}}\sum_{i=1}^{N_c}\mathbf{x}_{ci}\phi_i(s) = \frac{\partial g}{\partial \mathbf{u}}\frac{\partial \mathbf{u}}{\partial \mathbf{x}(s)}\phi_i(s) \tag{11.26}$$

for a performance function $g$ of only the displacement $\mathbf{u}$. If $N_{smax}$ nodes are placed on this spline, sensitivity of $g$ is thus computed by

$$\frac{D g}{D\mathbf{x}_{cj}}(\mathbf{u}) = \frac{\partial g}{\partial \mathbf{u}}\frac{\partial \mathbf{u}}{\partial \mathbf{x}_{cj}} = \frac{\partial g}{\partial \mathbf{u}}\sum_{i=1}^{N_{smax}}\frac{\partial \mathbf{u}}{\partial \mathbf{x}_i}\phi_j(s_i) \tag{11.27}$$

where $\mathbf{x}_i, i = 1,\dots,N_{smax}$ are the nodal coordinates of the finite element model on the design boundary segment, and $s_i$ are the parametric coordinates corresponding to $\mathbf{x}_i$. Therefore, we must compute the sensitivity of the displacement $\mathbf{u}$ with respect to all of the nodal coordinates on the design boundary segment of a finite element model. If a refined finite element model is used for stress analysis, computing sensitivity might become fairly large despite the small number of control points for the spline expression in the case that the analytical or semianalytical method is applied. It thus follows from this fact that design sensitivity for shape optimization should be effectively computed by finite difference methods by taking advantage of a few control points of the spline expression. Since a fairly small design perturbation is taken in the finite difference approximation for the sensitivity calculation, we need not regenerate the whole finite element model. We simply relocate the nodes on the design boundary segment to the curve perturbed, while the rest of the nodes in the model are fixed. Since the element connectivity and the total number of degrees of freedom of a finite element model are also fixed, sensitivity calculation by the finite difference method can be well justified.

Another issue we must discuss in shape optimization is the possible need for remeshing (or rezoning) of a finite element model, especially for the case of a large difference in the initial and optimum shape configurations. If the design change from the initial to the final is moderate, it is sufficient to adapt the finite element mesh by moving the nodes on the design boundary in the normal or specified direction, as desired above. However, if the design change becomes large, element distortion might not be reduced sufficiently by application of mesh smoothing schemes only. Excessively refined or coarse finite elements are generated in certain domains of the model, and then the quality of the analysis may deteriorate significantly. To avoid this, complete remeshing may be required; that is the whole finite element model must be completely regenerated by the current shape design. This remeshing concept is introduced by Bennett and Botkin (1985) as mentioned earlier.

## 11.5 IMPLEMENTATION OF STRUCTURAL OPTIMIZATION

On the basis of the concept described above, a prototype system has been implemented to solve sizing and shape optimization problems by integrating a generalized geometric modeler, an FEM analysis code, and an optimizer in the form small-size software driver. This system is flexible, easy to maintain, and can distribute the calculation to different types of computers, one to a graphic workstation with which the users are accustomed, and the other to faster computers with possibly a vector/parallel processor to carry out the analysis, as mentioned previously.

The system consists of (1) PATRAN as a geometric modeler and an automatic mesh generator, (2) the driver and file format converter written by PCL (Patran Command Language), (3) a general-purpose FEM analysis code, (4) a set of UNIX C-shell scripts to calculate sensitivity, (5) ADS for an optimizer, and (6) Mathematica to calculate the objective function value and derivatives. It is implemented on an APOLLO DN4000 (for man–machine interface and for system flow control) and a SONY NWS3710 (for finite element analysis) in the present study. The flowchart of this system is given in Fig. 11.7.

This prototype system can optimize (i.e. minimize) the volume or weight of a three-dimensional structure under constraints on the maximum principal stress, maximum strain energy density, maximum Mises equivalent stress, or maximum deformation, while physical dimensions of shells as well as their shape are the set of design variables. For example, an optimal design problem is formulated by

$$\text{minimize} \quad \text{weight}(\mathbf{d})$$

$$\text{subject to} \quad \frac{\text{stress}_{\text{maximum}}(\mathbf{d})}{\text{stress}_{\text{allowable}}} - 1 \leqslant 0$$

$$\underline{d}_i \leqslant d_i \leqslant \bar{d}_i, i = 1, 2, \dots, n \tag{11.28}$$

where $\underline{d}_i$ and $\bar{d}_i$ are the given lower and upper limit of each design variable. In this example, the weight is minimized in dimensions (sizes) of a structure, while the maximum deformation is restricted in the allowable range. In general, because the objective function and constraints are nonlinear functions of the design variable, the optimization problem may be solved by the sequential linear programming method. In this case, the derivatives of the constraints with respect to the design variables must be given to the system in the linearizing process to solve the problem.

Since the major portion of the optimization system consists of already existing (commercially available or public) codes, the only substantial portion of the present system is the main control flow routine written in UNIX C-shell scripts for sensitivity analysis and the optimization algorithm, while it calls the geometric

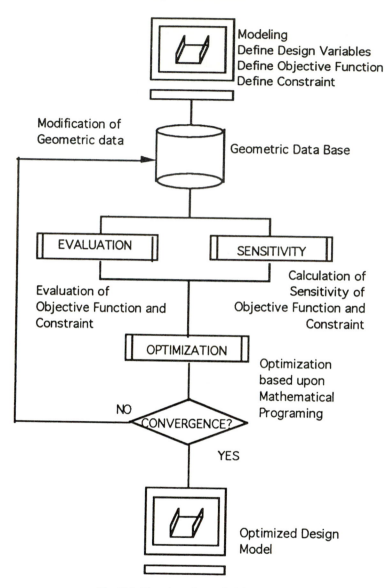

**Fig. 11.7** Flowchart for optimizing system.

modeler and finite element analysis code. Because of the choice of sensitivity calculation using the central finite difference scheme, whose size of perturbation can be controlled by the error analysis described in the previous section, the main routine becomes very simple. A sample program is given in Appendix 11A.

We simply call the geometric modeling module with automatic finite element mesh generation module, a finite element analysis code, and a program for the numerical optimization algorithm. If it is intended to use different geometric modules, analysis codes, and optimizers are in the system of optimization, we simply call the desired modules. Most of the rest of the program of the optimization is unchanged.

It is clear that if the total number of design variables is reasonably small, and if computing speed is not so critical, this simple and small design optimization system written in UNIX C-shell scripts can solve quite a large number of design optimization problems in both size and shape design.

## 11.6 EXAMPLES OF OPTIMIZATION BY THE PRESENT DESIGN SYSTEM

We shall give several examples of optimum structures using the above design system together with PATRAN's PCL and PATRAN to control geometric quantities for size and shape optimization and to develop finite element models automatically in the optimization system. As mentioned in the previous section, ADS, a public domain software package, is applied for optimization. More precisely, the method of feasible direction in ADS for constrained minimization is used as the optimizer, in which the golden method is used as the one-dimensional search.

### 11.6.1 Case of a folded plate

The geometry of a folded plate is given in Fig. 11.8. The objective of this example is to reduce the volume of the structure, while the maximum principal stress should not exceed the allowable value of the maximum principal stress by

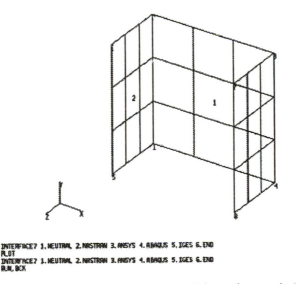

INTERFACE? 1. NEUTRAL 2. NASTRAN 3. ANSYS 4. ABAQUS 5. IGES 6. END
PLOT
INTERFACE? 1. NEUTRAL 2. NASTRAN 3. ANSYS 4. ABAQUS 5. IGES 6. END
RUN, BCK

**Fig. 11.8** Original model (patches in PATRAN for mesh generation).

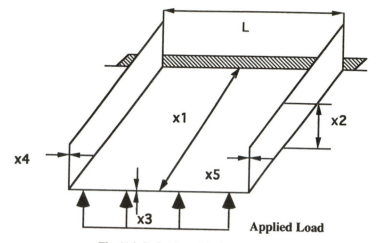

**Fig. 11.9** Definition of design variables.

reducing the weight. Design variables are shown in Fig. 11.9. While the width of the plate $L$(1.0 m) is fixed, design variables are the height of two folded portions, the length of the plate, and the thickness of the folded plate, and they are expressed by $x_i$, $i = 1, 2, \ldots, 5$. The upper and lower bounds of the design variables are also specified. This problem is mathematically formulated by

$$\text{minimize} \quad \text{volume}(\mathbf{x}) = Lx_1x_3 + 2x_1x_2(x_4 + x_5)$$

$$\text{subject to:} \quad \bar{\sigma}(r, x) \equiv \text{maximum principal stress}$$

$$\bar{\sigma}(r, x) \leqslant \bar{\sigma}_{\text{Allowable}}$$

$$0.1L \leqslant x_1 \leqslant 0.9L$$

$$0.4L \leqslant x_2 \leqslant 0.6L$$

$$0.03L \leqslant x_{3,4,5} \leqslant 0.07L \tag{11.29}$$

A finite element model of the initial design is shown in Fig. 11.10 with the boundary conditions, i.e. the one end is clamped and the other is subject to a uniformly distributed transverse traction (10 MPa).

The principal stress distribution of this design state is given in Fig. 11.11. The iteration process of the optimization method in ADS is described in Fig. 11.12 for the objective and the constraint on the maximum principal stress and in Fig. 11.13 for the design variables. It is clear that the convergence process is not monotonic at all except the vicinity of the optimum. This zigzag behavior of the convergence process of the optimization algorithm may be explained by the nonmonotonic variation of the objective function and its sensitivity with respect to the thickness of the plate (Fig. 11.14).

At the 13th iteration, ADS provides the convergent result with the optimal values of the design variables as shown in Table 11.1. The initial volume is reduced (by 20%) by changing the length of the folded plate to the lower limit while the height of the folded portions is also reduced to 0.48 m from 0.50 m without exceeding of the limit value of the maximum principal stress of the

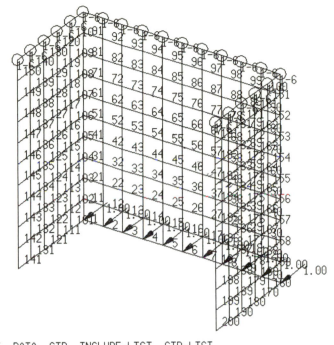

TYPE, DATA, SID, INCLUDE-LIST, CID-LIST

TYPE, DATA, SID, INCLUDE-LIST, CID-LIST

**Fig. 11.10** Mesh and boundary condition.

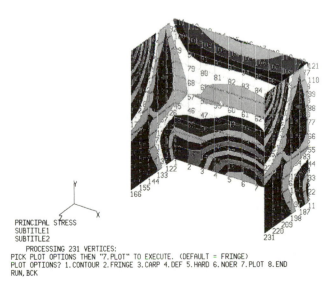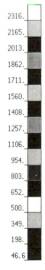

PRINCIPAL STRESS
SUBTITLE1
SUBTITLE2
 PROCESSING 231 VERTICES:
PICK PLOT OPTIONS THEN "7.PLOT" TO EXECUTE. (DEFAULT = FRINGE)
PLOT OPTIONS? 1.CONTOUR 2.FRINGE 3.CARP 4.DEF 5.HARD 6.NOER 7.PLOT 8.END
RUN,BCK

**Fig. 11.11** Initial distribution of principal stress.

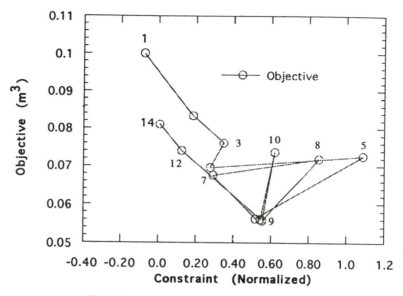

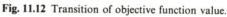

**Fig. 11.12** Transition of objective function value.

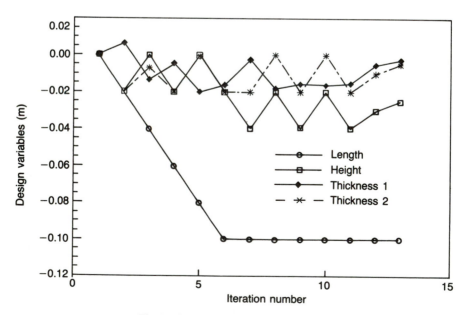

**Fig. 11.13** Transition of design variables.

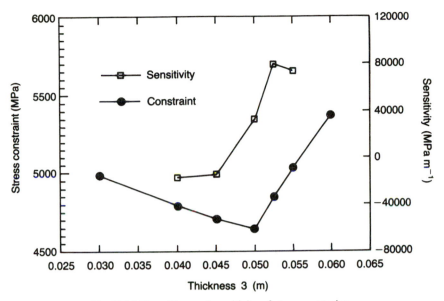

**Fig. 11.14** Transition and sensitivity of stress constraint.

**Table 11.1** Result of optimization example 1

|  | Initial | Lower | Upper | Optimum |
|---|---|---|---|---|
| Volume (m³) | 0.100 | | | 0.080 |
| Constraint | −0.070 | | | 0.010 |
| Design Var(m) 1 | 1.000 | 0.900 | 1.100 | 0.900 |
| 2 | 0.500 | 0.400 | 0.600 | 0.480 |
| 3 | 0.050 | 0.030 | 0.070 | 0.050 |
| 4 | 0.050 | 0.030 | 0.070 | 0.040 |
| 5 | 0.050 | 0.030 | 0.070 | 0.040 |

structure. The thickness of the folded portions is reduced to 0.04 m from 0.05 m, and the thickness of the base plate is unchanged although it changes in a zigzag manner in the optimization process. Figures 11.15 and 11.16 show the optimal model. Figure 11.17 shows the distribution of the maximum principal stress at the optimum. It is clear that the pattern of the distribution and the maximum value are unchanged from those of the initial design.

Since the height of the folded portions is a design variable, this problem possesses the nature of shape optimization as well as that of sizing optimization by modifying the thickness and the length of a plate. While the height of the folded portion is modified, the finite element must be regenerated. As shown in Figs. 11.10 and 11.16, the finite element model of the initial design is different from that of the optimum design; the number of elements has been changed from 200 and 180. In this case, PATRAN regenerates the finite element model with a fixed size of elements which we specified.

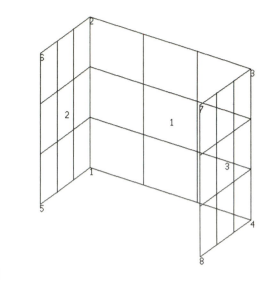

INPUT DIRECTIVE OR "END"
PLOT
INPUT DIRECTIVE OR "END"
RUN, BCK

**Fig. 11.15** Optimal model (patches in PATRAN for mesh generation).

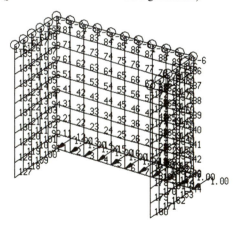

INPUT DIRECTIVE OR "END"
DISP, 1, PLOT
INPUT DIRECTIVE OR "END"
RUN, BCK

**Fig. 11.16** Optimal finite element model.

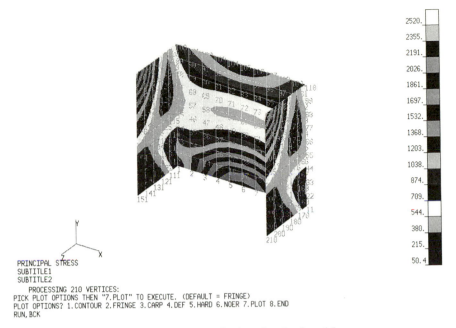

**Fig. 11.17** Stress distribution of optimal model.

## 11.6.2 Plate with a circular hole

We shall consider a shape design problem of a plate that is fixed along one edge and is subject to a uniform transverse traction along the opposite edge. The original geometry of the plate model is shown in Fig. 11.18. A circular hole is generated. In this example, the weight of the plate is minimized without increasing the maximum principal stress by changing the location of the center and the radius of the circular hole. Although the original shape of the circular hole is not modified in the optimization, the finite element model geometry must be changed because of modification to the location and radius of the hole. Thus, this must be classified as a shape design problem despite the definition of the design variables, the location and the size of a hole (Fig. 11.19). In order to maintain the symmetry of the structure, we shall place the hole along the center (horizontal) line. This problem is mathematically formulated as follows:

$$\text{minimize} \quad (L^2 - \pi r^2)t$$

$$\text{subject to} \quad \bar{\sigma}(r, x) \equiv \text{maximum principal stress}$$

$$\bar{\sigma}(r, x) \leqslant \bar{\sigma}_{\text{Initial}}$$

$$x + r \leqslant 0.9L$$

$$x - r \leqslant 0.1L \qquad\qquad (11.30)$$

$$0 \leqslant x \leqslant L$$

$$0 \leqslant r \leqslant 0.5L$$

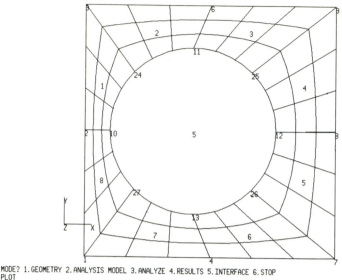

MODE? 1.GEOMETRY 2.ANALYSIS MODEL 3.ANALYZE 4.RESULTS 5.INTERFACE 6.STOP
PLOT
MODE? 1.GEOMETRY 2.ANALYSIS MODEL 3.ANALYZE 4.RESULTS 5.INTERFACE 6.STOP
RUN, BCK

**Fig. 11.18** Original model (patches in PATRAN for mesh generation).

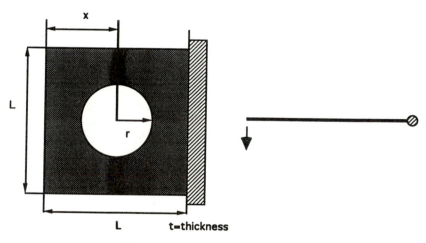

**Fig. 11.19** Definition of design variables.

where $L$ is the size of a square plate, $t$ is the plate thickness, $r$ is the radius of the hole, and $x$ is the (horizontal) location of the center of the hole. Here $L$ and $t$ are assumed to be fixed ($L = 1.0\,\mathrm{m}, t = 0.01\,\mathrm{m}$), i.e. only two design variables are defined in this shape design problem. The finite element model of the initial design state is shown in Fig. 11.20 with the boundary conditions. A small circle along the right edge indicates the fixed condition of a node, while the value '1.00'

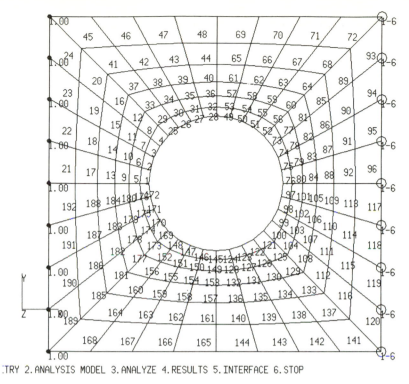

:TRY 2.ANALYSIS MODEL 3.ANALYZE 4.RESULTS 5.INTERFACE 6.STOP

:TRY 2.ANALYSIS MODEL 3.ANALYZE 4.RESULTS 5.INTERFACE 6.STOP

**Fig. 11.20** Mesh and boundary conditions.

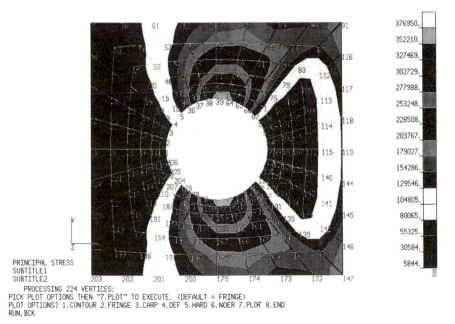

PRINCIPAL STRESS
SUBTITLE1
SUBTITLE2
PROCESSING 224 VERTICES:
PICK PLOT OPTIONS THEN "7.PLOT" TO EXECUTE. (DEFAULT = FRINGE)
PLOT OPTIONS? 1.CONTOUR 2.FRINGE 3.CARP 4.DEF 5.HARD 6.NOER 7.PLOT 8.END
RUN,BCK

**Fig. 11.21** Initial distribution of principal stress.

along the left edge shows the magnitude of the transverse force (1.00 Ma) applied on a node. The maximum principal stress of the initial design is given in Fig. 11.21.

Figures 11.22 and 11.23 show the iteration history of the optimization algorithm form the ADS program. Since lower stresses are generated in the left side of the

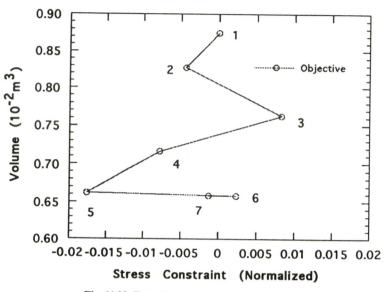

Fig. 11.22 Transition of objective function value.

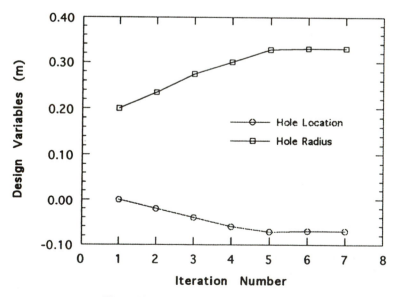

Fig. 11.23 Transition of design variables.

**Table 11.2** Result of optimization example 2

|  | Initial | Lower | Upper | Optimum |
|---|---|---|---|---|
| Volume ($10^{-2}\,\mathrm{m}^3$) | 0.8743 |  |  | 0.6582 |
| Stress constraint | 0.0000 |  |  | −0.0013 |
| Edge constraint | −2.0000 |  |  | −0.0008 |
| Hole location (m) | 0.5000 | 0.1000 | 0.9000 | 0.4299 |
| Hole radius (m) | 0.2000 | 0.1000 | 0.5000 | 0.3298 |

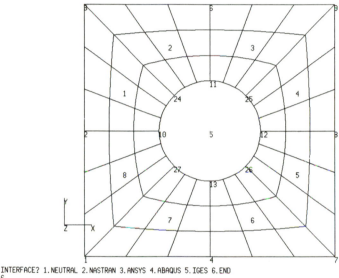

```
INTERFACE? 1.NEUTRAL 2.NASTRAN 3.ANSYS 4.ABAQUS 5.IGES 6.END
6
MODE? 1.GEOMETRY 2.ANALYSIS MODEL 3.ANALYZE 4.RESULTS 5.INTERFACE 6.STOP
RUN, BCK
```

**Fig. 11.24** Optimal model (patches in PATRAN for mesh generation).

plate, the hole location should be shifted to the left until it hits the limit $x - r \leqslant 0.1\,L$ (Fig. 11.23). As shown in Fig. 11.22, an unfeasible design is generated at the third and sixth iterations, but most of designs in the optimization process are feasible, i.e. do not violate the constraints we have set. Convergence is very monotonic. The volume is reduced to 0.6582 from 0.8743 by the optimum radius $r = 0.3298$ and the optimum location $x = 0.4299$ (Table 11.2). The geometric model of the optimum design is given in Fig. 11.24, and the distribution of the maximum principal stress is shown in Fig. 11.25. The overall pattern of the stress distribution is unchanged, although the stress in the left side of the hole becomes much larger than the original one.

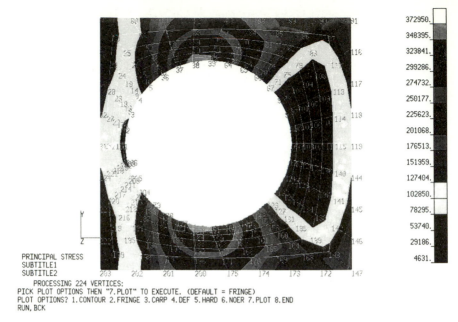

PRINCIPAL STRESS
SUBTITLE1
SUBTITLE2
PROCESSING 224 VERTICES:
PICK PLOT OPTIONS THEN "7.PLOT" TO EXECUTE. (DEFAULT = FRINGE)
PLOT OPTIONS? 1.CONTOUR 2.FRINGE 3.CARP 4.DEF 5.HARD 6.NOER 7.PLOT 8.END
RUN,BCK

**Fig. 11.25** Stress distribution of optimal model.

### 11.6.3 Plate with rib reinforcement

Let a square plate be reinforced by ribs as shown in Fig. 11.26. We shall consider a sizing optimization problem that minimizes the maximum deflection of the rib-reinforced plate by modifying the thickness of ribs without changing the rib

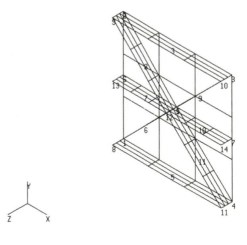

5.INTERFACE 6.STOP 7.(RESERVED) 8.(USER MENU)
RUN,BCK
CREATING IMAGE BACKUP FILE:
"patran.rst.3"

**Fig. 11.26** Original model (patches in PATRAN for mesh generation).

height. The total volume of the structure is also bounded to the one at the initial design. This problem is then mathematically defined by

minimize    $\bar{u}(\mathbf{x}) \equiv$ maximum deformation

subject to    $V = L^2 t + hL \left[ x_1 + x_2 + \dfrac{x_4 + x_7 + \sqrt{2}(x_3 + x_5 + x_6 + x_8)}{2} \right]$

$V \leqslant V_{\text{Initial}}$

$0.001 L \leqslant x \leqslant 0.1 L$    (11.31)

where $L$ is the size of a square plate, $t$ is the thickness of the plate, $h$ is the rib

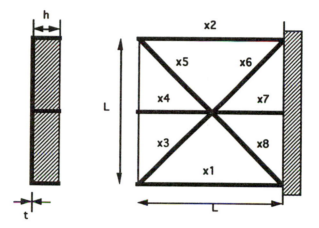

**Fig. 11.27** Definition of design variables.

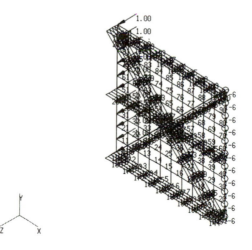

5. INTERFACE 6. STOP 7. (RESERVED) 8. (USER MENU)
RUN, BCK
CREATING IMAGE BACKUP FILE:
"patran.rst.4"

**Fig. 11.28** Mesh and boundary condition.

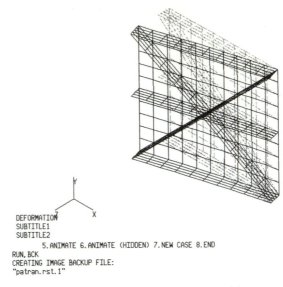

DEFORMATION
SUBTITLE1
SUBTITLE2
    5.ANIMATE 6.ANIMATE (HIDDEN) 7.NEW CASE 8.END
RUN,BCK
CREATING IMAGE BACKUP FILE:
"patran.rst.1"

**Fig. 11.29** Deformation of original model.

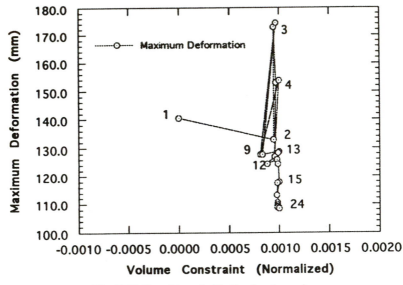

**Fig. 11.30** Transition of objective function value.

height, and $x_i$ are the thicknesses of the reinforced ribs (Fig. 11.27). The design variables in this problem are the rib thicknesses $x_i$, $i = 1, \ldots, 8$, while $L, h$, and $t$ are fixed. The finite element model and the boundary conditions are shown in Fig. 11.28. Reinforced ribs are modeled as plates, and are discretized by 4-node quadrilateral flat shell elements which are similar to QUAD4 elements in MSC/ NASTRAN. The right edge of the plate is fixed, while uniform transverse loads

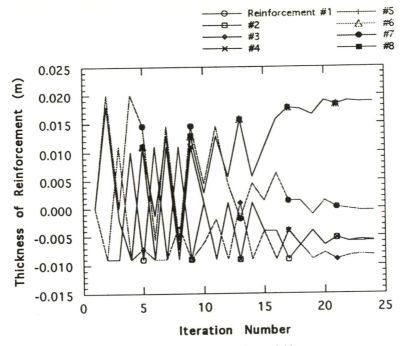

**Fig. 11.31** Transition of design variables.

**Table 11.3** Result of optimization example 3

|  |  | Initial | Lower | Upper | Optimum |
|---|---|---|---|---|---|
| Deformation (cm) |  | 1.404 |  |  | 1.085 |
| Volume |  | 0.000 |  |  | 0.001 |
| Thickness (m) | 1 | 0.010 | 0.001 | 0.100 | 0.004 |
|  | 2 | 0.010 | 0.001 | 0.100 | 0.004 |
|  | 3 | 0.010 | 0.001 | 0.100 | 0.002 |
|  | 4 | 0.010 | 0.001 | 0.100 | 0.004 |
|  | 5 | 0.010 | 0.001 | 0.100 | 0.002 |
|  | 6 | 0.010 | 0.001 | 0.100 | 0.029 |
|  | 7 | 0.010 | 0.001 | 0.100 | 0.010 |
|  | 8 | 0.010 | 0.001 | 0.100 | 0.029 |

are applied at every node along the left edge of the square plate. It is noted that the ribs are not subject to boundary conditions, i.e. the fixed condition is not imposed on the nodes of ribs. The deformed shape of the plate at the initial design is given in Fig. 11.29.

The iteration process of the optimization algorithm is shown in Figs. 11.30 and 11.31. The volume constraint is slightly violated, say about 1%, which is specified as allowable tolerance in ADS optimization code. As shown in Fig. 11.30, the volume constraint is always within the feasible range with a given allowable

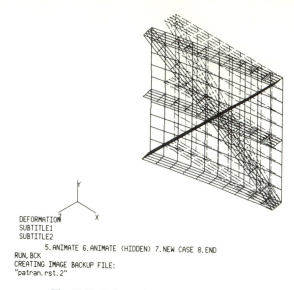

DEFORMATION
SUBTITLE1
SUBTITLE2
    5.ANIMATE 6.ANIMATE (HIDDEN) 7.NEW CASE 8.END
RUN, BCK
CREATING IMAGE BACKUP FILE:
"patran.rst.2"

**Fig. 11.32** Deformation of optimal model.

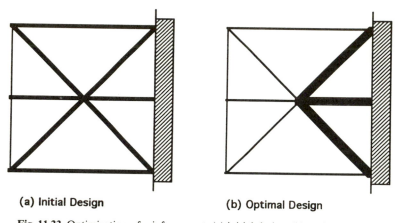

**(a) Initial Design**          **(b) Optimal Design**

**Fig. 11.33** Optimization of reinforcement: (a) initial design; (b) optimal design.

tolerance, while the objective function, the maximum deflection, is converging to the optimum in a zigzag manner. The behavior of the design variables is also very wild in most of the iteration history except in the vicinity of the optimum as shown in Fig. 11.31. The values of the maximum deflection and the design variables as given in Table 11.3. The deformed configuration of the rib-reinforced plate at the optimum is shown in Fig. 11.32. It is clear that both deformation and design maintain symmetry as shown in Fig. 11.33, which schematically represents the thickness of the ribs at both the initial and the optimum design. Thick ribs are formed in the vicinity of the fixed edge of the plate.

## 11.6.4 Optimum shell configuration

We shall present a capability of the design optimization system that can manage the shape change of a structure defined by a free-form Bezier surface in three-dimensional space. In this case we modify the shape of the Bezier surface by moving the control points. To do this we shall form the best shell configuration that minimizes the maximum principal stress under the volume constraint of the shell structure.

The original shape of a plate is defined by 16 control points and its patches are given in Fig. 11.34. As shown in Fig. 11.35, the size of the projected surface of a plate-shell structure onto the $x-y$ plane is fixed to be $L$, the thickness of the thin structure is also fixed to be $t$, while the $z$ coordinates of the four interior control points are varying in design. The objective is to reduce the maximum principal stress of the structure by increasing by 10% the original volume of the flat shell (i.e. plate) structure. This problem is mathematically represented by

$$\text{minimize} \quad \bar{\sigma}(\mathbf{x}) \equiv \text{maximum principal stress}$$

$$\text{subject to} \quad V = t \iint_\Omega \left| \frac{\partial B(\mathbf{x}, u, v)}{\partial u} \times \frac{\partial B(\mathbf{x}, u, v)}{\partial v} \right| du \, dv$$

$$V \leqslant 1.1 \, V_{\text{Initial}}$$

$$-\frac{L}{3} \leqslant x \leqslant \frac{L}{3} \tag{11.32}$$

Here the lower and upper bound of the $z$ coordinates of the four control points of the Bezier surface are also specified to be within $1/3$ of the size of the projected plane. The finite element model generated by the patch model in Fig. 11.34 in

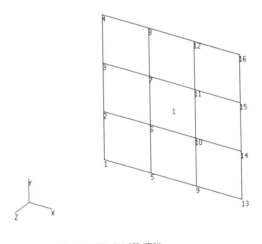

5. INTERFACE 6. STOP 7. (RESERVED) 8. (USER MENU)
RUN, BCK
CREATING IMAGE BACKUP FILE:
"patran.rst.4"

**Fig. 11.34** Original model (patches in PATRAN for mesh generation).

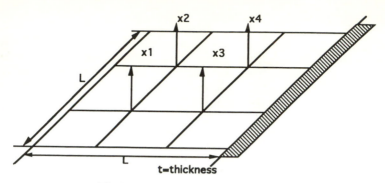

**Fig. 11.35** Definition of design variables.

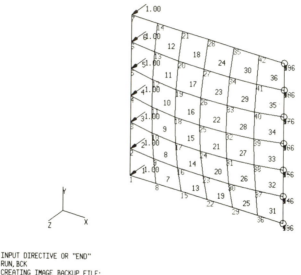

```
INPUT DIRECTIVE OR "END"
RUN, BCK
CREATING IMAGE BACKUP FILE:
"patran.rst.1"
```

**Fig. 11.36** Mesh and boundary condition.

PATRAN and the boundary conditions are given in Fig. 11.36. The right edge is fixed, and uniform transverse forces are applied at the nodes on the left edge. A small circle indicates the fixed condition at a node, and small arrows represent applied loads. The principal stress of the structure at the initial design is shown in Fig. 11.37.

Figures 11.38 and 11.39 show the iteration history of the optimization algorithm in ADS code. Convergence is very smooth and monotonic in this problem. The maximum principal stress is monotonically reduced by increasing the z coordinates of the control points. Because of the symmetry of this problem, four design variables have the same value in each design process (Table 11.4). Figure 11.40 shows the PATRAN patch model at the optimum. The four interior control

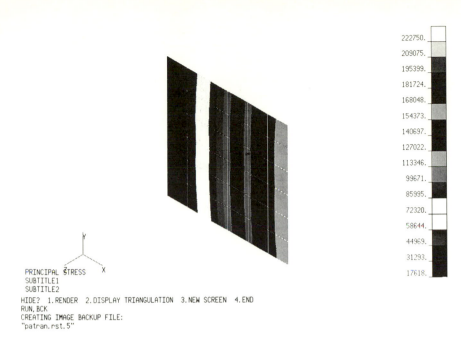

PRINCIPAL STRESS   X
SUBTITLE1
SUBTITLE2
HIDE?  1.RENDER  2.DISPLAY TRIANGULATION  3.NEW SCREEN  4.END
RUN,BCK
CREATING IMAGE BACKUP FILE:
"patran.rst.5"

222750.
209075.
195399.
181724.
168048.
154373.
140697.
127022.
113346.
99671.
85995.
72320.
58644.
44969.
31293.
17618.

**Fig. 11.37** Initial distribution of principal stress.

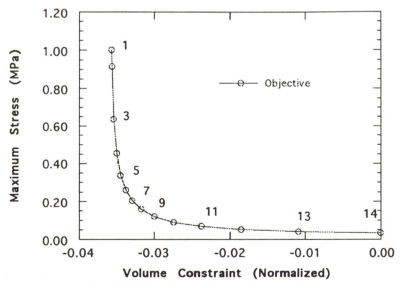

**Fig. 11.38** Transition of objective function value.

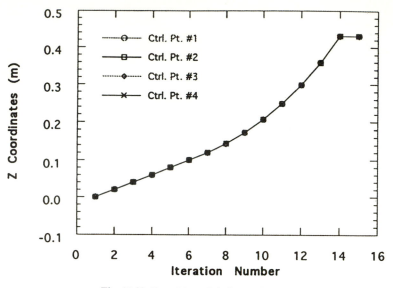

**Fig. 11.39** Transition of design variables.

**Table 11.4** Result of optimization example 4

|  | Initial | Lower | Upper | Optimum |
|---|---|---|---|---|
| Maxmum stress (MPa) | 1.0000 | | | 0.0340 |
| Volume constraint | −0.0357 | | | −0.0001 |
| Control point (m) 1 | 0.0000 | −0.5000 | 0.5000 | 0.4300 |
| 2 | 0.0000 | −0.5000 | 0.5000 | 0.4300 |
| 3 | 0.0000 | −0.5000 | 0.5000 | 0.4300 |
| 4 | 0.0000 | −0.5000 | 0.5000 | 0.4300 |

points are displaced to form a shell defined by the Bezier spline surface. The stress distribution at the optimum shell is given in Fig. 11.41.

### 11.6.5 Stiffener of the engine hood of a car

We shall consider a more practical example of structural optimization. A distributed load (10 kPa) is applied to the center of the engine hood of a car body (Fig. 11.42), and the hood is deformed as shown in Fig. 11.43. The objective of this example is to reduce the amount of deformation over the hood without increasing the volume. The shape and thickness of the hood cannot be modified because of the requirement of the design and manufacturing. Only the shape of the beam stiffener can vary. In this case, a beam stiffener with hat-shaped cross section is used (Fig. 11.44). Although any dimensions in the cross section can be

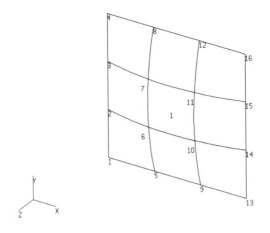

5. INTERFACE 6. STOP 7. (RESERVED) 8. (USER MENU)
RUN, BCK
CREATING IMAGE BACKUP FILE:
"patran.rst.3"

**Fig. 11.40** Optimal model.

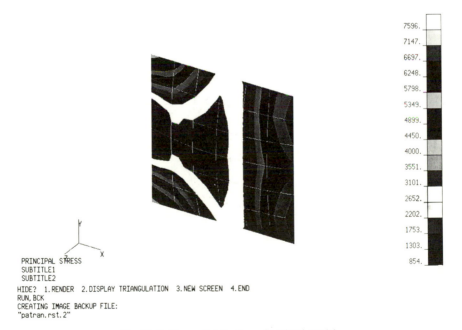

PRINCIPAL STRESS
SUBTITLE1
SUBTITLE2
HIDE? 1. RENDER 2. DISPLAY TRIANGULATION 3. NEW SCREEN 4. END
RUN, BCK
CREATING IMAGE BACKUP FILE:
"patran.rst.2"

**Fig. 11.41** Stress distribution of optimal model.

design variables in optimization, we shall fix the sizes of $f$ and $h$ for simplicity. From the symmetry of the structure, the beams on the left-hand side and the right-hand side should have the same cross section, and then $w$ and $t$ of the front and rear beam are considered to be design variables. In most finite element

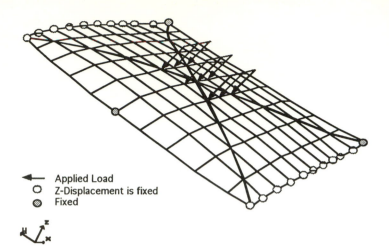

**Fig. 11.42** Discrete model and boundary conditions.

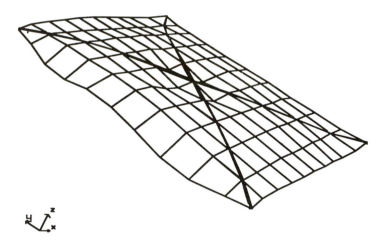

**Fig. 11.43** Deflection of the engine hood.

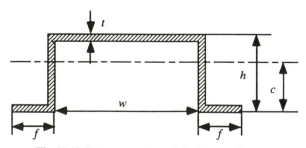

**Fig. 11.44** Hat cross section of the beam stiffener.

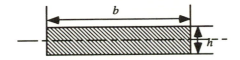

**Fig. 11.45** A rectangular cross section.

analysis codes, the moment of inertia of the cross section of the beam must be given as the element property. The neutral line is calculated from

$$\int_0^c dA = \int_c^h dA \Leftrightarrow c = \frac{w + 2h + 2t - 2f}{4} \qquad (11.33)$$

The moment of inertia is calculated using $c$ from

$$I_{hat} = \int_{-c}^{h-c} y^2 \, dA = \int_{-c}^{-c+t} 2fy^2 \, dy + \int_{-c+t}^{h-c} 2ty^2 \, dy + \int_{h-c-t}^{h-c} wy^2 \, dy \quad (11.34)$$

In this example, we use NIKE3D code from Lawrence Livermore National Laboratory to analyze the deformation of such a three-dimensional shell structure. NIKE3D does not have an input card to specify the moment of inertia but has two kinds of beam elements: one is for rectangular cross sections while the other is for pipe cross sections. If the effect due to the shear strain can be ignored, and if only pure bending of the beam is considered, the hat-type cross section beam can be described as a rectanglar beam (Fig. 11.45). The moment of inertia of the rectangular beam is given by

$$I_{rect} = \frac{bh^3}{12} \qquad (11.35)$$

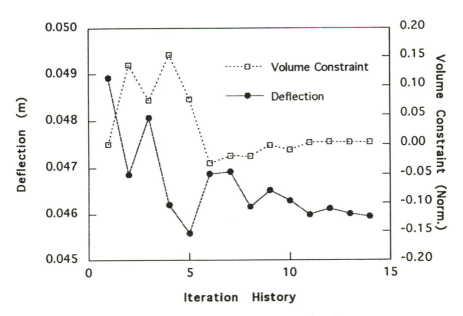

**Fig. 11.46** Iteration history of performance functions.

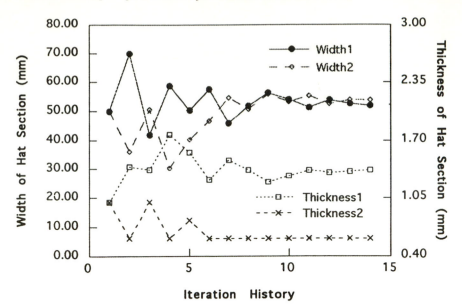

**Fig. 11.47** Iteration history of the dimensions of hat cross section.

**Table 11.5** Result of example of the engine hood

|               | Initial | Lower  | Upper  | Optimal |
|---------------|---------|--------|--------|---------|
| $w$ (front (mm) | 50.00   | 10.00  | 100.00 | 52.05   |
| $t$           | 1.00    | 0.50   | 10.00  | 1.37    |
| $w$ (rear)    | 50.00   | 10.00  | 100.00 | 53.94   |
| $t$           | 1.00    | 0.50   | 10.00  | 0.60    |
| Deflection (m) | 4.89   | –      | –      | 4.60    |
| Volume constraint [1] | 0.00 | –  | –      | 0.00    |

If the value $h$ for the rectangular beam is fixed, and if the shear strain effect is ignored, then $b$ may be a function of $w$ and $t$:

$$b \approx \frac{12}{h^3} I_{hat}(w, t) \tag{11.36}$$

Using this, the shell structure is analyzed and the cross section of beam stiffener for reinforcement of the shell is optimized. The iteration history of the performance function is given in Fig. 11.46 and dimensions of the cross section are given in Fig. 11.47. As the result of this optimization, the deflection over the structure is reduced to 4.60 from 4.89 without increasing the volume (Table 11.5).

## 11.7 REMARKS AND CONCLUSION

As shown above a very small system mostly written in UNIX C-shell scripts with PATRAN's PCL can solve design optimization problems which involve both size and shape of a structure by calling a finite element analysis program, optimization code, and symbolic manipulation software from the system. If the speed of the process is not too important, the present system can handle most of structural optimization problems. Furthermore, we have great flexibility in the choice of analysis and optimization codes, while commercially available design optimization systems are very sophisticated, but possess substantial restriction on these modules. In other words, our system can deal with a much wider class of design optimization problems than any other existing ones.

The key of the system is the capability of automatic mesh generation of the analysis model, since both sizing and shape of a structure are involved in most mechanical design problems. Thus the optimization system should be constructed by emphasizing the importance of the model development capability. Otherwise it will not be practical at all, no matter how sophisticated a sensitivity analysis is implemented. In spite of the restriction of PATRAN's automatic mesh generation capability, it can handle a fairly large class of problems. If a more capable full-scale automatic mesh generation capability that can control mesh density with simple commands is introduced, we can enhance the capability of the present design optimization system developed in this study. The disadvantage of the present method is the necessity of a large amount of processing time, because of the central difference method used to compute sensitivity of performance functions. It is certain that we have used other workstations to carry out the finite element analysis to reduce the overall processing time. However, still this may be insufficient if the number of design variables becomes large. At this moment, we must restrict the number of design variables to at most 20–30.

## ACKNOWLEDGEMENTS

During the present work, the first author was partially supported by Suzuki Corporation and RTB Corporation, while the second author was supported by Sony Corporation (through Dr K. Arita, Mr H. Harimaya and Mr K.C. Imaeda). They express sincere appreciation for this support.

Development of the optimization system was performed by using a computer system of Sony Corporation of America and the facilities of the Computer Aided Engineering Network, the College of Engineering, the University of Michigan, and using software supported by Mr J. Andusiak. Suggestions from Mr S. Ida, Mr M. Okamoto, Mr Fukushima and Mr Y. Park were very useful for implementing the prototype system.

## APPENDIX 11A SAMPLE PROGRAM

```
# !/bin/csh
#
# Optimization Procedure using UNIX C Shell Script
#
# set default file name and command aliases

set   PRM_FILE = parameters.dat
```

```
set   INFO_FILE = info.dat
set   BC = "bc -l"
set   ANALYZE = analysis

# Initialization of optimization parameters
# (Optimizer is called) and status is saved
set INFO = `optinit`

# set number of design variables/constraints
set NDV = `head -2 $PRM_FILE | tail -1 | awk '{print $1}'`
set NCON = `head -2 $PRM_FILE | tail -1 | awk '{print $2}'`

# Start iteration
set ino = 1
while ($INFO > 0)
    # Save current design variables into a file
    optdv > current.dv

    switch ($INFO)
        # When the values of performance functions are required
        case 1:
            # Generate geometric model
            model.sh `cat current.dv`

            Generate FE model with automatic mesh generator
            omesh > /dev/null

            # Convert FEM Model to the INPUT format of analyzer
            convert.sh meshgen.fem2

            # Analyze current design
            ${ANALYZE}

            # Calculate performance functions at the initial design
            #     Volume (Objective)
            volume.sh < current.dv > current.val

            #     Max Principal Stress (Constraint #1)
            #          (Calculate the value from OUTPUT file of analyzer)
            if  ($ino == 1) then
                awk -f max_ps.awk OUTFEM > normalizer
                set normal = `cat normalizer`
                echo "0.0" >> current.val
            else
                echo `awk -f max_ps.awk OUTFEM` / $norm1 - 1.0| \
                    $BC >> current.val
            endif

            #     Edge Constraint (Constraint #2)
            edge.sh < current.dv >> current.val

            # Reconstructing Parameter file
            optval current.val
            breaksw
```

```
# When the gradients of performance functions are required
case 2:
    # Calculate Perturbation Size
    set pertub = 0.001
    set ptmp0 = `perturb.sh current.dv`
    echo > temp.dat
    foreach p ($ptmp0)
        echo $p \* $pertub | $BC >> temp.dat
    end
    set psize = `cat temp.dat`
    echo > temp.dat
    foreach p ($psize)
        echo $p \* 2.0 | $BC >> temp.dat
    end
    set psize2 = `cat temp.dat`

    # Calculate Gradient for each design variables
    set i = 1
    while ($i < = $NDV)

        # Create perturbation model for design var #i
        sed $(i), \$d current.dv > temp.dat
        sed -n ${i}p current.dv |\
            awk '{print $1 + `$psize[$i]`}' >> temp.dat
        sed 1, ${i}d current.dv >> temp.dat

        # Calculate Performances at the Plus Perturbed Design
        model.sh `cat temp.dat`
        omesh > /dev/null
        convert.sh meshgen.fem2

        # Analyze perturbed model
        ${ANALYZE}

        # Calculate performances at the perturbed design
        volume.sh < temp.dat > obj + ${i}.val
        echo `awk -f max_ps.awk OUTFEM`/$norm1 |\
            $BC >> obj + ${i}.val
        edge.sh < temp.dat >> obj + ${i}.val

        # Create perturbation model for design var #i
        sed ${i}, \$d current.dv > temp.dat
        sed -n ${i}p current.dv |\
            awk '{print $1-`$psize[1]`}' >> temp.dat
        sed 1, ${i}d current.dv >> temp.dat
        # Calculate Performances at the Minus Perturbed Design
        model.sh `cat temp.dat`
        omesh > /dev/null
        convert.sh meshgen.fem2

        # Analyze perturbed model
        $ (ANALYZE}
```

```
        # Calculate Objectives at the perturbed design
        volume.sh < temp.dat > obj-${i}.val
        echo `awk -f max_ps.awk OUTFEM`/$norm1 | \
           $BC >> obj-${i}.val
        edge.sh ( temp.dat >> obj-${i}.val

        # Calculate Sensitivity
        set PERF_P = `cat obj + ${i}.val`
        set PERF_M = `cat obj - ${i}.val`
        set j = 1
        @  ONCON = $NCON + 1
        while ($j < = $ONCON)
           echo "(${PERF_P[$j]} - ${PERF_M[$j]}]/$psize2[$i]" | \
           $BC >> current.grad
           @ j+ +
        end
        @ i+ +
    end

    # Reconstruct parameter file
    optgrad < current.grad

    breaksw
  endsw

    # Calculate the next trial design variables
    #  (Optimizer is called) and save status
    optimize
    set INFO = `cat $INFO_-FILE`
    @ ino+ +
end

# Output the result of optimization

optout
```

## REFERENCES

Belegundu, A.D. and Zhang, S. (1992) Mesh distortion control in shape optimization. AIAA Proceedings of 33rd Structures, Structural Dynamics and Material Conference, Dallas, TX, April 13–15, 1992.

Bennet, J.A. and Botkin, M.E. (1985) Structural shape optimization with geometric problem description and adaptive mesh refinement. *AIIA Journal*, **23**(3), 458–64.

Bennet, J.A. and Botkin, M.E. (eds) (1986) *The Optimum Shape: Automated Structural Design*, GM Symposia Series, Plenum, New York and London.

Botkin, M.E. and Bennet, J.A. (1985) Shape optimization of three-dimensional folded plate structures. *AIAA Journal*, **23**(11), 1804–10.

Botkin, M.E., Yang, R.J. and Bennet, J.A. (1986) Shape optimization of three-dimensional stamped and solid automotive components, in *The Optimum Shape: Atuomated Structural Deisgn* (eds J.A. Bennet and M.E. Botkin), GM Symposia Series, Plenum, New York and London, pp. 235–61.

Braibant, V. and Fleury, C. (1984) Shape optimal design using B-splines. *Computer Methods in Applied Mechanical Engineering*, **44**, 247–67.

Choi, K.K. (1986) Shape design sensitivity analysis and optimal design of structural systems, in *Computer Aided Optimal Design: Structural and Mechanical Systems* (ed. C.A. Mota Soares), NATO ASI Series, Springer, pp. 439–92.

Choi, K.K. and Haug, E.J. (1983) Shape design sensitivity analysis of elastic structures. *Journal of Structural Mechanics*, **11**(2), 231–69.

Fleury, C. (1987) Computer aided optimal design of elsatic structures, in *Computer Aided Optimal Design: Structural and Mechanical Systems* (ed. C.A. Mota Soares), NATO ASI Series, Springer, pp. 831–900.

Haber, R.B. (1987) A new variational approach to structural shape design sensitivity analysis, in *Computer Aided Optimal Design: Structural and Mechanical Systems* (ed. C.A. Mota Soares), NATO ASI Series, Springer, pp. 573–88.

Haftka, R.T. and Grandhi, R.V. (1985) Structural shape optimization – a survey. The 26th AIAA SDM Conference, CP No. 85-0772, pp. 617–28.

Haftka, R.T. and Grandhi, R.V. (1986) Structural shape optimization – survey. *Computer Methods in Applied Mechanics and Engineering*, **57**(1), 91–106.

Haftka, R.T., Gurdal, Z. and Kamat, M.P. (1990) *Elements of Structural Optimization*, 2nd edition, Kluwer.

Haug, E.J., Choi, K.K., Hou, K.W. and Yoo, Y.M. (1983) A variational method for shape optimal design of elastic structures, in *New Directions in Optimum Structural Design* (eds E. Atrek, R.H. Gallagher, K.M. Ragsdell and O.C. Zienkiewicz), Wiley, New York, pp. 105–37.

Haug, E.J., Choi, K.K. and Komkov, V. (1985) *Design Sensitivity Analysis of Structural Systems*, Academic Press, New York.

Hörnlein, H.R.E.M. (1987) Take-off in optimum structural design, in *Computer Aided Optimal Design: Structural and Mechanical Systems* (ed C.A. Mota Soares), NATO ASI Series, Springer, 1987, pp. 901–19.

Kikuchi, N., Chung, K.Y., Torigaki, T. and Taylor, J.E. (1986) Adaptive finite element methods for shape optimization of linearly elastic structures, in *The Optimum Shape: Automated Structural Design* (eds. J.A. Bennet and M.E. Botkin), GM Symposia Series, Plenum, pp. 139–69.

Manicka, Y. and Belegundu, A.D. (1991) Analytical shape sensitivity by implicit differentiation for general velocity fields. *The First US National Congress on Computational Mechanics*, Chicago, IL, July 21–24.

Schnack, E. (1979) An optimization procedure for stress concentration by finite element technique. *International Journal of Numerical Methods in Engineering*, **14**, 115–24.

Zienkiewicz, O.C. and Campbell, J.S. (1973) Shape optimization and sequential linear programming, in *Optimal Structural Design* (eds R.H. Gallagher and O.C. Zienkiewicz), Wiley, London, Chapter 7.

Zienkiewicz, O.C. and Taylor, R.L. (1989) Mixed formulation and constraints – complete field methods, in *The Finite Element Method*, 4th edition, Volume 1, McGraw-Hill, London, Chapter 12.

# SOFTWARE BIBLIOGRAPHY

## ADS

Vanderplaats, G.N. (1985) *ADS: A FORTRAN Program for Automated Design Synthesis*, Engineering Design Optimization, Inc. Santa Bardara, CA.

Vanderplaats, G.N. (1987) *ADS Version 2.01 Manual*.

## ANSYS

*ANSYS User's Manual*, Swanson Analysis Systems Inc., Houston, PA.

## DOT

*DOT User's Manual Version 2.04*, VMA Engineering, Goleta, CA.


## ELFINI

Lecina, G. and Petiau, C. (1987) Advances in optimal design with composite materials, in *Computer Aided Optimal Design: Structural and Mechanical Systems* (ed. C.A. Mota Soares), NATO ASI Series, Springer, pp. 943–54.


## I-DEAS

*L-DEAS User's Guide Vol. I/II*, SDRC.
*L-DEAS Model Solution and Optimization User's Guide*, SDRC.
Ward, P. and Cobb, W.G.C. (1989) Application of I-DEAS optimization for the static and dynamic optimization of engineering structures, in *Computer Aided Optimum Design of Structures: Applications* (eds C.A. Brebbia and S. Hernandez), Computational Mechanics Publications, Springer, pp. 33–50.


## IDESIGN

Arora, J.S. and Haug, E.J. (1977) Methods of design sensitivity analysis in structural optimization. *AIAA Journal*, **17**(9), 970–4.
Arora, J.S. and Tseng, C.H. (1987) *User Manual for IDESIGN: Version 3.5*, Optimal Design Laboratory, College of Engineering, The University of Iowa, Iowa City, IA.
Haug, E.J. and Arora, J.S. (1979) *Applied Optimal Design*, Wiley, New York.
Lim, O.K. and Arora, J.S. (1986) An active set RQP algorithm for optimal design. *Computer Methods in Applied Mechanics and Engineering.*
Wu, C.C. and Arora, J.S. (1987) Design sensitivity analysis and of nonlinear structures, in *Computer Aided Optimal Design: Structural and Mechanical Systems* (ed. C.A. Mota Soares), NATO ASI Series, Springer, pp. 589–603.


## MATHEMATICA

Wolfram, S. (1991) *Mathematica*, 2nd edition, Addison-Wesley, Reading, MA.


## MECHANICA

*MECHANICA User's Manual*, Rasna Corp., San Jose, CA.


## NASTRAN

*MSC/NASTRAN User's and Application Manuals*, The MacNeal-Schwendler Corporation, 1984.
Fleury, C. and Liefooghe, D. (1986) Shape optimal design on an engineering workstation. *MSC/NASTRAN User's Conference*, University City, CA, March 20–21, 1986.
Nagendra, G.K. and Fleury, C. (1986) Sensitivity and optimization of composite structures using

MSC/NASTRAN. *NASA/VPI Symposium on Sensitivity Analysis in Engineering*, NASA Langley Research Center, Hampton, VA.

Vanderplaats, G.N., Miura, H., Nagendra, G. and Wallerstein, D. (1989) Optimization of large scale structures using MSC/NASTRAN, in *Computer Aided Optimum Design of Structures: Applications* (eds C.A. Brebbia and S. Hernandez), Computational Mechanics Publications, Springer, 1989, pp. 51–68.


## NISA II

*NISA II User's Manual*, EMRC, Troy, MI, 1988.


## NISAOPT

*NISAOPT User's Manual*, EMRC, Troy, MI, 1988.


## PATRAN

*PATRAN Plus User Manual*, PATRAN Division/PDA Engineering, July 1988.


## PCL

*PATRAN Command Language Guide*, PATRAN Division/PDA Engineering, September 1989.


## OASIS

Esping, B.J.D. (1985) A CAD approach to the minimum weight design problem. *International Journal of Numerical Methods in Engineering*, **21**, 1049–66.

Esping, B.J.P. (1986) The OASIS structural optimization system. *Computers and Structures*, **23**(3), 365–77.

Esping, B.J.D. and Holm, D. (1987) A CAD approach to structural optimization, in *Computer Aided Optimal Design: Structural and Mechanical Systems* (ed. C.A. Mota Soares), NATO ASI Series, Springer, pp. 987–1001.


## OPTSYS

Bråmå, T. (1989) Applications of structural optimization software in the design process, in *Computer Aided Optimum Design of Structures: Applications* (eds C.A. Brebbia and S. Hernandez), Computational Mechanics Publications, Springer, pp. 13–21.


## SAMCEF

*SAMCEF. Systeme d'Analyse des Milieux Continus per Elements Finis*, University of Liege, Belgium.

See also Fleury (1987) in the References.

## SHAPE

Atrek, E. (1989) SHAPE: a program for shape optimization of continuum structures, in *Computer Aided Optimum Design of Structures: Applications* (eds C.A. Brebbia and S. Hernandez), Computational Mechanics Publications, Springer, pp. 135–44.

## STROPT

Hariran, M., Paeng, J.K. and Belsare, S. (1988) STROPT – the structural optimization system. Proceedings of the 7th International Conference on Vehicle Structural Mechanics, Detroit, MI, April 11–13, 1988, SAE, pp. 37–47.

## STARS

Bartholomew, P. and Morris, A.J. (1984) STARS: a software package for structural optimization, in *New Directions in Optimum Structural Design* (eds E. Atrek *et al.*), Wiley, New York, 1984.

## VMCON

Crane, R.L., Hillstrom, K.E. and Minkoff, M. (1980) Solution of the general nonlinear programming problem with subroutine VMCON. *Report ANL-80-64*, Argonne National Laboratory, Argonne, Il.

# 12

# Practical optimization of structural steel frameworks

DONALD E. GRIERSON

## 12.1 INTRODUCTION

The means by which a structural design evolves from some initial form to a final form can be described as an optimization process whereby attempts are made to minimize features or to maximize benefits while satisfying design criteria reflecting performance and fabrication requirements. Many researchers over many years have pursued the quest to produce optimal designs. As early as the 1600s, Leonardo da Vinci and Galileo conducted design optimization through planned trials of models or actual full-scale structures. In the 1700s and 1800s, researchers such as Newton, Lagrange and Euler produced optimal designs using numeric calculations based on an optimality criterion that specified the strength of the structure to be uniform throughout all its parts. Maxwell in the late 1900s and Michell in the early part of the twentieth century altered structural form to achieve optimum designs having minimum structure weight. Since the advent of the computer in the last half-century there has been a veritable explosion of studies concerned with the optimal design of structural systems. One recent survey estimates that the modern literature has some 150 books and monographs and over 2500 research papers that deal with the topic of structure design optimization (Cohn, 1991). While some may argue that this long and extensive history of research has resulted in disappointingly few applications of optimization theory in professional practise, one notable exception in this regard concerns the design of structural steel building frameworks. In fact, today, there are a number of commercial software packages based on member-by-member (Computers and Structures, 1989; Research Engineers, 1987) and/or whole-structure (Grierson and Cameron, 1991) optimization that the professional designer may use to produce economical designs of such structures.

This chapter presents the details of a design synthesis strategy that employs member-by-member and whole-structure optimization in tandem to conduct computer-automated design of least-weight structural steel frameworks, where members are automatically sized using commercial standard steel sections in full conformance with governing steel design code provisions for strength–stability and stiffness. A number of design examples from professional practice are presented to illustrate the scope and effectiveness of the optimization technique.

The underlying features and functions of the corresponding computer software essential for its use in professional practice are not dealt with herein, but may be briefly summarized as follows: capability to conduct first-order and second-order analysis of planar and space frameworks; account for the strength–stability of individual members and the stiffness of the structure as a whole; clause-by-clause verification of the design in conformance with all of the relevant provisions of the governing steel design standard (e.g. American Institute of Steel Construction, 1989; Canadian Standards Association, 1989); automatic sizing of members using commercially available standard steel sections; account for a broad range of commercially available section shapes (e.g. W, T, hollow-box, single and double angle, etc.); determination of local-buckling classifications of member sections; calculation of effective length factors of members; calculation of the unbraced compression flange length of flexible members; account for fixed, pinned, spring and roller supports, as well as for pinned releases at connections and bolted connections; account for various gusset-plate thicknesses for back-to-back double-angle sections; account for strong-axis and weak-axis bending of member sections; capability to specify common section properties for groups of members, to impose limitations on section depths, and to fix the section properties of selected members; account for a variety of nodal and member load types, as well as for external effects due to temperature change and support settlement.

## 12.2 DESIGN FORMULATION

Design is herein defined as the detailed proportioning of the members of a structural steel framework to satisfy strength, stiffness and fabrication criteria. This definition assumes that the structural layout and loadings have been finalized and it only remains to size the structural components. Within this context the design process may be reduced to the following sequence of tasks: assume an initial trial design; analyze the trial structure to determine behavioral responses; compare these responses with design code specifications; select a new trial design either to remedy any code violations or to improve on the economy of the previous design (herein, structure weight is taken as the measure of design economy); reanalyze and continue the process until the design is deemed satisfactory. This iterative procedure has been traditionally carried out using trial and error, which often leads to a less-than-optimal final design after a tedious analysis–design process that owes its termination only to the skilled intuition and patience of the designer. Conducted in this way, the process becomes quite cumbersome because of the many facets of the design that must be coordinated. The initial design must be selected carefully because its quality has significant bearing on how quickly the design evolves to a satisfactory state. Design code verification is an exhaustive and time-consuming activity that requires selective application of critical code clauses at various stages of the synthesis process. The larger and more complicated the structure, the more difficult it becomes to have simultaneous concern for both design economy and design performance. On the other hand, contrary to the trial-and-error approach, computer-based optimization techniques can be effectively applied to overcome most if not all of these difficulties.

The design optimization seeks a least-weight structure while satisfying all strength and stiffness performance requirements simultaneously under static loading. For a structure having $i = 1, 2, \ldots, n$ members, and taking the cross-sectional area $a_i$ as the sizing variable for each member $i$, the general form of the design problem may be stated as

$$\text{minimize} \quad \sum_{i=1}^{n} w_i a_i \tag{12.1}$$

$$\text{subject to} \quad \underline{\delta}_j \leqslant \delta_j \leqslant \bar{\delta}_j \quad (j = 1, 2, \ldots, d) \tag{12.2}$$

$$\underline{\sigma}_k \leqslant \sigma_k \leqslant \bar{\sigma}_k \quad (k = 1, 2, \ldots, s) \tag{12.3}$$

$$a_i \in A_i \quad (i = 1, 2, \ldots, n) \tag{12.4}$$

Equation (12.1) defines the weight of the structure ($w_i$ is the weight coefficient for member $i$ given by material density × member length); equations (12.2) define $d$ constraints on displacements $\delta_j$ (under- and overscored quantities denote specified lower and upper bounds, respectively); equations (12.3) define $s$ constraints on stresses $\sigma_k$; equations (12.4) require each cross-sectional area $a_i$ to belong to the discrete set of areas $A_i \equiv \{a_1, a_2, \ldots\}_i$ prevailing for the section shape specified for member $i$ (e.g. W, T, hollow-box, double-angle, etc.).

In their present form, the displacement and stress constraints (12.2) and (12.3) are implicit nonlinear functions of the cross-sectional areas $a_i$. To facilitate computer solution, it is necessary to express these constraints as explicit functions of the sizing variables $a_i$. In fact, while introducing some approximations into the formulation, it is computationally advantageous to express these constraints as linear functions of the design variables. This suggests, then, to approximate each displacement and stress by its first-order Taylor series expansion. Moreover, since displacement and stresses in elastic structures under static loads generally vary inversely with member sizes $a_i$, it follows that better quality explicit linear constraint approximations result if the Taylor series expansions are expressed in terms of the reciprocal-sizing variables

$$x_i = \frac{1}{a_i} \quad (i = 1, 2, \ldots, n) \tag{12.5}$$

This is because displacements $\delta_j$ and stresses $\sigma_k$ generally vary directly with the $x_i$ variables and, therefore, the nonlinear relationships in $x_i$ space between $\delta_j$ and $x_i$, and between $\sigma_k$ and $x_i$, are shallower functions than are the corresponding $\delta_j$–$a_i$ and $\sigma_k$–$a_i$ relationships in $a_i$ space, i.e. the tangent-plane Taylor series expansions approximate the actual nonlinear functions more closely in $x_i$ space than in $a_i$ space.

The first-order Taylor series approximation of each displacement $\delta_j$ in terms of reciprocal sizing variables $x_i$ is

$$\delta_j = \delta_j^0 + \sum_{i=1}^{n} \left( \frac{\partial \delta_j}{\partial x_i} \right)^0 (x_i - x_i^0) \tag{12.6}$$

while that for each stress $\sigma_k$ is

$$\sigma_k = \sigma_k^0 + \sum_{i=1}^{n} \left( \frac{\partial \sigma_k}{\partial x_i} \right)^0 (x_i - x_i^0) \tag{12.7}$$

where the superscript (0) indicates known quantities for the current design stage (e.g. the initial 'trial' design). The quantities $(\partial \delta_j / \partial x_i)^0$ and $(\partial \sigma_k / \partial x_i)^0$ are displacement and stress gradients referenced to the current design, while the $x_i$ are the sizing variables for the next weight optimization. It is readily shown that

$$\delta_j^0 = \sum_{i=1}^{n} \left( \frac{\partial \delta_j}{\partial x_i} \right)^0 x_i^0 \tag{12.8}$$

$$\sigma_k^0 = \sum_{i=1}^{n} \left( \frac{\partial \sigma_k}{\partial x_i} \right)^0 x_i^0 \tag{12.9}$$

Therefore, from equations (12.1)–(12.9), the statement of the weight optimization problem becomes

$$\text{minimize} \quad \sum_{i=1}^{n} w_i / x_i \tag{12.10}$$

$$\text{subject to} \quad \underline{\delta}_j \leqslant \sum_{i=1}^{n} \left( \frac{\partial \delta_j}{\partial x_i} \right)^0 x_i \leqslant \bar{\delta}_j \quad (j = 1, 2, \ldots, d) \tag{12.11}$$

$$\underline{\sigma}_k \leqslant \sum_{i=1}^{n} \left( \frac{\partial \sigma_k}{\partial x_i} \right)^0 x_i \leqslant \bar{\sigma}_k \quad (k = 1, 2, \ldots, s) \tag{12.12}$$

$$x_i \in X_i \quad (i = 1, 2, \ldots, n) \tag{12.13}$$

where the elements of each discrete set $X_i = \{x_1, x_2, \ldots\}$ in equations (12.13) are the reciprocals of the cross-sectional areas constituting the corresponding discrete set $A_i$ in equations (12.4). The reciprocal areas $x_i$ are selected from the database of standard steel sections having the shape specified for member $i$. Table 12.1 presents the range of standard section shapes produced by North American steel

**Table 12.1** Standard steel sections

| Section shape | Section description | Number of sections | |
|---|---|---|---|
| | | USA | Canada |
| WWF | Welded wide flange | – | 83 |
| W | Wide flange | 187 | 196 |
| HP | Bearing piles | 15 | 18 |
| M | Miscellaneous beams and columns | 8 | 7 |
| S | Standard beams | 31 | 31 |
| C | Standard channels | 29 | 31 |
| MC | Miscellaneous channels | 40 | 36 |
| RHS | Rectangular hollow sections (tubing) | 127 | 65 |
| SHS | Square hollow sections (tubing) | 51 | 61 |
| CHS | Circular hollow sections (pipe) | 37 | 63 |
| WWT | Tees cut from WWF sections | – | 41 |
| WT | Tees cut from W sections | 187 | 190 |
| EL1L | Equal leg angles | 51 | 50 |
| UL1L | Unequal leg angles | 80 | 56 |
| EL2L | Two equal leg angles back to back | 40 | 50 |
| LL2L | Two unequal leg angles: long leg back to back | 48 | 56 |
| SL2L | Two unequal leg angles: short leg back to back | 48 | 56 |

mills, including the approximate number of sections available for each shape. Theoretically, the number of elements $x_i$ in each set $X_i$ is equal to the number of sections listed in Table 12.1 for the section shape prevailing for member $i$. Practically, however, this number is taken to be considerably smaller, for reasons that are discussed later.

The displacement and stress gradients in equations (12.11) and (12.12) may be evaluated using one of a number of different sensitivity analysis techniques. For example, a finite difference technique based on, say, a central difference scheme may be employed to approximate the gradients as

$$\frac{\partial \delta_j}{\partial x_i} = \frac{\delta_j(x_i + \Delta x_i) - \delta_j(x_i - \Delta x_i)}{2\Delta x_i} \tag{12.14}$$

$$\frac{\partial \sigma_k}{\partial x_i} = \frac{\sigma_k(x_i + \Delta x_i) - \sigma_k(x_i - \Delta x_i)}{2\Delta x_i} \tag{12.15}$$

Despite their apparent simplicity, however, equations (12.14) and (12.15) are computationally expensive to evaluate because they involve $2n$ analyses of the structure, where $n$ is the number of variables $x_i$ undergoing specified small perturbations $\Delta x_i$. Alternatively, an analytical technique may be employed to establish the displacement and stress gradients. To this end, displacement gradients are first considered in the following.

Each individual nodal displacement $\delta_j$ of concern to the design is related to the overall vector of nodal displacements $\mathbf{u}$ for the structure as

$$\delta_j = \mathbf{b}_j^T \mathbf{u} \tag{12.16}$$

where $\mathbf{b}_j$ is a Boolean vector that identifies $\delta_j$ from among all nodal displacements (e.g. if $\mathbf{b}_j^T = [1, 0 \ldots 0]$ then $\delta_j = u_1$) and, depending on the requirements of the design, the displacements $\mathbf{u}$ are found from either first-order or second-order analysis of the current structure by solving the equilibrium conditions

$$\mathbf{K}\mathbf{u} = \mathbf{P} \tag{12.17}$$

for each vector $\mathbf{P}$ of applied loads, where $\mathbf{K}$ is the structure stiffness matrix, to obtain

$$\mathbf{u} = \mathbf{K}^{-1}\mathbf{P} \tag{12.18}$$

Now, differentiate equation (12.16) with respect to each reciprocal-sizing variable $x_i$ to give

$$\frac{\partial \delta_j}{\partial x_i} = \mathbf{b}_j^T \frac{\partial \mathbf{u}}{\partial x_i} + \mathbf{u}^T \frac{\partial \mathbf{b}_j}{\partial x_i} = \mathbf{b}_j^T \frac{\partial \mathbf{u}}{\partial x_i} \tag{12.19}$$

which recognizes that $\partial \mathbf{b}_j / \partial x_i = 0$. Next, differentiate equation (12.17) w.r.t. $x_i$ to give

$$\mathbf{K}\frac{\partial \mathbf{u}}{\partial x_j} + \frac{\partial \mathbf{K}}{\partial x_i}\mathbf{u} = \frac{\partial \mathbf{P}}{\partial x_i} = 0 \tag{12.20}$$

which recognizes for the design of steel building frameworks that $\partial \mathbf{P}/\partial x_i = 0$ since applied loads $\mathbf{P}$ are generally taken independent of changes in member sizes $x_i$. Solve equation (12.20) to find

$$\frac{\partial \mathbf{u}}{\partial x_i} = - \mathbf{K}^{-1} \frac{\partial \mathbf{K}}{\partial x_i} \mathbf{u} \qquad (12.21)$$

The stiffness gradient $\partial \mathbf{K}/\partial x_i$ in equation (12.21) can be calculated using a finite difference approximation. Alternatively, an analytical approach can be taken to find $\partial \mathbf{K}/\partial x_i$ for steel building frameworks since for each member $i$ it is generally possible to express the member stiffness matrix $\mathbf{K}_i$ as a function of solely the reciprocal-sizing variable $x_i$, either exactly or instantaneous-approximately, such that the structure stiffness matrix can be expressed as

$$\mathbf{K} = \sum_{i=1}^{n} \mathbf{K}_i = \sum_{i=1}^{n} \frac{1}{x_i} \mathbf{K}_i^* \qquad (12.22)$$

where $\mathbf{K}_i^*$ is a constant matrix. Evidently, equation (12.22) is exact for pin-jointed truss structures that experience axial forces alone but only approximate for structures that also experience flexure, torsion and shear. In the latter case, the matrix $\mathbf{K}_i^*$ is updated from stage to stage of the design process to account for the changing relationships between the section area $a_i$ and the other section inertia parameters concerned with flexural, torsional and shear behavior. (Appendix 12A illustrates these relationships for several wide-flange sections that are typically used in steel building frameworks.)

Now, differentiate equation (12.22) w.r.t. $x_i$ to obtain the stiffness gradient as

$$\frac{\partial \mathbf{K}}{\partial x_i} = \frac{\partial}{\partial x_i} \left( \sum_{i=1}^{n} \frac{1}{x_i} \mathbf{K}_i^* \right) = \frac{\partial}{\partial x_i} \left( \frac{1}{x_i} \mathbf{K}_i^* \right) = - \frac{\mathbf{K}_i^*}{x_i^2} = - \frac{\mathbf{K}_i}{x_i} \qquad (12.23)$$

which recognizes that $\partial \mathbf{K}_r/\partial x_i = 0$ when $r \neq i$. Having $\partial \mathbf{K}/\partial x_i$ from equation (12.23), it is now possible to calculate the displacement gradient $\partial \delta_j/\partial x_i$ through equation (12.19) using either a pseudo-load or virtual-load method. The pseudo-load method proceeds by observing that the gradient vector $\partial \mathbf{u}/\partial x_i$ in equation (12.21) is actually the solution of the system of equilibrium equations for the structure subject to the vector of pseudo-loads $- (\partial \mathbf{K}/\partial x_i)\mathbf{u}$. The resulting vector $\partial \mathbf{u}/\partial x_i$ is then substituted into equation (12.19) along with the vector $\mathbf{b}_j$ to find the displacement gradient $\partial \delta_j/\partial x_i$. For $\partial \delta_j/\partial x_i$ to be found for all $i = 1, 2, \ldots, n$ sizing variables $x_i$, which is typically the case, this method involves analyzing the structure through equation (12.21) for $n$ different pseudo-load cases. Alternatively, the virtual-load method substitutes *a priori* for $\partial \mathbf{u}/\partial x_i$ from equation (12.21) into equation (12.19) to give

$$\frac{\partial \delta_j}{\partial x_i} = - \mathbf{b}_j^T \mathbf{K}^{-1} \frac{\partial \mathbf{K}}{\partial x_i} \mathbf{u} \qquad (12.24)$$

which becomes from equation (12.23)

$$\frac{\partial \delta_j}{\partial x_i} = \frac{1}{x_i} (\mathbf{b}_j^T \mathbf{K}^{-1} \mathbf{K}_i \mathbf{u}) \qquad (12.25)$$

It is then observed that if a unit virtual load is associated with the real displacement $\delta_j$ in equation (12.16), by the principle of virtual work equivalence the vector $\mathbf{b}_j$ may be viewed as a vector of virtual loads associated with the vector of real displacements **u** in equation (12.16). The vector of virtual nodal displacements $\mathbf{u}_j$ corresponding to the virtual loads $\mathbf{b}_j$ is found as

$$\mathbf{u}_j = \mathbf{K}^{-1}\mathbf{b}_j \tag{12.26}$$

and then substituted into equation (12.25) to find the displacement gradient as

$$\frac{\partial \delta_j}{\partial x_i} = \frac{1}{x_i}(\mathbf{u}_j^{\mathrm{T}}\mathbf{K}_i\mathbf{u}) \tag{12.27}$$

Contrary to the pseudo-load method, the virtual-load method only involves a single analysis of the structure for the virtual loads $\mathbf{b}_j$ to find the gradient $\partial \delta_j/\partial x_i$ for all $i = 1, 2, \ldots, n$ sizing variables $x_i$. On the other hand, if the gradients of $j = 1, 2, \ldots, m$ different displacements $\delta_j$ are to be found, which is typically the case, then the virtual-load method will involve $m$ analyses of the structure while the pseudo-load method will still only involve $n$ analyses regardless of the number of displacements of concern to the design of the structure. Therefore, assuming that the method requiring the fewest analyses of the structure is the most efficient, the pseudo-load method should be applied if the number $n$ of sizing variables is less than the number $m$ of response gradients. Otherwise, the virtual-load method should be applied if $n > m$. In actual fact, either method is quite efficient since each analysis of the structure to establish response gradients is really nothing more than a back-substitution operation for different pseudo- or virtual loads since the inverse or factorized stiffness matrix $\mathbf{K}^{-1}$ is already available from the basic analysis equation (12.18) that determines the displacements **u** due to the actual loading **P**. The virtual-load method (12.27) is adopted to calculate displacement gradients for the design examples presented later in the chapter.

Stress gradients are similarly calculated analytically using the virtual-load method described in the foregoing for displacement gradients. Here, each individual member stress $\sigma_k$ of concern to the design is related to the overall vector of nodal displacements **u** for the structure as

$$\sigma_k = \mathbf{t}_k^{\mathrm{T}}\mathbf{u} \tag{12.28}$$

where the vector $\mathbf{t}_k$ is row $k$ of the global-axis stress matrix for the member associated with the stress $\sigma_k$. The vector $\mathbf{t}_k$ is invariant for truss structures composed of axial members alone. For flexural structures, however, the vector $\mathbf{t}_k$ is a function of the neutral-axis position for the member section associated with the stress $\sigma_k$, and therefore it varies as the section design changes over the synthesis history. Upon viewing $\mathbf{t}_k$ as a vector of virtual loads applied to the structure to cause virtual nodal displacements

$$\mathbf{u}_k = \mathbf{K}^{-1}\mathbf{t}_k \tag{12.29}$$

the gradient of the stress $\sigma_k$ with respect to change in each reciprocal-sizing variable $x_i$ is found as

$$\frac{\partial \sigma_k}{\partial x_i} = \frac{1}{x_i}(\mathbf{u}_k^{\mathrm{T}}\mathbf{K}_i\mathbf{u}) \tag{12.30}$$

Equation (12.30) is based on the assumption that the vector $\mathbf{t}_k$ is invariant with changes in $x_i$ for the current design stage, for both truss and flexural structures. The error inherent in this assumption for flexural structures becomes negligible as the synthesis process converges to a final design state where no changes take place in the sizing variable $x_i$ for a number of successive design stages.

An illustration of the calculation of displacement and stress gradients using the virtual-load method is presented in Appendix 12B for a simple truss framework.

## 12.3 SYNTHESIS PROCESS

By virtue of the approximate nature of the performance constraints equations (12.11) and (12.12), the design synthesis is conducted through an iterative process that involves solving the weight optimization problem (12.10)–(12.13) during each design cycle. Attention is focused in the following on presenting the details of the iterative synthesis process that specifically pertain to the design of practical steel frameworks using standard steel sections in conformance with the provisions of a governing steel code (e.g. American Institute of Steel Construction, 1989; Canadian Standards Association, 1989).

The synthesis process begins by selecting an initial 'trial' design of the structure. There are several ways in which this can be done. One method is to employ some heuristics based on designer experience to derive an acceptable preliminary design. The advantage of this approach is that it provides an initial design that is reasonably well proportioned. A disadvantage, particularly for complex structures, is that it may not be readily possible to find an acceptable initial design using simple heuristics alone. Another approach is to identify the standard section shape for each member (e.g. W, T, etc.) and then to select the largest section from the corresponding database of available sections (i.e. as commercially supplied by steel mills). The advantage of this approach is that it is simple to apply to establish an initial design. Also, as the largest available sections are being used, this approach clearly signals that the design is probably ill-posed if the initial design is found to be infeasible for the performance requirements of the governing steel design code. A disadvantage of this approach is that the initial design is generally not very well proportioned. However, a reasonably proportioned design is generally found after one cycle of the synthesis process by solving the weight optimization problem for continuous-valued sizing variables by temporarily replacing the discrete sizing constraints (12.13) with the simple bounding conditions

$$\underline{x}_i \leqslant x_i \leqslant \bar{x}_i \quad (i = 1, 2, \dots, n) \tag{12.31}$$

where $\underline{x}_i = 1/\bar{a}_i$ and $\bar{x}_i = 1/\underline{a}_i$, in which $\bar{a}_i$ and $\underline{a}_i$ are the largest and smallest section areas available for member $i$, respectively. This latter approach to establish an initial design is adopted for the design examples presented later in the chapter.

Having a reasonably well-proportioned initial design, structural analysis is conducted to determine the corresponding elastic displacements and stresses for each load case. Either first-order or second-order analysis is performed depending on the requirements of the governing steel design code. If first-order analysis is the basis for the design calculations then effective length factors greater than unity are adopted for flexural members; if second-order analysis is employed then all member effective length factors are typically set equal to unity.

Having the stresses and displacements from structural analysis, the database of standard steel sections is searched, member by member, to find the least-weight sections that satisfy the stress-related strength–stability requirements of the governing steel design code while simultaneously ensuring that all displacement-related stiffness constraints are satisfied. This searching activity, hereafter referred to as member-by-member optimization, has two distinct aspects. The concern for strength–stability involves selecting standard sections taking into account axial tension and compression, bending, shear and torsion effects, and combinations of these effects, for individual members considered in isolation from the structure. This aspect of the search is somewhat more straightforward than that when displacements are of concern where member sections must be economically selected while at the same time ensuring that the stiffness of the assembled structure is satisfactory. To this latter end, an effective technique adopted herein is based on calculating displacement gradients (e.g. Appendix 12B) so as to establish the relative influence that individual members have on displacements, which then allows member sections to be selected while having concern for both displacements and economy (Cameron, 1989; Cameron *et al.*, 1992).

Having the member sections resulting from the member-by-member optimization procedure, structural analysis is again conducted and the strength–stability and stiffness requirements for the design are checked for the resulting stresses and displacements. If no design violations are detected, the synthesis process continues on to the next stage involving formal optimization. Otherwise, the design is infeasible and member-by-member optimization is conducted again.

The formal optimization stage of each cycle of the synthesis process involves formulating and solving the weight optimization problem posed by equations (12.10)–(12.13). With a view to computational efficiency, it is desirable to delete temporarily relatively inactive displacement and stress constraints from the formulation. For steel frameworks, this generally results in a significant reduction in the size of the active constraint set and thus allows either for faster computation for a certain size structure, or for the design of a larger structure than would be otherwise possible on a computer with fixed memory capacity. Deleted constraints are continuously monitored and added to the constraint set if and when they become active for a subsequent stage of the synthesis process. Moreover, to overcome possible design history oscillations that impede or thwart convergence of the synthesis process, it is recommended to retain continuously a constraint in the active set even if it subsequently becomes passive to the design (Grierson and Lee, 1984).

The weight coefficients $w_i$ in the objective function (12.10) remain constant throughout the synthesis process. This is also true for the displacement bounds $\bar{\delta}_j$ and $\underline{\delta}_j$ in equations (12.11). On the other hand, the allowable tensile and compressive stress bounds in equations (12.12) must be updated for each design cycle because corresponding axial, flexural, shear and torsional capacities are dependent on unbraced member lengths and the geometry of member sections, both of which vary over the synthesis history. For example, the lengths of unbraced compression flanges depend on the bending moment distribution, which changes with modifications in the stiffness distribution for the structure. Alternatively, the width–thickness ratios of section plate elements change as different standard sections are selected for a member from stage to stage of the synthesis process. The tensile stress bound $\bar{\sigma}_k$ accounts for the type of member connection (bolted, etc.), while the compressive stress bound $\underline{\sigma}_k$ depends on the stress state (axial, flexural, combined, etc.) and accounts for both local section buckling and overall member buckling.

For the current standard section sizes for the members of the structure, sensitivity analysis is conducted to establish the gradients $(\partial \delta_j / \partial x_i)^0$ and $(\partial \sigma_k / \partial x_i)^0$ of the active displacement and stress constraints in equations (12.11) and (12.12). It is important to note that the computational effort required to find these gradients is quite nominal since direct use is made of the inverse or factorized form of the structure stiffness matrix that has been previously established by the conventional analysis conducted for the current cycle of the synthesis process (e.g. Appendix 12B).

To complete the formulation of the weight optimization problem of equations (12.10)–(12.13) for the current design stage, it remains to update each set of discrete sizing variables $X_i$ in equations (12.13) from the corresponding database of standard steel sections to reflect only those sections having adequate strength–stability and stiffness properties for the current set of active stress and displacement constraints. In fact, the total number of sections reflected in each set $X_i$ is limited so as to restrict the extent of movement in the design space during solution of the weight optimization problem, thereby preserving the integrity of the first-order displacement and stress gradients in equations (12.11) and (12.12). The updated subset of standard sections for each member (or fabrication group of members) is arranged such as to bracket the current design section, thereby allowing for the selection of either a smaller or a larger section as the outcome of the weight optimization that solves equations (12.10)–(12.13).

## 12.4 OPTIMIZATION ALGORITHM

The solution of the least-weight design problem (12.10)–(12.13), hereafter referred to as the formal optimization stage of each design cycle, can be found using one of a number of different algorithms. An optimality criteria (OC) method involves temporarily replacing the discrete sizing constraints (12.13) with the simple bounding conditions (12.31) and deriving an optimality criterion based on the stationary conditions for the Lagrangian function involving equations (12.10)–(12.12). The Lagrange variables, which identify the active performance constraints for the design, are then determined along with the sizing variables $x_i$ through a recursive algorithm that converges after a number of cycles to a nonstandard design (i.e. a set of $x_i$ values that do not correspond to standard steel sections). A pseudo-discrete algorithm based on the optimality criterion is then applied to assign progressively a standard section to each of the members. This OC method is particularly well suited to the design of structures for which the number of performance constraints is small compared with the number of sizing variables, such as for the lateral stiffness design of tall steel frameworks subject to inter-story drift constraints (Chan, 1993).

Another optimization algorithm for solving the design problem (12.10)–(12.13) is a generalized optimality criteria (GOC) technique that is often referred to in the literature as the dual method (Schmit and Fleury, 1979; Fleury, 1979). The details of the GOC method, which is used for the design examples presented later in this chapter, are briefly developed in the following. To simplify the associated discussion it is convenient to express the optimization problem (12.10)–(12.13) in the following form:

$$\text{minimize} \quad \sum_{i=1}^{n} w_i/x_i \tag{12.32}$$

$$\text{subject to} \quad \sum_{i=1}^{n} c_{ir}x_i \leqslant \bar{g}_r \quad (r = 1, 2, \ldots, m) \tag{12.33}$$

$$x_i \in X_i \quad (i = 1, 2, \ldots, n) \tag{12.34}$$

where equations (12.33) are a concise expression of the displacement and stress constraints (12.11) and (12.12): the gradients $c_{ir} \equiv \pm(\partial \delta_j/\partial x_i)^0, \pm(\partial \sigma_k/\partial x_i)^0$; the bounds $\bar{g}_r \equiv \underline{\delta}_j, \bar{\delta}_j, \underline{\sigma}_k, \bar{\sigma}_k$; the total number of constraints $m = d + s$.

To commence the development of the GOC method, temporarily ignore the discrete sizing constraints (12.34) and formulate the Lagrangian function

$$\mathscr{L}(\mathbf{x}, \lambda) = \sum_{i=1}^{n} \frac{w_i}{x_i} + \sum_{r=1}^{m} \lambda_r \left( \sum_{i=1}^{n} c_{ir}x_i - \bar{g}_r \right) \tag{12.35}$$

where the Lagrange variables are sign restricted as

$$\lambda_r \geqslant 0 \quad (r = 1, 2, \ldots, m) \tag{12.36}$$

The GOC method proceeds by finding a saddle-point $(\mathbf{x}^*, \lambda^*)$ of equation (12.35) while accounting for the discrete sizing constraints (12.34), which in turn provides a solution $\mathbf{x}^*$ to the original optimization problem (12.32)–(12.34) (Zangwill, 1969). A saddle-point of equation (12.35) is given by a solution of the min–max problem

$$(\mathbf{x}^*, \lambda^*) = \max_{\lambda} \left[ \min_{\mathbf{x}} \mathscr{L}(\mathbf{x}, \lambda) \right] \tag{12.37}$$

The problem posed by equation (12.37) is solved by first minimizing the Lagrangian function over the $\mathbf{x}$ variables to establish a function in terms of $\lambda$ variables alone, i.e.

$$\mathscr{L}_m(\lambda) = \min_{\mathbf{x}} \mathscr{L}(\mathbf{x}, \lambda) = \min_{\mathbf{x}} \sum_{i=1}^{n} L_i(x_i, \lambda) \tag{12.38}$$

where, from equation (12.35), the functions

$$L_i(x_i, \lambda) = \frac{w_i}{x_i} + \sum_{r=1}^{m} \lambda_r(c_{ir}x_i - \bar{g}_r) \quad (i = 1, 2, \ldots, n) \tag{12.39}$$

are each expressed in terms of a single sizing variable $x_i$.

Equation (12.38), which is called the dual function, is readily established by separately minimizing each function $L_i(x_i, \lambda)$ in equations (12.39) over the corresponding discrete set of values $X_i$ for sizing variable $x_i$ and then summing, i.e.

$$\mathscr{L}_m(\lambda) = \sum_{i=1}^{n} \min_{x_i \in X_i} L_i(x_i, \lambda) \tag{12.40}$$

As the $\lambda$ variables vary in value, the value of the sizing variable $x_i$ for which the corresponding function $L_i(x_i, \lambda)$ is minimum may shift from one discrete value $x_i^k$ to the next discrete value $x_i^{k+1}$ in the set $X_i$. When this happens, continuity of the dual function $\mathscr{L}_m(\lambda)$ is maintained by requiring that

$$L_i(x_i^k, \lambda) = L_i(x_i^{k+1}, \lambda) \tag{12.41}$$

which reduces to, from equations (12.39),

$$\sum_{r=1}^{m} \lambda_r c_{ir} = \frac{w_i}{x_i^k x_i^{k+1}} \qquad (12.42)$$

From equation (12.42), the specific discrete value of each sizing variable $x_i$ for which the corresponding function $L_i(x_i, \lambda)$ is minimum is given by

$$x_i^0 = x_i^k \qquad (12.43)$$

if

$$\frac{w_i}{x_i^{k-1} x_i^k} < \sum_{r=1}^{m} \lambda_r^0 c_{ir} < \frac{w_i}{x_i^k x_i^{k+1}} \qquad (12.44)$$

where the values of the Lagrange variables $\lambda_r^0$ are initially assumed (e.g. if constraint $m$ of equations (12.33) is the most active for the initial design of the structure, set $\lambda_m^0 = 1$ and $\lambda_r^0 = 0\,(r = 1, 2, \ldots, m-1)$); thereafter, the $\lambda_r^0$ values are known from the previous iteration step of the maximization process conducted to solve equation (12.37), as outlined in the following.

Having established the dual function $\mathcal{L}_m(\lambda)$, the GOC method completes the solution of equation (12.37) by solving the maximization problem, from equations (12.36), (12.39) and (12.40),

$$\text{maximize} \quad \mathcal{L}_m(\lambda) = \sum_{i=1}^{n} \frac{w_i}{x_i^0} + \sum_{r=1}^{m} \lambda_r \left( \sum_{i=1}^{n} c_{ir} x_i^0 - \bar{g}_r \right) \qquad (12.45)$$

$$\text{subject to} \quad \lambda_r \geqslant 0 \quad (r = 1, 2, \ldots, m) \qquad (12.46)$$

where the $x_i^0$ values are known from equations (12.43) and (12.44). The problem posed by equations (12.45) and (12.46) is solved using a gradient projection method (Zangwill, 1969; Haftka and Gürdal, 1992) to update progressively the Lagrange variables as

$$\lambda = \lambda^0 + \alpha s \qquad (12.47)$$

where $s$ is the search direction, $\alpha$ the step length and $\lambda^0$ the variable values found at the end of the previous search step. Note from equations (12.43) and (12.44) that the $x_i^0$ values in equation (12.45) are subject to change as the $\lambda_r^0$ values change. The gradient projection search (12.47) progressively accounts for the bounds on the $\lambda_r$ variables, equations (12.46), and the discontinuities in the $x_i$ variables, equations (12.44), to find the solution point $(\mathbf{x}^*, \lambda^*)$ for which equation (12.45) is maximized. The point $\mathbf{x}^*$ is the solution of the design optimization problem (12.32)–(12.34).

The GOC method for discrete optimization is illustrated in Appendix 12C for the design of a simple truss structure.

## 12.5 EXAMPLE APPLICATIONS

The least-weight designs of five different steel frameworks are presented in the following. The designs are conducted using a software system based on the optimization technology described in the foregoing (Grierson and Cameron, 1991). The first design example has been selected more for its illustrative value than for its practicality and is presented in some detail. The remaining four design examples are presented somewhat more briefly and are representative of actual designs carried out in professional practice by structural engineering firms. (The names of the individuals and firms contributing particular design examples are acknowledged at the end of the chapter.)

EXAMPLE 12.1

The three-bay by two-bay by two-story steel framework in Fig. 12.1 is to be designed in accordance with the strength–stability and stiffness provisions of the American Allowable Stress Design steel code AISC-89 (American Institute of Steel Construction, 1989). The bay spacings and story heights in the $X, Z$ and $Y$ axis directions are 20, 16.4 and 10 feet, respectively.

First-order effects alone are to be accounted for, and sidesway is to be permitted. The in-plane and out-of-plane effective length factors $K_x$ and $K_y$ for the continuously connected $X$ direction girders and the $Y$ direction columns are to be calculated on the basis of their end stiffnesses (Johnston, 1976). The effective length factors for the pinned-end $Z$ direction girders and the diagonal bracing

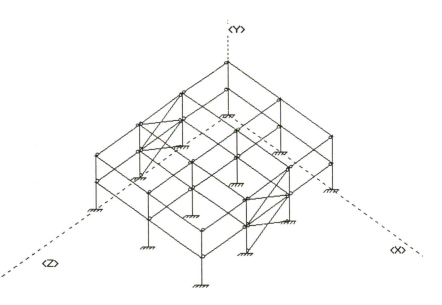

FABRICATION GROUPS

GX1-1 = First Story X-Girders
GX2-2 = Second Story X-Girders
GZ1-1 = First Story Z-Girders
GZ2-2 = Second Story Z-Girders

IC1-2 = Interior Columns
EC1-2 = Exterior Columns
BRACE = Diagonal Braces

**Fig. 12.1** Steel building framework.

members are specified to be $K_x = K_y = 1.0$. The maximum allowable slenderness ratios for each member are specified to be $KL/r = 200$ in compression and $KL/r = 300$ in tension (where $L$ is the member length and $r$ the section radius of gyration).

Member sections are to be selected from available AISC standard steel sections (American Institute of Steel Construction, 1989). All column and girder members are to have wide-flange (W) shapes, while all diagonal bracing members are to have square hollow section (SHS) shapes. For these specified section types, the Young's modulus for the steel is $29\,000\,\text{klbf in}^{-2}$, while the shear modulus is $11\,200\,\text{klbf in}^{-2}$, the yield stress is $36\,\text{klbf in}^{-2}$ and $50\,\text{klbf in}^{-2}$ for the W and SHS sections, respectively, and the ultimate stress is $58\,\text{klbf in}^{-2}$ and $62\,\text{klbf in}^{-2}$ for the W and SHS sections, respectively.

To satisfy practical fabrication requirements, the members of the structure are to be grouped as follows (Fig. 12.1). (a) The eight continuously connected girder members located in the $X$ axis direction at story level 1 (i.e. the second floor) are all to have the same W-shape section (the depth and size to be determined by the design). This fabrication group is designated as GX1-1. (b) The eight continuously connected girder members located in the $X$ axis direction at story level 2 (i.e. the roof) are all to have the same W-shape section. This fabrication group is designated as GX2-2. (c) The nine pinned-end girder members located in the $Z$ axis direction at story level 1 are all to have the W-shape section $W10 \times 33$ (i.e. fixed for the design). This fabrication group is designated as GZ1-1. (d) The nine pinned-end girder members located in the $Z$ axis direction at story level 2 are all to have the W-shape section $W10 \times 33$. This fabrication group is designated as GZ2-2. (e) The four interior column members in the $Y$ axis direction (i.e. two interior columns for each of the two story levels) are all to have the same W-shape section with a nominal depth of 12 in (the actual section size to be determined by the design). This fabrication group is designated as ICI-2. (f) The twenty exterior column members in the $Y$ axis direction (i.e. ten exterior columns for each of the two story levels) are all to have the same W-shape section with a nominal depth between 10 and 14 in (the appropriate depth and size to be determined by the design). This fabrication group is designated as EC1-2. (g) The eight pinned-end bracing members located in the two exterior $Z$–$Y$ faces of the structure are all to have the same SHS-shape section (the size to be determined by the design). This fabrication group is designated as BRACE.

Member cross-sections are to be oriented as follows (Fig. 12.1): (a) section webs for $X$ direction girders are to lie in a vertical $X$–$Y$ plane such that strong axis bending takes place about the $Z$ axis; (b) section webs for $Y$ direction columns are to lie in a vertical $X$–$Y$ plane such that strong-axis bending takes place about the $Z$ axis; (c) section webs for $Z$ direction girders are to lie in a vertical $Y$–$Z$ plane such that strong-axis bending takes place about the $X$ axis.

The service loading on the structure is (Fig. 12.1) as follows. (a) Wind loading acts horizontally in the positive $X$ axis direction of magnitude $11.0\,\text{klbf in}^{-2}$ at each exterior node of the $Z$–$Y$ face of the structure for which the $X$ coordinate is 0. This loading is designated as WIND:X. (b) Wind loading acts horizontally in the positive $Z$ axis direction of magnitude $9.0\,\text{klbf in}^{-2}$ at each exterior node of the $X$–$Y$ face of the structure for which the $Z$ coordinate is 0. This loading is designated as WIND:Z. (c) Gravity uniformly distributed loading acts vertically in the negative $Y$ axis direction of magnitude $1.4\,\text{klbf in}^{-2}\,\text{ft.}^{-1}$ on each of the $X$

axis girders in the second floor and the roof. This loading is designated as GRAVITY:X. (d) Gravity point loading acts vertically in the negative $Y$ axis direction of magnitude 14.5 klbf in$^{-2}$ at midspan of each of the $Z$ axis girders in the second floor and the roof. This loading is designated as GRAVITY:Z.

Two load combinations are considered for the design of the structure, as follows:

$$\text{GRAVITY:X} + \text{GRAVITY:Z} + \text{WIND:Z}$$
$$\text{GRAVITY:X} + \text{GRAVITY:Z} + \text{WIND:X}$$

Both load combinations define service loading schemes for which the stress related provisions of the AISC-ASD design code are to be satisfied (American Institute of Steel Construction, 1989). The second load combination also defines the service loading scheme for which the lateral sway in the $X$ axis direction is limited to 0.8 in at the roof level (i.e. $h/300$).

The least-weight design of the framework is conducted using the synthesis process involving formal and member-by-member optimization described earlier in this chapter. The iterative synthesis history and final design for the steel framework are shown in Fig. 12.2. The AISC standard steel section shapes and designations given in Fig. 12.2 apply for the seven fabrication groups defined in Fig. 12.1. The initial structure weight given in Fig. 12.2 refers to an arbitrarily

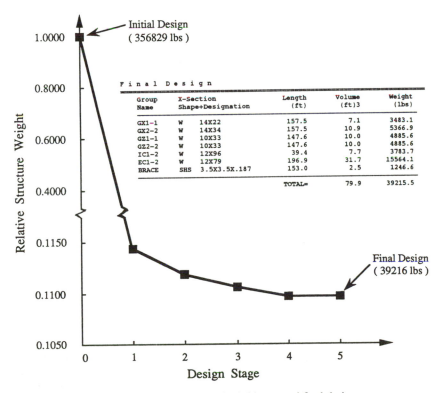

Fig. 12.2 Example 12.1: synthesis history and final design.

selected conservative design (i.e. largest section sizes satisfying fabrication requirements). As shown in Fig. 12.2, the optimization process converged after five iterations to a least-weight final design satisfying the imposed $X$ axis sway limitation and all relevant strength/stability requirements of the AISC–ASD steel design code (American Institute of Steel Construction, 1989; Cameron, Xu and Grierson, 1991).

The design optimization involved 7 sizing variables, 4 displacement constraints and 364 stress constraints, and required approximately 90 minutes execution time on a 386-based microcomputer. In fact, if formal optimization is excluded and member-by-member optimization is alone applied, a very similar final design is found after only approximately 5 minutes of computer time. However, this result is peculiar to this example as the use of formal optimization generally yields lower-weight designs compared with those produced by member-by-member optimization alone. In fact, as is demonstrated for the following example, the two optimization strategies generally work in tandem to produce a design that is an improvement on that produced by either strategy alone.

---

## EXAMPLE 12.2

The planar steel framework with trussed roof and pinned supports in Fig. 12.3 is part of a mill crane building and is to be designed in accordance with user-specified displacement constraints and the strength/stability provisions of the Canadian limit-states design steel design code CAN3-S16.1-M84 (Canadian Standards Association, 1989). Member sections are to be selected from the Canadian database of standard steel sections.

The crane framework is subject to 14 design load cases where, as indicated in Fig. 12.3, each load case is a particular combination of dead, live, wind and crane loading, coupled with a thermal effect caused by a 50 °C temperature change for the bottom chord members of the roof truss (due to elevated temperatures within the building enclosure). Load cases 1 to 6 are 1.25 dead + 1.5 live + temperature + 1.5 crane load at nodes 27 to 32; load cases 7 to 12 are 1.25 dead + 1.5 wind + temperature + 1.5 crane load at nodes 27 to 32; load case 13 is live + crane load at node 32; load case 14 is wind + crane load at node 32. Vertical displacement at node 32 is limited to 50 mm for load case 13, while horizontal displacement at node 26 is limited to 20 mm for load case 14.

Out-of-plane bracing is applied at all 56 nodes of the framework, including at the intermediate column nodes 51, 52 and 53 (Fig. 12.3). The framework has 103 members, consisting of 9 column members and 94 roof truss members. For the roof truss, the vertical and diagonal members are pin connected, while the members constituting the top and bottom chords are continuous connected over each of the two spans and pin connected only at the exterior and interior columns.

The column members and the top and bottom chord members in the roof truss are specified to have wide-flange (W) section shapes oriented such that bending takes place about the major axis and having compression flange bracing only at members ends. The vertical and diagonal members in the roof truss are

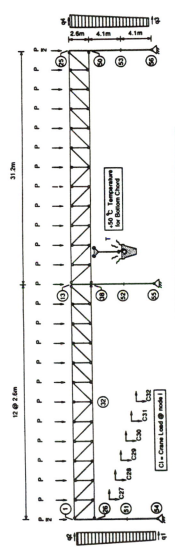

| LOAD CASE | P = Dead + Live | Crane | Wind = q1, q2 | Wind = q3, q4 | T (°C) |
|---|---|---|---|---|---|
| 1 | 1.25D + 1.5L | 1.5C27 | 1.5w1, 1.5w2 | 1.5w3, 1.5w4 | +50 |
| 2 | 1.25D + 1.5L | 1.5C28 | 1.5w1, 1.5w2 | 1.5w3, 1.5w4 | +50 |
| 3 | 1.25D + 1.5L | 1.5C29 | | | +50 |
| 4 | 1.25D + 1.5L | 1.5C30 | | | +50 |
| 5 | 1.25D + 1.5L | 1.5C31 | | | +50 |
| 6 | 1.25D + 1.5L | 1.5C32 | | | +50 |
| 7 | 1.25D | 1.5C27 | 1.5w1, 1.5w2 | 1.5w3, 1.5w4 | +50 |
| 8 | 1.25D | 1.5C28 | 1.5w1, 1.5w2 | 1.5w3, 1.5w4 | +50 |
| 9 | 1.25D | 1.5C29 | 1.5w1, 1.5w2 | 1.5w3, 1.5w4 | +50 |
| 10 | 1.25D | 1.5C30 | 1.5w1, 1.5w2 | 1.5w3, 1.5w4 | +50 |
| 11 | 1.25D | 1.5C31 | 1.5w1, 1.5w2 | 1.5w3, 1.5w4 | +50 |
| 12 | 1.25D | 1.5C32 | | | +50 |
| 13 | | C32 | | | |
| 14 | L | C32 | w1, w2 | w3, w4 | |

Where:  D = 40.0 kN,   L = 30.0 kN,   Crane = 80.0 kN (down)  & 15.0 kN (right)
w1 = 1.5 kN/m,   w2 = 2.5 kN/m,   w3 = 1.05 kN/m,   w4 = 1.75 kN/m
( 1 m = 3.28 feet,   1 kN = 0.225 kips,   1 kN/m = 68.5 plf )

+50 °C Temperature for Bottom Chord

Ci = Crane Load @ node i

**Fig. 12.3** Mill crane building framework.

specified to have double-angle section shapes with long legs back to back (⌐⌐)
separated by a 10 mm thick gusset plate.

To satisfy fabrication requirements, six groups of members are identified for
which all members in each group are specified to have common section properties.
The six fabrication groups are exterior column members (COLext), interior
column members (COLint), top chord truss members (CHtop), bottom chord
truss members (CHbot), vertical truss members (Vert), and diagonal truss
members (Diag).

The steel material properties for the design are as follows: Young's modulus,
200 000 MPa; shear modulus, 77 000 MPa; yield stress, 300 MPa; ultimate stress,
450 MPa; coefficient of thermal expansion, $0.117 \times 10^{-4}\,°C^{-1}$.

Second-order (P–delta) analysis is adopted as the underlying basis for the
design. The in-plane effective length factor is specified to be $K_x = 1.0$ for each
roof truss member and top column member, while $K_x = 2.0$ is specified for each
lower column member. The out-of-plane effective length factor is specified to be
$K_y = 1.0$ for each member. The maximum allowable slenderness ratios for each
member are specified to be $KL/r = 200$ in compression and $KL/r = 300$ in tension.

The least-weight design of the framework is conducted using the previously
described optimization process. Table 12.2 presents the design history results
from the initial design cycle 1 to the final design cycle 3 for the six member
fabrication groups for the framework (Grierson and Cameron, 1989). The
structure weight is given for each design stage along with the response ratios for
the most critical strength and displacement constraints for the corresponding
design. (The response ratio = actual/allowable response and, therefore, the maximum allowable response ratio for feasibility is unity.) The framework design
found at the beginning of design cycle 1 by formal optimization weighs 19 665 kg
but the response ratio is 1.292 for a strength constraint, which implies 29.2%
infeasibility. (Recall that the approximate nature of the constraints for the formal
optimization may result in designs that are infeasible relative to the actual
constraint functions.) The member-by-member optimization restores design
feasibility but, in the process, increases the structure weight to 19 725 kg. The
resulting design is nearly active for strength response (ratio = 0.978) but somewhat
inactive for displacement response (ratio = 0.638), suggesting that it may be
possible to redistribute member stiffnesses to achieve further weight reduction
without violating the displacement constraint. For the feasible design from cycle
1 as its starting basis, the formal optimization for cycle 2 determines a design
that weighs 17 110 kg but which is 13.8% infeasible for a displacement constraint.
The member-by-member optimization restores design feasibility but, in the
process, increases the structure weight to 17 500 kg. The resulting design is nearly
active for both strength response (ratio = 0.996) and displacement response
(ratio = 0.910), suggesting that the synthesis process is converging to a least-
weight structure. For the feasible design from cycle 2 as its starting basis, the
formal optimization for cycle 3 determines a design that weighs 16 795 kg but
which is 4.8% infeasible for a strength constraint. The member-by-member
optimization determines a heavier design weighing 17 190 kg, but, now, the critical
strength constraint is identically satisfied (ratio = 1.000) and the critical displacement constraint is nearly active (ratio = 0.969). The tight degree of satisfaction
of both the strength and displacement constraints indicates that the synthesis
process has converged to a least-weight feasible design of the framework, as given

**Table 12.2** Design history results for mill crane building framework

| Design cycle | 1 Section designation | | 2 Section designation | | 3 Section designation | |
|---|---|---|---|---|---|---|
| | Formal optimization | Member-by-member optimization | Formal optimization | Member-by-member optimization | Formal optimization | Member-by-member optimization |
| Member group | | | | | | |
| COLext | W610×174 | W610×174 | W610×113 | W610×140 | W610×125 | W610×125 |
| COLint | W610×195 | W610×195 | W530×123 | W610×125 | W610×125 | W610×125 |
| CHtop | W360×57 | W250×49 | W250×49 | W250×49 | W250×49 | W250×49 |
| CHbot | W410×85 | W310×97 | W250×89 | W310×86 | W250×80 | W310×86 |
| Vert | JL 100×90×10 | JL 125×90×10 | JL 125×90×10 | JL 125×90×10 | JL 125×90×10 | JL 125×90×10 |
| Diag | JL 150×100×10 | JL 100×75×13 | JL 125×90×10 | JL 125×90×10 | JL 125×90×10 | JL 125×90×10 |
| Structure weight (kg) | 19 665 | 19 725 | 17 110 | 17 500 | 16 795 | 17 190 |
| Critical strength response ratio | 1.292 | 0.978 | 0.994 | 0.996 | 1.048 | 1.000 |
| Critical displacement response ratio | 0.632 | 0.638 | 1.138 | 0.910 | 0.971 | 0.969 |

Note: W$d \times m \equiv$ wide-flange section; depth $d$ (mm) and mass $m$ (kg m$^{-1}$).
JL $l_1 \times l_2 \times t \equiv$ double-angle section with long legs back to back: long leg length $l_1$ (mm), short leg length $l_2$ (mm), and thickness $t$ (mm).

in the last column of Table 12.2. The iterative design process involved 2886 strength and displacement constraints and required approximately 120 minutes execution time on an IBM PS/2 Model 30 microcomputer.

As an illustration that formal optimization and member-by-member optimization together produce the lightest design, the design in the first column in Table 12.2 was taken as the starting basis for member-by-member optimization applied alone and the iterative synthesis process converged after three cycles to a feasible design weighing 18 070 kg. This design is 18 070–17 190 kg = 880 kg or 5.1% heavier that the final design in Table 12.2 found using formal and member-by-member optimization in tandem.

## EXAMPLE 12.3

This example concerns the design of an individual component part of the materials handling plant shown in Fig. 12.4. Material is input from the ground by conveyor

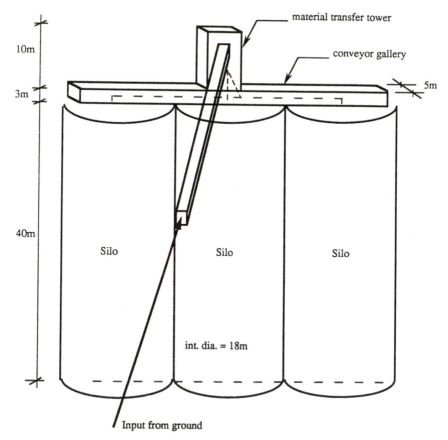

**Fig. 12.4** Materials handling plant.

to a transfer tower located above the central silo, where it is then distributed to the three silos through a conveyor gallery. The supporting roof system for the central silo shown in Fig. 12.5 consists of eleven stringers supported in span by two girders. As shown in Fig. 12.6, the roof girders are supported on the outer silo wall by short connector beams splice-connected at the girder ends.

It is required to design the silo roof girder and end-connectors shown in Fig. 12.6 in accordance with the provisions of the CAN3-S16.1-M89 steel design code (Canadian Standards Association, 1989). Both the girder and connectors are to have welded wide-flange (WWF) sections oriented such that bending takes place about the major axis. The sections are to be selected from the database of Canadian standard WWF steel sections. The girder is required to have a nominal depth between 1000 mm and 1200 mm, while each connector is required to have a nominal depth of 500 mm.

The steel material properties for the design of the girder and connectors are as follows: Youngs modulus, 200 000 MPa; shear modulus, 77 000 MPa; yield stress, 300 MPa; and ultimate stress, 450 MPa.

Both the girder and connectors are subject to uniformly distributed selfweight loading, while the girder is additionally subject to the selfweight of the stringers applied as point loads at the stringer locations (this loading is designated as

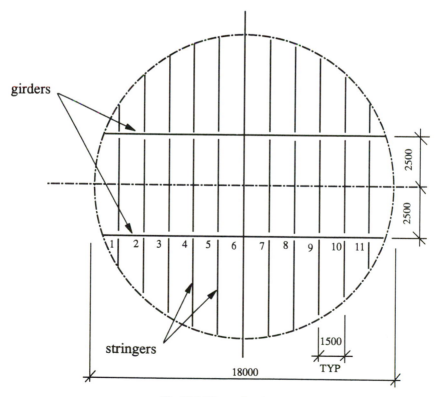

**Fig. 12.5** Silo roof system.

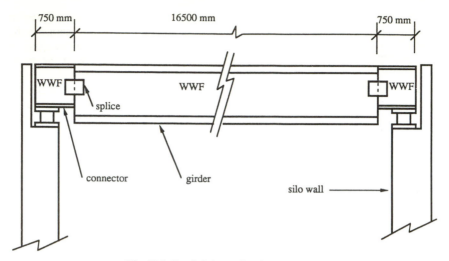

**Fig. 12.6** Roof girder and end-connectors.

DEAD). The girder is subject to live loading and snow loading applied as point loads at the stringer locations (these two loadings are designated as LIVE and SNOW). The girder is also subject to conveyor loading and wind loading applied, in each case, as two point loads at stringer locations 4 and 8 in Fig. 12.5 (these two loadings are designated as CONVEYOR and WIND). The roof girder and end-connectors are to be designed accounting for the following five load combinations:

> 1.25*DEAD + 1.5* (LIVE + CONVEYOR)
> 1.25* DEAD + 1.05* (LIVE + CONVEYOR + SNOW)
> 1.25* DEAD + 1.05* (LIVE + CONVEYOR + WIND)
> 1.25* DEAD + 0.9* (LIVE + CONVEYOR + SNOW + WIND)
> LIVE + CONVEYOR + SNOW + WIND

The strength–stability provisions of the CAN3-S16.1-M89 steel design code (Canadian Standards Association, 1989) are to be satisfied for the girder and connectors for the first four factored load combinations, while the vertical deflection at midspan of the girder is not to exceed 50 mm for the final unfactored service load combination.

The in-plane and out-of-plane effective length factors are specified to be $K_x = K_y = 1.0$ for the girder and connectors. The girder is specified to be fully braced against lateral-torsional buckling throughout its length, while the compression flanges of the end-connectors are considered to be unbraced. The maximum allowable slenderness ratios for both girder and connectors are specified to be $KL/r = 200$ in compression and $KL/r = 300$ in tension.

The design of the roof girder and end-connectors is conducted using the optimization process described earlier in this chapter. The iterative synthesis history and optimal WWF section designations are given in Fig. 12.7. The initial design weight given in Fig. 12.7 corresponds to the largest WWF sections satisfying the depth limitations specified for the girder and connectors. The

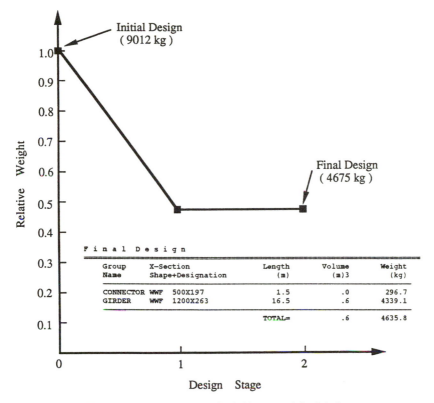

Fig. 12.7 Example 12.3: synthesis history and final design.

optimization process converged after two iterations to a least-weight final design satisfying all of the relevant strength–stability provisions of the CAN3-S16.1-M89 steel code (Canadian Standards Association, 1989) as well as the imposed limitation on the midspan deflection of the girder under service loads (in fact, this latter stiffness condition was the controlling design constraint). The design optimization required approximately 1 minute execution time on a 386-based microcomputer.

## EXAMPLE 12.4

The Warren pony truss structure shown in Fig. 12.8 is a pedestrian footbridge with a 5% camber that is to be designed in accordance with the provisions of the CAN3-S16.1-M89 steel design code (Canadian Standards Association, 1989), and the highway bridge code for Ontario, Canada (Ontario Ministry of Transport and Communications, 1983). The bridge has a span of 9.14 m (30 ft) and a width of 1.83 m (6 ft). The member joints are all moment-resisting rigid connections.

To satisfy practical construction requirements, the 47 members of the bridge

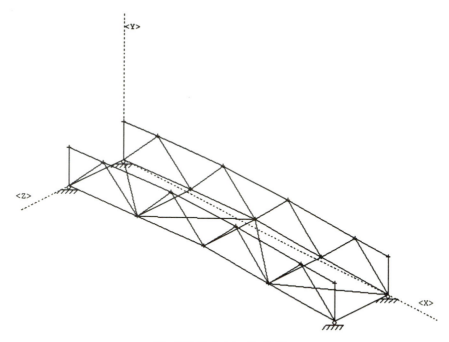

**Fig. 12.8** Pedestrian footbridge.

structure are collected into six fabrication groups: top chord members (T_CHORD); bottom chord members (B_CHORD); diagonal bracing members in bridge side walls (DIAG); diagonal bracing members in bridge floor (BRACE); parallel floor beams (BEAM); vertical end-members in side walls (VERT).

The in-plane and out-of-plane effective length factors for the T_CHORD and VERT members are taken to be $K_x = 2.0$ and $K_y = 1.0$, respectively, while those for the B_CHORD and BEAM members are taken to be $K_x = K_y = 1.0$. The effective length factors for the DIAG members are taken to be $K_x = 1.2$ and $K_y = 1.0$, while those for the BRACE members are taken to be $K_x = K_y = 0.2$. The maximum allowable slenderness ratios for each members are specified to be $KL/r = 200$ in compression and $KL/r = 300$ in tension.

The members of the bridge are to have rectangular hollow section (RHS) shapes, with the exception that the floor bracing members are to have square hollow section (SHS) shapes. Member sections are to be selected from the Canadian database of standard RHS and SHS steel sections: Young's modulus, 200 000 MPa; shear modulus, 77 000 MPa; yield stress, 350 MPa; ultimate stress, 450 MPa. Member sections are to be oriented such as to undergo strong-axis bending, with the exception that weak-axis bending is to be experienced by the sections of the beam members in the bridge floor.

The bridge is subject to gravity dead and live loading in the $Y$ axis direction (Fig. 12.8), wind loading in the $Y$ axis and $Z$ axis directions, rail loading on the top chord members in the $Z$ axis direction and temperature loading on all members. The bridge is to be designed for eight different factored combinations of these loads.

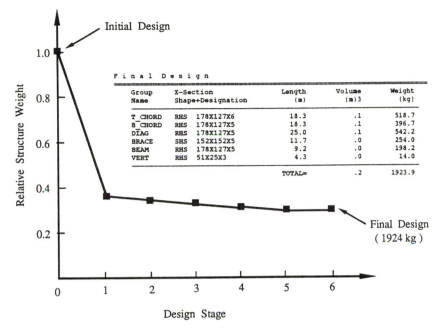

**Fig. 12.9** Example 12.4: synthesis history and final design.

The design of the pedestrian footbridge is conducted using the previously described optimization procedure. The iterative synthesis history and final design for the structure are shown in Fig. 12.9. The optimization process converged after six iterations to a least-weight final design satisfying all relevant strength–stability requirements of the governing design code (displacement constraints were not imposed for the design), and required approximately 8 minutes execution time on a 386-based microcomputer. In fact, an almost optimal design weighing only 2% heavier than the final design is found after completion of the first design stage in approximately 1.5 minutes execution time.

## EXAMPLE 12.5

The framework shown in Fig. 12.10 is part of a thermal power station boiler house structure that measures 138 ft × 167 ft in plan and requires approximately 2000 tons of structural steel. The boiler itself measures 75 ft × 79 ft × 187 ft high, weighs approximately 6000 tons and is suspended from large plate girders supported by twelve major columns. Other columns support the various components of the boiler house, such as elevators and stairs, access platforms, piping, air heaters and flue gas ductwork. The boiler house is required to withstand high seismic forces that are transmitted to the structure through bumpers at specific locations, and then to the foundations via a complex network of horizontal diaphragms and vertical bracing systems. The structure is designed

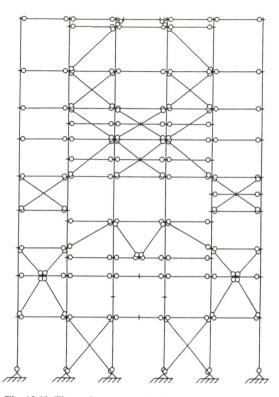

**Fig. 12.10** Thermal power station boiler house structure.

**Table 12.3** Example 12.5: final design

| Group name | X-section shape + designation | Length (ft) | Volume (ft³) | Weight (lb) |
|---|---|---|---|---|
| 1 | W 18 × 65 | 51.7 | 6.9 | 3367.9 |
| 2 | W 18 × 50 | 42.5 | 4.3 | 2128.1 |
| 3 | W 18 × 50 | 46.4 | 4.7 | 2322.5 |
| 4 | W 18 × 50 | 48.2 | 4.9 | 2415.7 |
| 5 | W 14 × 90 | 103.5 | 19.0 | 9345.5 |
| 6 | W 18 × 97 | 95.0 | 18.8 | 9225.2 |
| 7 | W 18 × 76 | 82.7 | 12.8 | 6285.1 |
| 8 | W 18 × 76 | 96.4 | 14.9 | 7329.3 |
| 9 | W 18 × 76 | 51.7 | 8.0 | 3932.2 |
| 10 | W 14 × 159 | 77.8 | 25.2 | 12374.2 |
| 11 | W 18 × 50 | 36.3 | 3.7 | 1820.5 |
| 12 | W 18 × 50 | 48.2 | 4.9 | 2415.7 |
| 13 | W 14 × 61 | 51.7 | 6.4 | 3156.3 |
| 14 | W 14 × 61 | 47.7 | 5.9 | 2910.5 |
| 15 | W 14 × 61 | 51.7 | 6.4 | 3156.3 |
| 16 | W 14 × 34 | 47.7 | 3.3 | 1626.0 |

**Table 12.3** (*Continued*)

| Group name | X-section shape + designation | Length (ft) | Volume (ft³) | Weight (lb) |
|---|---|---|---|---|
| 17 | W 10 × 39 | 107.8 | 8.6 | 4226.6 |
| 18 | W 12 × 30 | 202.0 | 12.3 | 6051.2 |
| 19 | W 12 × 65 | 25.3 | 3.3 | 1643.6 |
| 20 | W 16 × 31 | 75.8 | 4.8 | 2354.4 |
| 21 | W 14 × 30 | 25.3 | 1.6 | 761.6 |
| 22 | W 10 × 49 | 80.3 | 8.0 | 3940.3 |
| 23 | W 14 × 38 | 69.8 | 5.4 | 2666.0 |
| 24 | W 10 × 39 | 82.3 | 6.6 | 3225.3 |
| 25 | W 10 × 39 | 74.8 | 6.0 | 2933.1 |
| 26 | W 12 × 30 | 63.5 | 3.9 | 1902.3 |
| 27 | W 10 × 39 | 77.0 | 6.1 | 3017.2 |
| 28 | W 10 × 45 | 21.8 | 2.0 | 990.0 |
| 29 | W  8 × 40 | 21.9 | 1.8 | 871.9 |
| 30 | W 10 × 45 | 106.2 | 9.8 | 4812.9 |
| 31 | W 10 × 39 | 55.0 | 4.4 | 2157.4 |
| 32 | W 12 × 30 | 27.5 | 1.7 | 824.7 |
| 33 | W 14 × 43 | 34.4 | 3.0 | 1478.2 |
| 34 | W 12 × 30 | 17.6 | 1.1 | 527.6 |
| 35 | W 12 × 30 | 78.2 | 4.8 | 2343.8 |
| 36 | W 12 × 30 | 48.3 | 3.0 | 1448.3 |
| 37 | W 12 × 30 | 36.3 | 2.2 | 1088.6 |
| 38 | W 10 × 39 | 58.3 | 4.7 | 2285.3 |
| 39 | W 12 × 30 | 45.9 | 2.8 | 1373.8 |
| 40 | W 14 × 34 | 54.0 | 3.8 | 1840.3 |
| 41 | W 10 × 39 | 27.0 | 2.2 | 1058.2 |
| 42 | W 10 × 39 | 54.0 | 4.3 | 2116.4 |
| 43 | W 10 × 45 | 135.0 | 12.5 | 6119.1 |
| 44 | W 14 × 43 | 27.0 | 2.4 | 1159.4 |
| 45 | W 10 × 39 | 27.0 | 2.2 | 1058.2 |
| 46 | W 12 × 30 | 22.8 | 1.4 | 681.5 |
| 47 | W 12 × 30 | 22.8 | 1.4 | 681.5 |
| 48 | W 12 × 30 | 22.8 | 1.4 | 681.5 |
| 49 | W 12 × 30 | 45.5 | 2.8 | 1363.0 |
| 50 | W 24 × 55 | 45.5 | 5.1 | 2512.0 |
| 51 | W 24 × 55 | 68.3 | 7.7 | 3768.1 |
| 52 | W 12 × 30 | 22.8 | 1.4 | 681.5 |
| 53 | W 12 × 30 | 22.8 | 1.4 | 681.5 |
| 54 | W 12 × 30 | 18.9 | 1.2 | 565.4 |
| 55 | W 12 × 30 | 18.9 | 1.2 | 565.4 |
| 56 | W  8 × 31 | 18.9 | 1.2 | 587.3 |
| 57 | W 14 × 193 | 44.0 | 17.4 | 8517.3 |
| 58 | W 36 × 150 | 44.0 | 13.5 | 6627.9 |
| 59 | W 24 × 162 | 44.0 | 14.6 | 7152.7 |
| 60 | W 14 × 34 | 44.0 | 3.1 | 1499.5 |
| 61 | W 14 × 34 | 44.0 | 3.1 | 1499.5 |
| 62 | W 18 × 50 | 27.0 | 2.8 | 1352.6 |
| Total | | | 365.8 | 179505.0 |

as a series of planar frameworks, such as in Fig. 12.10, connected together by horizontal diaphragms.

The framework in Fig. 12.10 is to be designed in accordance with the strength–stability requirements of the AISC-89 steel design code (American Institute of Steel Construction, 1989), and user-specified horizontal displacement limitations are imposed at various heights above the foundation level. The various member joints are either moment-resisting welded connections or pin-jointed bolted connections (denoted by O in Fig. 12.10). The 181 members of the framework are collected into 62 fabrication groups, each one of which is specified to have a W-flange (W) section shape to be selected from the American database of standard W-shape steel sections. Maximum depth limitations are imposed on each of the 62 sections to satisfy practical construction considerations. The maximum allowable slenderness ratios for the members are specified to be 200 in compression and 300 in tension: Young's modulus, $29\,000\,\mathrm{klbf\,in}^{-2}$; shear modulus, $11\,200\,\mathrm{klbf\,in}^{-2}$; yield stress, $50\,\mathrm{klbf\,in}^{-2}$; and ultimate stress, $65\,\mathrm{klbf\,in}^{-2}$. The framework is subject to gravity dead and live loading, lateral wind loading and horizontal seismic loading (modeled as equivalent static

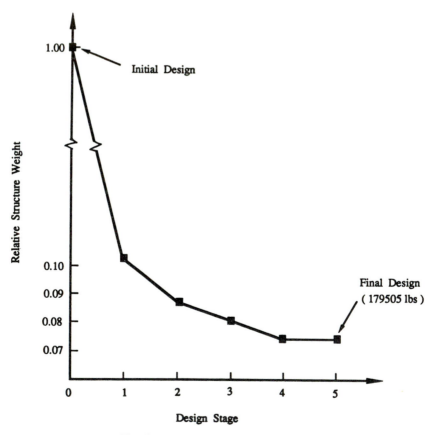

**Fig. 12.11** Example 12.5: synthesis history.

loading). The design is conducted accounting for ten different combinations of these loads.

The design optimization of the framework was carried out in two phases. Initially, the previously described optimization procedure was applied to establish a preliminary design that provided an estimate of the required steel tonnage and the foundation loads. Here, the effective length factors for all members were taken to be $K_x = K_y = 1.0$, and all beams were assumed to be fully laterally braced. As well, all loads were applied as nodal loading alone (minimum section sizes were imposed for the pin-jointed beam members to compensate for the absence of member loads). The framework was assumed to be sidesway prevented and second-order (P–delta) effects were not accounted for.

From the results of the preliminary design phase, member effective length factors and lateral bracing parameters were adjusted to reflect actual conditions. Actual nodal and member loads were then applied on the framework, and the optimization was conducted again to determine the final design listed in Table 12.3. The iterative synthesis history for the design is shown in Fig. 12.11. The optimization process converged after five iterations to a least-weight structure satisfying the imposed displacement constraints and all relevant strength–stability requirements of the AISC-89 steel design code (American Institute of Steel Construction, 1989) and required approximately 2.5 hours execution time on a 386-based microcomputer. A subsequent second-order analysis of the final design in Table 12.3 verified that all strength–stability and stiffness requirements are also satisfied when P–delta effects are accounted for.

## 12.6 CONCLUDING REMARKS

The practical design of structural steel frameworks is very effectively carried out using optimization techniques, such as those described in this chapter. Member-by-member and formal optimization employed in tandem permits the simultaneous consideration of both strength and stiffness aspects of the design, thereby eliminating much of the guesswork from the synthesis process. Ths use of optimization techniques results in reduced designer time and more economical designs compared to that when conventional trial-and-error techniques are employed. In fact, an optimal structural steel design typically represents a 3–10% saving in steel weight compared with conventional design, which for a large-scale structure can result in a significant monetary saving.

Much research is being carried out in many areas to improve further the applicability of computer-based optimization techniques to practical structural design. The body of work is too broad and varied to cite here, but listed in the following are some representative studies currently underway by the writer and his research associates at the University of Waterloo. The optimization strategy presented in this chapter is being extended to account for the cost and design of optimal member connections (Xu, 1993; Xu and Grierson, 1992). Optimality criteria techniques are being developed for the lateral-resistant design of tall three-dimensional steel building frameworks (Chan, 1993; Chan and Grierson, 1993), and for the strength and stiffness design of reinforced concrete building structures (Moharrami, 1993; Moharrami and Grierson, 1993). Genetic algorithms are being developed to establish not only optimal section sizes but also the optimal topological layout of the members for structural steel frameworks (Pak, 1993; Grierson and Pak, 1992). Finally, new programming paradigms are being employed to develop enhanced computer-based capabilities for structural design (Biedermann, 1993; Biedermann and Grierson, 1993).

## ACKNOWLEDGEMENTS

The writer wishes to thank Lei Xu, Chun-Man Chan and Nancy Simpson of the University of Waterloo for their help in preparing the text manuscript for this chapter. Gordon Cameron of Waterloo Engineering Software supplied the details for design example 12.2, as did freelance structural engineer Edward Greig for design examples 12.3 and 12.4. The writer is indebted to them, and to the anonymous-by-choice individual and firm that supplied the details for design example 12.5.

## APPENDIX 12A SECTION PROPERTY RELATIONSHIPS

Consider a structural steel member subject to a three-dimensional force field involving axial force, biaxial shear forces, torsional moment and biaxial bending moments, as indicated in Fig. 12A.1. The geometrical properties of the member cross section are $A_x$, $A_y$, $A_z$ the axial and shear areas, and $I_x$, $I_y$, $I_z$, the torsional and bending moments of inertia.

For commercial standard steel sections, each of the properties $A_y$, $A_z$, $I_x$, $I_y$ and $I_z$ may be expressed in terms of the axial area $A_x$ as follows (Chan, 1993):

$$
\begin{aligned}
1/A_Y &= C_{AY}(1/A_X) + C'_{AY} \\
1/A_Z &= C_{AZ}(1/A_X) + C'_{AZ} \\
1/I_X &= C_{IX}(1/A_X) + C'_{IX} \\
1/I_Y &= C_{IY}(1/A_X) + C'_{IY} \\
1/I_Z &= C_{IZ}(1/A_X) + C'_{IZ}
\end{aligned}
\tag{12A.1}
$$

where the coefficients $C$ and $C'$ are determined by linear regression analysis and have different values depending on the type and size of the section.

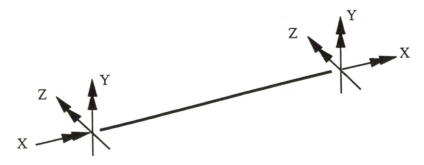

**Fig. 12A.1** Three-dimensional member force field.

**Table 12A.1** Relationships between cross section area $A_X$ and shear areas $A_Y$ and $A_Z$ for AISC-LRFD W14 and W24 sections

| Sections (Number of sections) | $\dfrac{1}{A_Y} = \dfrac{C_{AY}}{A_X} + C'_{AY}$ | | Maximum error (%) | $\dfrac{1}{A_Z} = \dfrac{C_{AZ}}{A_X} + C'_{AZ}$ | | Maximum error (%) |
|---|---|---|---|---|---|---|
| | $C_{AY}$ | $C'_{AY}$ | | $C_{AZ}$ | $C'_{AZ}$ | |
| W14 × 22–26 (2) | 1.435 326 | 0.095 276 | 0.00 | 2.561 815 | −0.096 225 | 0.00 |
| W14 × 30–38 (3) | 1.634 329 | 0.085 120 | 0.98 | 2.084 494 | −0.044 015 | 0.92 |
| W14 × 43–53 (3) | 3.004 888 | 0.001 081 | 0.26 | 1.572 869 | −0.006 856 | 0.02 |
| W14 × 61–82 (4) | 3.814 517 | −0.019 917 | 0.98 | 1.383 969 | 0.000 091 | 0.39 |
| W14 × 90–132 (5) | 4.725 603 | −0.015 799 | 0.86 | 1.293 418 | −0.000 474 | 0.34 |
| W14 × 145–176 (3) | 4.944 198 | −0.016 278 | 0.01 | 1.262 608 | −0.000 032 | 0.12 |
| W14 × 193–257 (4) | 4.641 405 | −0.009 561 | 0.60 | 1.276 141 | −0.000 330 | 0.16 |
| W14 × 283–426 (6) | 4.419 709 | −0.006 828 | 0.24 | 1.279 435 | −0.000 335 | 0.33 |
| W14 × 455–730 (6) | 4.108 586 | −0.004 616 | 0.24 | 1.263 193 | −0.000 185 | 0.14 |
| W24 × 55–62 (2) | 1.393 000 | 0.021 422 | 0.00 | 3.090 550 | −0.049 433 | 0.00 |
| W24 × 68–84 (3) | 1.428 414 | 0.030 837 | 0.43 | 2.517 252 | −0.030 232 | 0.43 |
| W24 × 94–103 (2) | 1.857 359 | 0.012 822 | 0.00 | 2.049 032 | −0.010 935 | 0.00 |
| W24 × 104–131 (3) | 2.328 042 | 0.007 155 | 0.15 | 1.755 494 | −0.005 079 | 0.01 |
| W24 × 146–192 (4) | 2.486 482 | 0.004 511 | 0.42 | 1.657 739 | −0.003 055 | 0.21 |
| W24 × 207–306 (5) | 2.897 758 | −0.003 009 | 0.24 | 1.518 109 | −0.000 560 | 0.15 |
| W24 × 335–492 (5) | 2.810 400 | −0.002 391 | 0.58 | 1.512 870 | −0.000 536 | 0.49 |

**Table 12.A2** Relationships between cross section area $A_X$ and moments of inertia $I_Z, I_Y$ and $I_X$ for AISC-LRFD W12 and W24 sections

| Sections (number of sections) | $\dfrac{1}{I_Z} = \dfrac{C_{IZ}}{A_X} + C'_{IX}$ | | | $\dfrac{1}{I_Y} = \dfrac{C_{IY}}{A_X} + C'_{IY}$ | | | $\dfrac{1}{I_X} = \dfrac{C_{IX}}{A_X} + C'_{IX}$ | | |
| --- | --- | --- | --- | --- | --- | --- | --- | --- | --- |
| | $C_{IZ}$ | $C'_{IZ}$ | Maximum error (%) | $C_{IY}$ | $C'_{IY}$ | Maximum error (%) | $C_{IX}$ | $C'_{IX}$ | Maximum error (%) |
| W14 × 22–26 (2) | 0.039 240 | −0.001 021 | 0.00 | 1.273 642 | −0.053 390 | 0.00 | 82.520 01 | −7.953 04 | 0.00 |
| W14 × 30–38 (3) | 0.035 327 | −0.000 573 | 0.63 | 0.570 689 | −0.013 801 | 0.81 | 56.961 14 | −3.878 09 | 3.50 |
| W14 × 43–53 (3) | 0.031 973 | −0.000 203 | 0.08 | 0.314 017 | −0.002 804 | 0.03 | 28.398 04 | −1.312 02 | 1.40 |
| W14 × 61–82 (4) | 0.029 882 | −0.000 110 | 0.31 | 0.181 081 | −0.000 790 | 0.50 | 17.591 99 | −0.540 04 | 3.65 |
| W14 × 90–132 (5) | 0.029 025 | −0.000 097 | 0.42 | 0.078 045 | −0.000 193 | 0.38 | 13.145 35 | −0.261 87 | 5.69 |
| W14 × 145–176 (3) | 0.028 522 | −0.000 084 | 0.11 | 0.068 920 | −0.000 138 | 0.07 | 6.705 84 | −0.092 28 | 1.58 |
| W14 × 193–257 (4) | 0.028 078 | −0.000 077 | 0.10 | 0.067 919 | −0.000 123 | 0.14 | 3.616 11 | −0.035 52 | 2.68 |
| W14 × 283–426 (6) | 0.027 137 | −0.000 065 | 0.31 | 0.067 382 | −0.000 115 | 0.13 | 1.566 43 | −0.009 66 | 5.12 |
| W14 × 455–730 (6) | 0.024 444 | −0.000 044 | 0.62 | 0.063 449 | −0.000 084 | 0.30 | 0.611 77 | −0.002 20 | 7.09 |
| W24 × 55–62 (2) | 0.014 090 | −0.000 129 | 0.00 | 0.792 936 | −0.014 582 | 0.00 | 38.721 68 | −1.542 77 | 0.00 |
| W24 × 68–84 (3) | 0.013 433 | −0.000 123 | 0.18 | 0.389 350 | −0.005 212 | 0.40 | 28.162 79 | −0.875 74 | 2.20 |
| W24 × 94–103 (2) | 0.011 956 | −0.000 061 | 0.00 | 0.248 872 | 0.000 190 | 0.00 | 15.968 55 | −0.386 37 | 0.00 |
| W24 × 104–131 (3) | 0.011 006 | −0.000 037 | 0.07 | 0.137 181 | −0.000 621 | 0.02 | 15.678 37 | −0.303 99 | 1.96 |
| W24 × 146–192 (4) | 0.010 668 | −0.000 030 | 0.16 | 0.120 064 | −0.000 247 | 0.58 | 7.577 72 | −0.103 49 | 3.69 |
| W24 × 207–306 (5) | 0.009 939 | −0.000 017 | 0.18 | 0.120 004 | −0.000 249 | 0.15 | 3.129 03 | −0.026 87 | 5.37 |
| W24 × 335–492 (5) | 0.009 775 | −0.000 016 | 0.41 | 0.114 721 | −0.000 201 | 0.67 | 1.268 54 | −0.006 74 | 5.90 |

Equations (12A.1) are illustrated in Tables 12A.1 and 12A.2 for W14 and W24 standard steel sections from the AISC-LRFD steel design code (American Institute of Steel Construction, 1986). Each set of coefficients $C$ and $C'$ applies for a group of from two to six sections for which the section properties $A_Y$, $A_Z$, $I_X$, $I_Y$ and $I_Z$ predicted by equations (12A.1), for given area $A_X$, approximate the actual section properties within small percentage error. Perhaps an exception is the prediction of the torsional moment of inertia $I_X$ for larger size sections, for which percentage errors up to 7% are noted to occur in Table 12A.2. However, this is not a significant shortcoming as torsional capacity requirements rarely govern the design of steel members of building frameworks. Similar results to those in Tables 12A.1 and 12A.2 may be derived for a broad range of commercially available standard steel sections.

Equations (12A.1) permit the stiffness matrix for a structural steel member to be instantaneously approximated in terms of its cross section area $A_X$ alone, thereby permitting the analytical derivation of stress and displacement gradients for the assembled framework of members (e.g. Appendix 12B).

## APPENDIX 12B DISPLACEMENT AND STRESS GRADIENTS

Consider the pin-jointed truss loaded as shown in Fig. 12B.1. The truss has six members of cross section area $a_i(i = 1, 2, \ldots, 6)$ and five independent nodal displacements $u_j(j = 1, 2, \ldots, 5)$. It is required to find the gradient (sensitivity) of the lateral displacement $u_3$ to changes in the reciprocal cross section areas $x_i = 1/a_i$ from (equation (12.27))

$$\frac{\partial u_3}{\partial x_i} = \frac{1}{x_i}(\mathbf{u}_3^T \mathbf{K}_i \mathbf{u}) \quad (i = 1, 2, \ldots, 6) \tag{12B.1}$$

Similarly, it is required to find the gradient of the axial stress $\sigma_5$ in member 5

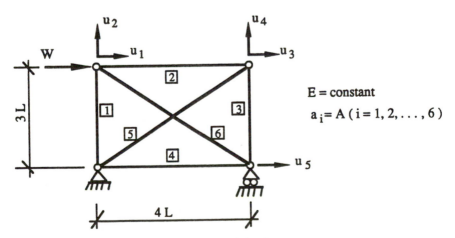

$$E = \text{constant}$$
$$a_i = A\,(i = 1, 2, \ldots, 6)$$

**Fig. 12B.1** Pin-jointed truss.

to changes in the reciprocal areas $x_i$ from (equation (12.30))

$$\frac{\partial \sigma_5}{\partial x_i} = \frac{1}{x_i}(\mathbf{u}_5^T \mathbf{K}_i \mathbf{u}) \quad (i = 1, 2, \ldots, 6) \tag{12B.2}$$

To evaluate equations (12B.1) and (12B.2), the nodal displacements are found from

$$\mathbf{u} = \mathbf{K}^{-1}\mathbf{P} \tag{12B.3}$$

which becomes for the truss in Fig. 12B.1,

$$
\begin{bmatrix} u_1 \\ u_2 \\ u_3 \\ u_4 \\ u_5 \end{bmatrix} = \frac{L}{EA}
\begin{bmatrix}
6.750 & & & & \\
1.125 & 2.813 & & \text{SYM} & \\
4.750 & 0.792 & 6.157 & & \\
-1.125 & -0.187 & -1.458 & 2.813 & \\
2.000 & -0.333 & 1.407 & -0.333 & 3.407
\end{bmatrix}
\begin{bmatrix} W \\ 0 \\ 0 \\ 0 \\ 0 \end{bmatrix}
$$

$$
= \frac{WL}{EA}
\begin{bmatrix}
6.750 \\
1.125 \\
4.750 \\
-1.125 \\
2.000
\end{bmatrix}
\tag{12B.4}
$$

where $E$ is Young's material modulus and, as indicated in Figure 12B.1, $L$ and $A$ are member length and section area parameters, respectively.

Having displacements $\mathbf{u}$ from equation (12B.4), the member axial stresses are found from

$$\sigma = \mathbf{T}\mathbf{u} \tag{12B.5}$$

where the transformation matrix $\mathbf{T}$ accounts for member stress–strain relations and the structure topology. Equation (12B.5) becomes for the truss in Fig. 12B.1

$$
\begin{bmatrix} \sigma_1 \\ \sigma_2 \\ \sigma_3 \\ \sigma_4 \\ \sigma_5 \\ \sigma_6 \end{bmatrix} = \frac{E}{L}
\begin{bmatrix}
0 & 1/3 & 0 & 0 & 0 \\
-1/4 & 0 & 1/4 & 0 & 0 \\
0 & 0 & 0 & 1/3 & 0 \\
0 & 0 & 0 & 0 & 1/4 \\
0 & 0 & 0.8/5 & 0.6/5 & 0 \\
-0.8/5 & 0.6/5 & 0 & 0 & 0.8/5
\end{bmatrix}
\frac{WL}{EA}
\begin{bmatrix}
6.750 \\
1.125 \\
4.750 \\
-1.125 \\
2.000
\end{bmatrix}
$$

$$
= \frac{W}{A}
\begin{bmatrix}
0.375 \\
-0.500 \\
-0.375 \\
0.500 \\
0.625 \\
-0.625
\end{bmatrix}
\tag{12B.6}
$$

The vector of virtual loads associated with displacement $u_3$ is $\mathbf{b}_3^T = [0\ 0\ 1\ 0\ 0]$, as shown in Fig. 12B.2, and the corresponding vector of virtual nodal displace-

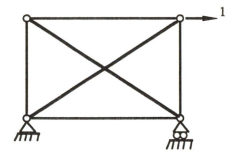

**Fig. 12B.2** Virtual loading $b_3$.

ments is found as

$$\mathbf{u}_3 = \mathbf{K}^{-1}\mathbf{b}_3 = \mathbf{K}^{-1}\begin{bmatrix} 0 \\ 0 \\ 1 \\ 0 \\ 0 \end{bmatrix} = \frac{L}{EA}\begin{bmatrix} 4.750 \\ 0.792 \\ 6.157 \\ -1.458 \\ 1.407 \end{bmatrix} \qquad (12B.7)$$

where the inverse stiffness matrix $\mathbf{K}^{-1}$ is as given in equation (12B.4).

The fifth row of the transformation matrix in equation (12B.6) identifies the vector of virtual loads associated with member stress $\sigma_5$ to be $\mathbf{t}_5^T = [0\ 0\ 0.8\ 0.6\ 0]E/5L$, as shown in Fig. 12B.3, and the corresponding vector of virtual nodal displacements is found as

$$\mathbf{u}_5 = \mathbf{K}^{-1}\mathbf{t}_5 = \mathbf{K}^{-1}\frac{E}{5L}\begin{bmatrix} 0 \\ 0 \\ 0.8 \\ 0.6 \\ 0 \end{bmatrix} = \frac{1}{A}\begin{bmatrix} 0.625 \\ 0.104 \\ 0.810 \\ 0.104 \\ 0.185 \end{bmatrix} \qquad (12B.8)$$

where, again, matrix $\mathbf{K}^{-1}$ is as given in equation (12B.4).

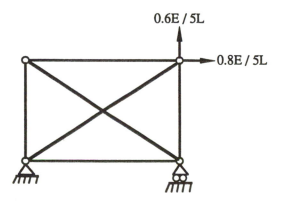

**Fig. 12B.3** Virtual loading $t_5$.

Having the actual and virtual displacements $\mathbf{u}$ and $\mathbf{u}_3$ from equations (12B.4) and (12B.7), respectively, the displacement gradient $\partial u_3/\partial x_i$ is readily calculated through equation (12B.1) upon identifying the (global-axis) stiffness matrix $\mathbf{K}_i$ for each member $i$. For example, the stiffness matrix for member $i = 6$ is

$$\mathbf{K}_6 = \frac{EA}{5L} \begin{bmatrix} 0.64 & -0.48 & 0 & 0 & -0.64 \\ -0.48 & 0.36 & 0 & 0 & 0.48 \\ 0 & 0 & 0 & 0 & 0 \\ 0 & 0 & 0 & 0 & 0 \\ -0.64 & 0.48 & 0 & 0 & 0.64 \end{bmatrix} \qquad (12\text{B}.9)$$

and, from Fig. 12B.1, the reciprocal area for member 6 is $x_6 = 1/a_6 = 1/A$. Therefore, from equations (12B.1), (12B.4), (12B.7) and (12B.9), the gradient of the displacement $u_3$ with respect to changes in the reciprocal-area $x_6$ for member 6 is

$$\frac{\partial u_3}{\partial x_6} = \frac{1}{x_6}(\mathbf{u}_3^{\mathrm{T}}\mathbf{K}_6\mathbf{u}) = +1.375\frac{WL}{A^2}$$

where the positive sign indicates that the nodal displacement $u_3$ varies directly with $x_6$ (which means that, as expected, the displacement $u_3$ varies inversely with changes in the actual section area $a_6 = 1/x_6$ for. member 6). The displacement gradient $\partial u_3/\partial x_i$ is similarly found for the remaining five members $i = 1, 2, \ldots, 5$, such that the vector of gradients for all six members is

$$\left[\frac{\partial u_3}{\partial \mathbf{x}}\right]^{\mathrm{T}} = [0.297 \ -0.704 \ 0.547 \ 0.704 \ 2.532 \ 1.375]WL/A^2$$

The negative sign for the gradient corresponding to member 2 indicates that the displacement $u_3$ varies inversely with the reciprocal area $x_2$ and, therefore, directly with the actual area $a_2$ for member 2 (i.e. $u_3$ increases as $a_2$ increases, and vice versa). This counterintuitive result is explained by the fact that since member 2 is in axial compression, for the loading shown in Fig. 12B.1, any increase in $a_2$ will tend to decrease the member axial shortening and, thereby, increase the nodal displacement $u_3$. Another interesting result can be derived from the fact that members 2 and 4 correspond to gradients that have equal magnitude but opposite sign. Namely, for the loading shown in Fig. 12B.1, any symmetrical change in the section areas of the top and bottom chord members 2 and 4 will have no effect on the magnitude of the nodal displacement $u_3$. In fact, in addition to providing the means to formulate explicit constraint approximations for computer-based design optimization, the gradient information itself provides valuable insight concerning the behavior of structures under loads.

Similarly, having the actual and virtual displacements $\mathbf{u}$ and $\mathbf{u}_5$ from equations (12B.4) and (12B.8), respectively, the stress gradient $\partial \sigma_5/\partial x_i$ associated with each member $i$ is readily calculated through equation (12B.2). For example, for member $i = 6$,

$$\frac{\partial \sigma_5}{\partial x_6} = \frac{1}{x_6}(\mathbf{u}_5^{\mathrm{T}}\mathbf{K}_6\mathbf{u}) = +0.181W$$

where the positive sign indicates that the axial stress $\sigma_5$ in member 5 varies directly with $x_6$, or that $\sigma_5$ varies inversely with changes in the actual section area $a_6 = 1/x_6$ for member 6. The stress gradient $\partial \sigma_5 / \partial x_i$ is similarly found for the remaining five members $i = 1, 2, \ldots, 5$, such that the vector of gradients for all six members is

$$\left[\frac{\partial \sigma_5}{\partial \mathbf{x}}\right]^{\mathrm{T}} = [0.039 \ -0.093 \ -0.039 \ 0.093 \ 0.444 \ 0.181]W$$

The negative signs for the gradients corresponding to members 2 and 3 indicate that the stress $\sigma_5$ in member 5 varies directly with changes in the section areas $a_2$ and $a_3$ for members 2 and 3. The fact that member pairs 1,3 and 2,4 each correspond to gradients that have equal magnitude but opposite sign means that any symmetrical change in the section areas of the vertical members 1,3 and/or the chord members 2,4 for the truss will have no effect on the magnitude of the axial stress $\sigma_5$ in member 5.

## APPENDIX 12C GOC METHOD FOR DISCRETE OPTIMIZATION

Consider the pin-jointed truss loaded as shown in Fig. 12C.1 (Schmit and Fleury, 1979; Haftka and Gürdal, 1992). The truss has two members of cross section area $a_1$ and $a_2$, and two independent nodal displacements $u_1$ and $u_2$. It is required to find the least-weight truss by selecting each of the section areas $a_1$ and $a_2$ from the discrete set of areas

$$A = \left\{1, \frac{3}{2}, 2\right\} \tag{12C.1}$$

while satisfying the displacement constraints

$$u_1 \leqslant 0.75PL/E; \quad u_2 \leqslant 0.25PL/E \tag{12C.2}$$

As the truss is statically determinate, the displacements are readily found to be

$$u_1 = \frac{PL}{2E}\left(\frac{1}{a_1} + \frac{1}{a_2}\right); \quad u_2 = \frac{PL}{2E}\left(\frac{1}{a_1} - \frac{1}{a_2}\right) \tag{12C.3}$$

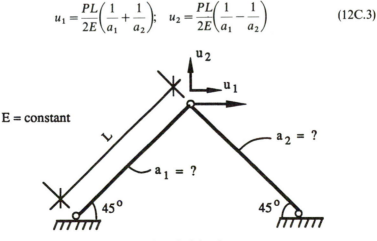

**Fig. 12C.1** Two-bar pin-jointed truss.

The weight of the truss is given by

$$W = \rho L(a_1 + a_2) \tag{12C.4}$$

where $L$ is the length of each members and $\rho$ is the material density.

From equations (12C.1)–(12C.4), the discrete optimization problem expressed in terms of reciprocal sizing variables $x_1 = 1/a_1$ and $x_2 = 1/a_2$ is

$$\text{minimize} \quad \frac{1}{x_1} + \frac{1}{x_2} \tag{12C.5}$$

$$\text{subject to} \quad x_1 + x_2 \leqslant 1.5 \tag{12C.6}$$

$$x_1 - x_2 \leqslant 0.5 \tag{12C.7}$$

$$x_1, x_2 \in \left\{ \frac{1}{2}, \frac{2}{3}, 1 \right\} \tag{12C.8}$$

The generalized optimality criteria (GOC) method is applied to solve the problem posed by equations (12C.5)–(12C.8). To that end, the Lagrangian function is formed as

$$\mathscr{L}(\mathbf{x}, \lambda) = \frac{1}{x_1} + \frac{1}{x_2} + \lambda_1(x_1 + x_2 - 1.5) + \lambda_2(x_1 - x_2 - 0.5) \tag{12C.9}$$

where the Lagrange variables are sign restricted as

$$\lambda_1 \geqslant 0; \quad \lambda_2 \geqslant 0 \tag{12C.10}$$

A saddle-point $(\mathbf{x}^*, \lambda^*)$ of equation (12C.9) is given by a solution of the min–max problem

$$(\mathbf{x}^*, \lambda^*) = \max_\lambda[\min_\mathbf{x} \mathscr{L}(\mathbf{x}, \lambda)] = \max_\lambda[\mathscr{L}_m(\lambda)] \tag{12C.11}$$

where the dual function is

$$\mathscr{L}_m(\lambda) = \sum_{i=1}^{2} \min_{x_i} L_i(x_i, \lambda) \tag{12C.12}$$

in which the individual functions of the variables $x_i$ are, from equation (12C.9),

$$L_1(x_1, \lambda) = \frac{1}{x_1} + (\lambda_1 + \lambda_2)x_1 - 1.5\lambda_1 - 0.5\lambda_2 \tag{12C.13}$$

$$L_2(x_2, \lambda) = \frac{1}{x_2} + (\lambda_1 - \lambda_2)x_2 \tag{12C.14}$$

When the dual function is maximized over the $\lambda$ variables in equation (12C.11), its continuity is maintained by requiring that

$$L_i(x_i^k, \lambda) = L_i(x_i^{k+1}, \lambda) \quad (i = 1, 2) \tag{12C.15}$$

where $x_i^k$ and $x_i^{k+1}$ are successive discrete values of each variable $x_i$ from equation (12C.8). Therefore, from equations (12C.8), (12C.13) and (12C.15), the boundaries for changes in the values of $x_1$ are

$$\frac{1}{1/2} + \frac{1}{2}(\lambda_1 + \lambda_2) = \frac{1}{2/3} + \frac{2}{3}(\lambda_1 + \lambda_2)$$

$$\frac{1}{2/3} + \frac{2}{3}(\lambda_1 + \lambda_2) = \frac{1}{1} + (\lambda_1 + \lambda_2)$$

which reduce to

$$\lambda_1 + \lambda_2 = 3; \quad \lambda_1 - \lambda_2 = 1.5 \tag{12C.16}$$

Similarly, from equations (12C.8), (12C.14) and (12C.15), the boundaries for changes in $x_2$ are

$$\lambda_1 - \lambda_2 = 3; \quad \lambda_1 - \lambda_2 = 1.5 \tag{12C.17}$$

As equation (12C.10) requires variables $\lambda_1$ and $\lambda_2$ to be nonnegative, equations (12C.16) and (12C.17) divide the positive quadrant of the $(\lambda_1, \lambda_2)$ plane into the six regions shown in Fig. 12C.2, where given in parentheses for each region are the fixed values of the variables $x_1, x_2$ prevailing for the dual function $\mathscr{L}_m(\lambda)$. For example, suppose that the search to maximize $\mathscr{L}_m(\lambda)$ in equation (12C.11) commences at the origin $\lambda = (0, 0)$. From equation (12C.9), the Lagrangian function $\mathscr{L}(\mathbf{x}, \lambda) = 1/x_1 + 1/x_2$, which is minimized for the discrete values $x_1 = x_2 = 1$, as indicated in Fig. 12C.2, such that the dual function is, from equations (12C.12)–(12C.14),

$$\mathscr{L}_m(\lambda) = 2 + 0.5\lambda_1 - 0.5\lambda_2 \tag{12C.18}$$

Since $\lambda_2$ cannot be reduced below zero, the right-hand side of equation (12C.18) is maximized by increasing $\lambda_1$ to the boundary of the region at $\lambda = (1.5, 0)$, at which point $\mathscr{L}_m(\lambda) = 2.75$. From Fig. 12C.2, it is now possible to move into one of the two regions for which $\mathbf{x} = (2/3, 1)$ or $\mathbf{x} = (2/3, 2/3)$ in a further attempt to maximize the dual function. In the former region, from equations (12C.12)–(12C.14), the dual function is

$$\mathscr{L}_m(\lambda) = 2.5 + \lambda_1/6 - 5\lambda_2/6 \tag{12C.19}$$

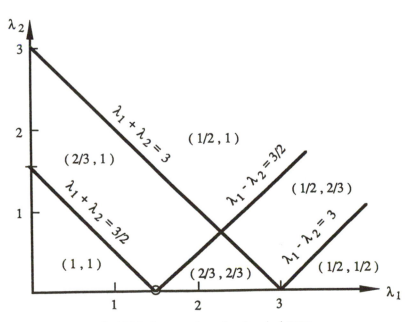

**Fig. 12C.2** Constant $(x_1, x_2)$ regions in $\lambda$ space.

while in the latter region it is

$$\mathscr{L}_m(X) = 3 - \lambda_1/6 - \lambda_2/2 \qquad (12\text{C}.20)$$

It is not possible to maximize further either of the dual functions (12C.19) and (12C.20) for nonnegative $\lambda$ values different from $\lambda^* = (1.5, 0)$, which means that the solution of the min–max problem (12C.11) occurs for any one of three pairs of $x_i$ values (1,1), (2/3, 1) or (2/3, 2/3). However, only $x^* = (2/3, 2/3)$ satisfies the constraints (12C.6) and (12C.7), for which the optimal cross section areas are $a_1 = a_2 = 3/2$ and the value of the objective function (12C.5) is 3.

A final comment is in order concerning the fact that the minimum value of the objective function (12C.5) is 3 while the maximum value of the corresponding dual function (12C.12) is 2.75. This difference serves as a reminder that the discrete optimization problem is not convex and, therefore, that there is no guarantee that the solution found is optimal and/or unique (as it happens, the solution point $x^* = (2/3, 2/3)$ found for the present example is one of two optima, the other being $x^* = (1/2, 1)$). The consequence of this for the design of structural steel frameworks is that the set of discrete standard sections found by the GOC method may correspond to but one of several near-optimal designs for the structure. Moreover, while all such designs are likely to be quite similar to each other from the viewpoint of both structure weight and standard section selection, it is theoretically possible that there may exist designs of similar weight but with radically different standard sections for the members.

# REFERENCES

American Institute of Steel Construction (1986) *Manual of Steel Construction: Load and Resistance Factor Design.*

American Institute of Steel Construction (1989) Specification for structural steel buildings: allowable stress design and plastic design.

Biedermann, J.D. (1992) Object-oriented computer-automated design of structures. PhD Thesis, Civil Engineering, University of Waterloo.

Biedermann, J.D. and Grierson, D.E. (1993) An object-oriented approach to detailed structural design. *Microcomputers in Civil Engineering*, **8**, 225–31.

Cameron, G.E. (1989) A knowledge-based expert system for structural steel design. PhD Thesis; Civil Engineering, University of Waterloo.

Cameron, G.E., Xu, L. and Grierson, D.E. (1991) Discrete optimal design of 3D frameworks. ASCE Structural Congress/10th Electronic Computation Conference, Indianapolis, IN.

Cameron, G.E., Chan, C.-M., Xu, L. and Grierson, D.E. (1992) Alternate methods for the optimal design of slender steel frameworks. *Computers and Structures*, **44**(4), 735–41.

Canadian Standards Association (1989) Steel structures for buildings (limit states design). *CAN3-S16.1M89.*

Chan, C.-M. (1992) An automatic resizing technique for the design of tall steel building frameworks. PhD Thesis, University of Waterloo.

Chan, C.-M. and Grierson, D.E. (1993) An efficient resizing technique for the design of tall steel buildings subject to multiple drift constraints. *Structural Design of Tall Buildings Journal*, **2**, 17–32.

Cohn, M.Z. (1991) Theory and practice of structural optimization. Proceedings of NATO–ASI Conference on Large-scale Structural Systems, Berchtesgaden.

Computers and Structures (1989) *ETABS, Building Analysis and Design*, Berkeley, CA.

Fleury, C. (1979) Structural weight optimization by dual methods of convex programming. *International Journal of Numerical Methods in Engineering*, **14**, 1761–83.

Grierson, D.E. and Cameron, G.E. (1989) Microcomputer-based optimization of steel structures in professional practice. *Microcomputers in Civil Engineering*, **4**(4), 289–96.

Grierson, D.E. and Cameron, G.E. (1991) *SODA: Structural Optimization Design and Analysis*, User Manual, Waterloo Engineering Software, Waterloo.

Grierson, D.E. and Lee, W.-H. (1984) Optimal synthesis of steel frameworks using standard sections. *Journal of Structural Mechanics*, **12**(3), 335–70.

Grierson, D.E. and Pak, W.H. (1992) Discrete optimal design using a genetic algorithm. Proceedings of NATO–ARW Conference on Topology Design of Structures, Sesimbra.

Haftka, R.T. and Gürdal, Z. (1992) *Elements of Structural Optimization*, 3rd edition, Kluwer, Dordrecht.

Johnston, B.G. (ed.) (1976) *Guide to Stability Design Criteria for Metal Structures*, 3rd edition, Wiley, New York.

Moharrami, H. (1993) Design optimization of reinforced concrete structures. PhD Thesis, Civil Engineering, University of Waterloo.

Moharrami, H. and Grierson, D.E. (1993) Computer-automated design of reinforced concrete frameworks. *ASCE Journal of Structural Engineering*, **119**(7) 2036–58.

Ontario Ministry of Transportation and Communications (1983) Ontario highway bridge design code.

Pak, W.H. (1993) Genetic algorithms in structural engineering. MASc Thesis, Civil Engineering, University of Waterloo.

Research Engineers (1987) *STAAD-III, Integrated Structural Design System*, New Jersey.

Schmit, L.A., Jr, and Fleury, C. (1979) An improved analysis/synthesis capability based on dual methods – ACCESS 3. Proceedings of AIAA/ASME/AHS 20th Structures, Structural Dynamics and Materials Conference, St. Louis, MO.

Xu, L. (1993) Optimal design of steel frameworks with semi-rigid connections. PhD Thesis, Civil Engineering, University of Waterloo.

Xu, L. and Grierson, D.E. (1992) Computer-automated design of semi-rigid steel frameworks. *ASCE Journal of Structural Engineering*, **119**(6) 1740–60

Zangwill, W.I. (1969) *Nonlinear Programming: A Unified Approach*, Prentice-Hall, Englewood Cliffs, NJ.

# 13

## Reliability-based structural optimization

### DAN M. FRANGOPOL and FRED MOSES

### 13.1 INTRODUCTION

The aim of structural optimization is to achieve the best possible design. Using deterministic concepts this has been interpreted in the following manner (Moses, 1973). The objective function, load conditions, design requirements, failure modes and design variables are all treated in a nonstatistical fashion. Furthermore, it is assumed that design codes provide adequate safety factors to cover any likelihood of failure. By using automated optimization procedures, member sizes are found so that the objective function (e.g. cost, weight) is minimized without violating any design requirements. In some instances deterministic optimization promotes structures with less redundancy and smaller ultimate overload margins than obtained with more conventional and conservative design procedures (Feng and Moses, 1986b; Frangopol and Klisinski, 1989a). Consequently, deterministic optimized structures will usually have higher failure probabilities than unoptimized structures. The reason for this is that the safety factors specified in design codes usually refer to the design of elements, such as beams and colums, and do not allow for a simultaneous occurrence of many failure modes in an optimized structure designed to its limits (Moses, 1973). Therefore, a balance must be developed between the safety needs of the structure and the aims of reducing its cost. Clearly, this requires the use of reliability-based design concepts in structural optimization.

Uncertainties are unavoidable in the design and evaluation of engineering structures. Structural reliability theory has been developed during the past three decades to handle problems involving such uncertainties. This theory has had considerable impact in recent years on structural design. For example, partial safety factors based on load and strength uncertainties have been proposed in the United States, Canada and Europe for building, bridge and offshore platform design codes. Today, the need for designing the best structures by balancing the conflicting goals of safety and economy provides the motivation for the application of structural reliability theory in structural optimization.

Reliability is a quantitative measure of confidence or belief that a system will successfully serve its intended purpose. Implicit in such a definition is the assumption that success and failure can be uniquely defined. With only two possible outcomes, reliability can be defined by the probability of success or the probability of failure, where

$$P_{\text{failure}} = 1.0 - P_{\text{success}} = 1 - \text{reliability} \qquad (13.1)$$

In most practical engineering undertakings, the probability of success is so close to 1.0 that it is customary to reference reliability by the probability of failure. The probability of failure can be interpreted as the number of failures that would occur given a large set of independent trials, i.e. 'a one in a thousand chance'. (In some instances, e.g. meeting structural serviceability criteria, there is no clear-cut distinction as to whether a system has passed or failed the requirements. In other words, the observation of the event even after the event has occurred is in dispute. A branch of probability known as fuzzy-set theory has recently arisen to deal with such problems in a decision mode. For the present chapter, the assumption is made of a crisp observation such that there is agreement regarding an outcome that is either a survival or a failure event.)

The aim of this chapter, which is partly based on previous work of the authors and their coworkers, is to present a brief state-of-the-art review of reliability-based structural optimization (RBSO) with applications. This review is divided into eight sections treating basic concepts and problem types, basic reliability modeling, element reliability-based optimization, design code optimization, analysis of system reliability, system reliability-based optimization, multiobjective reliability-based optimization, and residual and damage-oriented reliability-based optimization.


## 13.2 BASIC CONCEPTS AND PROBLEM TYPES

### 13.2.1 Concept development

Forsell (1924) is apparently the first who formulated the structural optimization problem as minimization of total expected cost. The term 'total cost' implies the initial cost (sum of all costs associated with erection and operation of the structure during its projected design life without failure) and the expected cost of failure (sum of all costs associated with the probability of failure, e.g. expected cost of repairs, expected cost of disruption of normal use, expected cost of loss of human life). This total cost criterion governs nearly all of the work that flourished during the 1950s and is recorded in Johnson (1953), Ferry-Borges (1954), Freudenthal (1956), and Paez and Torroja (1959), among others.

During the next decade (1960–1970) the importance of the Bayesian decision theory in structural optimization was recognized, and several devices aimed at improving the value of the optimum reliability-based solutions were introduced by Benjamin (1968), Cornell (1969a) and Turkstra (1967, 1970), among others. In the same decade, the need to reduce structural weight without compromising structural reliability, particularly in aerospace applications, was also recognized. The first major efforts along this line were led by Hilton and Feigen (1960). Subsequently, other efforts in this category were carried out by Kalaba (1962), Switzky (1964), and Moses and Kinser (1967), among others. The first survey of the optimum reliability-based design field has been written by Moses (1969); it was stated in this study, which focused on the relationship between reliability and optimization, that an optimization procedure which uses overall structural failure probability as the behavior constraint should produce more balanced designs, consistent with the development of rational safety. Furthermore, it was also stated that a truly optimum design should consider the behavior of the structure for various types of loading conditions as well as possible strength deteriorations.

By 1970 it had become apparent that the concepts and methods of probability are the bases for the development of optimum criteria for structural design (Turkstra, 1970). As a result, during the past two decades many investigators focused their efforts on developing the following:

- strategies for identifying failure modes for structural systems which incorporate brittle or ductile behavior, or both (Moses, 1977, 1982; Thoft-Christensen and Murotsu, 1986; Ishikawa and Iizuka, 1987; Zimmerman, Corotis and Ellis, 1991; among others);
- more sophisticated methods, taking failure mode correlation into account, for determining overall system reliability of both brittle and ductile structures (Vanmarcke, 1971; Ang and Ma, 1981; Chou, McIntosh and Corotis, 1983; Ditlevsen, 1979; Grimmelt and Schuëller, 1983; among others).
- random process and field structural reliability models (Vanmarcke, 1984; Madsen, Krenk and Lind, 1986; Melchers, 1987; Wen and Chen, 1989; among others), and simulation techniques (Deak, 1980; Bjerager, 1988; Bucher, 1988; Verma, Fu and Moses, 1990; among others).
- rational automatable structural design optimization techniques (Kirsch, 1981; Osyczka, 1984; Vanderplaats, 1984, 1986; Frangopol and Klisinski, 1989a; among others).
- automated RBSO procedures for minimum material cost with failure probability constraint, for maximum safety based on fixed weight or total expected cost, and for minimum total expected cost (Vanmarcke, 1971; Mau and Sexsmith, 1972; Moses, 1973; Frangopol, 1985b, c, 1986a; Feng and Moses, 1986a; Thoft-Christensen and Murotsu, 1986; Nakib and Frangopol, 1990a, b; among others).
- multiconstraint reliability-based optimal procedures for structural systems subject to probability requirements imposed at both serviceability and ultimate limit states (Cohn and Parimi, 1972; Parimi and Cohn, 1978; Frangopol, 1985b, 1986a, b, 1987; Frangopol and Fu, 1990; among others).
- damage-tolerant design and optimization of nondeterministic systems (Feng and Moses, 1986b; Frangopol and Fu, 1989; Frangopol, Klisinski and Iizuka, 1991; among others).
- reliability-based optimization of inspection and repair strategies (Sørensen and Thoft-Christensen, 1987; Fu, Liu and Moses, 1991; among others).
- time variant RBSO (Kim and Wen, 1987; Iizuka, 1991; among others).
- reliability-based shape optimzation of structural and material systems (Thoft-Christensen, 1987, 1991; Sørensen and Enevoldsen, 1989; Enevoldsen, Sørensen and Sigurdsson, 1990; Murotsu and Shao, 1990; Shao, 1991; among others).
- multiobjective structural optimization using reliability-based philosophy (Frangopol and Fu, 1989, 1990; Frangopol and Iizuka, 1991a, b; Fu and Frangopol, 1990a, b; Iizuka, 1991; among others).
- sensitivity studies and techniques to establish response changes of the RBSO

solutions to change in problem parameters (Moses, 1970; Frangopol, 1985a; Sørensen and Enevoldsen, 1989; among others).

The preceding developments allow reliability-based design concepts, which offer a fundamentally new approach to optimum design, to be used by structural engineers to a wide spectrum of design problems including trusses, frames, offshore platforms, highway bridges, shells and pressure vessels.

### 13.2.2 Problem types

Optimizing structural members and systems using reliability-based concepts and methods raises questions as to the meaning of an optimum design solution (Moses, 1969). Many objective functions have been proposed (Hilton and Feigen, 1960; Kalaba, 1962; Switzky, 1964; Moses and Kinser, 1967; Moses, 1969, 1970, 1973, 1977; Moses and Stevenson, 1970; Turkstra, 1970; Rosenblueth and Mendoza, 1971; Vanmarcke, 1971; Cohn and Parimi, 1972; Mau and Sexsmith, 1972; Frangopol and Rondal, 1976; Parimi and Cohn, 1978; Surahman and Rojiani, 1983; Rojiani and Bailey, 1984; Frangopol, 1984a, b, 1985b, c, 1986b, 1987; Feng and Moses, 1986b; Rosenblueth, 1986; Thoft-Christensen and Murotsu, 1986; Frangopol and Nakib, 1987; Ishikawa and Iizuka, 1987; Kim and Wen, 1987, 1990; Rackwitz and Cuntze, 1987; Sørensen, 1987, 1988; Sørensen and Thoft-Christensen, 1987; Thoft-Christensen, 1987, 1991; Sørensen and Enevoldsen, 1989; Frangopol and Fu, 1989, 1990; Enevoldsen, Sørensen and Sigurdsson, 1990; Murotsu and Shao, 1990; Nakib and Frangopol, 1990a, b; Fu and Frangopol 1990a, b; Frangopol and Iizuka, 1991a, b; Iizuka, 1991; Mahadevan and Haldar, 1991; Shao, 1991; among others).

These include cost and utility functions that should be minimized and maximized, respectively, as follows.

1. Minimization of the total expected cost of the structure $C = C(\mathbf{X})$ expressed as

$$C = C_0 + C_f P_f \tag{13.2}$$

where $C_0$ is the initial cost, which is a function of the vector of design variables $\mathbf{X}$, $C_f$ the cost of failure, and $P_f$ the probability of failure, which is also a function of $\mathbf{X}$.

2. Maximization of the total expected utility function, $U$, expressed in a monetary form as

$$U = B - C_0 - L \tag{13.3}$$

where $B$ is the benefit derived from the existence of the structure, and $L$ the expected loss due to failure. As shown by Rosenblueth (1986), in some cases $B$ can be a function of the design variables $\mathbf{X}$. All the quantities in equation (13.3) are expected present values. Conversion of future into present values is done by means of an exponential function (Rosenblueth, 1986).

Because of the difficulties in assuming monetary values to all failure consequences (e.g. placing monetary value on human life, environmental effects), the preceding (i.e. equations (13.2) and (13.3)) reliability-based unconstrained optimization formulations are of limited interest for practical purposes. As an alternative to placing a monetary value on failure consequences, Hilton and Feigen (1960) were the first to propose a reliability-based weight minimization formulation as follows.

3. Find the design variable vetcor

$$\mathbf{X} = (X_1 \, X_2 \cdots X_n)^t \tag{13.4}$$

that will minimize the objective function

$$W = W(\mathbf{X}) \tag{13.5}$$

subject to

$$P_f = P_f(\mathbf{X}) \leqslant P_f^0 \tag{13.6}$$

in which $W$ is the total structural weight, $P_f$ the probability of failure of the structure, and $P_f^0$ the allowable probability of failure. Side constraints of the form

$$X_i^l \leqslant X_i \leqslant X_i^u \qquad i = 1, 2, \ldots, n \tag{13.7}$$

can also be imposed, in which the superscripts l and u denote lower and upper bounds, respectively. In the above constrained optimization formulation (i.e. equations (13.4)–(13.7)) it is assumed that the measure of performance is given by the weight of the structure.

For the particular case of weakest-link structures (e.g. statically determinate systems) with independent failure modes under a single loading condition, Switzky (1964) showed that at the optimum (i.e. $P_f = P_f^0$) the following approximate linear relationship exists

$$\frac{W_i}{W} = \frac{P_{fi}}{P_f^0} \tag{13.8}$$

where $W_i$ and $P_{fi}$ are the weight and the probability of failure of member $i$, respectively. Other assumptions were used in the development of equation (13.8) including that the failure probability and weight functions are both linear, $P_f = \sum P_{fi}$ and $W = \sum W_i$, and that a change in $P_f^0$ does not affect the ratio $W_i/W$.

Another formulation of the optimum design problem in a reliability-based context is to minimize the probability of failure of a structure for a fixed structural weight (initial cost). In mathematical terms the formulation of this problem is as follows.

4. Find the design variable vector $\mathbf{X}$, that will minimize the objective function $P_f$, subject to

$$W(\mathbf{X}) \leqslant W^0 \tag{13.9}$$

where $W^0$ is the allowable total structural weight (initial cost).

The above four formulations are single-objective single-limit state RBSO problems. In practice there are many situations in which the designer wants to optimize a structure with regard to the occurrence of two or more limit states. In general, the RBSO formulation with collapse and loss of serviceability as the failure criteria may be stated in mathematical form as follows:

$$W = W(\mathbf{X}) \leqslant W^0 \tag{13.10}$$

$$P_{fcol} = P_{fcol}(\mathbf{X}) \leqslant P^0_{fcol} \tag{13.11}$$

$$P_{funs} = P_{funs}(\mathbf{X}) \leqslant P^0_{funs} \tag{13.12}$$

in which $P_{fcol}$ is the probability of plastic collapse, $P_{funs}$ the probability of unserviceability, and the superscript $^0$ denotes prescribed (allowable) values. Three single-objective biconstraint formulations could be used for finding the optimum solution as follows.

Find the design variable vector $\mathbf{X}$ which minimizes

5.
$$W(\mathbf{X}) \tag{13.13}$$

subject to
$$P_{fcol}(\mathbf{X}) \leqslant P^0_{fcol} \tag{13.14}$$

$$P_{funs}(\mathbf{X}) \leqslant P^0_{funs} \tag{13.15}$$

6.
$$P_{fcol}(\mathbf{X}) \tag{13.16}$$

subject to
$$W(\mathbf{X}) \leqslant W^0 \tag{13.17}$$

$$P_{funs}(\mathbf{X}) \leqslant P^0_{funs} \tag{13.18}$$

7.
$$P_{funs}(\mathbf{X}) \tag{13.19}$$

subject to
$$W(\mathbf{X}) \leqslant W^0 \tag{13.20}$$

$$P_{fcol}(\mathbf{X}) \leqslant P^0_{fcol} \tag{13.21}$$

The formulation 5 is useful when the optimization consists in finding the minimum weight (or initial cost) structure, given the maximum specifed values of the risk levels with regard to occurrence of both collapse and unserviceability. The formulations 6 and 7 are useful when the most reliable structure against the occurrence of collapse and unserviceability is, respectively, sought, given both the maximum specified total weight (or initial cost) of the structure and the preassigned maximum risk level against the occurrence of the other limit state of interest. For the particular case in which the cost is of no concern (i.e. $W^0$ is not specified) in formulations 6 and 7, the most reliable structure against the occurrence of collapse and unserviceability is, respectively, sought, given the allowable risk level against the occurrence of the other limit state of interest.

If $m$ limit states are of concern and the total weight is the objective to be minimized, the RBSO formulation may be stated in mathematical form as follows.

8. Find the design variable vector $\mathbf{X}$ that will minimize the total weight $W(\mathbf{X})$ subject to

$$P_{fj} = P_{fj}(\mathbf{X}) \leqslant P^0_{fj} \qquad j = 1, 2, \ldots, m \tag{13.22}$$

where $P_{fj}$ is the probability of occurrence of limit state $j$.

In RBSO, the selection of the objective function, the limit states and their allowable probability levels are extremely important considerations (Parimi and

Cohn, 1978; Frangopol, 1985c; Kim and Wen, 1987). The limit states in equation
(13.22) may be at the member level or system level, or both, and the allowable
probability levels should be chosen accordingly. Much effort has been devoted in
(a) developing reliability-based element (member)-level code formats (Ellingwood
*et al.*, 1980; Ellingwood and Galambos, 1982; Corotis, 1985; among others), (b)
improving the reliability bases of structural safety and design at both element
and system level (Ferry-Borges, 1954; Freudenthal, 1956; Cornell, 1967, 1969b;
Benjamin, 1968; Ang and Cornell, 1974; Moses, 1974, 1982, 1990; Ditlevsen, 1979;
Ang and Ma, 1981; Chou, McIntosh and Corotis, 1983; Shinozuka, 1983;
Vanmarcke, 1984; Madsen, Krenk and Lind, 1986; Thoft-Christensen and
Murotsu, 1986; Melchers, 1987; Bjerager, 1988; Bucher, 1988; Frangopol, 1989;
Wen and Chen, 1989; Frangopol and Corotis, 1990; Verma, Fu and Moses, 1990;
among others), (c) choosing an optimum level of the prescribed failure probabilities
$P_{fj}^0$ (Turkstra, 1967, 1970; Rosenblueth, 1986; among others), and (d) developing
computational optimization algorithms (Rosenbrock, 1960; Schmit, 1969; Kirsch,
1981; Duckstein, 1984; Koski, 1984; Osyczka, 1984; Vanderplaats, 1984, 1986;
Arora, 1989; Frangopol and Klisinski, 1989a; among others).

### 13.2.3 Hierarchy

The *n* design variables in the vector **X** (equation (13.4)) are varied during the
RBSO process in order to improve the design until the optimum is reached.
These variables can be divided into five groups (Schmit, 1969; Kirsch, 1981; Thoft-
Christensen, 1991):

- sizing design variables (e.g. the cross-sectional dimensions of structural members);
- shape (configuration, geometric layout) design variables (e.g. the coordinates
  of joints in a framed structure, the length of spans in a continuous bridge);
- material design variables (e.g. the types of mechanical and/or physical properties of the materials to be used);
- topological (general arrangement) design variables (e.g. the number of spans
  in a bridge);
- structural system types (e.g. truss structure, framed structure, plated structure).

Most RBSO studies deal exclusively with the first and lowest category in this
hierarchy of the design variables (i.e. the selection of cross-sectional sizes) because
of the relative simplicity of this problem. However, it should be noted that
extending optimization capabilities upward into the design variable hierarchy
(e.g. consideration of both shape and cross-sectional design variables) may
appreciably improve the optimum solution (Murotsu and Shao, 1990; Shao, 1991;
Thoft-Christensen, 1991). Some computational experience with the optimum fiber
orientation of a multiaxial fiber reinforced laminate under nondeterministic
conditions has been recently reported by Shao (1991). It is apparent that much
work remains to be done until the stage arrives when the designer will be able
to obtain for a specific structure a practical total optimum reliability-based
solution in which the type of structure, topology, material shape and member
sizes are all treated simultaneously as design variables.

## 13.3 BASIC RELIABILITY MODELING

### 13.3.1 Introduction

**Probability** deals with the future outcome of real or imagined experiments. If the outcome of the experiment was predetermined, then we are dealing with a deterministic and not a random phenomenon. It is the natural or uncontrolled variations as well as the lack of professional knowledge of all the important variables that lead to uncertainty in the result of an experiment.

The use of structural reliability concepts for structural safety requires descriptions of random phenomena and calculation of failure probability and risk. Because insufficient observational data exists for applying the methods of classical statistics, the tools of probability theory must be used. These can be applied in several ways to predict events.

- Fit existing probability functions such as normal, log-normal, or Weibull distributions to the phenomena (load, resistance, etc.) being described.
- Derive distribution functions from an underlying physical process such as a sum, product, repeated trial. For example, the maximum event in, say, 25 years may be derived from distribution of maximum annual events by a process which leads to an extremal distribution.
- Compute distributions derived from functions of more basic random variables as, for example, the wind-induced structure response being a function of the random variables associated with wind speed, drag coefficient, dynamic behavior, etc.

Considerable background material on probability applications in structures may be found in the books by Benjamin and Cornell (1970), and Ang and Tang (1975), or the more recent texts on structural reliability by Thoft-Christensen and Baker (1982), Ang and Tang (1984), Yao (1985), Madsen, Krenk and Lind (1986), Thoft-Christensen and Murotsu (1986), and Melchers (1987), among others. Several journals now deal extensively with the subject, including *Structural Safety* and *Probabilistic Engineering Mechanics*. Specialized conferences on structural reliability include the ASCE Speciality Conference and the international conferences ICASP (International Conference on Applications of Statistics and Probability in Civil Engineering) and ICOSSAR (International Conference on Structural Safety and Reliability), which are periodically held every four years in different countries.

### 13.3.2 Fundamental reliability problem

A basic design decision problem is now considered. In structural engineering this problem is called a fundamental reliability problem which considers the case where all the uncertainty in strength is associated with one capacity variable, $R$, and all the load variability with one demand variable, $S$. The fundamental structural reliability or $R$–$S$ problem has several aspects:

- calculation of failure probability, $P_f$, or probability that $S$ is greater than $R$ from the distributions of $R$ and $S$;

- the sensitivity of $P_f$ to assumptions in the model, distribution function, probability laws and parameters such as means, standard deviations, etc.;
- a subsequent decision problem is to optimize a component by selecting its safety margin for either a target failure probability or a minimization of total expected cost.

A precise calculation for risk may be illustrated by writing the exact integral for calculating failure probability, $P_f$, as

$$P_f = P[R < S] = \int_0^\infty F_R(t) f_S(t)\, dt \qquad (13.23)$$

where, by definition, the distribution function of resistance, $R$, can be written as

$$F_R(t) = P[R < t] \qquad (13.24)$$

$f_S(t)$ is the probability density of the demand random variable, $S$. As an alternative form, the risk may be written as

$$P_f = 1 - \int_0^\infty F_S(t) f_R(t)\, dt \qquad (13.25)$$

Figure 13.1 shows a pictorial representation of these distributions.

For an exact solution, such integrals must usually be evaluated numerically. Simplifications result if one of the two uncertainties dominate, i.e. has much larger dispersion. If the coefficient of variation of resistance, $V_R$, greatly exceeds the coefficient of variation of load, $V_S$, then

$$P_f \cong F_R(\bar{S}) \qquad (13.26)$$

where $\bar{S}$ is the mean value of $S$.

Some closed-form solutions of equation (13.23) are possible in certain cases.

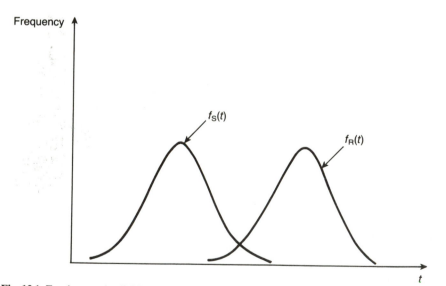

**Fig. 13.1** Fundamental reliability illustration showing distribution of load effect, S, and resistance, R.

Consider the case where both load and strength are normal (Gaussian) distributed variables.

Let $Z = R - S$, where $Z$ can be viewed as the margin of safety. $Z$ is also normal because $R$ and $S$ are normal. The statistical moments of $Z$ are the mean

$$\bar{Z} = \bar{R} - \bar{S} \tag{13.27}$$

and the variance (for independent $R$ and $S$)

$$\sigma_Z^2 = \sigma_R^2 + \sigma_S^2 \tag{13.28}$$

The failure probability is then

$$P_f = P[Z < 0] = F_U\left(\frac{0 - \bar{Z}}{\sigma_Z}\right) = F_U(-1/V_Z) \tag{13.29}$$

where $F_U$ is the standard normal form, and $V_Z$ the coefficient of variation (COV) of $Z$.

In terms of the central or mean safety factor, $\bar{n} = \bar{R}/\bar{S}$

$$P_f = 1 - F_U\left[\frac{\bar{R} - \bar{S}}{(\sigma_R^2 + \sigma_S^2)^{1/2}}\right] = 1 - F_U\left[\frac{\bar{n} - 1}{(\bar{n}^2 V_R^2 + V_S^2)^{1/2}}\right] \tag{13.30}$$

A similar closed-form expression for $P_f$ occurs when both $R$ and $S$ are log-normal variables. In that case, we obtain for values of $V_R$ and $V_S$ up to 30%

$$P_f = 1 - F_U[\ln(\bar{R}/\bar{S})/(V_R^2 + V_S^2)^{1/2}] \tag{13.31}$$

### 13.3.3 Simplified reliability design formats

The approach described above for calculating failure probabilities was developed by Freudenthal (1956) and Ang and Cornell (1974). Given a knowledge of the frequency distributions of load and strength, equation (13.23) can be used when sufficient data is available to determine the distribution functions for $R$ and $S$. It is thus a powerful analytical tool for predicting and controlling safety. It has severe limitations, however, both physical and psychological, which restrict its general application and acceptance. For example, the load random variable is not just a measurable environmental quantity, but also represents uncertainties in modeling load effects within the structure as well as in some of the basic limitations in the theory of mechanics. Similarly, the strength distribution depends not only on measurable material values, but also on assumptions of theoretical behavior of materials, quality of workmanship and fabrication control.

The major obstacles in accepting a fully probabilistic analytical treatment of safety directly in design can be summarized as follows.

- There are limitations in available data, particularly in a consistent form. This is combined with the low failure probabilities usually demanded of structures. A statistician, for example, would have to place a low confidence level on the results of most reliability predictions.
- Calculation of $P_f$ generally requires computer analysis programs that are independent of the usual design analysis programs. This makes it difficult and

inefficient to use during the design process since an iteration among programs would be required.

- Even if engineers felt it were possible to compute failure probabilities, the legal and professional responsibilities for establishing acceptable risks by consultants, owners, regulators, or even code writers, would present a severe problem.
- Safety factors in design are not always meant to protect against the phenomena being checked but rather to consider other unforeseen events such as possible future changes in function of the structure, blunders and accidents, serviceability requirements and possible rather gross approximate analysis and design assumptions.

Some of these objections to probabilistic analysis in structural engineering may not be as valid for such 'industrialized' structure types such as transmission lines or cranes. In cases where many structures of a basically similar design are to be constructed, this justifies a large engineering design and analysis cost to establish risk levels.

To overcome objections to explicit probability-based design procedures, there have been a number of attempts to introduce simplified formats. This work has been done by Cornell (1969b), Rosenblueth and Esteva (1972), Lind (1973) and Ang and Cornell (1974). It tries to incorporate as much as possible of the probabilistic basis while making the result appear in a format close to current design procedures.

For example, in the 'second moment format', the essential feature is that only mean and variance are used to describe a random variable. In order to have an explicit expression for the risk, an assumed probability law must be used. Alternatively, in place of a specified $P_f$ level, it is required that the mean safety margin must be $\beta$ standard deviations above zero, i.e.

$$\bar{Z} \geqslant \beta \sigma_Z \tag{13.32}$$

where $\beta$ is the **safety index** (also called **reliability index**). In terms of load and strength values, the safety criterion is met if

$$\bar{R} \geqslant \bar{S} + \beta(\sigma_R^2 + \sigma_S^2)^{1/2} \tag{13.33}$$

The safety index, $\beta$, may be related to risk levels if the fundamental reliability problem is used with a normal distribution of load and strength, as follows:

| $\beta$ | $P_f$ |
|---|---|
| 0 | 0.5 |
| 1 | 0.158655 |
| 2 | 0.022750 |
| 2.5 | 0.006210 |
| 3 | 0.001350 |
| 4 | $0.316712 \times 10^{-4}$ |

Alternatively, a log-normal assumption for $R$ and $S$ leads to the safety expression

$$\bar{R} \geqslant \bar{S} \exp[\beta(V_R^2 + V_S^2)^{1/2}] \tag{13.34}$$

Both equations (13.33) and (13.34) can easily be used to derive safety factors for use in design checking. Such factors depend on the desired safety index, $\beta$, and the uncertainty level expressed by $V_R$ and $V_S$.

### 13.3.4 Advanced reliability design formats

The first step in component reliability modeling is to define the limit state condition. This is usually written in the form of a limit state requirement:

$$g(X_1, X_2, X_3, \ldots) = 0 \qquad (13.35)$$

The $X_i$ are the random variables (e.g. load, resistance, dimensional, material). $g$ is defined such that a realization of the performance function $g$ which is positive means the component has not failed.

The function $g$ is basically deterministic in nature and expresses the mechanics of the strength check. A performance function (also called state or failure function) must be clear in the sense that if values of all the random variables were precisely known, then the engineer would know whether the structure failed (i.e. $g < 0$) or not ($g > 0$). In this approach, analysis or equation approximations can be treated as additional random variables. However, if after knowing the magnitudes of each random variable, the engineer is still not certain regarding failure or survival, it means the mechanics are not well understood. Such problems are obviously not ripe enough for reliability analysis when even the deterministic solutions are not known.

The first-order second moment approximation to reliability fits a mean and variance to the function $g$. The best approximation to the mean, $\bar{g}$, is

$$\bar{g} = g(\bar{X}_1, \bar{X}_2, \bar{X}_3, \ldots) \qquad (13.36)$$

i.e. set all the variables equal to the mean value and compute the corresponding $g$.

Approximating the performance function $g$ by a first-order Taylor series, expanded at the mean value of the variables, $X_i$, gives for the variance of $g$, $\sigma_g^2$, the following expression:

$$\sigma_g^2 = \sum (\partial g / \partial X_i)^2 \sigma_{X_i}^2 \qquad (13.37)$$

provided that the random variables, $X_i$, are uncorrelated. Knowing the mean and variance of $g$ allows the safety index to be computed:

$$\beta = \bar{g} / \sigma_g \qquad (13.38)$$

If $g$ were in fact normal, $\beta$ could provide an exact value for the failure probability $P_f$ from a normal table. By the central limit theorem, $g$ would be normal if the performance function were linear and all the $X_i$ were also normal. This first-order second moment expression for $\beta$ looks attractive since $\beta$ can easily be found using finite difference approximation for the derivative of $g$. A major difficulty arose with the expression, however, soon after it was proposed, which is called the lack of invariance. That is, one would like the same result regardless of mathematical form of $g$ as long as the physical models were the same. This turns out not to be the case because of the approximation of the derivative at the mean value.

This 'lack of invariance' was solved by the Hasofer–Lind (Hasofer and Lind, 1978) reliability criterion which recognized that the linearized approximation to the variance of $g$ should not be done at the mean, $\bar{g}$, but rather at a point on the limit state surface (i.e. $g = 0$). The 'best' point recommended to do this linearization would be the point (known as the **design point**) on the limit state surface $g = 0$ which is closest to the mean value. The reason for selecting this closest point is that it is likely to have the highest probability density. This design point can be found in several ways by iteration (algorithms of Rackwitz and Fiessler (1978), Baker (1977) or Wirshing (1985)) or by minimization (Shinozuka, 1983). The design point is the point on the limit state surface closest to the mean value (i.e. shortest distance to the mean). Mathematically, we now expand $g$ about a point on the limit state surface (also called failure surface) rather than the mean. The problem is finding the point, say $(X_1^*, X_2^*, \ldots)$ identically satisfying $g(\mathbf{X}^*) = 0$. This leads to an iterative calculation for the safety index.

This iteration provides the safety index for a particular design. The inverse design problem is to find the safety factors so that a target safety index is reached. It can be shown that if the function $g$ is linear in $\mathbf{X}$ and the variables $X_i$ are all normal, $\beta$ correlates to an exact probability of failure, $P_f$, using a normal table.

The AFOSM or advanced first-order second moment method goes another step further by approximating the 'true' nonnormal distribution of a variable $X_i$ by replacing it by a normal distribution (with appropriate mean and variance) obtained by matching at the design point $X_i^*$. Thus, the equivalent or approximating normal distribution will have (a) the same density and (b) the same cumulative probability as the actual nonnormal distribution at the point $X_i^*$. In fact the method can go further and treat correlated random variables by first performing an uncoupling that produces an independent set of variables. The transformation of the performance function to these new uncorrelated variables must also be carried out. Computer programs for automatically performing such safety index calculations for general performance functions of correlated variables with nonnormal distributions are generally available. In this approach, the safety index is uniquely defined as the shortest distance from the origin (normalized mean value) to a point on the limit state surface (Hasofer–Lind definition of safety or reliability index (Hasofer and Lind, 1978)). These AFOSM methods are all related to finding the safety index, or the probability of failure of individual components provided that a performance function and distributions for the random variables are both given. Only one failure mode is usually considered.

## 13.4 ELEMENT RELIABILITY-BASED OPTIMIZATION

### 13.4.1 General

Reliability-based specifications and design procedures are usually aimed at design of individual elements such as beams, columns and connections (Moses, 1974). This reflects the activities of designers who look on design as satisfying individual component safety checks. Thus, there evolved reliability-oriented prediction programs to estimate the probability of failure of single components. To apply such probabilistic results, design methods were developed so that the designers could be satisfied that a target element reliability index was realized (Moses, 1990).

### 13.4.2 Optimization formulations

The reliability-based optimization problem at the element level may be formulated in various ways, as follows.

1. Minimization of the total expected cost of the element $C_{el} = C_{el}(\mathbf{X})$ expressed as a function of the probability of occurrence of the dominant element limit state (e.g. collapse):

$$C_{el} = C_{o,el} + C_{fd,el}P_{fd,el} = C_{o,el} + C_{Fd,el} \qquad (13.39)$$

where $C_{o,el}$ is the element initial cost, which is a function of the vector of design variables $\mathbf{X} = (X_1 \, X_2 \cdots X_n)^t$, $C_{fd,el}$ the cost of element failure due to the occurrence of the dominant limit state, and $P_{fd,el}$ the probability of failure (i.e. occurrence of the dominant limit state of interest) of the element, which is also a function of $\mathbf{X}$.

   An illustration of the variation of total, initial and failure costs as a function of the risk $P_{fd,el}$ is shown in Fig. 13.2 (Moses, 1977). Increased initial cost causes a decreased risk and a reduction in the (expected) failure cost $C_{Fd,el}$. An optimum design (i.e. minimum expected total cost) is reached when an increase of initial cost $C_{o,el}$ is balanced by a reduction in the (expected) failure cost $C_{Fd,el}$.

2. Minimization of the total expected cost of the element $C_{el} = C_{el}(\mathbf{X})$ expressed as a function of the probabilities of occurrence of dominant ultimate (e.g. collapse) and serviceability (e.g. excessive elastic deformation) limit states $P_{f1,el}$ and $P_{f2,el}$ respectively:

$$C_{el} = C_{o,el} + C_{f1,el}P_{f1,el} + C_{f2,el}P_{f2,el} \qquad (13.40)$$

where $C_{f1,el}$ and $C_{f2,el}$ are the costs of failure of the element by occurrence of dominant ultimate and serviceability limit states, respectively. Both $P_{f1,el}$ and $P_{f2,el}$ are functions of the design variable vector $\mathbf{X}$.

3. Minimization of the total expected cost of the element $C_{el} = C_{el}(\mathbf{X})$ taking

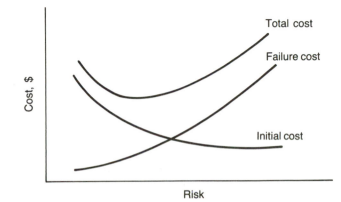

**Fig. 13.2** Total cost, failure cost and initial cost versus risk (Moses, 1977).

into account all possible ultimate and serviceability limit states:

$$C_{el} = C_{o,el} + \sum_{j=1}^{m} C_{fj,el} P_{fj,el} \quad j = 1, 2, ..., m \tag{13.41}$$

where $m$ is the total number of limit states of the element.

4. Minimization of the total weight of the element:

$$W_{el} = W_{el}(\mathbf{X}) \tag{13.42}$$

subject to

$$P_{fj,el}(\mathbf{X}) \leqslant P_{fj,el}^{o} \quad j = 1, 2, ..., m \tag{13.43a}$$

or, alternatively,

$$\beta_{j,el}(\mathbf{X}) \geqslant \beta_{j,el}^{o} \quad j = 1, 2, ..., m \tag{13.43b}$$

where

$$\beta_{j,el}(\mathbf{X}) = \Phi^{-1}[1 - P_{fj,el}(\mathbf{X})] \tag{13.44}$$

is the reliability index with respect to the occurrence of limit state $j$, $\Phi^{-1}(a)$ is the value of the standard nominal variate at the probability level $a$, $P_{fj,el}^{o}$ and $\beta_{j,el}^{o}$ are allowable values of $P_{fj,el}$ and $\beta_{j,el}$, respectively. Fabrication or other side constraints can also be imposed in the form given by equation (13.7).

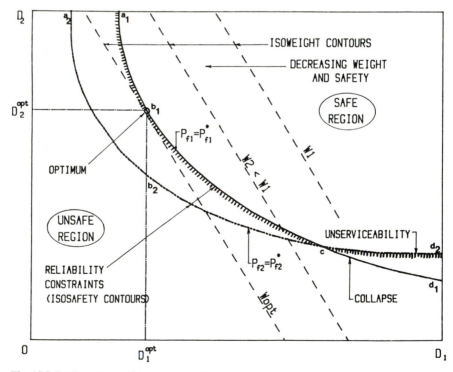

**Fig. 13.3** Design space with two reliability constraints: collapse critical at the optimum (Frangopol, 1986a).

5. Minimization of the probability of occurrence of a specified limit state $k$:

$$P_{fk,el} = P_{fk,el}(\mathbf{X}) \qquad (13.45a)$$

subject to

$$W_{el}(\mathbf{X}) \leqslant W^o \qquad (13.46)$$

$$P_{fj,el}(\mathbf{X}) \leqslant P^o_{fj,el} \quad j = 1, 2, \ldots, m \quad j \neq k \qquad (13.47a)$$

Alternatively, using reliability indices described in the previous section instead of failure probabilities the above formulation could be viewed as the maximization of the reliability index with respect to the occurrence of a specified limit state $k$:

$$\beta_{k,el} = \beta_{k,el}(\mathbf{X}) \qquad (13.45b)$$

subject to the constraints given by both equation (13.46) and the following $m-1$ inequalities:

$$\beta_{j,el}(\mathbf{X}) \geqslant \beta^o_{j,el} \quad j = 1, 2, \ldots, m \quad j \neq k \qquad (13.47b)$$

Of course, side constraints can be added to this formulation in the form given by equation (13.7).

Figures 13.3, 13.4 and 13.5 (Frangopol, 1986a) are useful in understanding the solution of the formulation 4 for the particular case of the reliability-based

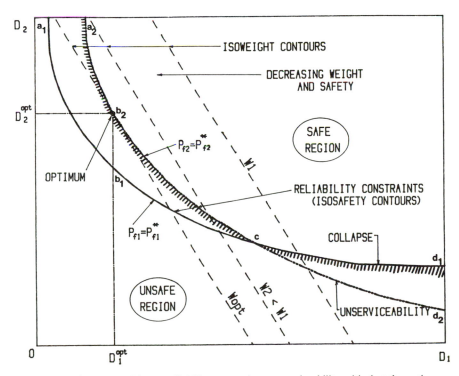

**Fig. 13.4** Design space with two reliability constraints: unserviceability critical at the optimum (Frangopol, 1986a).

optimization of an element with two limit states: collapse (e.g. formation of one mechanism) and unserviceability (e.g. first plastic hinge occurrence). In Figs 13.3–13.5 the design variable vector **X** has two components ($X_1 = D_1$ and $X_2 = D_2$) and the active collapse and loss of serviceability constraints are denoted as $P_{f1} = P_{f1}^*$ and $P_{f2} = P_{f2}^*$, respectively, where $P_{f1}^* = P_{f1}^o$ and $P_{f2}^* = P_{f2}^o$ are allowable failure probabilities.

The boundary of the acceptable design space (which contains points with acceptable probabilities of plastic collapse and first plastic hinge occurrences) depends on the relative magnitudes of the acceptable risk levels $P_{f1}^*$ and $P_{f2}^*$. In Fig. 13.3 the collapse constraint is the only one which is critical at the optimum (point $b_1$ in Fig. 13.3), while in Fig. 13.4 the unserviceability constraint is the only one which is critical at the optimum (point $b_2$ in Fig. 13.4).

It is interesting to note that (a) the points lying on some regions of the isosafety contours with regard to loss of serviceability limit state (e.g. regions $a_2b_2c$ in Fig. 13.3, $cd_2$ in Fig. 13.4, and $bc_2d$ in Fig. 13.5) have unacceptable reliabilities with regard to plastic collapse ($P_{f1} > P_{f1}^*$), (b) the points lying on some regions of the isosafety contours with regard to collapse (e.g. regions $cd_1$ in Fig. 13.3, $a_1b_1c$ in Fig. 13.4, and $a_1b$ and $de_1$ in Fig. 13.5) have unacceptable reliabilities with regard to loss of serviceability ($P_{f2} > P_{f2}^*$), and (c) both reliability constraints are active (i.e. satisfied as equalities) at one (e.g. point $c$ in Figs 13.3 and 13.4)

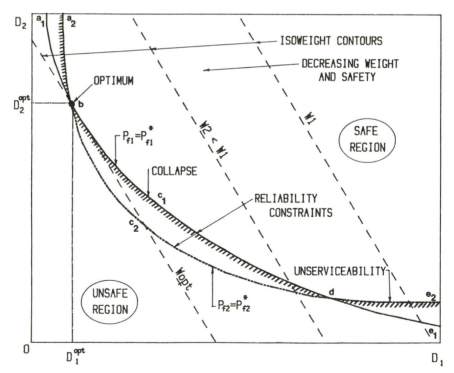

**Fig. 13.5** Design space with two reliability constraints: collapse and unserviceability critical at the optimum (Frangopol, 1986a).

or two (e.g. points b and d in Fig. 13.5) design points. It is also relevant to note that only one constraint (e.g. reliability against collapse) may govern the boundary of the acceptable design space; consequently, in this case the two reliability constraints have no common points. Another interesting remark is that one or both reliability constraints may be satisfied as equalities at the optimum. For example, the minimum weight solution is governed by the collapse constraint in Fig. 13.3 (optimum point $b_1$), by the unserviceability constraint in Fig. 13.4 (optimum point $b_2$), and by both constraints satisfied as equalities in Fig. 13.5 (optimum point b).

### 13.4.3 Applications

From the above formulations it is clear that an important step in element reliability-based optimization is the evaluation of risk associated with various element failure modes. If proper failure modes are not selected, the solution will not be useful. Another step in achieving the optimum solution is to use a reliable and efficient optimization algorithm. The illustrative examples presented in this section show only some of the possible applications of reliability-based optimization methods in elements design.

---

### EXAMPLE 13.1 STEEL BEAM-COLUMN (Iizuka, 1991)

A new approach for reliability analysis of steel beam-columns was recently presented by Iizuka (1991) using the theorem of total probability. In this approach both beam-column instability and material yielding are considered as limit states and both loading and material properties are considered as random variables. Figure 13.6 (Iizuka, 1991) displays the variations of the reliability indices against both material yielding and instability with the change in the cross-sectional area of the beam-column. A discrete constrained minimization programming method, considering standard steel beam-column selection based on minimum area design (equations (13.42) and (13.43b)), will permit optimum selection of nondeterministic beam-columns.

---

### EXAMPLE 13.2 STEEL BRIDGE GIRDER DESIGN FOR FATIGUE (Moses, 1977)

In highway bridges the design fatigue stress is a function of truck volume (load cycles), specific weld category life and shape of the random load histogram. The applied stress (load effect) is a function of truck bending moment, section modulus, impact factor and girder load distribution. These random variables have been studied in great detail in a number of field measurement, laboratory and analytical studies. These studies have evaluated the means and variances of these random

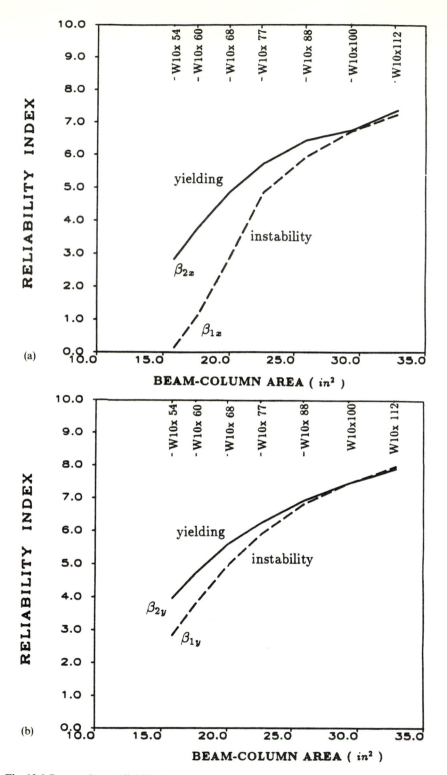

**Fig. 13.6** Beam-column reliability index: (a) instability can only occur about $x$ axis; (b) instability can only occur about $y$ axis (Iizuka, 1991).

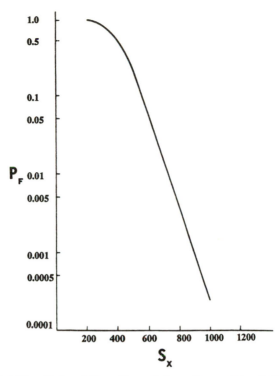

Fig. 13.7 Probability of fatigue failure, $P_f$, vs girder section modulus, $S_x$ (Moses, 1977).

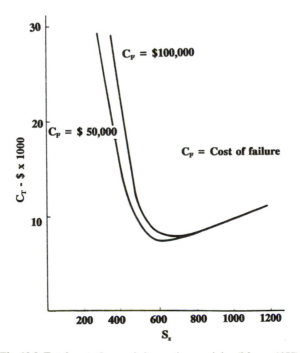

Fig. 13.8 Total cost, $C_T$, vs girder section modulus (Moses, 1977).

variables. The design variable in this case may be taken as simply the section modulus $S_x$ since the fatigue life distributions are not known as a function of say girder thickness or depth but only of weld category. Figure 13.7 (Moses, 1977) shows for a specific girder design the probability of fatigue failure (during the bridge lifetime) vs section modulus. By adding to initial cost $C_0 = C_{0,el}$ the expected cost of failure $C_F = C_{Fd,el}$ (equation (13.39)) we obtain the curves in Fig. 13.8 (Moses, 1977) which show the total expected cost $C_T = C_{el}$ (equation (13.39)) vs the girder section modulus $S_x$ for two values of the cost of failure $C_F = C_{fd,el}$.

---

## EXAMPLE 13.3 REINFORCED CONCRETE BEAM
(Moses, 1973)

Means and standard deviations of reinforced concrete beams resistances in several failure modes were provided in Moses (1973), Ellingwood and Ang (1974), and Surahman and Rojiani (1983), among others. Ellingwood and Ang (1974), working within the reliability-based design philosophy, evaluated the risks associated with existing design procedures for reinforced concrete beams in both flexure and

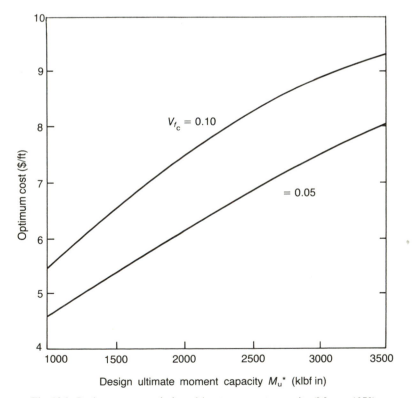

**Fig. 13.9** Optimum cost vs design ultimate moment capacity (Moses, 1973).

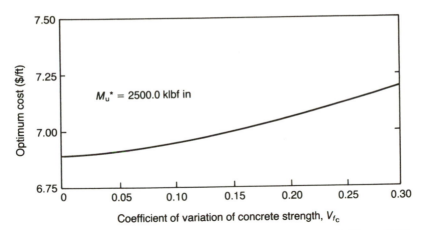

**Fig. 13.10** Optimum cost vs coefficient of variation of concrete strength (Moses, 1973).

shear. They were able to calculate the probability of failure in flexure and shear of simply supported beams. Load, strength and model prediction uncertainties were taken into account. Although Ellingwood and Ang's (1974) study does not include optimization, it is important because the risk of unfavorable performance of reinforced concrete beams commensurate with the level of uncertainty provides a consistent measure of design safety and thus furnishes a systematic measure for comparing design alternatives.

Some examples of reliability-based under-reinforced concrete beam optimization results are presented in Figs 13.9 and 13.10 (Moses, 1973), showing the variation of the optimum cost with design ultimate moment capacity, $M_u^*$, and coefficient of variation of concrete strength, $V_{f_c}$. The total (expected) cost includes concrete, steel and forming costs. The design problem is to find the mean values of the steel area, depth and width of the rectangular cross section, steel bar strength, and concrete strength to minimize the total cost subject to five constraints including both strength and side requirements.

Other optimum element reliability-based design examples under both time invariant and time varying loads have been presented (Iizuka, 1991; among others). In all reliability-based optimization applications, it is necessary to subject the results to sensitivity analysis to determine the influence of input parameters, including distribution functions and coefficients of variation, on optimum solutions (Moses, 1970, 1973; Frangopol, 1985a; among others).

## 13.5 DESIGN CODE OPTIMIZATION

### 13.5.1 Introduction

The implementation of reliability concepts into engineering practice may occur in several ways, for example, as an aid to evaluation of alternative design concepts or by establishing routine design criteria for checking structures. The most broadly acceptable implementation, however, will be by introducing a reliability basis into documented design checking codes such as in the steel, concrete or other code practices. Such industry-wide design checking formulae represent an industry consensus based on theory, practice and experience. Once it is in the book (blue, white, red, or whatever the corresponding cover), it will have to be considered by all designers in their routine operations.

Consequently, much of recent structural reliability research has been in the area of code implementation, that is, to provide the methodology by which design codes can introduce the benefits and advantages of reliability-based design. This is not to say that existing codes were not earlier influenced by reliability concepts since, in fact, codes have historically been modified in a quasi-Bayesian manner. This is apparent when one examines, for example, the reductions in safety factors for steel beams which have occurred over the last century. These reductions reflect the low failure rate or high reliability experienced by these components. A lower safety factor presumably provides a more economical design decision. Similarly, the recent code increases in seismic factors reflect the large damages that did occur during some earthquakes and the need to increase reliability of seismic resistant designs.

The difference between a historical or quasi-Bayesian evolution of codes and the present reliability developments is the latter formulation of the safety decision problem in a comprehensive statistical decision-based framework. These include the various types of load intensities such as dead, live, wind and seismic, analysis uncertainties, material properties, strength formula and failure consequences. The aim is to treat all the uncertainties consistently and the corresponding limit states and to derive design checking criteria which will lead to acceptable and economical reliability levels.

Despite some slow initial progress in introducing changes, the development of reliability-based checking formats has advanced rapidly in recent years. The new codes are often called limit-state formats but in reality they are independent of both the particular methodology for analyzing the force distribution within the structure and even the maximum acceptable strength definition for failure. Conventional linear structural analyses are in fact mostly used in code applications described as limit state design. The approximations inherent in an elastic force distribution are only one of the uncertainties in predicting the load effect and resistance.

### 13.5.2 Design checking formats

Traditional design checking procedures are based on allowable or working stress design (WSD) methods. In this format, the maximum design stress due to applied loads is restricted to be equal to the nominal yield divided by a safety factor. That is,

$$R/\text{SF} = D + L + W + \cdots \qquad (13.48)$$

where $R$ is the nominal resistance or component strength, $D$ the dead load effect, $L$ the live load effect, $W$ the environmental load effect, and SF the safety factor. This approach, if precisely followed, would lead to obvious contradictions. For example, environmental load is more uncertain than live load which, in turn, is usually more uncertain than dead load. Also, the combination of different loading effects makes it less likely that the full design value will be reached compared with situations in which only one load effect is present. Similarly, some components, e.g. members in compression or connections, may often require higher safety factors than members in tension because of their failure consequences.

Over the years, however, many of the obvious disadvantages of the WSD format have been indirectly eliminated. For example, the one-third increase in allowable stresses for some load combinations or the deliberate conservative reserves in connection formulae do, in fact, accomplish some of the goals of reliability-based formats. The limitations of WSD, however, become apparent in several distinct areas.

- It is difficult to distinguish the overall safety margin in WSD since safety is enhanced by the safety factor as well as deliberate but unquantified bias in strength formula, analysis procedures, etc.
- It is difficult to develop consistent safety levels when extending the provisions to new material applications or environmental exposures.
- It lacks flexibility in approaching nontraditional applications such as evaluating remedial construction, repair, rehabilitation, or short-term load exposures.
- Allowable stresses are awkward when dealing with nonlinear structure behavior problems such as seismic response, foundation flexibility, or dynamic response to ocean waves.

Reliability procedures have been shown to offer a comprehensive guideline for code writing. The first step is a code checking format followed by optimization of the checking factors.

### 13.5.3 Load and resistance factor design (LRFD) calibration

The terminology denoted as 'load and resistance factor design' has been generally adopted to implement reliability-based design practices in the United States. It is known in Canada and the United Kingdom as limit-state design while in other parts of Europe it is called partial safety factor design. Similarly, in concrete design practice it has been called ultimate strength design, while in highway bridge design it is called load factor design. The basic format is repeated here as:

$$\phi_i R = \gamma_D D + \gamma_L L + \gamma_W W + \cdots \tag{13.49}$$

where $R$, $D$, $L$ and $W$ retain the same definitions as in WSD, $\phi_i$ is the resistance factor on the $i$th component type, e.g. beam bending, shear, $\gamma_D$ the load factor applied to dead load effect, $\gamma_L$ the load factor applied to live load effect, and $\gamma_W$ the load factor applied to environmental load effect.

In checking, there may be more than one load combination used for the right-hand side in equation (13.49), for example, gravity alone, gravity plus wind, gravity plus seismic, overturning, etc. This is similar to the formulation in a WSD code. Each partial safety factor in equation (13.49), i.e. $\phi$ on the resistance and $\gamma$ on the load, is associated with its corresponding quantity.

The load and resistance factors depend on several items.

- The bias of that variable, i.e. ratio of mean value to nominal code recommended formula.
- Corresponding uncertainty of that variable.
- An overall calibration exercise which varies the partial load and resistance factors in order to come as close as possible to the target safety index over a range or possible parameters. The latter exercise is known as code calibration or code optimization and is described in more detail in the next section.

The LRFD format in equation (13.49) is not unique, since additional factors may also be used to augment the $\phi$ and $\gamma$ shown. For example, a system factor may reflect consequences of component failure such that

$$\phi_i = \phi_{i,R}\phi_S \tag{13.50}$$

where $\phi_S$ is the system factor and $\phi_{i,R}$ the resistance factor, for a component in the absence of consideration of failure consequences. The system factor can depend on redundancy, ductility, and consequences of component failure. Similarly the $\gamma$ factor on load effects can be made a function of analysis procedures, such as

$$\gamma_L = \gamma_1\gamma_A \tag{13.51}$$

where $\gamma_A$ is a factor based on analysis uncertainty including modeling details and any field performance proof testing of the analysis method, $\gamma_1$ a factor based on environmental load intensity uncertainty, and $\gamma_L$ the overall load factor.

Additional partial safety factors can lead to a finer 'tuning' of the code safety margin to the specific application that the user is designing. This approach is similar to European practices such as in the CEB code which contain several additional partial factors such as separate factors for workmanship and quality control procedures. Many of these latter applications are more appropriate for concrete structures, rather than steel structures, but there are instances where such factors apply also in steel.

There are obvious advantages to having more factors in attaining uniform reliability over different applications. The disadvantages, however, include the added effort required by the designer in applying more factors. In particular, there may be subjective interpretations in using such factors as workmanship, ductility, or failure consequences. There is the further limitation that adding more factors implies that the statistical data is sufficiently refined to reflect a breakdown of uncertainties into a greater number of variables. One possible area that benefits from more factors is the application to reevaluation of structures. For example, a recent code developed for evaluating existing bridges considers a variety of load and resistance factors based on the level of effort (i) to inspect the existing structure, (ii) to improve maintenance to avoid further deterioration, (iii) to perform accurate rather than routine approximate analysis, and (iv) to limit or enforce the external operational loading (Verma and Moses, 1989).

### 13.5.4 Code optimization

The process by which load and resistance factors are derived for codes has been denoted as **code calibration**. A direct reliability approach would be to analyze each component using the reliability and safety index calculation. Rather than applying equation (13.49), the member checking criterion would be that a design component is acceptable if the safety index exceeds the required target value. This approach is sometimes called a level II approach. A level III approach would use an integration or Monte Carlo simulation to evaluate the risk directly to avoid any approximations that are inherent in the AFOSM safety index method (section 13.3.4). The so-called level I procedure is generally used in US–LRFD codes and requires the partial factors to be specified by the code writers in such a way that they produce, on the average, the desired target safety index. For designers, the checking procedures are carried out in a deterministic manner.

In the code calibration, the following steps are needed.

1. Assemble a representative population of components that will be checked by the code. These components should be generally acceptable to industry and code writers. That is, they were checked by existing (WSD) codes or are known historically to have acceptable performance.
2. Develop a reliability formulation including a failure function ($g$), and statistical parameters and distributions for each of the random variables, including load, analysis, and strength.
3. Compute safety indices for each component by the AFOSM method.
4. Average these values and produce a target safety index, $\beta^{\mathrm{T}}$.
5. Find the load and resistance factors ($\gamma$ and $\phi$) appearing in the LRFD format (equation (13.49)) which minimize the following functions and satisfy the constraints:

$$\text{find} \qquad \gamma_i, \phi_i$$

$$\text{minimize} \quad \sum_i W_i (\beta_i - \beta^{\mathrm{T}})^2 \qquad (13.52)$$

$$\text{subject to} \quad \sum_i W_i = 1.0 \qquad (13.53)$$

where $W_i$ is the weighting factor assigned to design component $i$ in the population, $\beta_i$ the computed reliability for component $i$ using parameters $\gamma_i$ and $\phi_\gamma$, and $\beta^{\mathrm{T}}$ the target reliability.

In addition, another constraint in the optimization may be imposed to ensure that the average risk for all components equals the risk target, $P_{\mathrm{f}}^{\mathrm{T}}$, or

$$\sum_i W_i P_{\mathrm{f}i} = P_{\mathrm{f}}^{\mathrm{T}} \qquad (13.54)$$

where $P_{\mathrm{f}i}$ is the risk of failure or inverse function of safety index $\beta_i$ and $P_{\mathrm{f}}^{\mathrm{T}}$ the probability of failure equivalent to target reliability, $\beta^{\mathrm{T}}$.

In general, most studies have found that the safety indices found with WSD will have unexplainable variation. This is a result of the fact that design margins were not historically derived with any reliability models and further the presence of only a single safety factor in the format is too restrictive to achieve uniform indices ($\beta_i$). The LRFD format is more flexible simply by virtue of having more factors and hence can reduce the scatter in $\beta$ and produce components with more uniform reliability. This type of code optimization is well researched and documented. Standard minimization algorithms may be used to solve equations (13.52)–(13.54).

## 13.6 ANALYSIS OF SYSTEM RELIABILITY

### 13.6.1 Introduction

What is meant by the conventional design of structures is the design checking of individual components or members (beams, columns, piles, connections, etc.). Codes do not usually check the overall structure capacity. In some types of structures, several components must simultaneously fail in sequence before there is overall structure damage or failure. On the other hand, in other types of structures there may be a multitude of critical members; the failure of any single one of the members may lead to catastrophic conditions.

This discussion suggests that there is an area of investigation related to reliability of the overall structure system as distinct from the reliability of components. This is important since what is meant by reliability in an owner's or the public's mind is not related to component, but rather to the system reliability with respect to some specific consequence of failure.

Another limitation in code checking is that the target $\beta$ values discussed in section 13.5 refer to notional probabilities of component failure due to occurrence of overload and understrength. Most reported structural failures are due to accidents, fabrication mistakes, or human error which are not always influenced by code-specified safety margins. Rather, these types of failures must be controlled by tighter design review and inspection procedures which reveal such gross errors before they cause failure. System reliability can, in many cases, identify whether there are critical locations from which failure may initiate and assist in the best utilization of resources to promote better design accuracy, ductility, redundancy, or inspection. In-service assessments of existing structures are another area where comparative system reliability estimates can be used to recommend levels of in-depth inspection, analysis, testing, or repair.

It is important in some instances for engineers to understand both the limitations of present reliability-based code procedures as well as the potential usefulness of system reliability application. Examples include remedial studies of existing structures, comparison of new alternative structure concepts and geometries, and calibration of actual system reserve levels. The considerable research in recent years in structure system reliability is due to interest in nuclear plants, bridges, long span roofs, and offshore platforms where ultimate capacity rather than component serviceability is important and where identification of critical members for inspection, review, and strengthening is an option. In the future, more attention is expected to focus on system reliability, both in code work and in specific structural investigations for criteria selection, concept evaluation, or remedial re-evaluation.

### 13.6.2 Basic system models

The basic problem of system reliability is to estimate the system reliability from the reliability of the components. Several approaches have been proposed including direct methods (e.g. Monte Carlo simulation, point estimate, and response surface) and failure mode analysis.

The direct methods which are simple to apply have been generally thought to be too time consuming and offer little insight into the contributing factors in the system behavior. Recent developments, however, in simulation, such as importance sampling or directional simulation, have offered improvements in computation efficiencies while removing some of the limitations in other analysis approaches (Hasofer, 1987; Moses, Fu and Verma, 1989). Expected further cost reductions in computer time suggest that in the future there will be further advances in the direct methods.

The failure mode analysis will be emphasized here because of its wider usage in the structure application and the fact that it more easily lends itself to an optimization formulation. Three characteristics of a structure model greatly affecting its reliability performance are (i) the material behavior, (ii) the geometry, and (iii) the statistical correlation (Moses, 1982).

A structural system is an assemblage of components. Both the reliability and behavior of the components affect system reliability. Various types of material models are used for component behavior – these include elastic brittle, elastic–plastic (ductile), strain hardening in which loads increase after component yielding, and semibrittle in which loads decrease after yielding. Ductile components continue to carry load after failure. Brittle components shed all of their load to remaining components upon failure. Because of the load redistribution that occurs after component failure, the system reliability must be closely connected to the structural analysis.

The second system characteristic of importance is the geometry. Series and parallel geometries are idealized models that represent the two extremes of system behavior. However, most structures contain some combination, and it is instructive initially to consider these two limiting cases.

A **series system** fails if any component fails, and is analogous to a 'weakest link' chain which fails if any link fails. Statically determinate structures, and independent subsystems such as superstructure, and foundation, are examples of series systems. The reliability model of series systems is the same for both ductile and brittle components.

The resistance, $R$, of ideal series systems can be found from the $n$ component resistances:

$$R = \min_{i=1}^{n} (R_i) \tag{13.55}$$

where $R_i$ is the resistance of the $i$th component.

If the component resistances are independent, the distribution function, $F_R(r)$, of a series system can be written:

$$F_R(r) = 1 - \prod_{i=1}^{n} [1 - F_{Ri}(r)] \tag{13.56}$$

Series systems may have very poor reliability characteristics. The reliability of a series system will always be equal to or less than the reliability of the least reliable element. The more elements are added to the series system, the worse its reliability becomes.

The third characteristic of a system is the statistical correlation between component failure events. Correlation between component resistances indicates that there is some dependence between their strengths, so that if one element is above its mean strength, there is a high probability that the other strengths also exceed their mean. Such correlation may arise as a result of common material sources, similar fabrication, inspection, and control methods, and perhaps a uniform interpretation of the strength formulae by the designer. Independence of element resistance means that the strengths are not statistically correlated. More importantly, in cases of environmental loadings, correlation of component failure events results from the fact that the same external loading affects the reliability of each component.

For a series system, increasing the correlation between element resistances decreases the probability of failure (or increases the reliability) of the system.

Independence of element strengths increases the probability that one of the elements (the 'weakest link') will fail, and hence the system will fail. Increasing statistical correlation improves the reliability of a series system, i.e. common or identical material supply, fabrication, inspection, testing, etc. increases reliability.

Unlike a series system, a **parallel system** allows for redistribution of load following an element failure. As the failure of a single element is not tantamount to failure of the system, parallel systems are often referred to as 'fail-safe' systems. In a true parallel system, all elements must fail for the system to fail.

The reliability characteristics of parallel systems depend upon the element failure mode. Parallel systems can be considered for both ductile elements and brittle elements.

Ductile parallel systems have good reliability characteristics. The reliability is at least as good as the most reliable element. Increasing the number of parallel elements increases the reliability of the system.

The coefficient of variation of system resistance $R$ is increased by any positive correlation between elements. Thus, increasing the correlation between component resistances increases the uncertainty in the system resistance, and increases the probability of failure (or decreases reliability). Increasing correlation reduces the reliability of a parallel system.

Brittle elements that fail carry little or no load thereafter. In parallel structures, the load that was carried by the failed element must be distributed to the remaining elements. This greatly complicates the reliability analysis. The results depend on the safety factor levels used. If a member adjacent to a failed element has sufficient reserve capacity, it can pick up the redistributed load. Otherwise, it will fail and a cascading effect of failed members may result.

Several factors suggest that except for highly redundant systems (e.g. yarn-type systems), the reliability of brittle parallel systems can be approximated by a 'weakest link' analysis. Unless strength variability or reserve margins are large, failure of one element will usually trigger consecutive failures in other elements following load redistribution. This failure pattern is often termed 'progressive collapse'.

### 13.6.3 Failure modes analysis

A failure mode is any distinct sequence of component failures that causes the system to fail. For example, in a simple frame, plastic hinges can form leading to beam, sway, and combined types of plastic mechanisms. In typical structural frames or trusses, there is a very large number of possible failure modes (some are more probable than others). Failure mode analysis methods seek the modes that are most likely to occur, and use these modes to estimate the system reliability. The individual failure modes are considerably less complex to analyze than the failure of the system.

The modal failure probability, $P(g_i)$, for mode $i$ can be written in terms of the modal safety margin, $g_i$:

$$P(g_i) = \text{prob}(g_i \leqslant 0) \qquad (13.57)$$

and can be calculated from a knowledge of the distribution of load and component resistances. For a particular failure mode to occur, all of the components in that mode must fail. Thus, a failure mode can be modeled as a parallel system. The structure fails if any failure mode occurs. The overall structure can be modeled as a system with the failure modes in series, and the structure can be thought of as a series–parallel system.

The reliability of a structural system can be improved by increasing the length of the failure path, i.e. increasing the number of components that must fail before the structure fails (increasing the number of components in parallel). The length of the failure path can be increased by increasing the structural redundancy (i.e. the degree of static indeterminacy). However, increasing the structural redundancy by increasing the number of members also increases the number of possible failure modes. Increasing the number of possible failure modes (a series system) decreases system reliability. Some initial calculations by Gorman (1985) show that the greatest marginal benefit is achieved in going from zero redundancy (i.e. statically determinate) to two and three degrees of redundancy.

For the system, we then have

$$P_f(\text{system}) = \text{prob}(\text{any } g_i \leqslant 0) \qquad (13.58)$$

### (a) Correlation between modes

In general, there will be correlation between the failure modes, i.e. given that one mode has a high safety margin there is a greater probability that another mode will also have a high safety margin. The modal correlation arises from the following: (a) correlation between the component strengths due to common material, design methods, fabrication, etc.; (b) a single component may appear in several different modes and create correlation between the modes; (c) most importantly, there is a common source of loading that affects the safety of each mode.

The correlation plays an important role on the system reliability. If, for example, load coefficient of variation (COV) greatly exceeds strength COV, there may be high modal correlation. In this case, the most critical mode (highest probability of occurrence) may be close to the overall system reliability. If strength COV exceeds load COV, the correlation between modes may be small. In this case, it is necessary to consider in the system model all modes with significant probabilities of occurrence and the system reliability may act like a chain or weakest link model.

### (b) Large systems

For application to large-scale realistic structures, the system analysis should be divided into two parts: (a) engineering modeling, which means identification, description and enumeration of the statistically significant collapse modes, and (b) probabilistic calculations, to determine individual mode failure probabilities (or safety indices) and then combine them into an overall system assessment. For simple structures, such as one-story frames (plastic analysis), the identification and enumeration of collapse modes in terms of loads and component resistances is straightforward. For large structural frameworks, the search for critical modes is difficult and it is not possible to assure that all significant failure modes have been found (Moses and Rashedi, 1983).

Collapse mode events are correlated through loading and resistances, so an exact solution to equation (13.58) is usually impossible. Several investigations considered this combination problem by finding either approximate solutions or bounds (Nordal, Cornell and Karamchandani, 1987). Recently, there have been considerable improvements in system reliability bounds. These upper and lower bounds on system reliability include correlation. Monte Carlo simulation is another precise method to combine the statistics of each collapse mode to estimate the system reliability (Fu and Moses, 1988). Improvements in importance sampling techniques have made such approaches highly efficient.

### 13.6.4 A practical expression of system reliability

A conclusion drawn from the idealized behavior studies (e.g. series, parallel, ductile, brittle, correlated, independent) is that there is a multiplicity of possible modes which are interconnected in geometry, material behavior and correlation assumptions. This suggests that detailed failure mode analyses are needed for practical systems which can be studied to define realistic reliability ranges within the parametric values likely to be encountered.

Gorman and Moses (1979) had initiated studies to develop an incremental loading method. This was extended by Moses and Stahl (1979) to offshore platform frameworks. The method leads to a failure mode equation and second-order modeling (mean and variance) of structure systems appropriate for computing safety indices.

For general member behavior, the enumeration of significant modes is difficult and various fault tree search procedures are necessary. The object is to express the reserve margins, $g$, for each failure mode in terms of basic load and element resistance variables. This simplifies the subsequent computation of modal safety indices. The combination of these failure modes into a system reliability index is also complex.

To investigate a general system reliability approach, consider the incremental analysis technique. It identifies a failure mode by following a load path from initial component failure to system collapse and leads to a linear failure expression. The basic steps of the incremental loading method are quite simple. A much greater problem is the detailed 'book-keeping' that is involved in generating and keeping track of each significant possible failure alternative in a large realistically modeled system.

Gorman (1979) used heuristic techniques and Monte Carlo simulation to cause different failure sequences to appear. The latter were encouraged by using artificially high strength coefficients of variation. Linear programming and nonlinear programming techniques have also been used for finding ductile collapse modes for frames first analyzed for basic plastic mechanisms (Rashedi and Moses, 1986). More general approaches for systems with different member behavior have used safety index criteria to enumerate significant failure modes. It is important for large frameworks to reduce the number of structural reanalysis.

The incremental procedure begins with critical components and the possible existence of redundant load paths. For design and quality assurance, it also provides the overall importance of a member to the load carrying capacity of the system under a particular loading condition. The steps are outlined as follows:

1. Analyze the structural system intact with no members failed.
2. Identify critical components. This may be based on (a) mean strength as done above, (b) a safety index, or (c) other measures of importance based on professional judgment.
3. Allow a critical component to fail and recognize its ductile, brittle or semi-brittle behavior.
4. Examine the changes in component utilization ratios as candidates for a failure sequence.
5. For each component identified in step 4, extrapolate the load factor to find its failure value. Ignore sequences which have high load factors.
6. Remove in succession components identified in steps 4 and 5 as remaining important.
7. The procedure is continued until either the load needed to cause additional failures is larger and hence the path can be ignored or else several components have failed with little change in load factor suggesting that the load is flattening out.
8. Continue the analysis (steps 3–7) with other critical members.

This strategy may be viewed with a failure tree (Moses and Rashedi, 1983). Each branch leads to a failure mode expression. Some paths may be similar so failure mode correlation in assessing system reliability is important. It is often found that only two or three failed components may provide a sufficiently accurate reliability. Further, changes in utilization ratios for typical frameworks suggest that a component failure significantly affects only members located adjacent to it or in a parallel load path arrangement.

The identification strategy can be automated for analyzing large frameworks (Bjerager, Karamchandani and Cornell, 1987). Moses and Rashedi (1983) extended the basic incremental loading model to include brittle and semibrittle components, multiple loads, significant mode identification and even partial failure models. Applications included both evaluation of design concepts (e.g. structure topology and redundancy) and inspection criteria. The latter utilize the tree search to identify which members are critical to the system and should require additional quality review in design, construction and maintenance. The difficult task of combining the failure mode expressions into an overall system risk was also studied by Gorman (1979) who used Monte Carlo simulation. Fu and Moses (1988) found a new approach in importance sampling that both provided good accuracy and high efficiency. The major recent extensions of the failure tree system modeling were done by Cornell and his associates (Bjerager, Karamchandani and Cornell, 1987; Nordal, Cornell and Karamchandani, 1987). They have developed programs for practical modeling of offshore platform frameworks. Among the advances were improved assessments and comparisons of alternate structure geometries and topologies, effects of redundancy, simulated damage conditions and detailed investigation of structure redistribution capabilities, improved reanalysis, more accurate modal combination probabilities and more realistic post-failure behavior models. Further work was identified to include more detailed and realistic post-failure element models including combined loading cases and more intelligent generation of failure paths.

## 13.7 SYSTEM RELIABILITY-BASED OPTIMIZATION

### 13.7.1 General

In the previous section, various methods and techniques for reliability analysis of structural systems have been considered. Depending on the type and topology of the structure, material, configuration, sizing of members, and types of connections, as well as on the loading acting on the structure, the reliability of the system can be vastly different. The results of structural system reliability analysis may be used to assess the adequacy and relative merits of alternative system designs with respect to established objectives and to find the optimum solution. This section is devoted to this subject.

### 13.7.2 Unconstrained and single-constrained formulations

A summary of the classical (i.e. single-objective) system reliability-based optimization formulations was given by Parimi and Cohn (1978) and Frangopol (1987). The context of these classical formulations is provided by (a) the philosophy of limit states design for both steel and reinforced concrete structures, and (b) the structural system reliability and optimization approaches. The classical system reliability-based optimization formulations may be classified into three groups, namely

1. minimizing the total expected cost, which includes (a) the initial cost as a function of design variables, and (b) the expected loss in case of failure (equation (13.2)),
2. minimizing the initial cost (or structure weight), with system failure probability specified as a constraint, and
3. minimizing the system failure probability, with initial cost or weight specified as a constraint.

As mentioned previously, in the unconstrained formulation 1 some of the expected loss such as human life may not be easily converted into monetary value. Therefore, such a formulation is difficult to implement. Contributions to this formulation and to its applications have been made by Turkstra (1970), Rosenblueth and Mendoza (1971), Mau and Sexsmith (1972), Frangopol and Rondal (1976), Surahman and Rojiani (1983), Rojiani and Bailey (1984), and Rosenblueth (1986). Formulation 2 implicitly assumes a lifetime expected cost translated to the allowable value of the system failure probability.

Among the most used single-objective single-constraint system reliability-based optimization formulations pertaining to groups 2 and 3 above it is worthwhile mentioning the following. For group 2,

1. minimizing the initial cost (weight) $W$, when the allowable probability of collapse of the system $P^o_{fcol}$ is prescribed (the probability of loss of system serviceability $P_{funs}$ is of no concern):

$$\min W(\mathbf{X}) \tag{13.59}$$

$$P_{fcol}(\mathbf{X}) \leqslant P^o_{fcol} \tag{13.60}$$

where $\mathbf{X} = (X_1 \, X_2 \cdots X_n)^t$ is the design variable vector;

2. minimizing the initial cost (weight) $W$, when the allowable probability of unserviceability of the system $P^o_{funs}$ is prescribed (the probability of system collapse $P_{fcol}$ is of no concern):

$$\min W(\mathbf{X}) \tag{13.61}$$

$$P_{funs}(\mathbf{X}) \leqslant P^o_{funs} \tag{13.62}$$

For group 3,

1. minimizing the probability of collapse of the system $P_{fcol}$, when the allowable initial cost (weight) $W^o$ is prescribed (the probability of system unserviceability is of no concern):

$$\min P_{fcol}(\mathbf{X}) \tag{13.63}$$

$$W(\mathbf{X}) \leqslant W^o \tag{13.64}$$

2. minimizing the probability of system unserviceability, $P_{funs}$, when the allowable initial cost (weight) $W^o$ is prescribed (the probability of system collapse is of no concern):

$$\min P_{funs}(\mathbf{X}) \tag{13.65}$$

$$W(\mathbf{X}) \leqslant W^o \tag{13.66}$$

Fabrication or other side constraints can also be added to the above four formulations in the form given by equation (13.7).

### 13.7.3 Multiconstrained formulations

In practice there are many situations in which the designer wants to optimize a structure with regard to the occurrence of two or more limit states. In general, the problem of optimization under uncertainty with three limit states $(i, j, k)$ as the failure criteria may be stated in mathematical programming form as follows (Frangopol, 1987):

$$W = W(\mathbf{X}) \leqslant W^o \quad \text{cost (weight) constraint} \tag{13.67}$$

$$P_{fi} = P_{fi}(\mathbf{X}) \leqslant P^0_{fi} \quad \text{reliability constraint with respect to limit state } i \tag{13.68}$$

$$P_{fj} = P_{fj}(\mathbf{X}) \leqslant P^0_{fj} \quad \text{reliability constraint with respect to limit state } j \tag{13.69}$$

$$P_{fk} = P_{fk}(\mathbf{X}) \leqslant P^0_{fk} \quad \text{reliability constraint with respect to limit state } k \tag{13.70}$$

in which $\mathbf{X}$ is the vector of design variables, $W$ the initial cost (or weight) of the structure, $P_{fi}$ the probability of occurrence of limit state $i$ (e.g. excessive deformation), $P_{fj}$ the probability of occurrence of limit state $j$ (e.g. instability), $P_{fk}$ the probability of occurrence of limit state $k$ (e.g. plastic collapse), and the superscript $^0$ denotes prescribed (allowable) values.

Sixteen different single-objective multiconstrained optimum design problems could be obtained from the combinations of equations (13.67)–(13.70) as follows:

$$\min W, \text{ subject to constraints (13.68), (13.69) and (13.70)} \qquad (13.71)$$

$$\min P_{fi}, \text{ subject to constraints (13.67), (13.69) and (13.70)} \qquad (13.72)$$

$$\min P_{fj}, \text{ subject to constraints (13.67), (13.68) and (13.70)} \qquad (13.73)$$

$$\min P_{fk}, \text{ subject to constraints (13.67), (13.68) and (13.69)} \qquad (13.74)$$

$$\min W, \text{ subject to constraints (13.68) and (13.69)} \qquad (13.75)$$

$$\min W, \text{ subject to constraints (13.68) and (13.70)} \qquad (13.76)$$

$$\min W, \text{ subject to constraints (13.69) and (13.70)} \qquad (13.77)$$

$$\min P_{fi}, \text{ subject to constraints (13.67) and (13.69)} \qquad (13.78)$$

$$\min P_{fi}, \text{ subject to constraints (13.67) and (13.70)} \qquad (13.79)$$

$$\min P_{fi}, \text{ subject to constraints (13.69) and (13.70)} \qquad (13.80)$$

$$\min P_{fj}, \text{ subject to constraints (13.67) and (13.68)} \qquad (13.81)$$

$$\min P_{fj}, \text{ subject to constraints (13.67) and (13.70)} \qquad (13.82)$$

$$\min P_{fj}, \text{ subject to constraints (13.68) and (13.70)} \qquad (13.83)$$

$$\min P_{fk}, \text{ subject to constraints (13.67) and (13.68)} \qquad (13.84)$$

$$\min P_{fk}, \text{ subject to constraints (13.67) and (13.69)} \qquad (13.85)$$

$$\min P_{fk}, \text{ subject to constraints (13.68) and (13.69)} \qquad (13.86)$$

The formulations given by equations (13.71)–(13.86) are multiconstraint system RBSO problems with three (equations (13.71)–(13.74) or two (equations (13.75)–(13.86)) constraints. The formulations (13.71), (13.75)–(13.77) are useful when the optimization consists of finding the minimum weight structure, given the allowable risk levels with regard to the occurrence of three (formulation (13.71)) or two (formulations (13.75)–(13.77)) limit states. The multiconstraint formulations given by equations (13.72)–(13.74) and (13.78)–(13.86) are useful when the most reliable structure with respect to the occurrence of a limit state is sought, given (a) the allowable cost (formulations (13.72)–(13.74), (13.78), (13.79), (13.81), (13.82), (13.84) and (13.85) and the preassigned values of the risk levels with respect to the occurrence of the other limit state(s) of concern, or (b) the preassigned values of the risk levels with respect to the occurrence of two other limit states concern (formulations (13.80), (13.83) and (13.86)).

It is interesting to note that when only two of the four equations (13.67)–(13.70) are chosen in the system reliability-based optimization process, twelve different formulations are possible, as follows:

$$\min W, \text{ subject to constraint (13.68)} \qquad (13.87)$$

$$\min W, \text{ subject to constraint (13.69)} \qquad (13.88)$$

$$\text{min } W, \text{ subject to constraint (13.70)} \qquad (13.89)$$

$$\text{min } P_{fi}, \text{ subject to constraint (13.67)} \qquad (13.90)$$

$$\text{min } P_{fi}, \text{ subject to constraint (13.69)} \qquad (13.91)$$

$$\text{min } P_{fi}, \text{ subject to constraint (13.70)} \qquad (13.92)$$

$$\text{min } P_{fj}, \text{ subject to constraint (13.67)} \qquad (13.93)$$

$$\text{min } P_{fj}, \text{ subject to constraint (13.68)} \qquad (13.94)$$

$$\text{min } P_{fj}, \text{ subject to constraint (13.70)} \qquad (13.95)$$

$$\text{min } P_{fk}, \text{ subject to constraint (13.67)} \qquad (13.96)$$

$$\text{min } P_{fk}, \text{ subject to constraint (13.68)} \qquad (13.97)$$

$$\text{min } P_{fk}, \text{ subject to constraint (13.69)} \qquad (13.98)$$

The above single-constrained formulations are useful when the system optimization consists of determining (a) the minimum weight structure, subject to a reliability constraint with respect to occurrence of a single limit state (formulations (13.87)–(13.89)) (also the formulations of equations (13.59) and (13.60) and of equation (13.61) and (13.62)), or (b) the most reliable structure with respect to the occurrence of a single limit state, given constraints on weight (formulations (13.90), (13.93) and (13.96)) or on risk of occurrence of another limit state (formulations (13.91), (13.92), (13.94), (13.95), (13.97) and (13.98)).

The selection of the objective function, allowable cost (weight) level, limit states and associated risk levels are crucial considerations in system RBSO. As shown previously, a variety of objectives can be chosen in the optimization process. Consequently, an optimum design solution is always relative. The selection of allowable cost and/or risk levels in RBSO is a problem beyond the scope of this chapter. According to limit states design specifications (e.g. Ellingwood *et al.*, 1980; Ellingwood and Galambos, 1982; Corotis, 1985) a moderate (say, $10^{-2} \geqslant P_f^0 \geqslant 10^{-3}$) and a low (say, $10^{-4} \geqslant P_f^0 \geqslant 10^{-6}$) risk level has to be required for serviceability and ultimate limit states, respectively.

As outlined in section 13.4, present reliability-based limit states design codes are almost exclusively aimed towards an element-level reliability format. This is achieved through the use of load and resistance factor design checking equations at the element level. Design according to these equations may not satisfy the required system reliability level. Conversely, an optimum design based on system reliability requirements only may not guarantee the element-level reliability requirements in the current codes. Consequently, RBSO with consideration of both system-level and element-level reliability requirements may at present produce an attractive balanced (system–element) risk consistent design approach. Efforts in this direction were made by Kim and Wen (1987), Nakib and Frangopol (1990a), and Mahadevan and Haldar (1991), among others.

### 13.7.4 Applications

The examples used herein to illustrate the applications of the system reliability-based optimization approach cover both steel and reinforced concrete structures. In addition to these examples, numerous additional applications are described by Cohn and Parimi (1972), Frangopol (1984b, 1986b, c), Ishikawa and Iizuka (1987), Kim and Wen (1987), Mahadevan and Haldar (1991), Moses (1969, 1970), Moses and Kinser (1967), Nakib and Frangopol (1990a), Parimi and Cohn (1978), Rackwitz and Cuntze (1987), Rojiani and Bailey (1984), Sørensen (1987, 1988), Stevenson and Moses (1970), Surahman and Rojiani (1983), Thoft-Christensen (1987), Thoft-Christensen and Murotsu (1986), and Vanmarcke (1971). The following single-limit state (Examples 13.4 to 13.6) and multilimit state (Example 13.7) system reliability-based optimization examples are only briefly presented. The interested reader will find the detailed data and explanation of these examples in the original references from which they were excerpted.

---

### EXAMPLE 13.4 TWO-STORY TWO-BAY FRAME
(Moses and Stevenson, 1970; Moses, 1973)

The nondeterministic two-story two-bay steel frame under the mean random loads shown in Fig. 13.11 was optimized by Moses and Stevenson (1970) for plastic collapse failure using the system reliability-based optimization formulation of equations (13.59) and (13.60). There are six mean moment capacity design variables, including three for beams and three for columns, and as many as fifty-three collapse modes. A deterministic collapse analysis using linear programming would indicate at least six simultaneous active collapse mechanisms at the optimum. Table 13.1 (Moses, 1973) shows some examples of system RBSO optima. Moses and Stevenson (1970) pointed out that only one collapse mode

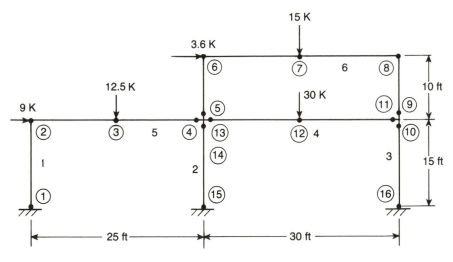

Mean loads are shown

1    Member number

①    Critical joint number

**Fig. 13.11**  Two-story two-bay frame (Moses and Stevenson, 1970).

**Table 13.1** Optimum-design results of two-story two-bay frame shown in Fig. 13.11 (Moses, 1973)

| Example number | Optimum mean moment capacities (klbf ft) | | | | | | COV | | Allowable $P_f$ | Weight function | Frequency distribution |
|---|---|---|---|---|---|---|---|---|---|---|---|
| | $M_1$ | $M_2$ | $M_3$ | $M_4$ | $M_5$ | $M_6$ | Moment capacity | Load | | | |
| 1 | 29.2 | 95.8 | 84.4 | 175.0 | 73.2 | 74.4 | 0.10 | 0.20 | 7.78(−2)* | 312.47 | Normal |
| 2 | 27.8 | 96.3 | 84.4 | 173.8 | 72.0 | 77.9 | 0.10 | 0.20 | 7.80(−2) | 312.89 | Log-normal |
| | Monte Carlo value of $P_f$ (9500 trials) | | | | | | | | 7.59(−2) | | |
| 3 | 28.0 | 78.7 | 71.0 | 170.9 | 69.4 | 74.9 | 0.20 | 0.10 | 7.72(−2) | 297.26 | Normal |
| 4 | 27.3 | 78.3 | 71.3 | 166.4 | 65.1 | 74.9 | 0.20 | 0.10 | 7.16(−2) | 293.53 | Log-normal |
| 5 | 29.1 | 87.8 | 72.3 | 170.3 | 68.0 | 74.1 | 0.15 | 0.15 | 7.52(−2) | 300.56 | Normal |
| | Monte Carlo value of $P_f$ (7000 trials) | | | | | | | | 7.50(−2) | | |

*Exponents of failure probability are shown in parentheses $(m)$ and should be read as $10^m$.

dominates, in contributing most of the system failure probability to the allowable value. A check on the system reliability analysis was obtained for two cases in Table 13.1 by comparison with a Monte Carlo simulation of the system failure probability. It is interesting to note that the central-safety-factor checks alone, as used in conventional deterministic optimization methods, were not good indications of system failure probability. In fact, it was shown that some collapse modes of the frame with higher safety factors also had higher occurrence probabilities.

---

## EXAMPLE 13.5 TWO-STORY ONE-BAY FRAME
### (Frangopol, 1986b)

The constrained optimization formulation aimed at minimizing the probability of system collapse $P_f = P_{fcol}$ when the allowable weight $W = W^0$ is prescribed (equations (13.63) and (13.64)) is illustrated in Fig. 13.12(b) for the two-story one-bay nondeterministic steel plastic frame in Fig. 13.12(a) (Frangopol, 1986b). Figure 13.12(b) shows isosafety contours $P_f = P_{fcol}$ of the frame with respect to plastic collapse, obtained using sixty possible collapse modes, for two values of the prescribed weight function $W = 7000\,\mathrm{kN\,m^2}$ and $W = 6500\,\mathrm{kN\,m^2}$. The load and strength variabilities and correlations are included in the design. The design variables are the three mean values of the beams and columns plastic moments $\bar{M}_1$, $\bar{M}_2$ and $\bar{M}_3$ indicated in Fig. 13.12(a). As expected, a higher allowable weight means a lower optimum probability of collapse.

---

Both the two-story two-bay frame (Moses and Stevenson, 1970) and the two-story one-bay frame (Frangopol, 1986b) showed that the optimum proportioning of a nondeterministic structure with many collapse modes for minimum weight (Moses and Stevenson, 1970) or for minimum collapse failure probability (Frangopol, 1986b) is a complex interplay of members participating in a different collapse mode. The computer-based optimization is needed in both the system reliability analysis and the mathematical programming methods for finding the minimum weight or the maximum safety design.

---

## EXAMPLE 13.6 STEEL PORTAL FRAMES (Frangopol, 1985a;
### Nakib and Frangopol, 1990b)

• The nondeterministic steel plastic frame in Fig. 13.13(a) was optimized for minimum weight, with a prescribed collapse failure probability, $P_f = 10^{-5}$ (equations (13.59) and (13.60)). The two loads and plastic moments at the seven critical sections are assumed random variables with prescribed variabilities and correlations (Frangopol, 1985a).

The purpose here is to study the sensitivity of the optimum solution of the frame shown in Fig. 13.13(a) to various types of plastic moment correlation

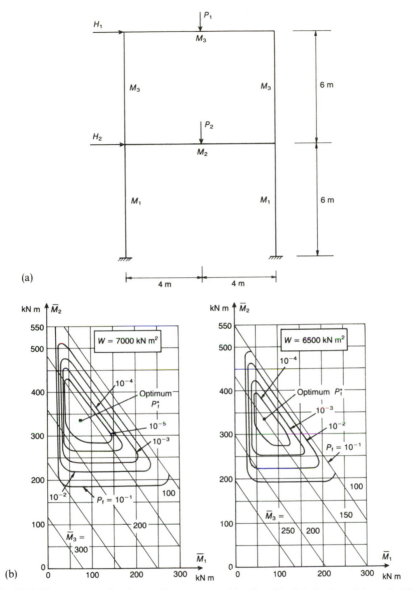

(a)

(b)

**Fig. 13.12** Two-story one-bay frame: (a) geometry and loading; (b) minimization of the probability of collapse failure for a specified total weight, $W$ (Frangopol, 1986b).

and various methods for determining the overall probability of plastic collapse of the structure. In order to study this sensitivity five different cases of resistance correlation are considered as follows (Frangopol, 1985a): (a) independence among all plastic moments; (b) perfect within columns and column–column correlation, and independence among the other plastic moments; (c) perfect

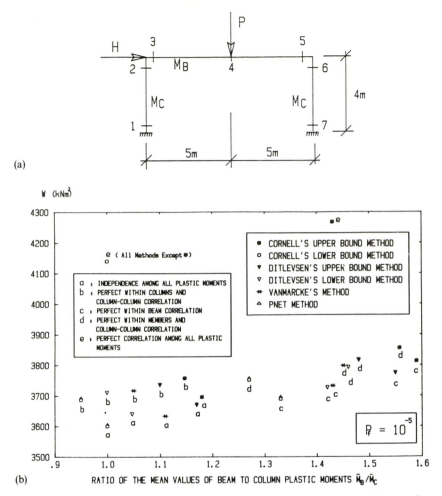

**Fig. 13.13** Portal frame: (a) geometry and loading; (b) sensitivity of optimum solution to changes in strength correlation and in method for system reliability evaluation (Frangopol, 1985a).

within beam correlation, and independence among the other plastic moments; (d) perfect within members and column–column correlation, and independence among the other plastic moments; (e) perfect correlation among all plastic moments.

For each one of the five cases of correlation among plastic moments just mentioned, six different methods for determining the overall probability of plastic collapse failure of the frame are used: Cornell's lower and upper bounds method (Cornell, 1967); Ditlevsen's lower and upper bounds method (Ditlevsen, 1979); Vanmarcke's method (Vamarcke, 1971); PNET method, choosing the value 0.8 for the demarcating correlation coefficient (Ang and Ma, 1981). Figure 13.13(b) provides information on the effect of varying both the statistical correlation of the plastic moments and the method for determining the overall

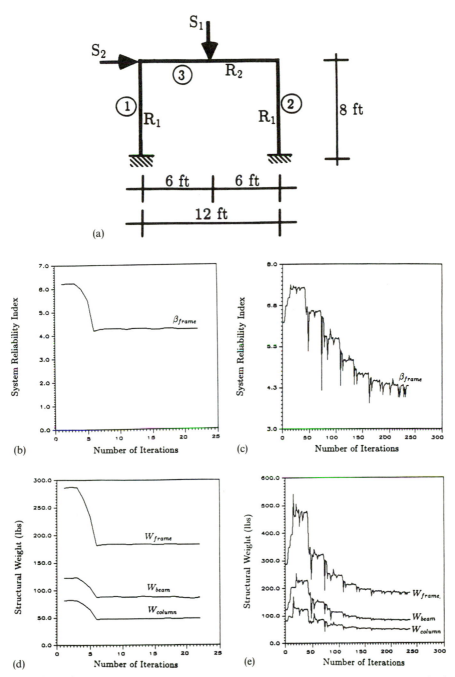

**Fig. 13.14** Portal frame: (a) geometry and loading; (b) reliability index iteration history (method of feasible directions); (c) reliability index iteration history (interior penalty method); (d) weight iteration history (method of feasible directions); (e) weight iteration history (interior penalty method) (Nakib and Frangopol, 1990b).

probability of plastic collapse of the frame on the optimum solution (Frangopol, 1985a).

It should be emphasized that for this particular example the dominant failure modes that correspond to the optimum solution are always beam modes. For this reason the correlation coefficient among loads as well as some of the correlation coefficients among plastic moments have no effect on the optimum solution.

As shown in Fig. 13.13(b), plastic moment correlation can have a considerable effect on the optimum solution for each of the six methods considered.

The information provided by this sensitivity analysis allows for identification of those factors with the greatest influence on optimum solutions. It appears from the results presented that strength correlation can play an important role in the accuracy achieved in reliability-based optimization calculations. A further point is that these results shows that their sensitivity effect to strength correlation may exceed the sensitivity effect to changes in the method for system reliability evaluation.

- Another example of a nondeterministic steel portal frame optimized by Nakib and Frangopol (1990b) is shown in Fig. 13.14(a). In the optimization process, the weight of the structure is minimized with a constraint on the frame reliability index with respect to plastic collapse occurrence. The optimum is obtained when the minimum allowable reliability index of the frame $\beta^0_{col} = \Phi^{-1}(1 - P^0_{fcol}) = 4.265$ is reached (i.e. $P^0_{fcol} = 10^{-5}$). Using an interactive graphics environment, the optimum obtained by using both the feasible directions (FD) and the interior penalty function (IPF) methods are compared in Figs 13.14(b) to 13.14(e) with regard to convergence histories of both the system reliability index (Figs 13.14(b) and 13.14(c)) and weight (Figs. 13.14(d) and 13.14(e)). Despite the fact that both optimization methods achieve approximately the same optimum solution, the feasible directions method offers a significant advantage with regard to both the number of convergence cycles and the computing time. For example, the number of convergence cycles is 21 and 236 for FD and IPF, respectively, while their associated CPU times are 5.83 s and 48.18 s (Nakib and Frangopol, 1990b).

## EXAMPLE 13.6 REINFORCED CONCRETE PORTAL FRAME (Frangopol, 1985d)

Attempts to optimize nondeterministic reinforced concrete framed structures subjected to random loads with respect to plastic collapse occurrence were made by Cohn and Parimi (1972), Parimi and Cohn (1978), and Frangopol (1985d), among others. The object of reliability-based plastic optimization of reinforced concrete framed structures is usually to find the concrete cross-sectional dimensions and the corresponding amount of reinforcing steel so that a specified reliability level against plastic collapse can be provided and an adopted objective be minimized. In general, it is assumed that the minimization of the total cost of concrete and longitudinal steel is the design objective. If the concrete section sizes are assigned from reliability-based elastic design computations, the reliability-

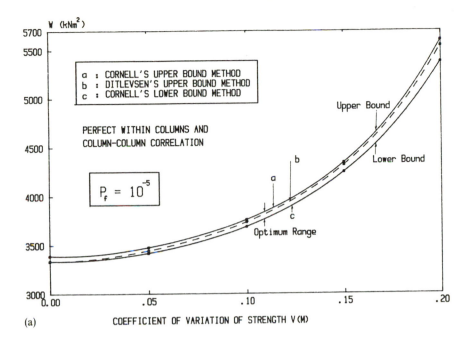

(a)

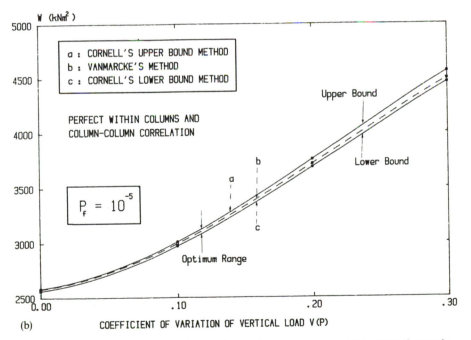

(b)

**Fig. 13.15** Sensitivity of optimum solution of a reinforced concrete portal frame to changes in the coefficients of variation of (a) plastic moment (strength) and (b) vertical load (Frangopol, 1985d).

based plastic design objective is simpler: the minimization of the longitudinal steel weight (or volume). Using this assumption, the plastic optimization problem can be formulated as follows:

$$\min W = \min \sum_{j=1}^{s} (l_j^+ \bar{M}_j^+ + l_j^- \bar{M}_j^-) \qquad (13.99)$$

subject to

$$P_f(\bar{\mathbf{M}}) \leqslant P_f^0 \qquad (13.100)$$

in which $P_f^0$ is the allowable probability of plastic collapse, and $\bar{\mathbf{M}}$ is the optimum design vector defined by

$$\bar{\mathbf{M}} = \{\bar{M}_1^+ \bar{M}_1^- \cdots \bar{M}_s^+ \bar{M}_s^-\}^t \qquad (13.101)$$

where $\bar{M}_j^+$ and $\bar{M}_j^-$ are the mean values of the positive and negative plastic moments associated with the $j$th critical section. The moments $M_j^+$ and $M_j^-$ are assumed to govern over the prescribed lengths $l_j^+$ and $l_j^-$ along the structure and, consequently, the longitudinal steel areas $A_j^+$ and $A_j^-$ are constant over the lengths associated with the critical section $j$.

The optimum reliability-based solution of the nondeterministic reinforced concrete frame with the geometry and loading shown in Fig. 13.13(a) was obtained using the assumptions presented in Frangopol (1985d). Figures 13.15(a) and 13.15(b) show the sensitivity of the optimum objective (equation (13.99)) to changes in the coefficients of variation of strength (plastic moment) and vertical load, respectively. As expected, the optimum weight increases with increasing uncertainties in strength and/or loading.

---

## EXAMPLE 13.7 MULTILIMIT STATE RELIABILITY ANALYSIS AND OPTIMIZATION OF A STEEL PORTAL FRAME (Frangopol and Nakib, 1987; Frangopol, 1985c)

Practical design is much more complex than the single-level system reliability approach illustrated in the previous examples. Usually, a structure is proportioned for conformance with a set of limit states including both serviceability and ultimate requirements. Therefore, to evaluate the reliability of a structure a multilevel system reliability approach has to be used. In order to accomplish this, an incremental mean loading technique was formulated for evaluating the risks associated with various limit states of a structural sytem (Moses, 1982; Frangopol and Nakib, 1986). This technique was applied in the following multilimit state reliability analysis and optimization examples.

- For a nondeterministic steel portal frame under two random loads (i.e. lateral load $H$ and gravity load $P$), Fig. 13.16 (Frangopol and Nakib, 1987) indicates isosafety mean loading functions for four limit states including excessive deflection, first yield, first plastic hinge, and plastic collapse. All points on the three serviceability limit state curves (i.e. deflection, first yield, first plastic hinge) represent different combinations of mean loads leading to the same

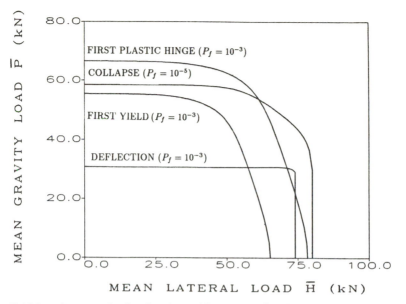

**Fig. 13.16** Isosafety mean loading functions with respect to four limit states of a portal frame (Frangopol and Nakib, 1987).

value of the overall probability of frame unreliability (i.e. $P_f = 10^{-3}$, or $\beta = 3.09$), and all points on the ultimate limit state curve (i.e. plastic collapse) represent different combinations of mean loads leading to the same value of the overall probability of plastic collapse (i.e. $P_f = 10^{-5}$, or $\beta = 4.265$).

Figures 13.17 and 13.18 (Frangopol and Nakib, 1987) show the system reliability levels (i.e. system safety index $\beta$) with respect to three limit states (i.e. first yield, first plastic hinge, and plastic collapse in Fig. 13.17, and excessive deflection, first yield and first plastic hinge in Fig. 13.18) corresponding to all possible combinations of mean loads represented by the isosafety mean loading functions with respect to excessive deflection and plastic collapse, respectively. It is important to note that using the isosafety loading function against excessive deflection as the only serviceability limit state (Fig. 13.17) does not protect against unreliability by occurrence of other limit states for large values of the mean load ratio $\bar{H}/\bar{P}$ (i.e. say $> 2.0$). Also, using the isosafety loading function against plastic collapse (Fig. 13.18) does not always protect against unreliability by loss of serviceability. Consequently, the serviceability and ultimate limit states have to be considered simultaneously in the reliability-based analysis and optimization processes. This latter case is demonstrated in the following example.

- This example is the rectangular fix-ended single-story rigid portal frame with deterministic geometry and random strengths ($M_B$ is the plastic moment of the critical sections of the beam and $M_C$ the plastic moment of the critical sections of the columns) shown in Fig. 13.13(a). The frame is subjected to two random concentrated loads $H$ and $P$. The statistical information is given in Frangopol (1985c). The total weight of the frame is approximated by the linear

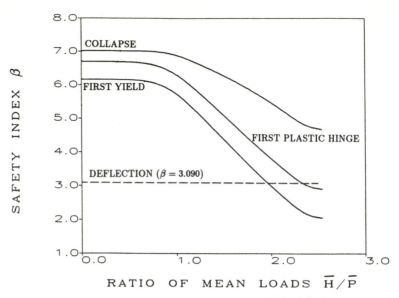

**Fig. 13.17** Isosafety mean loading function with respect to limit deflection occurrence and associated reliability levels of three other limit states (Frangopol and Nakib, 1987).

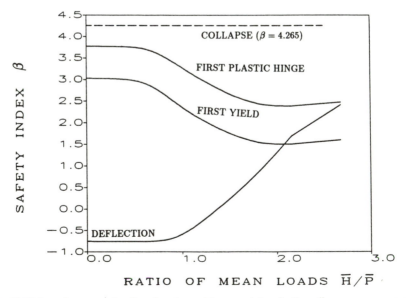

**Fig. 13.18** Isosafety mean loading function with respect to plastic collapse occurrence and associated reliability levels of three other limit states (Frangopol and Nakib, 1987).

function $W = l\bar{M}_B + 2h\bar{M}_C$, in which $l = 10\,\text{m}$ (beam span) and $h = 4\,\text{m}$ (column height).

Figure 13.19 (Frangopol, 1985c) contains the design space with the two isoreliability contours with respect to plastic collapse and loss of serviceability: $P_{f1}^0 = 10^{-5}$ and $P_{f2}^0 = 10^{-2}$, respectively. All points on the curves $P_{f1}^0 = 10^{-5}$ and $P_{f2}^0 = 10^{-2}$ have the probabilities of plastic collapse and unserviceability equal to the prescribed values $10^{-5}$ and $10^{-2}$, respectively. Also shown is the linear objective function, $W$, that should be minimized. The boundary of the feasible design space (which contains points with acceptable probabilities of failure) depends on the optimization formulation as follows: boundary $a_2bc_1de_2$ corresponds to the biconstraint (collapse and unserviceability) formulation given by equation (13.75) and boundaries $a_1bc_1de_1$ and $a_2bc_2de_2$ correspond to the single-constraint optimization formulations with reliability against collapse only (equations (13.59) and (13.60)) and unserviceability only (equations (13.61) and (13.62)), respectively. Consequently, if the biconstraint formulation given by equation (13.75) or the single-constraint formulation given by equations (13.59) and (13.60) is chosen, the optimum solution (the point at which $W$ is tangent to the $P_f$ constraint) is the design point $c_1$ (collapse

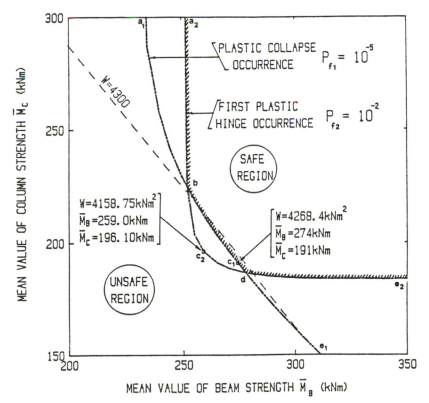

**Fig. 13.19** Design space with two system reliability constraints used in minimum weight optimization of a steel portal frame (Frangopol, 1985c).

constraint is critical at the optimum). On the other hand, if the single-constraint formulation given by equations (13.59) and (13.60) is chosen, the optimum solution is the design point $c_2$. It is interesting to note that (a) both reliability constraints are satisfied as equalities at two design points (b and d), (b) the points lying on the curve $bc_2d$ have unacceptable reliabilities with regard to plastic collapse ($P_{f1} > 10^{-5}$), and (c) the points lying on the curves $a_1b$ and $de_1$ have unacceptable reliabilities with regard to loss of serviceability ($P_{f2} > 10^{-2}$).

Figures 13.20 and 13.21 (Frangopol, 1985c) show minimum weight solutions for various sets of specified risk levels against plastic collapse and unserviceability, respectively. Figure 13.20 presents solutions for different values of $P_{f1}^0$ when $P_{f2}^0$ is kept constant, while Fig. 13.21 presents solutions for different values of $P_{f2}^0$ when $P_{f1}^0$ is kept constant.

The results in Figs 13.20 and 13.21 indicate the possible fallacy in a single-limit-state optimization approach. For example, when a biconstraint optimization formulation is used the solutions $d_1$ in Fig. 13.20 and $c_2$ and $d_2$ in Fig. 13.21 have unacceptable reliabilities with regard to unserviceability and plastic collapse, respectively.

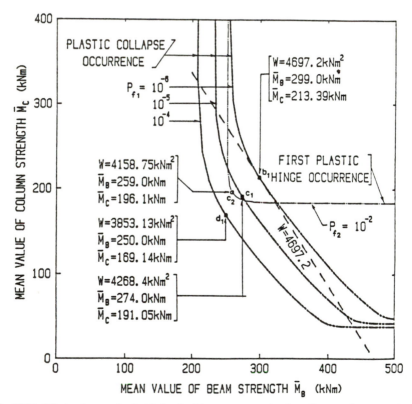

**Fig. 13.20** Effect of prescribed risk level with respect to plastic collapse, $P_{f1}^0$, on optimum solution (Frangopol, 1985c).

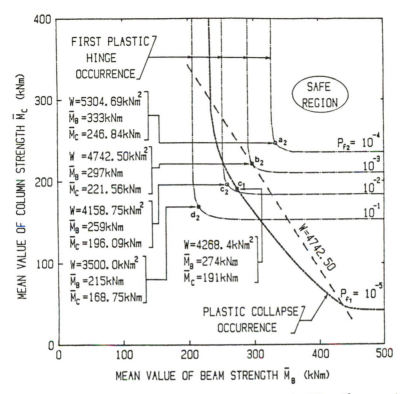

**Fig. 13.21** Effect of prescribed risk level with respect to unserviceability, $P_{f_2}^0$, on optimum solution (Frangopol, 1985c).

## 13.8 MULTIOBJECTIVE RELIABILITY-BASED OPTIMIZATION

### 13.8.1 General

Single-objective optimization has been the basic approach in most previous work on the design of structural systems. The purpose was to seek optimal values of design variables which minimize or maximize a specific single quantity termed the objective function, while satisfying a variety of behavioral and geometrical conditions, termed constraints. In this definition of structural optimization, the quality of a structural system is evaluated using a single criterion (e.g. total expected cost, total weight, system reliability). For most design problems, however, engineers are confronted with alternatives that are conflicting in nature. For these problems, where the quality of a design is evaluated using several competing criteria simultaneously, multiobjective optimization (also called vector, multicriterion or Pareto optimization) should be used.

Multiobjective optimization of structural systems is an important idea that has only recently been brought to the attention of the structural optimization community by Duckstein (1984), Koski (1984), Osyczka (1984), Frangopol and Klisinski (1989b, 1991), and Fu and Frangopol (1990a, b), among others. It was shown that there are many structural design situations in which several conflicting objectives should be considered. For example, a structural system is expected to be designed such that both its total weight and maximum displacements be minimized. In such a situation, the designer's goal is to minimize not a single objective but several objectives simultaneously.

The main difference between single- and multiobjective optimization is that the latter almost always is characterized not by a single solution but by a set of solutions. These solutions are called Pareto optimum, noninferior or nondominated solutions. If a point belongs to a Pareto set there is no way of improving any objective without worsening at least another one. The advantage of the multiobjective optimization is that it allows one to choose between different results and to find possibly the best compromise. It requires, however, a considerable computation effort. In this section which is highly based on the results obtained during the past four years at the University of Colorado (Frangopol and Fu, 1989, 1990; Frangopol and Iizuka, 1991a, b; Frangopol and Klisinski, 1989b, 1991; Fu and Frangopol, 1990a, b), the mathematical formulation and the solution methods used in multiobjective structural optimization are briefly reviewed and examples of solutions for both nondeterministic truss and frame systems are presented.

### 13.8.2 Mathematical formulation

A multiobjective optimization problem can be formulated in the following way (Duckstein, 1984; Koski, 1984):

$$\min \mathbf{F}(\mathbf{X}) \tag{13.102}$$

where $\mathbf{X} \in \Omega$ and $\mathbf{F}: \Omega \to R^m$ is a vector objective function given by

$$\mathbf{F}(\mathbf{X}) = [F_1(\mathbf{X}) \, F_2(\mathbf{X}) \cdots F_m(\mathbf{X})]^t \tag{13.103}$$

Each component of this vector describes a single objective

$$F_i: \Omega \to R \qquad i = 1, 2, \ldots, m \tag{13.104}$$

The design variable vector $\mathbf{X}$ belongs to the feasible set $\Omega$ defined by equality and inequality constraints as follows:

$$\Omega = (\mathbf{X} \in R^n : \mathbf{H}(\mathbf{X}) = \mathbf{0}, \mathbf{G}(\mathbf{X}) \leqslant \mathbf{0}) \tag{13.105}$$

The image of the feasible set $\Omega$ in the objective function space is denoted by $\mathbf{F}(\Omega)$.

In a multiobjective RBSO problem, the components of the objective vector $\mathbf{F}(\mathbf{X})$ are certain characteristic properties of the structural system to be designed, such as failure cost, material volume or weight, system probabilities of ultimate collapse, first yielding, or plastic hinge occurrences, among others. The components of the design variable vector $\mathbf{X}$ are parameters to be detemined in the optimization process, such as cross-sectional dimensions or mean plastic moments. Some of the objective vectors used in RBSO were specified by Frangopol and Fu (1989, 1990), Frangopol and Iizuka (1991a, b), Fu and Frangopol (1990a, b), and

Iizuka (1991), as follows:

$$\mathbf{F}_1(\mathbf{X}) = (V(\mathbf{X}) \quad P_{\text{fcol}}(\mathbf{X}) \quad P_{\text{fyld}}(\mathbf{X}) \quad P_{\text{fdfm}}(\mathbf{X}))^t \qquad (13.106)$$

$$\mathbf{F}_2(\mathbf{X}) = (V(\mathbf{X}) \quad P_{\text{fint}}(\mathbf{X}) \quad P_{\text{fdmg}}(\mathbf{X}))^t \qquad (13.107)$$

$$\mathbf{F}_3(\mathbf{X}) = (V(\mathbf{X}) \quad P_{\text{fint}}(\mathbf{X}) \quad -R_1(\mathbf{X}))^t \qquad (13.108)$$

$$\mathbf{F}_4(\mathbf{X}) = (V(\mathbf{X}) \quad -\beta_{\text{yld}}(\mathbf{X}) \quad -R_2(\mathbf{X}))^t \qquad (13.109)$$

in which $V$ is material volume of the structure, $P_{\text{fcol}}$ the system probability of plastic collapse, $P_{\text{fyld}}$ the system probability of first yielding, $P_{\text{fdfm}}$ the system probability of excessive elastic deformation, $P_{\text{fint}}$ the system probability of collapse initiated from the intact state (i.e. $P_{\text{fint}} = P_{\text{fcol}}$), $P_{\text{fdmg}}$ the system probability of collapse initiated from a given damaged state (e.g. the system probability of collapse given that one element is completely removed), $R_1$ the redundancy factor of the system given by the ratio (Fu and Frangopol, 1990a)

$$R_1 = (P_{\text{fdmg}} - P_{\text{fint}})/P_{\text{fint}} \qquad (13.110)$$

and $R_2$ the redundancy factor of the system given by the difference (Frangopol and Iizuka, 1991b)

$$R_2 = \beta_{\text{col}} - \beta_{\text{yld}} \qquad (13.111)$$

in which $\beta_{\text{col}}$ is the system reliability index with respect to plastic collapse (i.e. $\Phi^{-1}(1 - P_{\text{fcol}})$) and $\beta_{\text{yld}}$ the system reliability index with respect to first yielding (i.e. $\Phi^{-1}(1 - P_{\text{fyld}})$). Other definitions of redundancy factors are given in Fu, Liu and Moses (1991), and Iizuka (1991), among others.

If the components of the objective vector $\mathbf{F}(\mathbf{X})$ are not fully independent there is not usually a unique point at which all the objective functions reach their minima simultaneously. As previously mentioned, the solution of a multi-objective optimization problem is called Pareto optimum or nondominated solution. Two types of nondominated solutions can be defined as follows (Duckstein, 1984; Koski, 1984):

- a point $\mathbf{X}_0 \in \Omega$ is weakly nondominated solution if and only if there is no $\mathbf{X} \in \Omega$ such that

$$\mathbf{F}(\mathbf{X}) < \mathbf{F}(\mathbf{X}_0) \qquad (13.112)$$

- a point $\mathbf{X}_0 \in \Omega$ is a strongly nondominated solution if there is no $\mathbf{X} \in \Omega$ such that

$$F(\mathbf{X}) \leqslant F(\mathbf{X}_0) \qquad (13.113)$$

and for at least one component $i$

$$F_i(\mathbf{X}) \leqslant F_i(\mathbf{X}_0) \qquad i = 1, 2, \ldots, m \qquad (13.114)$$

If the solution is strongly nondominated it is also weakly nondominated. Such a strongly nondominated solution is also called Pareto optimum. In other words, the above definitions state that if the solution is strongly nondominated, no one of the objective functions can be decreased without causing a simultaneous increase in at least one other objective.

The main task of multiobjective optimization is to find the set of strongly nondominated solutions (also called Pareto optimal objectives or minimal curve) in the objective function space and the corresponding values of the design

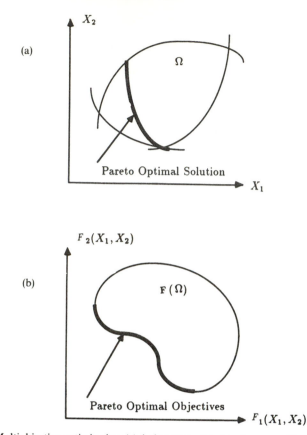

**Fig. 13.22** Multiobjective optimization: (a) design space; (b) objective space (Fu and Frangopol, 1990b).

variables (Pareto optimal solution) in the feasible region space. Figure 13.22 (Fu and Frangopol, 1990b) qualitatively displays the feasible space $\Omega$, the Pareto optimal solution, and its corresponding Pareto optimal objectives in a biobjective optimization problem with two design variables.

### 13.8.3 Solution techniques

There are a number of multiobjective optimization solution techniques described in the literature (Duckstein, 1984; Koski, 1984; Osyczka, 1984; among others). Not all of them are, however, suitable for reliability-based structural optimization. Three solution methods which may be applied to multiobjective reliability-based structural optimization are briefly described herein: the weighting method, the minimax method and the constraint method. For a more complete description of these methods the interested reader is referred to Duckstein (1984), Koski (1984), and Osyczka (1984), among others.

- The basic idea of the **weighting method** is to define the objective function $F$ as a scalar product of the weight vector $\mathbf{W} = [W_1 \ W_2 \cdots W_m]^t$ and the objective vector $\mathbf{F}$, as follows:

$$F = \mathbf{W} \cdot \mathbf{F} \tag{13.115}$$

Without loss of generality the vector $\mathbf{W}$ can be nomalized. The Pareto optimal objective set can be theoretically obtained by varying the weight vector $\mathbf{W}$. In the objective function space the single objective function $F$ is linear. For a fixed vector $\mathbf{W}$, the optimization process results in a point at which a hyperplane representing the single objective function is tangential to the image of the feasible set $\mathbf{F}(\Omega)$. Only if the set $\mathbf{F}(\Omega)$ is convex is the weighting method able to generate all the strongly nondominated solutions. The second drawback of this technique is the difficulty involved in choosing the proper weight factors. Since the shape of the image of the feasible set $\mathbf{F}(\Omega)$ is generally unknown it is almost impossible to predict where the solution will be located. Sometimes the problem is also very sensitive to the variation of weights. In such a case the weighting approach can prove to be unsatisfactory.

- Another method for solving vector optimization problems is the **minimax method** described in Koski (1984), among others. This method introduces distance norms into the objective function space. In this method the reference point from which distances are measured is the so-called ideal solution. This solution can be described by the vector

$$\mathbf{F}^* = [F_1^* \ F_2^* \cdots F_m^*]^t \tag{13.116}$$

where all its components are obtained as the solutions of $m$ independent single-objective optimization problems:

$$F_i^* = \min_{\mathbf{X} \in \Omega} F_i(\mathbf{X}) \tag{13.117}$$

Generally, the ideal solution is not feasible, so it does not belong to the set $\mathbf{F}(\Omega)$.

In the minimax method the norm is defined as follows:

$$^-\max_i W_i(F_i - F_i^*) \qquad i = 1, 2, \ldots, m \tag{13.118}$$

where the components $W_i$ and $F_i$ of the vectors $\mathbf{W}$ and $\mathbf{F}$, respectively, were previously defined. For a given vector $\mathbf{W}$, the norm associated with equation (13.118) corresponds to the search in a direction of some line starting from the ideal solution. If the other norms are used, this approach generates the entire family of nondominated solutions.

The minimax approach eliminates one drawback of the weighting method, because it is also suitable for nonconvex problems. It may also, however, be sensitive to the values of the weight factors. The prediction where the solution will be located is improved. To use this method it is necessary to know first the ideal solution, which calls for solving $m$ scalar optimization problems: it is the price to pay for using this method.

- Another alternative to the previous methods is the **constraint method**. The main idea of this method is to convert $m - 1$ objective functions into constraints. This can be obtained by the assumption that the values below some prescribed levels for all these functions are satisfactory. Without loss of generality it may be assumed that the components $F_2, F_3, \ldots, F_m$ of the objective vector will be constrained and only $F_1$ will be minimized.

In mathematical terms, the constraint method is formulated according to

Duckstein (1984) and Koski (1984) as follows. The original minimization problem (equations (13.102–13.105)) is replaced by

$$\min_{\Omega \cap \Omega_0} F_1(\mathbf{X}) \tag{13.119}$$

where

$$\mathbf{X} \in \Omega \cap \Omega_0 \text{ and } \Omega_0 = (\mathbf{X}: F_i(\mathbf{X}) \leqslant E_i, i = 2, \ldots, m) \tag{13.120}$$

and

$$\mathbf{E} = [E_2 \ E_3 \cdots E_m]^t \tag{13.121}$$

The entire Pareto set can be obtained by varying the **E** vector. The constraint method applies to nonconvex problems and does not require any additional computations.

The computational experience gained at the University of Colorado has indicated that the constraint method is the most appropriate technique for solving both deterministic (Frangopol and Klisinski, 1989b, 1991) and reliability-based (Fu and Frangopol, 1990b; Frangopol and Iizuka, 1991a, b; Iizuka, 1991) multiobjective structural optimization problems. For this reason, the constraint method may be treated as a basic numerical technique in multi-objective RBSO.

There also many other multiobjective optimization techniques (e.g. Duckstein, 1984; Koski, 1984; Osyczka, 1984), but they are less or not suitable for multi-objective RBSO.

### 13.8.4 Applications

The solution of a multiobjective RBSO formulation provides alternatives to the decision maker which are optimal in the Pareto (noninferior) sense. As the numerical result of this optimization process, a space of Pareto optimal objectives (decision support space) can be offered to the designers (Fig. 13.22(b)). This space contains optimal combinations of the individual objective functions. These combinations are optimal in the sense that none of the objectives can be further reduced without increasing at least one of the others. According to specific conditions encountered in design practice, the decision makers can choose from the decision support space one of the alternatives as the final preferred solution. For example, system reliabilities with respect to different limit states may be given different preference levels in order to reach a balanced design. Trade-off techniques for determining the preferred solution from a decision support space have been discussed in the literature by Osyczka (1984), among others. In the following numerical examples decision support spaces are provided for both nondeterministic truss and frame systems. The interested reader will find the detailed data and explanation of these examples in the original references from which they were excerpted.

---

## EXAMPLE 13.8 FIVE-BAR TRUSS (Fu and Frangopol, 1990a, b)

Consider the simple steel truss structure shown in Fig. 13.23 (Fu and Frangopol, 1990b). This structural system consists of five uniaxial components and is acted on by the load $S$. The deterministic cross-sectional areas of the truss members constitute the design variable vector $\mathbf{A} = \mathbf{X}$. The three cross-sectional areas to be determined are $A_1$, $A_2$, and $A_3$ for vertical, horizontal, and diagonal members, respectively.

All members are assumed to exhibit ductile (elastic–perfectly plastic) behavior. The random variables considered in this problem are the three uniaxial capacities ($C_1$, $C_2$, and $C_3$) and the load $S$. The component capacities are related to component areas by $C_i = A_i \sigma_{yi}$ ($i = 1, 2, 3$), where $\sigma_{yi}$ is the random yielding stress of component $i$; it contributes randomness to the component capacity $C_i$. The mean values of $\sigma_{yi}$ for tension and compression are $25\,\text{kN cm}^{-2}$ and $12.5\,\text{kN cm}^{-2}$, respectively. The mean value of $S$ is $23.8\,\text{kN}$ and the coefficients of variation of $\sigma_{yi}$ and $S$ are 0.10 and 0.20, respectively. In the following, the five-bar truss is optimized successively considering the objective vectors $\mathbf{F}_2(\mathbf{X})$ and $\mathbf{F}_3(\mathbf{X})$ given by equations (13.107) and (13.108), respectively.

● *Objective vector* $\mathbf{F}_2(\mathbf{X})$(*Fu and Frangopol, 1990b*). In this case, the truss in Fig. 13.23 is optimized to satisfy the conditions

$$\min[V(\mathbf{A}), P_{\text{fint}}(\mathbf{A}), P_{\text{fdmg}}(\mathbf{A})] \qquad (13.122)$$

subject to

$$\mathbf{A} \in \Omega = (\mathbf{A} | A_i \geqslant 2\,\text{cm}^2) \qquad i = 1, 2, 3 \qquad (13.123)$$

where $V(\mathbf{A})$ is the total material volume, $P_{\text{fint}}(\mathbf{A})$ the probability of system collapse initiated from intact state, and $P_{\text{fdmg}}(\mathbf{A})$ the probability of system collapse initiated from damaged state in which any one and only one of the five members of the truss is removed. The system failure probabilities $P_{\text{fint}}$ and $P_{\text{fdmg}}$ are evaluated by Ditlevsen's upper bound method (Ditlevsen, 1979).

Figure 13.24 (Fu and Frangopol, 1990b) displays the decision support space (Pareto optimal objectives) for the five-bar truss shown in Fig. 13.23, which is obtained by conducting the multiobjective optimization given by equations (13.122) and (13.123) with the constraint method described earlier. This space indicates trade-offs among the three objectives, $V$, $P_{\text{fint}}$ and $P_{\text{fdmg}}$. A point on the surface shown in Fig. 13.24 represents a combination of the objectives

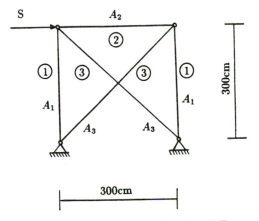

**Fig. 13.23** Five-bar truss: geometry and loading (Fu and Frangopol, 1990b).

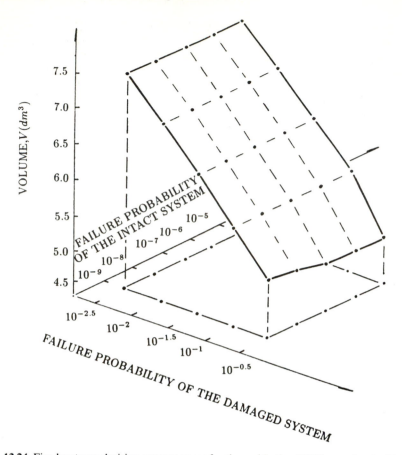

**Fig. 13.24** Five-bar truss: decision support space for three-objective RBSO associated with the objective vector $F_2(X)$ (Fu and Frangopol, 1990b).

which is optimal in the Pareto sense. These points provide alternatives to decision makers. $V$ and $P_{fdmg}$ are the dominant objectives for this problem in most of the region considered. This is because one component removed from the five-bar truss in Fig. 13.23 results in a statically determinate structural system which has a much lower reliability level than the intact system. This low level of residual reliability of the indeterminate five-bar system is taken into account in the optimal design searching. It is important to note that for high redundant truss systems, where removal of any member will not drastically affect the reliability of the system, $V$ and $P_{fint}$ could be the dominant objectives.

• *Objective vector $F_3(X)$(Fu and Frangopol, 1990a)*. In this case, the truss in Fig. 13.23 is optimized to satisfy the conditions

$$\min[V(\mathbf{A}), P_{fint}(\mathbf{A}), -R_1(\mathbf{A})]$$

subject to

$$\mathbf{A} \in \Omega = (\mathbf{A} \mid A_i \geqslant 2 \, \text{cm}^2) \qquad i = 1, 2, 3 \qquad (13.125)$$

where the redundancy factor $R_1$ is given by equation (13.110).

(13.124)

The trade-offs among the three objectives in equation (13.124) are described by the decision support space shown in Fig. 13.25 (Fu and Frangopol, 1990a). It is important to notice that considering only the system reliability in structural optimization does not always result in a rational design in the sense of redundancy gaining. For example, Figure 13.25 shows $\ln(R_1) = 8$ when $V$ is minimized to $5\,\text{dm}^3$ under the constraint $P_{\text{fint}} \leqslant 10^{-8}$. However, in this case, the decision space in Fig. 13.25 indicates that by consuming about the same amount of material under the same system reliability requirement the system redundancy can be substantially increased to $\ln(R_1) = 11$. This is reached by including both reliability and redundancy in the optimization process.

---

## EXAMPLE 13.9 ONE-STORY TWO-BAY FRAME
### (Fu and Frangopol, 1990b)

A nondeterministic steel frame under three random loads ($S_i$, $i = 1, 2, 3$) is considered in this example as shown in Fig. 13.26 (Fu and Frangopol, 1990b). This single-story two-bay structure has five random bending components ($M_i$, $i = 1, 2, \ldots, 5$). The design variables are the deterministic cross-sectional areas $\mathbf{A} = \mathbf{X}$ of those components. Three cross-sectional areas are to be determined, namely left-hand beam $A_2$, right-hand beam $A_3$ and columns $A_1$.

The frame in Fig. 13.26 is optimized considering the objective vector $\mathbf{F}_1(\mathbf{X})$ given by equation (13.106). The formulation is as follows:

$$\min[V(\mathbf{A}), P_{\text{fcol}}(\mathbf{A}), P_{\text{fyld}}(\mathbf{A}), P_{\text{fdfm}}(\mathbf{A})] \tag{13.126}$$

subject to

$$\mathbf{A} \in \Omega = (\mathbf{A} \,|\, A_i \geqslant 2.5\,\text{cm}^2) \qquad i = 1, 2, 3 \tag{13.127}$$

Twenty-eight failure modes are considered for computing $P_{\text{fcol}}$. The evaluation of $P_{\text{fyld}}$ involves consideration of the elastic moments at twelve critical sections (six column sections and six beam sections) of the frame shown in Fig. 13.26. They are either at a connection or at a load acting section. The system failure due to yielding occurrence is evaluated by considering that at least one of those twelve moments may exceed the elastic moment capacity of the corresponding critical section. The linear elastic deformations at three sections are considered for the evaluation of $P_{\text{fdfm}}$, namely section 7 (in the horizontal direction), and sections 3 and 6 (both in the vertical direction). They are regarded as excessive if 4 cm, 4 cm, and 6 cm are exceeded, respectively.

Figure 13.27 (Fu and Frangopol, 1990b) presents the decision support space (i.e. Pareto optimal objectives) for the four-objective vector optimization problem associated with the frame in Fig. 13.26. It exhibits a group of constant volume surfaces subject to the three failure probabilities of collapse, first yielding and deformation. These surfaces display the interaction among the four objectives in equation (13.126). A point on these isovolume surfaces is defined by its coordinates in the three-dimensional unreliability decision space. It can be seen in Fig. 13.27 that the collapse probability is not interactive in most cases. Its effect increases when this probability is very low (close to $10^{-9}$) and the other two failure probabilities are relatively high (say, $\geqslant 10^{-3}$). This can be observed in the lower

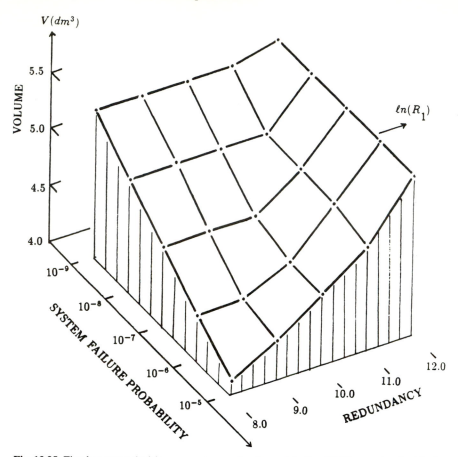

**Fig. 13.25** Five-bar truss: decision support space for three-objective RBSO associated with the objective vector $F_3(X)$ (Fu and Frangopol, 1990a).

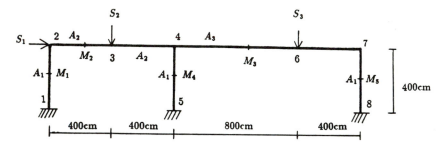

**Fig. 13.26** One-story two-bay frame: geometry and loading (Fu and Frangopol, 1990b).

right-hand part of the decision space shown in Fig. 13.27. It is then possible to consider only the other two failure probabilities in certain regions for final decision making. This observation also indicates that $P_{fcol}$ may not be the dominant factor to be taken into account in design. Considering only the collapse risk in the

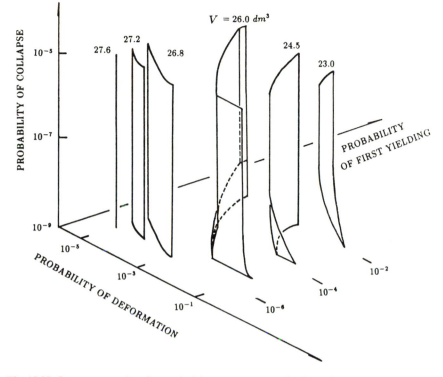

**Fig. 13.27** One-story two-bay frame: decision support space for four-objective RBSO (Fu and Frangopol, 1990b).

optimization process may implicitly sacrifice reliability with respect to excessive deformation or to other limit states of concern.

EXAMPLE 13.10 TRUSS BRIDGE (Frangopol and Fu, 1989)

The multiobjective optimization of the nondeterministic truss bridge in Fig. 13.28 (Frangopol and Fu, 1989) is presented here for illustration. The material to be used for uniaxial components is assumed ductile (elastic–perfectly plastic). The design variables are chosen as the cross-sectional areas of these uniaxial components, namely $\mathbf{A} = \mathbf{X}$.

Considering the symmetric vertical (vehicle) load $S$ as shown in Fig. 13.28, symmetric components are required. The members are classified in four deterministic design variables (i.e. cross-sectional areas) as follows: vertical members $(A_1)$, lower chord members $(A_2)$, diagonal members $(A_3)$, and upper chord members $(A_4)$. They are indicated also in Fig. 13.28.

The statistical parameters of the random variables (i.e. load $S$ and resistances $C_1$, $C_2$, $C_3$, $C_4$ of the four types of members) are given in Frangopol and Fu

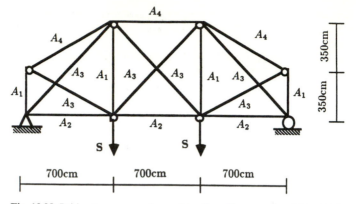

Fig. 13.28 Bridge truss: geometry and loading (Frangopol and Fu, 1989).

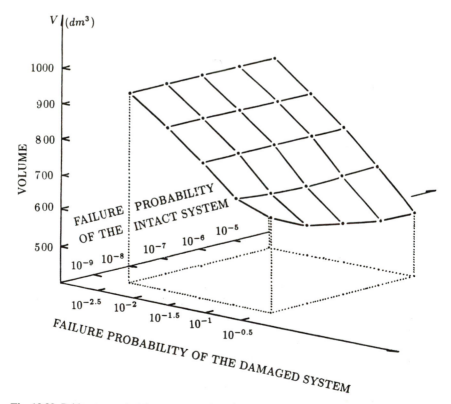

Fig. 13.29 Bridge truss: decision support space for three-objective RBSO associated with the objective vector $\mathbf{F}_2(\mathbf{X})$ (Frangopol and Fu, 1989).

(1989). It is noted that the compression capacities of the truss members take into account buckling effects indirectly. Consequently, they are different from the tension capacities.

The truss bridge is optimized considering the objective vector $\mathbf{F}_2(\mathbf{X})$ given by equation (13.107). The formulation is as follows:

$$\min[V(\mathbf{A}), P_{\text{fint}}(\mathbf{A}), P_{\text{fdmg}}(\mathbf{A})] \tag{13.128}$$

subject to

$$\mathbf{A} \in \Omega = (\mathbf{A} \mid A_i \geqslant 25\,\text{cm}^2) \qquad i = 1, 2, 3, 4 \tag{13.129}$$

In equation (13.128), $P_{\text{fint}}$ and $P_{\text{fdmg}}$ have the same meanings as in the five-bar truss example (equation (13.122)) optimized previously.

Figure 13.29 (Frangopol and Fu, 1989) shows the decision support space obtained by solving the problem defined by equations (13.128) and (13.129). It provides Pareto optimal objectives to the decision maker for the final decision. It is easy to recognize that $V$ and $P_{\text{fdmg}}$ are dominant figures for the final decision making. Consequently, the final design can be determined based only on the trade-offs between $V$ and $P_{\text{fdmg}}$, and $P_{\text{fint}}$ will be automatically kept at a satisfactorily low level.

---

## EXAMPLE 13.11 TWO-STORY FOUR-BAY FRAME (Frangopol and Iizuka, 1991a, b)

Consider the nondeterministic two-story four-bay steel frame shown in Fig. 13.30 (Frangopol and Iizuka, 1991a). The geometry of the frame ($L_1 = 3$ m, $L_2 = 4$ m, $H_1 = 6$ m, $H_2 = 4.8$ m) and the cross-sectional area design vector $\mathbf{X} = [A_1$ (bottom columns) $A_2$ (bottom beams) $A_3$ (top columns) $A_4$ (top beams)$]^t$ are deterministic. The loads and the yield stress at each section are random variables with distributions, mean values, coefficients of variation, and correlations given in Iizuka (1991).

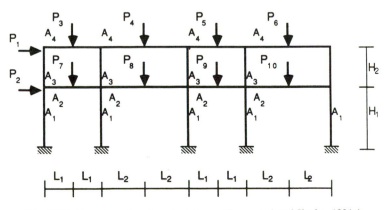

**Fig. 13.30** Two-story four-bay steel frame (Frangopol and Iizuka, 1991a).

The two-story four-bay frame is optimized successively considering the objective vectors $\mathbf{F}_1(\mathbf{X})$ and $\mathbf{F}_4(\mathbf{X})$ given by equations (13.106) and (13.109), respectively,

● *Objective vector* $\mathbf{F}_1(\mathbf{X})$ *(Frangopol and Iizuka, 1991a)*. In this case, the frame in Fig. 13.30 is optimized to satisfy the conditions

$$\min[V(\mathbf{A}), P_{\text{fcol}}(\mathbf{A}), P_{\text{fyld}}(\mathbf{A}), P_{\text{fdfm}}(\mathbf{A})] \tag{13.130}$$

subject to

$$\mathbf{A} \in \Omega = (\mathbf{A} | A_i \geqslant 0) \qquad i = 1, 2, 3, 4 \tag{13.131}$$

The system probabilities of collapse and unserviceability are computed using the method proposed by Ang and Ma (1981) and modified by Ishikawa and Iizuka (1987). The allowable vertical and horizontal elastic displacements of the frame in Fig. 13.30 are given in Iizuka (1991). The probabilities $P_{\text{fyld}}$ and $P_{\text{fdef}}$ are calculated using forty-four critical sections (i.e. top and bottom for each column, and two joints and the midspan for each beam) and ten reference sections (i.e., one for each load), respectively.

Figures 13.31 and 13.32 (Iizuka, 1991) show different representations of decision support spaces for the Pareto optimal objectives associated with the frame in Fig. 13.30. For example, in Fig. 13.31 the isoprobability of collapse surfaces are drawn in the three-dimensional space defined by the structural volume and probabilities of deformation and yielding. This space is more convenient than that of Fig. 13.32, when the designer's purpose is to balance cost and serviceability requirements given the collapse reliability level. Consequently,

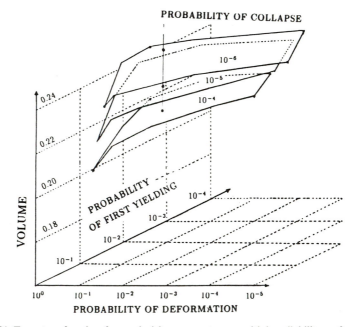

**Fig. 13.31** Two-story four-bay frame: decision support space with isoreliability surfaces against collapse for four-objective RBSO associated with the objective vector $\mathbf{F}_1(\mathbf{X})$ (Iizuka, 1991).

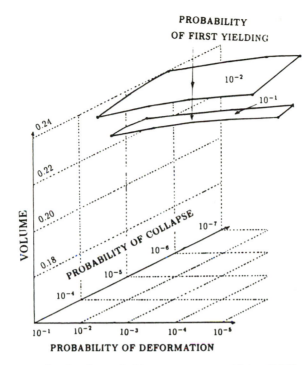

**Fig. 13.32** Two-story four-bay frame: decision support space with isoreliability surfaces against first yielding for four-objective RBSO associated with the objective vector $\mathbf{F}_1(\mathbf{X})$ (Iizuka, 1991).

appropriate decision support spaces must be chosen according to specified purposes.

- *Objective vector* $\mathbf{F}_4(\mathbf{X})$ *(Frangopol and Iizuka, 1991b)*. In this case, the frame in Fig. 13.30 is optimized to satisfy the conditions

$$\min[V(\mathbf{A}), -\beta_{\text{yld}}(\mathbf{A}), -R_2(\mathbf{A})] \qquad (13.132)$$

subject to

$$\mathbf{A} \in \Omega = (\mathbf{A} \mid A_i \geqslant 0) \qquad i = 1, 2, 3, 4 \qquad (13.133)$$

where the reliability index with respect to first yielding is $\beta_{\text{yld}} = \Phi^{-1}(1 - P_{\text{fyld}})$, and the redundancy factor $R_2$ is given by equation (13.111). Consequently, the purpose is to obtain Pareto optimum solutions by considering the weight (material volume) of the structure, the system reliability with respect to first yielding occurrence (i.e. system serviceability requirement), and system redundancy as three equally important goals of the design process.

Figure 13.33 (Frangopol and Iizuka, 1991b) displays the Pareto optimal objectives in the space of system yielding reliability index $\beta_{\text{yld}}$ and system redundancy factor $R_2$. In this space five isovolume contours and the two borders which enclose the domain of strongly nondominated solutions are indicated. The variations of Pareto optimal objectives and solutions along the isovolume contour A–B in Fig. 13.33 are shown in Figs 13.34(a) and 13.34(b), respectively. Several trade-offs can be observed.

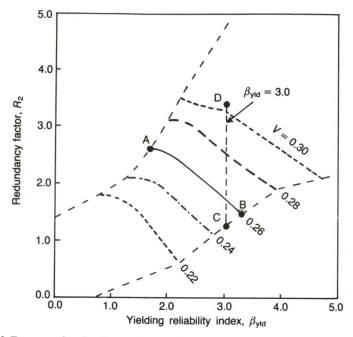

**Fig. 13.33** Two-story four-bay frame: Pareto objective set with isovolume contours for three-objective RBSO associated with the objective vector $F_4(X)$ (Frangopol and Iizuka, 1991b).

Figures 13.35(a) and 13.35(b) show the variations of Pareto optimal objectives and solutions, respectively, along the isoreliability contour C–D shown in Fig. 13.33. From Fig. 13.35(a) it can be observed that a small increase in volume causes great improvement of redundancy under a constant reliability index with respect to first yielding. The conclusion is obvious: it is surely worthwhile to add so little material to improve system performance so much as a result of higher redundancy.

## 13.9 RESIDUAL AND DAMAGE-ORIENTED RELIABILITY-BASED OPTIMIZATION

### 13.9.1 Introduction

A major limitation in many structural optimization programs is the use of deterministic constraints on stresses and performance. Overall safety should consider the statistical properties of loads, materials and analysis variables. System reliability analysis has improved as tools for predicting the overall safety under extreme load conditions. This section presents the optimization of structures with system reliability constraints imposed on both the performance of the original intact structure (reserve reliability) as well as the structure response after specified accident or damage scenarios (residual reliability). Several examples (in

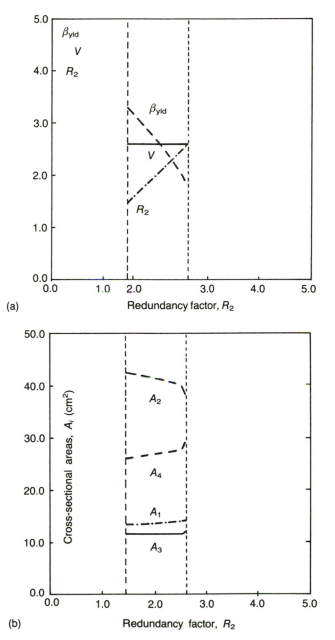

(a)

(b)

**Fig. 13.34** Variation along the isovolume contour $A{-}B$ in Fig. 13.33 of (a) Pareto objective set and (b) Pareto solution set.

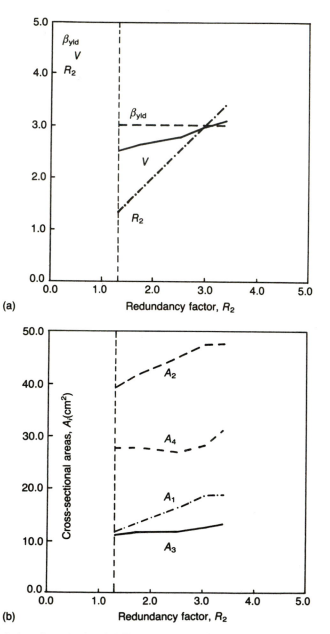

**Fig. 13.35** Variation along the isoreliability contour $C-D$ in Fig. 13.33 of (a) Pareto objective set and (b) Pareto solution set.

addition to those presented in Section 13.8) show the benefits of such optimization for selecting both member sizes as well as structure topologies. Different behavior models include ductile and brittle behavior and degrees of strength correlation. The methods allow the designers to specify different accident scenarios and their corresponding target reliability levels as well as the target reliability for the intact geometry.

During the expected lifetime of a structure, various types of deterioration or damage may occur as a result of corrosion, fatigue and/or accidental loss of structural member capacity. Therefore, if the design only optimized the system reliability for the initial intact structure, this still does not guarantee that the structure is safe over its lifetime.

The method considers system reliability constraints for system reserve (for the initial installed structure) and system residual reliability corresponding to a damaged structural condition. These methods are illustrated for a truss structural optimization example.

In order to select a practical system reliability analysis, several methods including bounding and simulation were compared. The study found that the average value of the well-known Ditlevsen bounds on system reliability is an adequate approach owing to its accuracy, efficiency and especially its stable sensitivity analysis. The optimization algorithm uses a modified method of feasible directions for minimization.

Several examples are included to illustrate the minimum weight of structures subject to overall system reliability constraints both for the intact structure and for life cycle performance including damaged conditions.

### 13.9.2 System reliability formulations

Using the incremental loading methods of section 13.6, linear failure mode expressions (FME) are derived in the form of safety margins as follows:

$$g_j = \sum_j c_{ji} R_i - \sum_k d_{jk} S_k \quad (j = 1, 2, \ldots, M) \tag{13.134}$$

where $g_j$ is the $j$th failure mode margin of safety, $R_i$ is the component strength $i$ in mode $j$, $S_k$ are the loads acting in mode $j$, and $c_{ji}$ and $d_{jk}$ are constant coefficients representing the participation of strength and load variables in a failure mode. System failure occurs if any $g_j$ is less than zero. $M$ is the total number of significant failure modes identified. If the load and material statistical data is given, then the number of modes, $M$, that must be examined is fixed by the accuracy requirements. During the structure optimization, $M$ may be changed as a result of the relative contributions of different collapse modes to the system reliability.

### (a) Reliability bounding methods

The simplest system reliability bounds are of the following form (Cornell, 1967):

$$P_f(\text{sys}) \begin{cases} \leqslant \sum_{m=1}^{M} P_{f,j} \\ \\ \geqslant \max_j P_{f,j} \end{cases} \tag{13.135}$$

where $P_f(\text{sys})$ is the system failure probability and $P_{f,i}$ the failure probability of any single mode of collapse. Alternate and more exact system reliability bounds are available (Ditlevsen, 1979):

$$P_f(\text{sys}) \begin{cases} \leqslant \sum_{j=1}^{M} P_{f,j} - \sum_{j=2}^{M} \max_{i<j} P_{f,ij} \\ \geqslant P_{f,1} + \sum_{j=2}^{M} \max\left( 0; P_{f,j} - \sum_{i=1}^{j-1} P_{f,ij} \right) \end{cases} \tag{13.136}$$

where the doubled subscripts of $P_f$ indicate simultaneous occurrence of two collapse modes, and $P_{f,1}$ is the probability of occurrence of the dominant failure mode.

Several examples have demonstrated that the use of the bounds provides accurate and stable calculation of system reliability applicable to structures which typically have a multitude of significant collapse modes. The studies included brittle and ductile components, correlated and independent strength members and a range of strength and load variables.

### 13.9.3 Optimization formulation for truss structures

(a) Case 1: deterministic optimization for intact truss structures

The structural optimization considered is restricted to weight minimization of trusses with fixed geometry and material properties. The cross-sectional areas of structural members are the design variables, $A_i$. Consider the 15-bar truss (shown in Fig. 13.36) (strength correlation coefficient $\rho = 0.5$). The deterministic fully

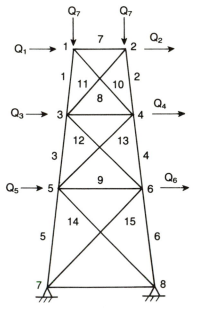

**Fig. 13.36** 15-bar truss example.

stressed design (FSD) for this truss structure, $\sigma = 16\,000\,\text{kN}\,\text{m}^{-2}$, corresponds to a total truss weight $W = 52.83$.

The reliability analysis for the FSD truss gave the following results for the system reliability index, $\beta_s$: for ductile material behavior,

$$\text{FSD truss} \quad \beta_s = 2.86$$

$$(\text{SF} = 2.0)$$

for brittle material behavior,

$$\text{FSD truss} \quad \beta_s = 1.72$$

$$(\text{SF} = 2.0)$$

It was found that the weight of structure is evidently reduced from that of the original design, by using a deterministic optimization program, but the safety level is also evidently reduced and therefore the failure probability is increased. It was also seen that while the original design has similar system reliability $\beta_s$ for ductile and brittle behavior (3.79 and 3.41 respectively), the FSD $\beta_s$ is much lower for brittle compared with ductile behavior. Thus, structures designed by deterministic optimization may not be optimal structures in a safety sense and, in some instances, these structures can be unsafe structures.

## (b) Case 2: reliability optimization for intact truss structures

In order to overcome shortcomings in deterministic structure optimization, the reliability optimization for intact truss structures is compared. Find member areas ($A$), to

$$\text{min weight}(A)$$

$$\text{subject to } \beta_s(A) \geqslant \beta_s^* \tag{13.137}$$

where $\beta_s^*$ is the target structural system reliability index required by the designer and $\beta_s(A)$ is the truss structural reliability index when design variables are $A$ (the value used is $\beta_M$, the average of the Ditlevsen bounds).

Using this optimization formulation, the truss structure weight can be minimized while providing the safety levels required for the original intact structures. For example, with this formulation for the 15-bar truss and $\beta_s^* = 4.0$, the following results are found for the ductile case:

$$\text{original} \quad W = 86.43, \beta_s = 3.79$$

$$\text{optimum} \quad W = 76.16, \beta_s = 4.00$$

Thus, in this optimal formulation, the truss weight is actually reduced from 86.43 to 76.16 while at the same time the structure safety level is increased. Other optimization results were presented showing optimal weights for different target system $\beta_s$.

## (c) Case 3: truss optimization under reserve and residual reliability constraints

During the expected lifetime of a structure, various types of deterioration or damage often occur as a result of corrosion, fatigue, fracture, and/or accidental loss of structural members. Therefore, sufficient reserve based on the system reliability constraints for the intact structure cannot ensure that the optimal structure is also safe during its expected or damaged lifetime.

After optimization, the weight of the 15-bar truss in Fig. 13.36 is 76.16 with the reliability (reserve or intact system) index $\beta_s = 4.0$. If the top horizontal bar (bar 7) is suddenly lost because of damage, then the system reliability $\beta_s$ of the damaged structure is reduced to 1.15. This corresponds to a failure probability increased to 0.1251 from the intact value of $0.3167 \times 10^{-4}$ or a ratio of 3950. The occurrence of such a lost member is not an uncommon event, say in marine offshore structures, due to fatigue, boat collision or dropped objects during operations.

Thus, all possible damage situations or accident scenarios during the expected lifetime of the structure should be considered. According to this safety philosophy, the following formulation is used for considering both system reserve and system residual reliability requirements simultaneously.

Find the vector $A$, denoted by $(A)$, such that:

$$\min W(A)$$

subject to:

$$\beta_s(A) \geqslant \beta_s^*$$
$$\beta_k(A) \geqslant \beta_k^* \quad (k = 1, 2, \ldots, P) \tag{13.138}$$

where $\beta_s^*$ is the target structural reliability index required for intact structure, $\beta_k^*$ is the target structural reliability index required for the $k$th residual structure (possible damage scenarios must be defined), $\beta_s(A)$ is the intact truss structure reliability index, when design variables are $A$, $\beta_k(A)$ is the $k$th residual truss structure reliability index, also when design variables are $A$, and $P$ is the number of residual or damage cases considered.

This optimal formulation can ensure a tolerable safety level for a structure during its expected lifetime for all possible damage cases. For example, with this optimal formulation consider the 15-bar truss structures with intact and lost bars again. Set $\beta_s^* = 4.0$ for the intact system and $\beta_7^* = 3.5$ for the scenario in which truss bar 7 accidentally fails. For this case, the new optimal structure weight is increased to 76.81 from $W = 76.16$ which is the optimal result for intact structure with $\beta_s^* = 4.0$. At the same time, $\beta_s = 4.0$ and $\beta_7$ is increased to 3.5 from $\beta = 1.15$. Thus, only a slight weight change makes the structure satisfactory.

As another example, consider the 13-bar truss shown in Fig. 13.37. Again, set $\beta_s^* = 4.0$ for the intact truss and $\beta_7^* = 3.5$ for a constraint for the truss without bar 7. Using only the intact system constraint ($\beta_s^* = 4.0$) the structure weight is 77.23 kN but $\beta_7 = -1.57$. That is, this truss is almost certain to fail if an accident damages member 7. This low $\beta$ value in this case is due to the less redundant geometry in the 13-member truss compared with the 15-member truss considered above. Using both intact $\beta_s^* = 4.0$ and residual $\beta_7^* = 3.5$ constraints, the structure weight is increased to 85.21 from $W = 77.23$. This represents a weight increase of about 10%. Similarly, looking at other potential accident scenarios would also have the effect of raising the weight.

Comparisons of this nature also help to compare overall topologies such as the 15-bar vs 13-bar designs. In general, deterministic optimization will often lead to nonredundant deterministic structures. The approach herein accounts for both intact and possible damage scenarios in fixing member sizes.

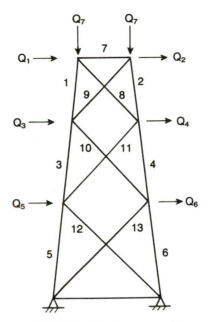

**Fig. 13.37** 13-bar truss example.

### 13.9.4 System reliability with a damage distribution

In some applications, a partial damage such as corrosion loss of strength should be considered in the optimization. The results above can be extended to a distribution of member damage with a random variable $D$.

The system probability of failure $P_{RSD}$ can be found by a conditional integration over $D$ as follows:

$$P_{RSD} = \int_0^1 P_{f_{(D=x)}} f_D(x)\,dx \qquad (13.139)$$

This formulation was given in Liu and Moses (1991) for different member damage formulations, and residual reliability targets. Results show weight increases with expected damage level. If damages in more than one member simultaneously are considered, then the optimal formulation can be expanded. Some illustrations were given in Liu and Moses (1991) for different member-group damages and residual reliability levels. In actual practice, uniform damage in all members is likely to be due to corrosion and is likely to be small. Large damage in single members will probably be repaired. In general, the damage distributions will have to be correlated to inspection intervals and maintenance practices.

### 13.10 CONCLUSIONS

- In order to obtain a consistent approach to safety in an optimization, the criteria should be based on a probabilistic formulation of failure events.

- Reliability-based optimization permits specification of either load or risk constraints and illustrates the interplay beteween cost allocation to specific components and overall risk.
- System reliability is needed to account properly for the large number of potential failure modes and their complex interaction in terms of common loadings, strength correlation and relative occurrence risks.
- System optimization can be cast in a risk formulation which has evolved to a stage which allows multiple collapse modes to be identified and overall risk to be computed.
- Reliability-based design has reached a stage where both serviceability and ultimate limit states constraints can be incorporated in the same probabilistic framework. This is especially important in certain structures such as reinforced concrete frames where serviceability limits may control the optimization.
- Formulation of system constraints for large structures has been accomplished using incremental loading schemes with fault tree logic. Realistic applications to weight and cost optimization with risk constraints have been reported.
- Further work is needed to make the models more accessible to designers and to provide a consistent database acceptable for making risk–benefit trade-offs.

## ACKNOWLEDGEMENTS

Several postdoctoral researchers at both the University of Colorado and Case Western Reserve University including G. Fu, M. Klisinski, Y. Liu and D. Verma, and a number of former graduate students including M. Iizuka, M. Gorman, Y.-H. Lee, R. Nakib, J. Stevenson, R. Rashedi and K. Yoshida contributed to the results presented in this chapter. Their contributions and assistance are greatly appreciated. Support for part of the reliability-based structural optimization research at the University of Colorado has mainly come from the National Science Foundation under grants MSM-8618108, MSM-8800882 and MSM-9013017, and at Case Western Reserve University, also from the National Science Foundation grants and also research contracts with Transportation Research Board, Federal Highway Administration and Ohio Department of Transportation. This support is gratefully acknowledged.

## REFERENCES

Ang, A. H.-S. and Cornell, C.A. (1974) Reliability bases of structural safety and design. *Journal of the Structural Division, ASCE,* **100**(ST9), 1755–69.

Ang, A. H.-S. and Ma, H.-F. (1981) On the reliability of structural systems, in Proceedings of the Third International Conference on Structural Safety and Reliability: ICOSSAR'81, Trondheim, pp. 295–314.

Ang. A. H.-S. and Tang, W.H. (1975) *Probability Concepts in Engineering Planning and Design,* Vol. I, Wiley, New York.

Ang, A. H.-S. and Tang, W.H. (1984) *Probability Concepts in Engineering Planning and Design,* Vol. II, Wiley, New York.

Arora, J.S. (1989) *Indroduction to Optimum Design,* McGraw-Hill, New York.

Baker, M. (1977) Rationalization of safety and serviceability factors in structural codes. *CIRIA Report No. 63,* Construction Industry Research and Information Association, London.

Benjamin, J.R. (1968) Probabilistic structural analysis and design. *Journal of the Structural Division, ASCE*, **94**(ST7), 1665–79.

Benjamin, J.R. and Cornell, C.A. (1970) *Probability, Statistics, and Decision for Civil Engineers*, McGraw-Hill, New York.

Bjerager, P. (1988) Prbability integration by numerical simulation. *Journal of Engineering Mechanics, ASCE*, **114**(8), 1285–302.

Bjerager, P., Karamchandani, A. and Cornell, C.A. (1987) Failure tree analysis in structural system reliability, in *Reliability and Risk Analysis in Civil Engineering* (ed. N.C. Lind), Vol. 2, Proceedings of ICASP 5, Vancouver, pp. 985–96.

Bucher, C.G. (1988) Adaptive sampling: an iterative fast Monte-Carlo procedure. *Structural Safety*, **5**(2), 119–26.

Chou, K.C., McIntosh, C. and Corotis, R.B. (1983) Observations on structural system reliability and the role of modal correlations. *Structural Safety*, **1**, 189–98.

Cohn, M.Z. and Parimi, S.R. (1972) Multi-criteria probabilistic design of reinforced concrete structures, in *Inelasticity and Non-linearity in Structural Concrete* (ed. M.Z. Cohn), SM Study No. 8, University of Waterloo, Waterloo, Ontario, pp. 471–92.

Cornell, C.A. (1967) Bounds on the reliability of structural systems. *Journal of the Structural Division, ASCE*, **93**(ST1), 171–200.

Cornell, C.A. (1969a) Bayesian statistical decision theory and reliability-based design, in Proceedings of the International Conference on Structural Safety and Reliability, Washington, DC, pp. 47–66.

Cornell, C.A. (1969b) A probability-based structural code. *Journal of the American Concrete Institute, ACI*, **66**(12), 974–85.

Corotis, R.B. (1985) Probability-based design codes. *Concrete International*, **7**, 42–9.

Deak, I. (1980) Three digit accurate multiple normal probabilities. *Numerische Mathematik*, **35**, 369–80.

Ditlevsen, O. (1979) Narrow reliability bounds for structural systems. *Journal of Structural Mechanics*, **7**(4), 453–72.

Duckstein, L. (1984) Multiobjective optimization in structural design: the model choice problem, in *New Directions in Optimum Structural Design* (eds. E. Atrek, R.H. Gallagher, K.M. Ragsdell and O.C. Zienkiewicz), Wiley, Chichester, pp. 459–81.

Ellingwood, B. and Ang, A. H.-S. (1974) Reliability bases of structural safety and design. *Journal of the Structural Division, ASCE*, **100**(ST9), 1771–88.

Ellingwood, B. and Galambos, T.V. (1982) Probability based criteria for structural design. *Structural Safety*, **1**, 15–26.

Ellingwood, B., Galambos, T.V., MacGregor, J.G. and Cornell, C.A. (1980) Development of a probability based load criterion for American National Standard A58. *NBS Special Publication 577*, Washington, DC.

Enevoldsen, I., Sørensen, J.D. and Sigurdsson, G. (1990) Reliability-based shape optimization using stochastic finite element methods. *Structural Reliability Theory Paper No. 73*, Institute of Building Technology and Structural Engineering, Aalborg University, Aalborg.

Feng, Y.S. and Moses, F. (1986a) A method of structural optimization based on structural system reliability. *Journal of Structural Mechanics*, **14**(4), 437–53.

Feng, Y.S. and Moses, F. (1986b) Optimum design, redundancy and reliability of structural systems. *Computers and Structures*, **24**(2), 239–51.

Ferry-Borges, J. (1954) *O Dimensionamento de Estruturas*, Ministry of Public Works, National Laboratory of Civil Engineering, **54**, Lisbon.

Forssell, C. (1924) Ekonomi och byggnadsvasen (economy and construction), *Sunt Fornoft*, **4**, 74–7 (in Swedish). (Translated in excerpts in Lind, S.M. (1970) *Structural Reliability and Codified Design*, SM study No. 3, Solid Mechanics Division, University of Waterloo, Waterloo.)

Frangopol, D.M. (1984a) A reliability-based optimization technique for automatic plastic design. *Computer Methods in Applied Mechanics and Engineering*, **44**, 105–17.

Frangopol, D.M. (1984b) Interactive reliability-based structural optimization. *Computers and Structures*, **19**(4), 559–63.

Frangopol, D.M. (1985a) Sensitivity of reliability-based optimum design. *Journal of Structural Engineering, ASCE*, **111**(8), 1703–21.

Frangopol, D.M. (1985b) Multicriteria reliability-based optimum design. *Structural Safety*, **3**(1), 23–8.

Frangopol, D.M. (1985c) Structural optimization using reliability concepts. *Journal of Structural Engineering, ASCE*, **111**(11), 2288–301.

Frangopol, D.M. (1985d) Towards reliability-based computer aided optimization of reinforced concrete structures. *Engineering Optimization*, **8**(4), 301–13.

Frangopol, D.M. (1986a) Computer-automated design of structural systems under reliability-based performance constraints. *Engineering Computations*, **3**(2), 109–15.

Frangopol, D.M. (1986b) Structural optimization under conditions of uncertainty, with reference to serviceability and ultimate limit states, in *Recent Developments in Structural Optimization* (ed. F.Y. Cheng), ASCE, New York, pp. 54–71.

Frangopol, D.M. (1986c) Computer-automated sensitivity analysis in reliability-based plastic design. *Computers and Structures*, **22**(1), 63–75.

Frangopol, D.M. (1987) Unified approach to reliability-based structural optimization, in *Dynamics of Structures* (ed. J.M. Roesset), ASCE, New York, pp. 156–67.

Frangopol, D.M. (ed.) (1989) *New Directions in Structural System Reliability*, University of Colorado, Boulder, CO.

Frangopol, D.M. and Corotis, R.B. (eds) (1990) System reliability in structural analysis, design and optimization. *Structural Safety*, **7**(2–4), 83–312.

Frangopol, D.M. and Fu, G. (1989) Optimization of structural systems under reserve and residual reliability requirements, in *Lecture Notes in Engineering* (eds C.A. Brebbia and S.A. Orszag), Vol. 48 (ed. P. Thoft-Christensen), Springer, Berlin, pp. 135–45.

Frangopol, D.M. and Fu, G. (1990) Limit states reliability interaction in optimum design of structural systems, in *Structural Safety and Reliability* (eds. A. H-S. Ang, M. Shinozuka and G.I. Schuëller), Vol. III, ASCE, New York, pp. 1879–86.

Frangopol, D.M. and Iizuka, M. (1991a) Multiobjective decision support spaces for optimum design of nondeterministic structural systems, in *Probabilistic Safety Assessment and Management* (ed. G. Apostolakis), Vol. 2, Elsevier, Amsterdam, pp. 977–82.

Frangopol, D.M. and Iizuka, M. (1991b) Pareto optimum solutions for nondeterministic systems, in Proceedings of the Sixth International Conference on Applications of Statistics and Probability in Civil Engineering, ICASP6 (eds L. Esteva and S.E. Ruis), Mexico City, Vol. 1, pp. 216–23.

Frangopol, D.M. and Klisinski, M. (1989a) Material behavior and optimum design of structural systems. *Journal of Structural Engineering, ASCE*, **115**(5), 1054–75.

Frangopol, D.M. and Klisinski, M. (1989b) Vector optimization of structural systems, in *Computer Utilization in Structural Engineering* (ed. J.K. Nelson), ASCE, New York, pp. 490–9.

Frangopol, D.M. and Klisinski, M. (1991) Computational experience with vector optimization techniques for structural systems, in *Lecture Notes in Engineering* (eds C.A. Brebbia and S.A. Orszag), Vol. 61 (eds A. Der Kiureghian and P. Thoft-Christensen), Springer, Berlin, pp. 99–111.

Frangopol, D.M. and Nakib, R. (1986) Isosafety loading functions in system reliability analysis. *Computers and Structures*, **24**(3), 425–36.

Frangopol, D.M. and Nakib, R. (1987) Reliability of structural systems with multiple limit states, in *Materials and Member Behavior* (ed. D.S. Ellifritt), ASCE, New York, 1987, pp. 638–46.

Frangopol, D.M. and Rondal, J. (1976) Considerations on optimum combination of safety and economy, in Final Report, Tenth Congress of the International Association for Bridge and Structural Engineering, Tokyo, pp. 45–8.

Frangopol, D.M., Klisinski, M. and Iizuka, M. (1991) Computational experience with damage-tolerant optimization of structural systems, in Proceedings of the 1st International Conference on Computational Stochastic Mechanics (eds P.D. Spanos and C.A. Brebbia),

Computational Mechanics Publications, Southampton, and Elsevier Applied Science, London, pp. 199–210.

Freudenthal, A.M. (1956) Safety and the probability of structural failure. *Transactions, ASCE,* **121,** 1337–75.

Fu, G. and Frangopol, D.M. (1990a) Balancing weight, system reliability and redundancy in a multiobjective optimization framework. *Structural Safety,* **7**(2–4), 165–75.

Fu, G. and Frangopol, D.M. (1990b) Reliability-based vector optimization of structural systems. *Journal of Structural Engineering, ASCE,* **116**(8), 2141–61.

Fu, G. and Moses, F. (1988) Importance sampling in structural system reliability, in *Probabilistic Methods in Civil Engineering* (ed. P.D. Spanos). Proceedings of the 5th ASCE Specialty Conference on Probabilistic Methods, Blacksburg, VA, pp. 340–3.

Fu, G., Liu, Y. and Moses, F. (1991) Management of structural system reliability, in *Lecture Notes in Engineering* (eds C.A. Brebbia and S.A. Orszag), Vol. 61 (eds A. Der Kiureghian and P. Thoft-Christensen), Springer, Berlin, pp. 113–28.

Gorman, M.R. (1979) Reliability of structural systems. *Report CE 79-2,* Civil Engineering Department, Case Western Reserve University, Cleveland, OH, May 1979.

Gorman, M.R. (1985) Resistance modeling. *ASCE Short Course Notes on Structural Reliability Analysis of Offshore Platforms,* May 1985.

Gorman, M.R. and Moses, F. (1979) Direct estimate of structural system reliability. ASCE Seventh Electronic Computation Conference, St. Louis, MO.

Grimmelt, M.J. and Schuëller, G.I. (1983) Benchmark study on methods to determine collapse failure probabilities of redundant structures. *Structural Safety,* **1**(2), 93–106.

Hasofer, A.M. (1987) Directional simulation with applications to outcrossings of Gaussian processes, in *Reliability and Risk Analysis in Civil Engineering* (ed. N.C. Lind), Vol. 1, Proceedings ICASP 5, Vancouver, pp. 35–45.

Hasofer, A.M. and Lind, N.C. (1978) Exact and invariant second-moment code format. *Journal of the Engineering Mechanics Division, ASCE,* **100**(EM1), 829–44.

Hilton, H.H. and Feigen, M. (1960) Minimum weight analysis based on structural reliability. *Journal of the Aerospace Sciences,* **27,** 641–53.

Iizuka, M. (1991) Time invariant and time variant reliability analysis and optimization of structural systems. Ph.D. Thesis, Department of Civil Engineering, University of Colorado, Boulder, CO.

Ishikawa, N. and Iizuka, M. (1987) Optimum reliability-based design of large framed structures. *Engineering Optimization,* **10**(4), 245–61.

Johnson, A.I. (1953) *Strength, Safety and Economical Dimensions of Structures,* Division of Building Statistics and Structural Engineering, Royal Institute of Technology, **12,** Stockholm.

Kalaba, R.E. (1962) Design of minimum weight structures given reliability and cost. *Journal of the Aerospace Sciences,* **29,** 355–6.

Kim, S.H. and Wen, Y.K. (1987) Reliability-based structural optimization under stochastic time varying loads. *Civil Engineering Studies, Structural Research Series No. 533,* University of Illinois, Urbana, IL.

Kim, S.H. and Wen, Y.K. (1990) Optimization of structures under stochastic loads. *Structural Safety,* **7**(2–4), 177–90.

Kirsch, U. (1981) *Optimum Structural Design,* McGraw-Hill, New York.

Koski, J. (1984) Multicriterion optimization in structural design, in *New Directions in Optimum Structural Design* (eds E. Atrek, R.H. Ragsdell and O.C. Zienkiewicz), Wiley, Chicester, pp. 483–503.

Lind, N.C. (1973) The design of structural design norms. *Journal of Structural Mechanics,* **1**(3), 357–70.

Liu, Y. and Moses, F. (1991) Bridge design with reserve and residual reliability constraints. *Structural Safety,* **11**(1), 29–42.

Madsen, H.O., Krenk, S. and Lind, N.C. (1986) *Methods of Structural Safety,* Prentice-Hall, Englewood Cliffs, NJ.

Mahadevan, S. and Haldar, A. (1991) Reliability-based optimization using SFEM, In *Lecture*

*Notes in Engineering* (eds C.A. Brebbia and S.A. Orszag), Vol. 61 (eds A. Der Kiureghian and P. Thoft-Christensen), Springer, Berlin, pp. 241–50.

Mau, S.-T. and Sexsmith, R.G. (1972) Minimum expected cost optimization. *Journal of the Structural Division, ASCE*, **98**(ST9), 2043–58.

Melchers, R.E. (1987) *Structural Reliability Analysis and Prediction*, Ellis Horwood, Chichester.

Moses, F. (1969) Approaches to structural reliability and optimization, in *An Introduction to Structural Optimization* (ed. M.Z. Cohn), SM Study No. 1, Solid Mechanics Division, University of Waterloo, Waterloo, Ontario, pp. 81–120.

Moses, F. (1970) Sensitivity studies in structural reliability, in *Structural Reliability and Codified Design* (ed. N.C. Lind), SM Study No. 3, Solid Mechanics Division, University of Waterloo, Waterloo, Ontario, pp. 1–18.

Moses, F. (1973) Design for reliability – concepts and applications, in *Optimum Structural Design* (eds R.H. Gallagher and O.C. Zienkiewicz), Wiley, New York, pp. 241–65.

Moses, F. (1974) Reliability of structural systems. *Journal of the Structural Division, ASCE*, **100**(ST9), 1813–20.

Moses, F. (1977) Structural system reliability and optimization. *Computers and Structures*, **7**, 283–90.

Moses, F. (1982) System reliability developments in structural engineering. *Structural Safety*, **1**(1), 3–13.

Moses, F. (1990) New directions and research needs in system reliability research. *Structural Safety*, **7**(2–4), 93–100.

Moses, F. and Kinser, D.E. (1967) Optimum structural design with failure probability constraints. *AIAA Journal*, **5**(6), 1152–8.

Moses, F. and Rashedi, M.R. (1983) The application of system reliability to structural safety, in *Applications of Statistics and Probability in Soil and Structural Engineering* (eds G. Augusti, A. Borri and G. Vannucchi), Proceedings ICASP 4, Vol. 1, Florence, pp. 573–84.

Moses, F. and Stahl, B. (1979) Reliability analysis format for offshore structures. *Journal of Petroleum Technology*, (March), 347–54.

Moses, F. and Stevenson, J.D. (1970) Reliability-based structural design. *Journal of the Structural Division, ASCE*, **96**(ST2), 221–44.

Moses, F., Fu, G. and Verma, D. (1989) Advanced simulation methods in system reliability, in *Computational Mechanics of Reliability Analysis* (eds W.K. Liu and T. Belytschko), Elme, Lausanne.

Murotsu, Y. and Shao, S. (1990) Optimum shape design of truss structures based on reliability. *Structural Optimization*, **2**(2), 65–76.

Nakib, R. and Frangopol, D.M. (1990a) RSBA and RSBA-OPT: two computer programs for structural system reliability analysis and optimization. *Computers and Structures*, **36**(1), 13–27.

Nakib, R. and Frangopol, D.M. (1990b) Reliability-based structural optimization using interactive graphics. *Computers and Structures*, **37**(1), 27–34.

Nordal, H., Cornell, C.A. and Karamchandani, A. (1987) A structural system reliability case study of an eight leg steel jacket offshore production platform, in Proceedings of the Marine Structural Reliability Symposium, SNAME, Arlington, VA, October 1987.

Osyczka, A. (1984) *Multicriterion Optimization in Engineering*, Ellis Horwood, Chichester.

Paez, A. and Torroja, E. (1959) *La Determination del Coeficiente de Seguridad en las Distintas Obras*, Instituto Technico de la Construccion y del Cemento, Madrid.

Parimi, S.R. and Cohn, M.Z. (1978) Optimum solutions in probabilistic structural design. *Journal of Applied Mechanics*, **2**(1), 47–92.

Rackwitz, R. and Cuntze, R. (1987) Formulations of reliability-oriented optimization. *Engineering Optimization*, **11**(1, 2), 69–76.

Rackwitz, R. and Fiessler, B. (1978) Structural reliability under combined random load sequences. *Computers and Structures*, **9**, 489–94.

Rashedi, R. and Moses, F. (1986) Application of linear programming to structural system reliability. *Computers and Structures*, **24**(3), 375–84.

Rojiani, K.B. and Bailey, G.L. (1984) Reliability-based optimum design of steel structures, in *New Directions in Optimum Structural Design* (eds E. Atrek, R.H. Gallagher, K.M. Ragsdell and O.C. Zienkiewicz), Wiley, Chichester, pp. 443–57.

Rosenblueth, E. (1986) Optimum reliabilities and optimum design. *Structural Saftey*, **3**(1), 69–83.

Rosenblueth, E. and Esteva, L. (1972) Reliability basis for some mexican codes, in *Probabilistic Design of Reinforced Concrete Buildings*, ACI Publication SP-31, pp. 1–41.

Rosenbleuth, E. and Mendoza, E. (1971) Reliability optimization in isostatic structures. *Journal of the Engineering Mechanics Division, ASCE*, **97**(EM6), 1625–40.

Rosenbrock, H.H. (1960) An automatic method for finding the greatest or least value of a function. *Computer Journal*, **3**, 175–84.

Schmit, L.A. (1969) Problem formulation, methods and solutions in the optimum design of structures, in *An Introduction to Structural Optimization* (ed. M.Z. Cohn), SM Study No. 1, Solid Mechanics Division, University of Waterloo, Waterloo, Ontario, pp. 19–46.

Shao, S. (1991) *Reliability-based shape optimization of structural and material systems*. Ph.D. Thesis, Division of Engineering, University of Osaka Prefecture, Osaka.

Shinozuka, M. (1983) Basic analysis of structural safety. *Journal of Structural Engineering, ASCE*, **109**(3), 721–40.

Sørensen, J.D. (1987) Reliability-based optimization of structural systems. *Structural Reliability Theory Paper No. 32*, Institute of Building Technology and Structural Engineering, Aalborg University, Aalborg.

Sørensen, J.D. (1988) Optimal design with reliability constraints. *Structural Reliability Theory Paper No. 45*, Institute of Building Technology and Structural Engineering. Aalborg University, Aalborg.

Sørensen, J.D. and Enevoldsen, I. (1989) Sensitivity analysis in reliability-based shape optimization. *Structural Reliability Theory Paper No. 69*, Institute of Building Technology and Structural Engineering, Aalborg University, Aalborg.

Sørensen, J.D. and Thoft-Christensen, P. (1987) Integrated reliability-based optimal design of structures. *Structural Reliability Theory Paper No. 29*, Institute of Building Technology and Structural Engineering, Aalborg University, Aalborg.

Stevenson, J. and Moses, F. (1970) Reliability analysis of frame structures. *Journal of the Structural Division, ASCE*, **96**(ST11), 2409–27.

Surahman, A. and Rojiani, K.B. (1983) Reliability-based optimum design of concrete frames. *Journal of Structural Engineering, ASCE*, **109**(3), 741–57.

Switzky, H. (1964) Minimum weight design with structural reliability, in Proceedings of the Fifth Annual Structures and Materials Conference, pp. 316–22.

Thoft-Christensen, P. (1987) Application of optimization methods in structural systems reliability theory. *Structural Reliability Theory Paper No. 33*, Institute of Building Technology and Structural Engineering, Aalborg University, Aalborg.

Thoft-Christensen, P. (1991) On reliability-based structural optimization, in *Lecture Notes in Engineering* (eds C.A. Brebbia and S.A. Orszag), Vol. 61 (eds A. Der Kiureghian and P. Thoft-Christensen), Springer, Berlin, pp. 387–402.

Thoft-Christensen, P. and Baker, M.J. (1982) *Structural Reliability Theory and Its Applications*, Springer, Berlin.

Thoft-Christensen, P. and Murotsu, Y. (1986) *Applications of Structural Systems Relaibility Theory*, Springer, Berlin.

Turkstra, C.J. (1967) Choice of failure probabilities. *Journal of the Structural Division, ASCE*, **93**(ST6), 189–200.

Turkstra, C.J. (1970) *Theory of Structural Design Decisions* (ed. N.C. Lind), SM Study No. 2, Solid Mechanics Division, University of Waterloo, Waterloo, Ontario.

Vanderplaats, G.N. (1984) *Numerical Optimization Techniques for Engineering Design: With Applications*, McGraw-Hill, New York.

Vanderplaats, G.N. (1986) *ADS – A Fortran Program for Automated Design Synthesis*, Version 1.10. Engineering Design Optimization, Inc. Santa Barbara, CA.

Vanmarcke, E. (1971) Matrix formulation of reliability analysis and reliability-based design. *Computers and Structures*, **3**, 757–70.

Vanmarcke, E. (1984) *Random Fields: Analysis and Synthesis.* The MIT Press, Cambridge, MA.

Verma, D., Fu, G. and Moses, F. (1990) Efficient structural system reliability assessment by Monte-Carlo methods, in *Structural Safety and Reliability* (eds A. H-S. Ang, M. Shinozuka and G.I. Schuëller), Vol. II, ASCE, New York, pp. 895–901.

Verma, D. and Moses, F. (1989) Calibration of bridge-strength evaluation code. *Journal of Structural Engineering, ASCE*, **115**(8), 1538–54.

Wen, Y.K. and Chen, H-C. (1989) System reliability under time varying loads: I. *Journal of Engineering Mechanics, ASCE*, **115**(4), 808–23.

Wirshing, P. (1985) Reliability methods. *ASCE Short Course Notes on Structural Reliability Analysis of Offshore Platforms*, May, 1985.

Yao, J.T.P. (1985) *Safety and Reliability of Existing Structures*, Pitman, Boston.

Zimmerman, J.J., Corotis, R.B. and Ellis, J.H. (1991) Stochastic programs for identifying significant collapse modes in structural systems, in *Lecture Notes in Engineering* (eds C.A. Brebbia and S.A. Orszag), Vol. 61 (eds A. Der Kiureghian and P. Thoft-Christensen), Springer, Berlin, pp. 359–67.

# Index